GRAPES & WINES

GRAPES & WINES

A COMPREHENSIVE GUIDE TO VARIETIES & FLAVOURS

STERLING EPICURE
New York

An Imprint of Sterling Publishing
1166 Avenue of the Americas
New York, NY 10036

STERLING EPICURE is a trademark of Sterling Publishing Co., Inc.
The distinctive Sterling logo is a registered trademark of Sterling Publishing Co., Inc.

This edition first published in the United States in 2015 by Sterling Publishing Co. Inc.

First published in the United Kingdom in 2015 by Pavilion

ISBN: 978-1-4549-1598-0

Distributed in Canada by Sterling Publishing
c/o Canadian Manda Group,
664 Annette Street
Toronto, Ontario, Canada M6S 2C8

For information about custom editions, special sales, and premium and corporate purchases, please contact
Sterling Special Sales at 800-805-5489 or specialsales@sterlingpublishing.com.

Manufactured in China

10 9 8 7 6 5 4 3 2 1

www.sterlingpublishing.com

For wine recommendations, special offers and news on Oz's upcoming books and events visit www.ozclarke.com and sign up to the Oz Clarke newsletter.
Follow Oz on Twitter@OzClarke

Page 1: Cabernet Sauvignon at budbreak in early spring.
Page 2: Bunches of Merlot grapes ready for harvest in the Sonoma Valley, California.
Page 3: Zinfandel has been adopted by the Californians as their very own classic grape.
Page 4: Tinto Fino (better known as Tempranillo) grapes on a 60-year-old vine at Dominio de Pingus, Ribera del Duero, Spain.

CONTENTS

INTRODUCTION

So what is a grape, then? Well, it's juice and flesh, obviously. It's skin, obviously. It's pips, and I suppose it could be the stalks as well. And then what? And then everything, that's what.

If we have any interest in wine and in flavours, we have to be interested in the grape variety itself. If we have any interest in how a wine matures and changes with age, we have to know about the potential of the particular grape. If we care about the style of a wine, whether it should be sweet or dry, fizzy or fortified or still, each different grape variety's peculiar talents will be of prime importance. Do we like the flavour of oak barrel aging in our wine? Some grapes take to oak, again some don't – it's vital to know which ones love the kiss of oak and which ones loathe its hot embrace. And are we fascinated by how completely different wines taste when they come from different countries and from different regions within those countries? Without the consistent character of each different grape variety to use as a measuring point, mere comparison of place would be meaningless. However far we delve into all the things that influence the flavours of our wine, it all comes back to the grape.

Above: Chardonnay and Pinot Noir are just about two of the most famous classic grape varieties in the world. Both hail from Burgundy and are responsible for two of the world's most magical wine styles – Pinot Noir for haunting sensuous reds and Chardonnay for stunningly rich, honeyed, nutty, oatmealy whites. Yet isn't it strange that neither red or white Burgundy trumpets the grape names on the label? Facing page: Double magnums stored away in the vintage bottle cellar at Château Canon-la-Gaffelière in St-Émilion, Bordeaux.

I mean, think about it. I give you a glass of pale golden green wine. It's got a wonderful pungent scent of gooseberry and passion fruit and lime. You taste it and the acidity crackles against your teeth, the exhilarating attack of citrus fruit scours your palate clean and makes your mouth drool with desire for food. Who made the wine? No idea. Where does it come from? It could be the Loire Valley in France. But it could also be South Africa or Chile, it could be Spain or northern Italy. And it could certainly be New Zealand. So. The four corners of the earth, really. But the grape variety? When the wine smells and tastes like that, you know it is Sauvignon Blanc. The unique, brilliantly recognizable character of the wine is down to the grape variety – Sauvignon Blanc – above all else. It is refined by the relative talent of the men and women who grow the grape and vinify the wine. It is modified or intensified by the local conditions under which it grows. But the core of the flavour comes from the grape.

Now, Sauvignon Blanc is a very dramatic grape. But so is Viognier with its powerful scent of apricot and mayblossom. So is Gewürztraminer with its explosion of musky rose petals and lychees. So is Muscat with its overpowering aroma of hothouse grapes. Riesling is more subtle, but the unmistakeable balance of high acidity with floral notes and citrus fruit is unique to the grape. Chardonnay's nutty, oatmealy ripeness is created with the help of oak barrel aging, but no other grape achieves quite that taste, however similarly you treat it, wherever it is grown.

Red wine grapes are frequently less outspoken, and just at present the obsession with overzealous use of new oak to age wines is spoiling the thrilling individuality of many grapes' flavours – but good varieties still shine through. Tannic sturdiness and blackcurrant fruit mark out Cabernet Sauvignon in a way no other grape can replicate. The ethereal scent and strawberry/cherry fruit of Pinot Noir, the damson fruit and violet perfume of Malbec, the sour cherry and herbal rasp of Sangiovese, the brilliant chocolate and smoky black plum blast of Shiraz – all of these experiences and many more are above all else due to the particular characteristics of the grape variety.

It is remarkable how, over the years, grapes seem to have been relegated to a subordinate role in wine books when they are so evidently of such massive importance. Well, one of the reasons has to be that until the advent of modern 'New World' winemaking techniques that allow the winemaker to pinpoint the potential flavour of the grape and then maximize it, I suspect that few people – winemakers, wine writers and wine drinkers all – actually had much idea of what a grape variety was supposed to taste like. It was easier to say that a wine's particular taste derived from where it was grown, that it tasted of what the French call '*terroir*'. Indeed, the wines often did have a minerally or earthy flavour which probably did emanate from the vineyard and from old-fashioned winemaking, rather than the actual grape itself. That's why, until recently, many experts and critics were obsessed with the minutiae of a wine's birthplace rather than its chief component – the juice of the grape itself.

But when the New World producers brashly barged their way through into our wine consciousness, everything changed. The Australians and Californians, New Zealanders, South Africans and Chileans didn't have much of a story to tell when it came to the traditions and historical importance of their vineyards – many of these selfsame vineyards had only just been planted. The one story they could tell, and the one their ultra-modern winemaking allowed them to tell, was that of the grape itself and the flavour it imparted to the wine. Varietal labelling – labelling a wine according to its grape variety – was one of the most revolutionary moves in wine of recent times, a brilliant stroke of democratization, opening up the wonderful world of flavour to millions previously bamboozled and excluded by the obscurity and unhelpfulness of most wine labels. It allows all of us to make informed choices about what we like and don't like, and makes it so easy to extend our knowledge and experience and pleasure. The grape variety blazes the way: everything else follows.

And there's never been a time when a book recognizing grape varieties as the most important factor in the flavour of a wine has been needed more. The 21st century has seen an explosion of interest, not just in the main varieties, but in the local grapes of every area that grows wine, the ancient varieties facing extinction, the ones previously thought of as dullards, or as too much trouble to grow. Suddenly we're interested in all of these. Social media spreads the word in seconds. Winemakers (wherever they are) have no qualms about planting experimental varieties. Indeed, it's almost a badge of honour to have something unusual and unexpected growing in your vineyard. And always with the objective of creating new, exciting, often unexpected flavours in your wine.

We are now seeing a progression as even the most futuristic of New Wave winemakers begin to search out the special characteristics of different patches of land – a return to *terroir*, I suppose. So does the grape variety simply become a vehicle for the flavour of place? Well, try planting Muscat or Malvasia or Marsanne in Meursault and see if you capture the compelling flavours of its vineyards. You won't. Chardonnay is indeed the vehicle for Meursault's unique personality and indispensable as such. But the grapes' unique characteristics are just as important as those of the vineyard – and a great deal more adaptable, since they will express themselves in recognizable ways wherever the vine is planted.

Along with this move towards a greater understanding of each vineyard and its potential is a move away from relying solely on the best known grapes. Eighty per cent of the world's wine comes from a mere 20 grape varieties. We cover all these in great detail, but also examine hundreds more from every corner of the wine world, so that when you're ready to move out of the comfort zone, this book will be there for you, to encourage, to explain, to lead you on to the fabulous adventure of new flavours, new sensations, new pleasure. The world of wine with the wine grape at its heart – that's what this book is about.

While planning the project, I realized I simply couldn't handle it all by myself. I knew I needed someone to work with who was a first class wine intellectual who could shoulder the bulk of the research, make the best of the brilliant, up-to-date and minutely detailed data that we would obtain from every important vineyard area worldwide, and then put masses of often quite indigestible source material into a highly readable form. I also needed someone who, when I was flying off on my frequent flights of fancy, could rap me firmly on the knuckles and say 'Come on, Oz. Back to earth. This is a serious book of research as much as a joyous celebration of grape varieties in all their multi-flavoured glory'. Well, she didn't quite say that; she'd have edited out most of the final sentence for a start. But the person who shared in the creation of this book with me is Margaret Rand. And without her talent and determination we'd never have got it done.

Oz Clarke

THE STORY OF THE VINE

The story of human cultivation of the vines is the story of the spread of civilization and the movement of populations. All cultivated vines – all 10,000 or so varieties of grapevine – are the descendants of wild vines. And while there doesn't (so far) seem to be an Eve-vine theory, to parallel the way that the study of mitochondrial DNA has hypothesized a single female ancestor for humanity, DNA studies seem to have pinpointed where the first vine was cultivated.

Or at least, they have pinpointed the two most likely places. Southeast Anatolia is one; Transcaucasia the other. The upper reaches of the Tigris and Euphrates rivers, in the Taurus mountains of eastern Turkey, are the favourite candidate. It's partly the enormous range of cultivated vines native to these regions that give the clue, and partly the close relationships between the DNA of local wild vines (which still proliferate here) and that of the local cultivated varieties. When did it happen? Presumably some while after humans discovered that wine – which will develop naturally, provided that ripe grapes are left in warm conditions for a few days – had an agreeable flavour and could cheer you up after a hard day. Wild grapes would have been plentiful, and would have been a useful part of the diet. From grapes to wine is a very small step.

The step from wild vines to cultivated ones is bigger. Wild vines come in male and female versions; cultivated vines are hermaphrodite. The difference in the shape of the seeds is a useful clue for archeologists trying to determine if the inhabitants of a particular early settlement were cultivating the vine or not; but why should vines make this jump?

A tiny percentage of wild vines, maybe two or three per cent, are hermaphrodite. Hermaphrodite vines would be easier to cultivate and would produce a more certain crop, giving them an obvious advantage. So it is reasonable to assume that hermaphrodite wild vines are the probable forebears of our modern vines. Vines can be cultivated from seed, but every time a berry from an early vine fell to the ground and the seed germinated and took root, a new variety of vine would be born. Some would have found favour with early farmers for their sweetness and flavour and the attractiveness of their wine; others would have been dismissed. It took the introduction of propagation by taking cuttings – in this way, every vine can be identical to its parents – for the cultivated vine to begin to squeeze out and dominate its wild family relations. Now we are on the way to today's 10,000 or so known varieties. How did *Vitis vinifera* come to be the chosen species? Well, probably a good old Darwinian process of its wines tasting better.

Surely humans in different parts of the globe would have discovered the same sort of thing independently? Quite possibly; but at the moment it looks as if other centres of early vine cultivation – the Iberian Peninsula was one, as was Sardinia – were not starting from scratch, because there doesn't seem to be any DNA relationship between early cultivated vines here, and wild ones.

Putting a date on the first cultivation of the vine, or the first wine, is a matter of guesswork at the moment, or romance. There is evidence of both wild and cultivated vines in sites dating from the Neolithic era (8500–4000 BC) in southeast Anatolia and Transcaucasia. The remains of liquid containing tartaric acid, identified by infrared spectrometry, have been found in Neolithic settlements from the sixth millenium BC in northern Iran; the presence of tartaric acid is a certain pointer to grapes. The liquid had been in stoppered jars, laid on their

This copy of a wall painting from the tomb of Kha'emwese at Thebes, c.1450 BC, is both a technical aide-mémoire, ensuring that the departed will be well supplied with wine in the next world, and perhaps an expression of pleasure in all the different stages of the cultivation of the vine and the making of wine. For us it is also invaluable documentary evidence of how grapes were grown and wine made in an early but extremely sophisticated civilization. The viti- and vinicultural techniques shown here are far from primitive, even if the wines that resulted might not be to our modern taste.

sides, and terebinth tree resin had been added as a preservative. Would they have gone to this trouble just to preserve grape juice? Maybe. But would they have been able to stop the grape juice beginning to ferment? That's what grape juice will always do, if you don't stop it, so isn't it more likely that these stoppered jars held this weird and wonderful nectar called wine, the bringer of joy and laughter and good times? If so, who was first? Was it a mistake that turned out brilliantly? Well, an awful lot of the seminal moments in the development of all alcoholic drinks are 'happy mistakes'. And the chance fermentation of some wild grape juice was probably the first of the lot, lost in the mists of alcoholic time.

It really shouldn't come as a surprise that the grapevine was first cultivated around the headwaters of the Tigris and Euphrates. The Fertile Crescent was where many 'founder crops' were first cultivated, those from which agriculture throughout the world is derived: chickpeas, lentils, rye, peas, emmer and einkorn wheat come from here. As do languages: all the Indo-European tongues which dominate Western European culture – and consequently much of the rest of the world – have been shown to have originated here. To say that wine and the vine are part of the warp and weft of human civilization is not an exaggeration.

Vitis vinifera and wine

All vines belong to the genus *Vitis*, which in turn belongs to the Vitaceae (formerly Ampelidaceae) family (see chart right). The *Vitis* genus includes around 60 species and is generally divided into two sections: *Euvites*, which contains nearly all the American, Asian and European vine species, including the European wine vine, *Vitis vinifera*; and *Muscadiniae*, which is sometimes considered a separate genus. This book focuses on *Vitis vinifera*; other species of vine, like *Vitis labrusca*, *Vitis riparia* or *Vitis berlandieri*, are important to wine largely because they provide, either directly or by way of crossings, the rootstocks on to which *Vitis vinifera* vines are grafted. (See page 22.)

There are also hybrids: vines whose parents are of different vine species, as opposed to crosses, which have both parents of the same species.

The usual object of breeding hybrids is to combine some genetic advantage of a *labrusca* or *rupestris* vine with the better wine flavours obtained from *vinifera*: non-*vinifera* varieties give pungently scented wine often described

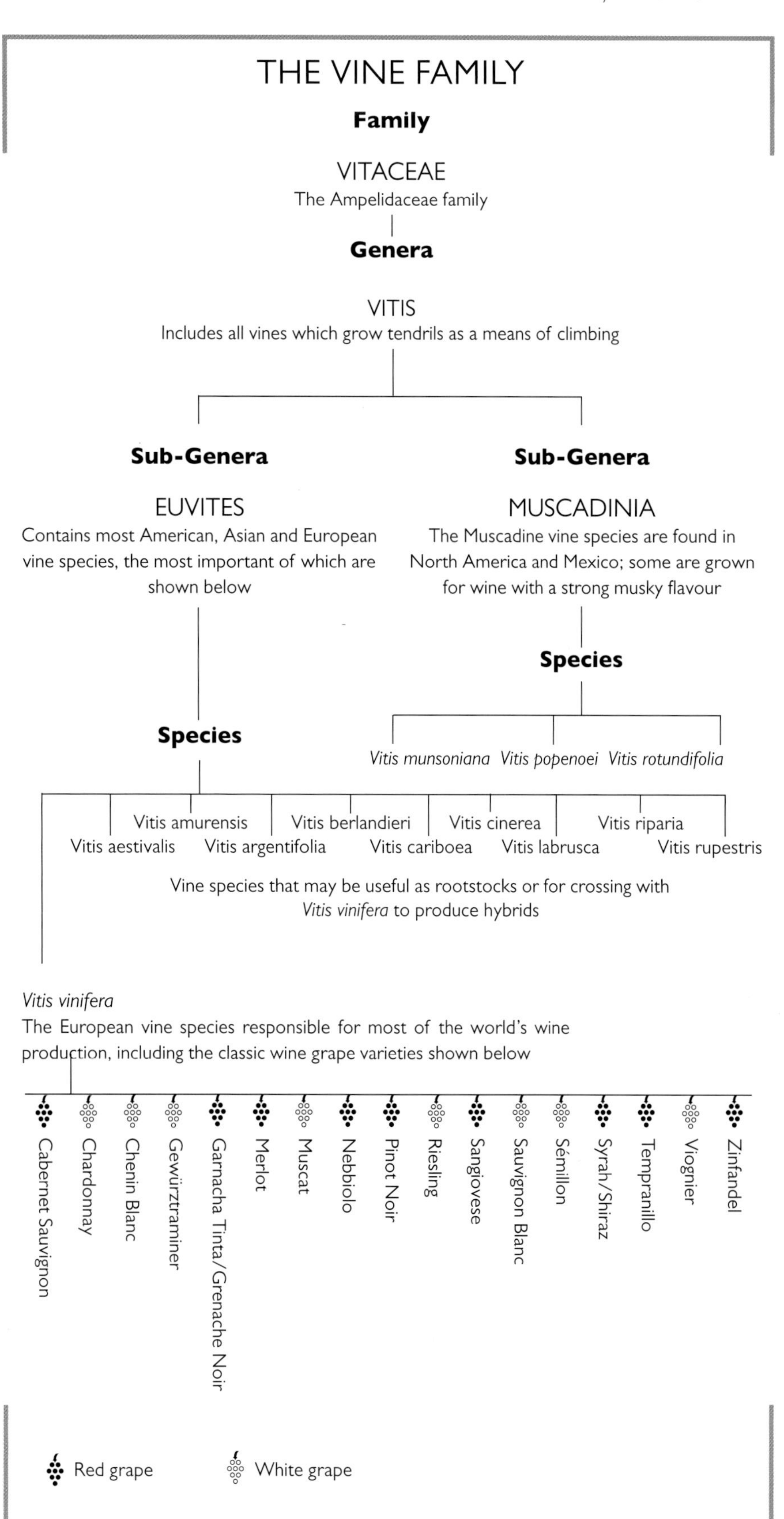

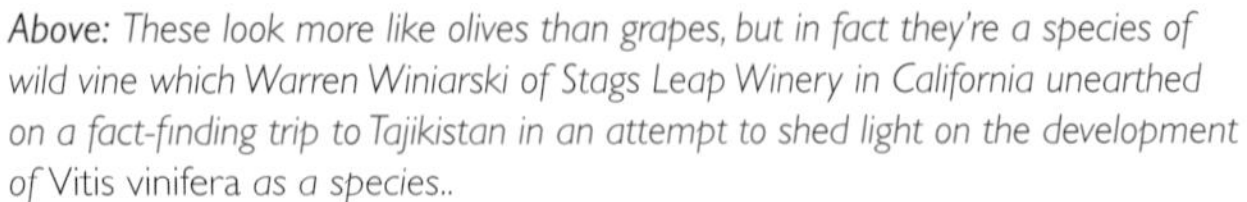

Above: *These look more like olives than grapes, but in fact they're a species of wild vine which Warren Winiarski of Stags Leap Winery in California unearthed on a fact-finding trip to Tajikistan in an attempt to shed light on the development of* Vitis vinifera *as a species..*

Above: A 400-year-old tree vine of the White Horse Breast variety. No, I'd never heard of it either, but the ancient vine lands are full of survivors like this, and each one might have a fascinating story to tell, or part to play in the evolution of modern vines. This one is in the Samtskhe-Javakheti region of Georgia.

as 'foxy', though the aroma is in fact more reminiscent of mayblossom or nail varnish. Maybe that's why winemaking was slow to catch on among the early settlers in North America whose native vines were all non-*vinifera*. But these vines may have resistance to disease (particularly phylloxera) or cold: hybrids of American vines with *vinifera* are widely planted in many more northerly states in the USA because *vinifera* vines find it difficult to withstand the winters. Seyval Blanc, a so-called French hybrid (French hybrids are a group of hybrid vines bred in France in the late 19th and early 20th centuries in an attempt to find a solution to phylloxera), was much planted in England until the focus of English growers changed to sparkling wine.

Hybrids are becoming increasingly sophisticated. There are some now being bred at research stations in Germany that are said to produce wine indistinguishable from the wine of certain *vinifera* varieties. These latest breedings involve not simply a crossing of an American vine with a *vinifera* vine, but numerous back-crossings to stabilize the European character. There is one, for example, that is said to be very close in flavour to Pinot Blanc, but which has no Pinot Blanc genes; another is said to be indistinguishable from Riesling, and does have some Riesling somewhere in its ancestry. The best red hybrids resemble Merlot, with good colour, low acidity, and blackberry-cherry fruit.

At the moment hybrids are not permitted for the production of quality wine in the European Union. And anyway, what the world seems to want at the moment is older varieties, preferably ones that are almost extinct, and can be rescued and propagated from three scrubby old vines growing in a corner of an untended vineyard 10km (6 miles) from the nearest road. The speed at which old varieties are being rediscovered is astonishing: most, truth to tell, were probably growing unacknowledged in old mixed vineyards, and not all are wonderful. Some will undoubtedly prove to be genetically the same as other varieties grown elsewhere; some may turn out to be missing links in stories of genetic relationships which are still incomplete. Ten years ago, the story of the vine seemed to be shrinking down to a tail-end of popular varieties, supplemented by new crossings; now we're rediscovering where we've come from.

But some just might completely transform our experience of what is possible in the spectrum of flavours. The guys who first planted the remote Waitaki area in New Zealand's South Island with vines said the flavours were so different from what had gone before, it was as if they'd discovered a new colour.

VINES TODAY

Of the 10,000 or so varieties of grapevine, not all are cultivated for wine. Some are used for table grapes, some for raisins, and some are wild and shown no sign of pining for the benefits of civilization. DNA profiling has established that some of the vines thought to be different are in fact genetically the same; and some thought to be the same are different. Vines are not quite what we thought. Fewer than 400 of those 10,000 are responsible for most of the wine we drink.

*To you, me and most growers – and indeed to every bureaucrat in charge of enforcing wine laws – Pinot Noir (**left**) and Pinot Gris (**right**) are different varieties. Try putting Pinot Noir on your label when you used Pinot Gris, and you won't get far. Yet to a geneticist they are the same, because they have the same DNA.*

The vine, *Vitis* – whether *vinifera* or another type – has been on the planet a long, long time. It has had ample time to cross-breed. It also mutates constantly and unpredictably. It used to be thought that some vines, Pinot and Muscat being the ones usually cited, were more prone to mutation than others; now it's thought that the great age of these vines – they're among the oldest we have – accounts for their numerous variations. When it comes to mutation, *Vitis vinifera* can outdo most fruits: its heterozygous nature (meaning that its already complex gene pattern can be readily rearranged) enables it to adapt itself with surprising ease to new circumstances.

Which leads us to another question: at what point is a variation on an existing vine considered to be a separate vine variety? This is a complicated matter, and a grower's answer is different to the answer of an ampelographer (or vine geneticist).

To a grower, a vine is a different variety if it looks different. The points of difference that matter are the shape and appearance of the leaves – the depth of the indentations, the appearance of the veins – the hairiness of the shoot tips, the shape and tightness or looseness of the bunches, the colour of the berries, the appearance of the canes, buds, flowers and seeds. Thus, to a grower, Pinot Noir and Pinot Gris are separate varieties.

To vine geneticists like Dr Carole Meredith of the University of California at Davis (UCD), however, Pinot Noir and Pinot Gris are merely colour forms of the same variety. They have the same DNA, therefore they are the same vine. 'In genetic terms,' she says, 'each variety is descended from a single unique original seedling. All vines of that variety can be traced back to that original seedling and have been derived from it by cuttings or buds.'

Pinot Noir is notorious for its variation: some strains grow upright and produce large, loose bunches; others do the opposite. They can be so different in appearance, and the wine can be so different in quality and flavour, that as consumers we might be justified in wondering if all these different strains should really all bear the same name. Yes, says Meredith, they should. Their DNA is the same. To produce a new variety you have to produce a seedling. You can do this by crossing two separate varieties, but even a seedling of the same variety (*Vitis vinifera* vines, remember, are hermaphrodite) is unlikely to be true to its parent. (The propensity of *Vitis vinifera* to produce seedlings that are different to its parent is the reason why vine propagation is done with cuttings.) Vines may mutate so that they look different and even behave differently, but they are still the same variety, and always will be; and this ability to mutate is one reason why the same variety can succeed in different climates and circumstances. And it is the basis of clonal selection.

What is a clone?

Good question. 'Clone' is a word that is going to crop up a lot in this book. If you take a cutting of a plant – it might be a vine, it might be a pelargonium – you have a clone. Any plant you grow from that cutting will be genetically identical to the first. Any properly observant grower will be aware of which vines in his vineyard are the best. They may be the healthiest, or they may give grapes that ripen a little earlier or a little later, or have particularly good flavours. Whatever their positive attributes, by propagating cuttings from those vines rather than others, that grower is practising a form of clonal selection.

Massal versus clonal selection

This 'home-made' selection is known as massal selection, which means a 'mass' of different cuttings from a vineyard are planted out, usually to encourage diversity, and complexity of flavours – wine growers in traditional areas have been doing this for centuries. The alternative, clonal selection proper, is done in the laboratory: it involves taking cuttings from single vines that have been observed to be especially good in some way, and producing successive generations of identical vines from this lone parent. The object is not only to

Tissue culturing is a crucial part of the virus elimination process. The plantlet on the left is in a rooting medium to form roots, while the one on the right is now in a soil medium.

Meristem culture – the tissue culture of the 1mm tip of a vine shoot – allows the propagation of large numbers of plantlets hopefully free of virus.

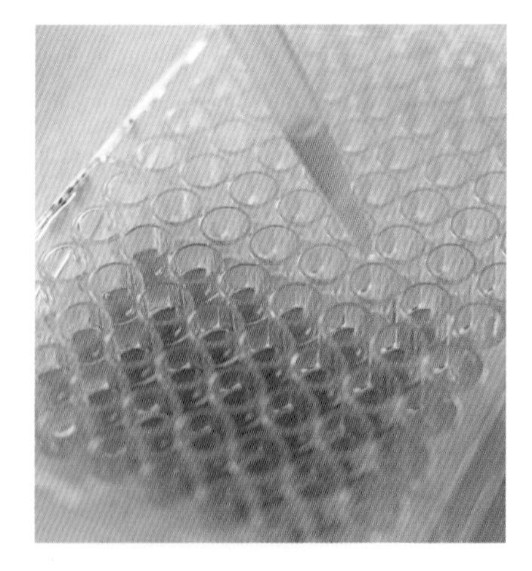

Elisa indexing is a serological test used to check vine pathogens and see if the virus elimination process was successful.

reproduce their good qualities, but also to eradicate their virus diseases.

Some virus diseases may eventually kill the vine; others may make it chronically weak and underproducing. There is no easy way of curing virus diseases once a vineyard is infected: the viruses live on pieces of root in the soil and are spread by nematode worms. By growing cuttings at high temperatures (around 38°C/100°F) and then taking a cutting of just the very point of the growing tip it is possible, since the spread of viruses through the vine is always just a fraction behind the growth of the vine, to have a piece of vine that is free of viruses. From that you grow your new vine. (There is some debate about how much clonal variation is itself the result of varying viral incidence.)

Clones take a long time – often around 15 years – to be made commercially available to growers: the plants must be selected over several generations to ensure that they stay true to type. (Mutations may become more stable later on – or less stable.) Their resistance to disease and poor weather, and their suitability to different soil types must be tested; and, most importantly of all, the quality of their wine must be judged. Many clones produced in the 1970s produced high yields but poor quality wine; much current clonal research is aimed at putting right those mistakes.

GM vines: getting closer

In 2007 scientists at the Institut National de la Recherche Agronomique reported that they had cracked the genetic code of *Vitis vinifera*. Research into genetic modification of grapevines had already been started; field trials of GM vines have been underway in many places for some years now. But this unravelling of the whole genetic code opens the door to wider changes.

Do we want GM vines? What if it were possible to produce wine without chemical sprays, because they were immune to disease and rot? Resistance to mildew, both powdery and downy, is one of the first targets. Resistance to cold is another.

The French team has isolated large families of genes linked with the production of tannins and terpenes – the latter give flavours and aromas. What if they come up with grapes that produce the best-tasting wine the world has ever seen? Well, I'd certainly want to taste it. Wouldn't you? What if you could raise the resveratrol content of red wine, thus perhaps increasing its anti-cancer properties? GM vines are not going to go away, and such pressing issues as climate change and the spread of so-far incurable vine infections are going to fuel the arguments. Sometimes it's amazing how quickly one can change one's mind.

Grapevines are not the same as maize or soya, and are of mercifully less interest to big agro-industrial companies. Once planted, vines generally remain in their vineyards for 30 years or more, unlike other crops which must be replanted from seed every year. With maize or soya there is the double profitability of every year selling seed genetically modified to resist a particular weedkiller, and then selling the weedkiller to go with it; that is not the case with vines.

But one big danger of GM vines is that of ever greater uniformity. The vineyards of the world are a huge resource of genetic variation: this new edition of *Grapes & Vines* includes numerous grapes that had been forgotten about, but which have come to prominence again as enterprising growers have rediscovered them. And that's without considering the genetic variation in vines like Pinot Noir. If GM vines turn out to be as useful, as economical, as labour-saving as they could be, will growers continue to bother with these curiosities? Will massal selection become a thing of the past? And will we be confined to a diet of standardized flavours?

Genetically modified yeasts also exist, though they're used mostly by the brewing and distilling industries so far. The aim of such yeasts is more efficient conversion of sugar to alcohol, and perhaps some built-in antimicrobial action. But what if GM yeasts could be used to reduce the alcohol level in wine? That's becoming an obsession with a lot of winemakers now. So-called 'lazy yeasts' producing as much as one degree less of alcohol from a given level of sugar in the grape juice have already been isolated. If all the brainpower that has gone into making yeasts more efficient could be employed to make yeasts less efficient, and our wines far lower in alcohol but just as tasty – isn't that extremely tempting, GM or not? These are issues we have to think about seriously – and not too emotionally.

Vine confusion: a question of identity

The identification of vines is a fascinating mixture of science and happenstance. Throughout human history people have traded and migrated; and they've taken their vines with them. With those vines have gone their local names; but new local names might later have been appended. So the same vine may

This is how infant vines are brought into the world at Yalumba's nursery in Australia. Each is a cutting of a parent vine, and so genetically identical to its parent, and indeed to its siblings. Nevertheless, when sold to different growers and planted in different soils, climates and conditions, each will behave slightly differently. Planting laboratory-produced clones is a way of weeding out poor quality strains of a vine, but it need not produce identikit wines.

Omega grafting – named after the shape of the graft cut. The scion material is on the left, the rootstock on the right.

Newly grafted Pinot Noir vines. The red wax is for protection in the early weeks. The choice of rootstock is crucial to the quality of the future wine.

be known by many different names in many different countries or regions. Conversely, the same name might be given to many different vines, even in the same region.

Chile has provided several of our most recent identity crises. Vines were brought out from Bordeaux in the 19th century and labelled haphazardly – one might suggest they were labelled according to what the market seemed to want. So Carmenère – a remarkable Bordeaux variety – was labelled by the easier to sell title of Merlot. Sauvignonasse – a rather neutral Bordeaux variety related to Sauvignon Blanc – was nonetheless called Sauvignon Blanc, because, well, which one would you buy? The search for the true Zinfandel has led us a merry dance through California, New York, Vienna, the heel of Italy and various tiny patches of vines in southern Croatia and produced a fistful of alternative names. Tribidrag seems to be the most ancient name for Zinfandel – so far. DNA studies have also revealed, for example, that Vermentino, Favorita and Pigato are all the same vine.

The results of DNA studies have been fascinating. Jancis Robinson, Julia Harding and José Vouillamoz (*Wine Grapes*, 2012) argue in favour of a small number of founder varieties responsible via natural crosses for the huge range of vines we have today: such founder varieties would include the ancestor(s) of Pinot and Savagnin, for example, which now head a pedigree involving no fewer than 165 varieties, including such well-known vines as Grüner Veltliner, Cabernet Sauvignon and Syrah, and such obscure ones as Raffiat de Moncade and Timpurie – neither of which could be squeezed into this book. Everything, it would seem, is interrelated.

Thousands of waxed and grafted vinifera *vines on American rootstock at Chateau Junding nursery in Shandong Province, China. Demand for healthy plant material is huge in China.*

THE GLOBAL PICTURE

First of all, let's get our bearings. This map shows where wine is grown: it gives the bare facts of latitude. It shows that most wine is grown between about 32° and 52° north, and about 26° and 45° south. If you're thinking of making wine, that's where you should stick your pin. After that, though, things get a little more complicated.

Northern hemisphere vineyards, as you can see from the figures above, tend to be at higher latitudes than those in the southern hemisphere. Partly this is for the very simple reason that most land masses in the southern hemisphere don't extend all that far south: only the curving tail of South American goes past the 50° mark. That South Africa doesn't extend any further south is certainly a great source of frustration to some of the producers there, who have difficulties in finding sites cool enough for some varieties – Pinot Noir, for example.

But the question of latitude is not just a simple one of temperature. The northern hemisphere is generally warmer than the southern at the same latitude, both in the growing season and over the course of the whole year. This is partly because of the greater land mass, and partly (as far as western Europe is concerned) because of the warming effect of the Gulf Stream. So comparing a northern latitude with a southern one can be misleading. Because it is possible (just) to ripen grapes successfully in southern England does not mean that the same grapes would ripen at the same latitude in Chile or Argentina. They wouldn't.

The effect of latitude on wine flavour is enormous. Lower latitudes have early springs and late autumns, and thus provide the long ripening season that winemakers the world over covet: grapes that ripen slowly on the vine have time to develop plenty of flavour and aroma as well as just sugar. But on the other hand, low latitudes tend to mean high temperatures, so that grapes quickly reach high sugar levels, and lose acidity, which is the opposite of what growers want.

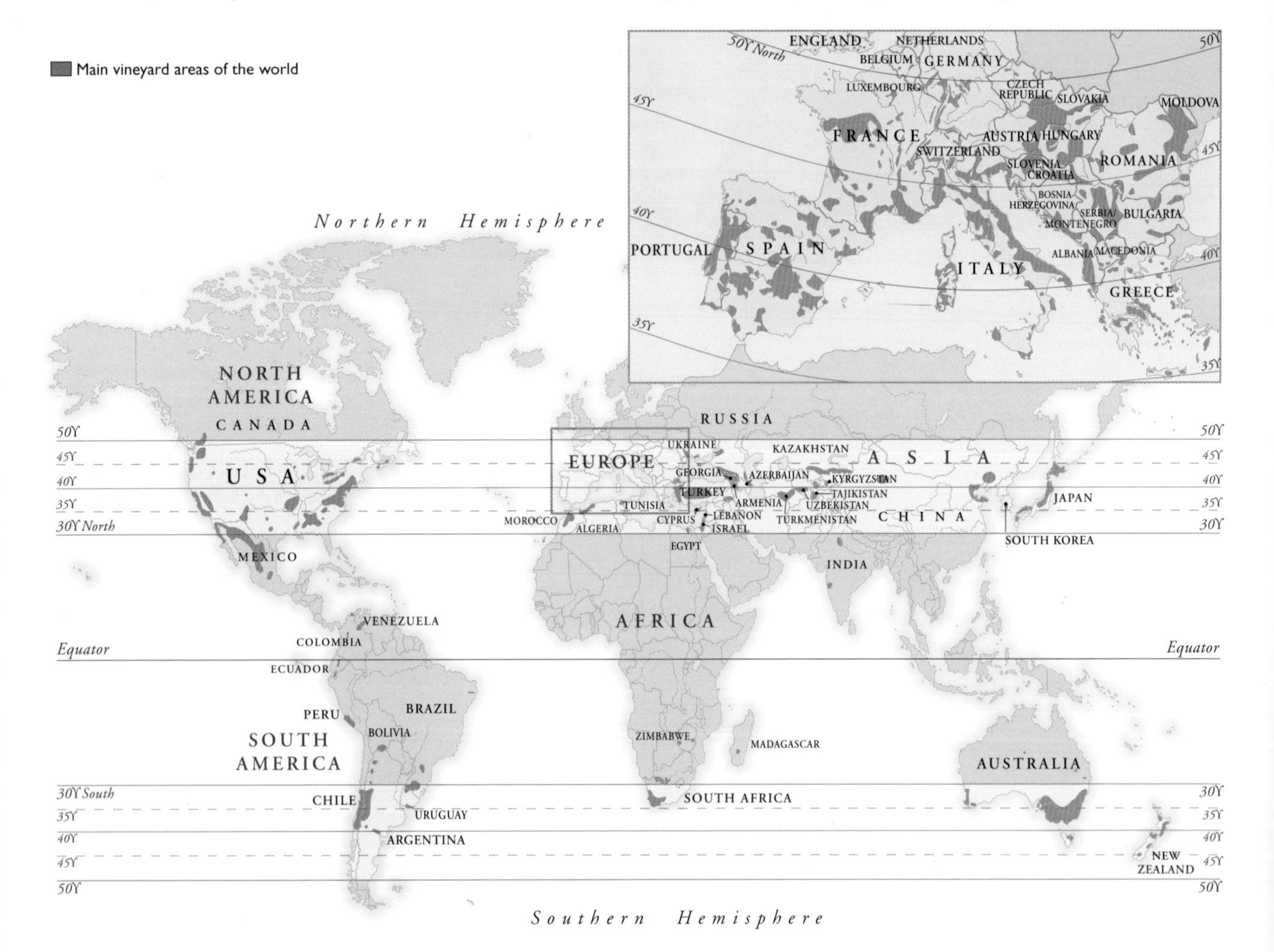

By contrast, higher latitudes mean shorter, cooler summers, but longer hours of daylight. More hours of daylight, and sunshine that is less fierce but lasts longer during the day, means more efficient photosynthesis, better retention of acidity and better development of flavour and aroma. Grapes ripen relatively faster at higher latitudes – in Bordeaux they reckon to have about 100 days between flowering and picking, whereas on California's Central Coast the figure is more like 150 – but if the temperature is too cool there is nevertheless a risk that the grapes won't ripen fully before autumn sets in.

Ideally, therefore, high latitudes need to be tempered with warm climates, and vice versa. We'll be looking at climate more closely on page 17 onwards. The third element in the equation – and the element that balances the other two – is the grape variety.

What grows where

Contrary to expectations, the world total amount of land under vine has decreased since the 1950s. The 1951 figure of 8,845,130 hectares (ha) of vines peaked to 10,213,000ha in the late 1970s; then vine-pull schemes in Europe and the USSR started reducing the total. The aim in Europe was to drain the wine lake; the aim in the USSR was to reduce alcohol abuse. By 1998 both these initiatives were effectively over; the European one, at least, had succeeded in its aim. 1998 saw a world level of 7,716,000ha. Since 1998 there have been local adjustments and in 2012 (the most recent year for which figures exist), the world vineyard area totalled 7,528,000ha.

In the southern hemisphere, South Africa, New Zealand, Chile and Argentina have all shown significant percentage growths in the period since 2000 – New Zealand showed a staggering increase of 170% between 2000 and 2015. With an increasing emphasis on producing quality wines for the export market, this isn't all made up of new plantings: many of these new vineyards are, in fact, replantings with better quality 'international' varieties.

In the same period China planted a further 270,000ha of vines, bringing its total to 570,000ha in 2012. Although the majority of these produce table grapes and raisins, plantings of wine grapes (mainly international grapes with some traditional Chinese, German and Russian varieties) are expanding rapidly. Despite the vine-pull schemes, Europe's vineyards in 2011 still covered 4,253,000ha, which makes it far and away the world's biggest grape-producing area. To put that in perspective, Australia had around 170,000, while in France, Languedoc-Roussillon alone had 193,748ha.

(Below) The starkest statistics here are those for the European powerhouses of Spain, France and Italy. Their longterm decline continues relentlessly and, pointedly, their wine consumption is slumping too. Hungary has also suffered a dramatic loss of vineyards, and only Romania, of the major producers, is reasonably steady. The figures for the USA and the Southern Hemisphere are more encouraging, and Australia's slight decline is more a correction than a trend. Chile registers a 4% rise in vineyard area, but production has shot ahead by over 40%. New Zealand saw a 7% rise, but the rise since 2000 is 168%, and most suitable vineyard land is now planted. In China, accurate statistics are hard to come by, but there's no doubt that the world's biggest vineyard expansion is taking place in China – up 19% since 2008, and up 90% since 2000.

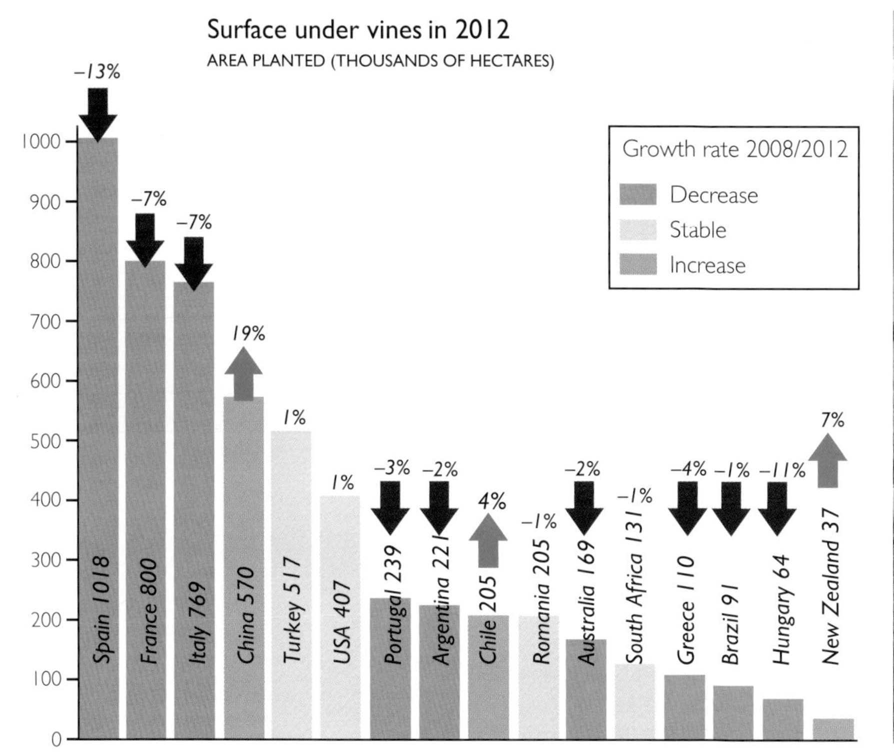

(Below) This is Lake Wanaka in Central Otago in New Zealand's South island. Vineyard latitudes in the Southern Hemisphere don't get much higher than Central Otago, at around 45° south, and according to climatic statistics it should be impossible to grow grapes here. In fact, though, the summer's extremely long sunshine hours, and the almost complete lack of summer cloud cover combine with favourable local mesoclimates to produce conditions suitable for both Chardonnay and Pinot Noir.

WHERE GRAPES GROW

If climate and latitude alone determined where vines grew, life would be so much simpler. You would simply look at the sunshine, the temperature, the rainfall and the chance of frost, and work out what to plant where. Hey presto: great wine, every time. Unfortunately, it's not quite that simple. Weather is only one of many factors that determine the quality and style of wine. When considering where grapes grow, it's worth remembering that European vineyards were seldom planted after close analysis of the weather or the soil. If by chance they turned out to be great vineyards, then people tried to work out why. It's been a long process of discovery, and it's not over yet.

Winemakers have been studying Burgundy's Côte d'Or for years, but we still don't know precisely what it is about this little stretch of French vineyard that produces such marvellous Pinot Noir and Chardonnay. And if people can't agree on that, it's not surprising that they also can't agree on which of the attributes of the Côte d'Or you should try to imitate if you want to make great Pinot elsewhere, or how closely you should imitate them. Should you find somewhere that mimics the Burgundian climate? Or is the climate in fact a disadvantage? If you think that, then you'll seek somewhere warmer and drier. Should you be trying to copy the soil? And if so, should you be looking at its structure, its mineral content, or what?

Vineyard dirt is receiving intense study from winemakers everywhere. Even those few who seem perversely determined to eradicate all traces of *terroir* in their wines by overripeness, massive extraction and as much manipulation as possible in the winery (added tannins, added acidity, added colour) still take pride in having good or even great *terroir*. *Terroir* has swung back into fashion, big time. In Bordeaux, top châteaux divide their parcels of vines ever more precisely according to which bits ripen together – a parcel might be one hectare, or half a row; there's really no limit to how precise you can be – and then redesign their wineries to suit, with one vat, appropriately sized, for each parcel. Suddenly they're able to ferment individual parcels that have never before been fermented separately, and the effect on the wine, in terms of precision and accuracy, can be dramatic. Accuracy in the wine comes from accuracy in the winery, which comes directly from accuracy in the vineyard – via, it has to be said, a substantial investment budget.

Yet as we're seeing in Champagne, while it wrestles with the complexities of bringing into the appellation those bits of land judged to be at least as good as those already in it, analyzing *terroir* is one thing. Proving that one *terroir* is better or worse than another is quite another, though it provides a gratifying amount of work for m'learned friends, since these matters always seem to end up in court.

If it weren't for the water, there wouldn't be any wine here. The Finger Lakes in New York State have a relatively small surface area but are extremely deep and seldom freeze in winter, so act as a very efficient heat store. Vineyards on the surrounding hillsides are protected against extreme winter temperatures and late spring frosts, and enjoy milder autumns than areas further away.

Terroir

This is still a relatively poorly understood concept – but it is important to realize that '*terroir*' does not mean 'soil'. The *terroir* of a vineyard is the sum of all its parts: its geology, its climate, its topology, its water-holding ability and the amount of sun it receives, and the effect of man. Without human intervention there wouldn't be vineyards in the first place. So the soil, both topsoil and subsoil, is important, as are the mineral components of the soil. How fertile or infertile it is, and its depth and structure, which affects how well or poorly drained it is, are also factors. Altitude, steepness of slope and exposure to the sun all matter, as does the mesoclimate, or climate particular to that vineyard. From the French point of view (and it is most of all a French concept), it is the *terroir* that makes each vineyard different. It underpins the Appellation Contrôlée system not least because, as the underlying factor behind wine quality and style, it should show in the wine no matter who the winemaker is or what he or she does to the wine. Winemakers come and go; the *terroir* remains.

However, good viticulture and winemaking can permit the expression of the *terroir* while bad viticulture and winemaking can mask it. And since good vineyard practice can mean installing drainage where necessary, thereby changing the soil's character, and since good winemaking in northern Europe does not exclude chaptalization (the addition of sugar before or during fermentation to increase a wine's alcoholic strength; see page 34), growers are in practice not absolute slaves to what their *terroir* dictates.

Climate

It's not surprising that so many winegrowers pay more attention to climate than to any other factor: its effects are obvious and undeniable. Even the most dedicated *terroir*iste is likely to blame the weather rather than the *terroir* when his vines are hit by spring frost (although those with vineyards sited outside frost pockets may well smugly praise their *terroir* when they escape). Rain or high winds during flowering; drought in late summer, causing photosynthesis, and thus the ripening process, to stop; summer rain that provides the humidity in which rot can attack the grapes; rain at harvest which dilutes the juice; hail at any time: the climatic hazards faced by the grower are endless. In Europe, growers are largely stuck with existing appellation boundaries and the weather conditions that go with them. In the New World you can plant a vineyard where you like – which in practice means, since the perfect vineyard site does not exist, that up to a point you choose your problems rather than inherit them.

Nevertheless, the weather during the growing season does not make much difference to the quality of the vintage until *véraison* (the onset of ripening, when the hard, green immature fruit softens and takes on the colour of the ripe grape) takes place. Spring frosts or excessive water stress (i.e. drought) may reduce quantity, but as far as quality is concerned it is the last couple of months before the harvest that really matter.

Cool or warm

There is no absolute definition of what constitutes a cool, warm or hot climate for viticulture: what feels cool to a grower in California's Napa Valley might seem warm to a grower in the Saar Valley of Germany. A cool climate is usually understood to mean one where only early-ripening grape varieties will ripen: grapes like Pinot Noir, Riesling, Chardonnay or Gewürztraminer. An intermediate climate will ripen later-ripening varieties like Merlot, Cabernet Sauvignon and Syrah. In warm climates you get very late-ripening grapes like Mourvèdre, Grenache and Touriga Nacional, but also earlier-ripening varieties such as Muscat if the aim is sweet fortified wines. Carneros and Sonoma Coast in California, Oregon, Australia's Mornington Peninsula, New Zealand's South Island, Burgundy, Germany and, obviously, England come under the heading of cool climate; Bordeaux, Tuscany, California's Napa Valley, Spain's Rioja, Chile's Maipo and Australia's Coonawarra and Margaret River are intermediate. The south of France, the Douro Valley in Portugal, Jerez in Spain, most of Mendoza in Argentina and McLaren Vale in Australia are warm.

Hot climates grow grapes mostly for raisins or table grapes, though wine is made nevertheless in some tropical climates in South America and elsewhere.

Mediterranean or Continental

A Mediterranean climate has the sort of mild winters and long, warm summers that prevail around the Mediterranean. A Continental climate is more extreme, with colder winters and hotter, sometimes shorter, summers. More inland parts of Europe, such as Germany, Austria and Hungary, have a Continental climate. Regions near the sea, like Bordeaux, have a maritime climate: the proximity of a large body of water tempers extremes of heat and cold, but also brings the danger of storms and rain.

Having an expanse of water near a vineyard can, in fact, be vital in marginal climates (a marginal climate is one where a given grape variety will only just ripen: marginal climates often produce the most complex, elegant wines – in years when there's enough sun, that is. The downside of the equation is raw, green wine when the sun doesn't shine). A large river or lake, by acting as a heat store, can raise the average temperature by a vital degree or two. It's one of the reasons why Riesling will ripen on the southeast- to southwest-facing banks of the Mosel, but not further away from the river's edge.

Getting warmer?

It's a question that preoccupies many growers: how to deal with climate change? In Australia, for example, they're increasingly looking at vines that can survive heat and need little irrigation. Champagne growers have been seen prowling round the chalk slopes of southern England. And China, where there is great scope for planting further north, and further north again, could be a net beneficiary.

There's no question that the climate is changing, says Giuseppe Rizzardi of Guerrieri-Rizzardi, producer of Bardolino and Valpolicella: 'the grapes tell us so'. Grapes are the ultimate, unarguable voice of climate change. They don't tell you what's causing it; all they tell you is whether they like it or not.

Interestingly, grapes in many European regions do like it, so far. German Riesling has moved into its comfort zone: until the 1980s catastrophic vintages were not uncommon, and it was legal for growers to add water to

THE WINKLER AND AMERINE HEAT SUMMATION SCALE

It is only right to give a proper account of the Winkler and Amerine Degree Day System, since it governed so much of what was planted and where in California from the 1940s on up until relatively recently.

Albert Julius Winkler (1894–1989) was a scientist at the University of California at Davis (UCD), the leading US viticultural research institute as well as the leading US college for aspiring winegrowers and makers. He and Maynard Amerine (1911–98), also of UCD, devised a system of classifying the regions of California by heat summations. They assumed a growing season that extended from 1 April to 31 October, and worked out the number of 'degree days' in different regions. This was done by calculating the amount by which the temperature exceeded 50°F/10°C, which is the lowest temperature at which the vine will grow. They divided California into five different regions, with Region I, the coolest, having fewer than 2500 degree days, to Region V, with more than 4000. Specific grape varieties were then recommended for each region.

When applied to California the system is broadly reliable, since it so happens that in California temperature is a pretty good guide to overall climate. Elsewhere it works less well: applied to Australia, for example, it would deny the possibility of making good Semillon in New South Wales's Hunter Valley – yet Hunter Semillon is one of the classic wine styles of Australia.

In a very hot country you may have to go very, very high up to find a suitable climate for growing vines. In Catamarca in Argentina, Michel Torino has planted vines at around 2400m (7880ft): luckily the Andes happen to be there to provide such sites.

their wines to reduce the acidity. Bordeaux and Burgundy have more good vintages than poor ones now. Champagne is on a roll, and vintage wines are now made almost every year, though not always by the same houses.

What we don't yet necessarily know is how great each vine's comfort zone – its plasticity, or its ability to adapt – is. Chardonnay is pretty flexible: it can grow almost anywhere. For Riesling, we will eventually find out.

Defining ripeness

What exactly is ripeness? On the face of it, it seems a silly question. We all know the difference between a ripe plum and an unripe one. But take the analogy of bananas. Some people consider a banana to be ripe when the skin is yellow all over. To others, a ripe banana has a skin speckled with black. In wine, a grape that is considered ripe in Champagne would be thought unripe in the Napa Valley. The high acid/low sugar 'ripeness' that is crucial for making sparkling wines of balance and vivacity in Champagne makes painfully acidic table wines. Napa's sun-soaked grapes make full-bodied table wines but would make fat, flabby fizz.

High sugar levels do not, on their own, constitute ripeness. A ripe grape has brown seeds, not green, and the stalk is lignified to the first joint of the bunch. When you taste it (and all good wine producers judge ripeness by taste first, and by analysis second), a red grape should have minimal vegetal green bean flavour, and its skin tannins should be soft and velvety. In other words, the aim is physiological ripeness, and in warmer climates this occurs at higher sugar levels than in cooler climates, simply because it cannot be hurried. Heat will speed up sugar formation; but for tannin ripeness you simply have to wait. This is why ripe grapes in Australia's Barossa will tend to produce more alcoholic wines than grapes of similar physiological ripeness in Bordeaux. The battle in warm regions is to attain full physiological ripeness without head-banging levels of alcohol: get the equation wrong and you end up with over 15 per cent alcohol and you still get unripe flavours.

The ideal is to have sugar ripeness and tannin ripeness at the same time. We would argue that this is one of the criteria of a great *terroir*. We would even question whether places that find this impossible, at least in the majority of years, can be considered great *terroirs*. To have sugar ripeness hugely out of sync with tannin ripeness makes it impossible to have balanced wines without regular intervention in the winery, whether it's to add acidity or to add sugar. Is this too demanding an approach? Since it would rule out most of Western Europe at least until the 1980s, and much of the New World now, it perhaps is. But if were are talking about a great *terroir* – no, it's not too demanding. And whereas until the 1980s the wine world's problems were mostly about trying to make up for lack of ripeness, climate change means that an increasing number of our problems nowadays are to do with overripeness and loss of any 'typical' flavours from a vineyard. Grapes will have to be grown differently, and, if necessary, different varieties will need to be planted, ones which can express their sense of place in our warmer world. However you look at it, climate is an integral part of *terroir*.

Picking grapes too early can be unavoidable, if rain is on the way, or if, in cooler areas, a bad summer or autumn has meant that the grapes can't ripen. In white grapes this means high levels of sharp-tasting malic acid; leaving the grapes on the vine for longer would mean lower levels of malic acid, which falls as the grapes ripen, and higher levels of the riper-tasting tartaric acid.

In red grapes, there is the additional factor of tannins, which should be silky. Green tannins range from dry to bitter and unpleasant, but a touch of greenness in flavour can add some balancing freshness without being in the least frightening; most great Bordeaux of the 20th century had a streak of greenness in them. Some producers, however, particularly in California's Napa Valley, but also in various other New World areas as well as parts of Spain and Italy and even in patches of Bordeaux itself, dread the thought of green tannins so much that they are apt to leave the grapes on the vine until they are shrivelled and are heading for raisin status. By this time sugar levels are sky-high and acidity is falling faster than stock markets after a Euro crisis. The wines may have tannins so soft you could wrap a baby in them, or they may have just got fiercer and fiercer, but either way, the wines have lost most of the freshness that makes you want to drink them. And as for the alcohol, well, either it's very high, or has been reduced from 'very high' by perfectly legal but hardly desirable methods involving dilution or manipulation of the grape juice or wine. It is perfectly possible to produce ripe, silky tannins without destroying the wine's freshness, but you need the magic combination of a good ripening season in a relatively cool climate that will hold the sugar level back and prevent the grapes shrivelling to raisins while you wait for the tannins to ripen. Bordeaux at its finest is the best example of this.

How much alcohol?

If you can taste the alcohol on the palate, as a burn on the finish or, even worse, all the way through the wine, the wine is not in balance. It may be a good wine in all other respects, but alcohol that sticks out is a fault. Balance can be found in wines of 14.5 or 15 per cent; unbalanced wines can be found at 13.5 per cent. Alcohol levels are not a problem in themselves; it is lack of balance that is the problem. But it has to be said that high alcohol tends to go hand in hand with lack of aroma, lack of freshness and general soupiness because the more you search for higher sugars, the more acidity the grape will lose. Good acidity balances high alcohol and inadequate acidity will taste jammy, soupy and tiring.

There are many reasons for the apparently inexorable rise of alcohol in recent years. Warmer summers mean more sugar in the grapes; the desire for riper tannins means longer hang times; vigorous, virus-free clones also put in their ha'porth and the pruning of the vine to produce self-defeatingly small crops of grapes per plant whacks up the sugar. Ways of bringing alcohol down, or at least getting it under control, include long-term strategies like planting at higher altitudes and higher latitudes, and shorter-term ones like the development of less efficient yeasts; de-alcoholization of wine, where it is allowed; adding water to the must, where that is allowed(!); open-topped fermenters that allow some alcohol to evaporate during fermentation; less leaf-pulling and more shading of clusters; and greater selection in vineyard and winery.

Château Margaux brought the alcohol in its white wine, Pavillon Blanc, down from more than 15 per cent in 2007 to 13.4 per cent in 2012 by, among other measures, using only the first part of the juice to be pressed off the skins, which has less sugar and more acidity, and rejecting the later part, which is closer to the skins, lower in acidity and higher in sugar. In the case of Pavillon Blanc, these measures may mean rejecting 60 per cent of the crop. As with restructuring your winery to suit your vineyard and thus reflect your *terroir*, it helps not to be skint.

COMMON VINEYARD SOILS

What you see at the surface of the vineyard may be material deposited by rivers, glaciers or erosion, or it may be the topsoil: the weathered bedrock combined with organic matter. If this layer is too fertile the vine will spread its roots sideways, rather than sending them deep – sometimes as much as 30m (100ft) – into the subsoil to establish a stable foundation. The subsoil, or pure weathered bedrock, may be relatively uniform or may consist of several strata.

Alluvial: soil deposited by rivers. Alluvial soils can be very fertile, and contain sand, silt and gravel. **Argillaceous:** includes clays, marls, shales and others. **Calcareous:** contains calcium and magnesium carbonates. These are cool, water-retaining soils; usually alkaline, they give grapes with high acidity. **Chalk:** a type of limestone. It is cool and alkaline, and combines good drainage with sufficient moisture retention. **Clay:** cold, acid, poorly drained soil. Solid clay is difficult for vines, but clay mixed with other soils, in particular calcareous soils, can be excellent for vine growing. **Clayey-loam:** even more fertile than loam on its own. Can drain poorly in wet weather. **Ferruginous clay:** clay with an admixture of iron. **Gneiss:** a type of granite. **Granite:** warm, mineral-rich soil that tends to produce grapes low in acidity. **Gravel:** pebbly, well-drained soil, generally infertile. It is acid, so produces grapes with low acidity. Gravel over the alkaline limestone gives wines with more acidity than gravel over clay. **Limestone:** carbonate-rich alkaline soil. There are various sorts of limestone, of which chalk is one. Different limestones have different water retention. Limestone soils generally give grapes with high acidity. The Kimmeridgian soil of Chablis is a calcareous clay containing Kimmeridgian limestone. **Loam:** warm, fertile soil – generally too fertile, in fact, for fine wine. Loam contains clay, silt and sand. **Marl:** calcareous clay. Marl is cold soil that holds back ripening and gives wines with high acidity. **Sand:** fine-grained, warm soil that drains freely. **Sandstone:** rock with sand-sized particles bound together by minerals or forced together by pressure. **Sandy-loam:** loam with a large admixture of sand. Sandy loam is warm and well drained. **Schist:** warm, splintery, crystalline rock derived from shale, sandstone or granite. **Shale:** warm, reasonably fertile soil. **Slate:** warm, well-drained soil. Slate soils like those in the Mosel can consist almost entirely of thinly split fragments of rock.

Large stones like these at Ata Rangi in Martinborough, New Zealand, can act as a heat store, soaking up heat during the day to advance ripening in what is a cool, marginal climate.

The deep beds of gravel here in Pessac-Léognan are warm and provide good drainage, which is vital in Bordeaux's damp climate, and can make all the difference between good and mediocre wine.

Again, good drainage and warmth are the key advantages of the slate of Germany's Middle Mosel region. There is so little actual soil here that one wonders how vines live at all.

The Elqui Valley in Chile should really be too warm for growing good grapes, bordering on the Atacama Desert at 29° South. But a mixture of high altitudes – grapes are grown at up to 2000m (6600ft) altitude – and keen icy winds drawn up the valley from a Pacific shore cooled down by the Antarctic Humboldt Current has seen exciting wines produced, particularly from the Syrah grape. Indeed, these steep granite slopes bear some resemblance to Syrah's homeland – the Northern Rhône Valley in France.

The rise of balance

This is the cause of much cheering in the wine world. Excess alcohol looks dated, and – hooray! – so does excess oak. Regions that had previously made wines in which you could stand a spoon up are now making wines that offer a sense of place instead of a sense of prunes in brandy. Priorat is fresher, more appetizing and far less intimidating than it was, without being a jot less impressive. Barossa is becoming far too aware of its varied *terroir* to want to cover it up. Australian Chardonnay is now crisp and elegant and the delights of cool climate Syrah (not Shiraz) are increasingly being lauded. Chile is now looking to newer, cooler regions like Elqui and Limarí for its stylistic guides, and even Central Valley wines are looking lighter. Producers in Argentina are pushing higher and higher up in the Andes. Winemakers, having discovered how to do ripeness, are now having fun doing restraint.

Fashion is moving away from both overripeness and overextraction, but it's important to differentiate between them. We've discussed the former; the latter happens in the winery. You can have overextraction without overripeness, and hard, dry tannins are the result. You can have overripeness without overextraction: you get a soft jamminess. But because alcohol is such a good extractor of tannins, the first usually leads to the second.

Climate and photosynthesis

Vines produce sugar by photosynthesis, the process in which sunlight is used to combine water with atmospheric carbon dioxide. The sugar thus made in the leaves is transported to the grapes. At low temperatures photosynthesis, and thus the ripening process, stops. But it also stops at high temperatures, and in drought conditions. In very hot, dry spells, if no irrigation is available, grapes may shrivel, but they won't ripen. This is why it's perfectly possible to get overripe, jammy flavours hand in hand with green tannins.

The heights and the slopes

One of the few incontrovertible statements one can make about vineyards is that higher altitudes are cooler. In California they reckon that the temperature cools by 2.2°C (3.96°F) for every 329m (1080ft) of altitude. This may not be a conveniently round figure, but it shows that if you want the longer summer that goes with lower latitude, you can get a cooler climate – and thus get the possibility of very long hang time, with all the benefits to ripeness and flavour that that implies – by heading up into the mountains. It's worth remembering that 19th-century vineyard pioneers in California chose the hillsides, not the valley floors.

Go too high, of course, and you may find that your grapes won't ripen properly. On the Côte d'Or and in Hungary's Tokaj region it is the middle of the slopes that produce the best, ripest grapes. In Portugal's Douro Valley the highest vineyards may ripen two to three weeks after those nearer the river, and may indeed be too cool for port; but they come into their own for unfortified table wine, which needs less sugar but more acidity.

But slopes have another advantage, too: they get more direct sunshine – crucially when the sun rises lower in the sky as late summer heads in to autumn/fall. In cool climates that means better ripeness. They also tend to have poorer, thinner, better drained, less fertile soil, which acts as a natural brake on the vigour of the vine; this in turn helps to control the amount of grapes the vine produces, and gives more concentrated flavours.

Planting on valley floors is certainly desirable if you want to produce large quantities of inexpensive wine. But this is not to suggest that only sloping vineyards are good. In areas that seem flat, minor gradations of only a few metres can dramatically change the quality of the soil. This is particularly true in gravelly regions, where ridges of gravel give well-drained infertile soil conditions, yet the troughs between the ridges are fuller of more fertile, damper clay soils, likely to give bigger crops of slower-ripening fruit. And in some of these seemingly flat areas, the changes can happen several times across 100m (330ft). In marginal and cool climate regions, this has a dramatic effect on the final flavour of the wine. The gravelly Médoc looks remarkably flat to the untutored eye, and the stony vineyards of Marlborough in New Zealand, seem to be about as flat as land can be, but are in fact full of gentle rises and falls, with gravel on the ridges and clays in the troughs. The floor of the Napa Valley seems to be ultra-flat, but underneath the surface all kinds of intricate geological patterns are woven.

Back to the soil

One of the current buzzwords in wine is 'minerality'. Ask 10 people to define it and you'll get 10 different definitions; ours would be that minerality is a certain saltiness, savouriness or stoniness – sometimes even a metallic quality – on the palate that is not fruitiness and not precisely earthiness, either.

Where does it come from, and how does it get into the wine? Is it a function of ripeness or lack of ripeness? Is it low-level reduction, or is it a genuine soil flavour?

Let's dispose of the second suggestions first. There seems to be a correlation between certain soil and certain characteristics in wine: the slate of the Mosel is instantly recognizable as a smoky tang in the valley's Riesling; limestone gives liveliness, clay gives weight and solidity; and volcanic soil gives a certain fieriness, a sparkiness, which sounds like a bad case of auto-suggestion, but which is recognizable. But these notes do not result from some movement of slate or clay or volcanic molecules through the vine, into the grapes and, through fermentation, into the wine. Or, at least, no pathway for such movement has yet been identified by plant biologists. Until such a pathway is discovered we should assume that while Chablis might have a trademark flavour that is like licking limestone, what we are tasting is not actually limestone – not yet, anyway.

Some authorities suggest that what we term minerality may derive from esters, which are created when alcohol reacts with organic acids, and which vaporize easily. In addition, the gunflint note associated with some Sauvignon Blanc comes from a volatile thiol, Benzenemethanethiol (BMT); volatile thiols are associated with low levels of reduction or sulphides, and it is possible that sometimes 'minerality' of this 'gunflint' variety may be the result of bottling under screwcap.

Growers who have minerality in their wines, along with those who covet it, maintain that it comes from old, deep-rooted vines with the ability to extract nutrients from way down in the soil, aided by the presence of mycorrhizal fungi in the soil. It's true that minerality has been more associated with the Old World than the New, where irrigation tends to make life in the top 50cm (20in) of soil much more enticing for a vine; at the same time the popularity of screwcaps does seem to have made minerality much more discernible in New Zealand Sauvignon Blanc. In addition, some vintages in Chablis show more minerality than others and a cooler Mosel vintage may seem more slatey than a warmer one, suggesting that ripeness levels can be a factor, with less ripe vintages displaying more minerality.

Chardonnay vines growing on the pale chalky slopes of Le Mesnil-sur-Oger, a Grand Cru village famous for its Chardonnay. The belemnite chalk spreads right down the Côte des Blancs; its whiteness reflects sunlight back on the vines, yet it retains some heat and just enough moisture, and its alkalinity is crucial in creating the acidity in the Chardonnay juice so prized by winemakers.

More research is needed. And when we know what it is, it should be a lot easier to agree on a definition of what it tastes like. But, I don't know, I think it'll remain subjective for a fair while yet.

The physical structure of soil should, however, be differentiated from its chemical attributes. Most authorities agree that the most important aspect of soil is its water-holding capacity, and how easily the vine can access that water. What vines like best is well-drained soils that nevertheless have an adequate, but not too high, supply of water. The chalk of Champagne and southern England is an example: chalk drains freely yet maintains a high water table, and the water is accessible to the vine. Clay, by contrast, drains poorly, yet can be so dense that much of its water content may be inaccessible to the vine's roots. Wet soils are cold soils; well-drained soils are warmer. Well-drained soils are thus likely to encourage budbreak earlier in the season, and aid ripening; they can be suitable for later-ripening varieties in marginal climates because they speed them up. A cold soil might be better suited to an earlier ripening variety. In Bordeaux, for example, Merlot does better on the clay soils of Pomerol and St-Émilion, where the later-ripening Cabernet Sauvignon would not ripen. Cabernet needs the warmer, freely draining gravel soil of the Médoc, where Merlot might overripen in hot years. Dark soils are warmer; light-coloured soils are colder. And very stony soils, such as those found in parts of Châteauneuf-du-Pape, conduct and hold the heat well, again aiding ripening of the grapes.

Some aspects of the chemical and mineral content of soil are more clearly understood and can be related to other crops. There is certainly a direct relationship between excess nitrogen and excess growth: vines getting too much nitrogen produce enormous quantities of leaves that in turn shade the grapes and hinder ripening. Too much potassium in the soil can reduce the acidity of wine. (Potassium-based fertilizers, very popular in the 1970s, are out of fashion.) High levels of organic matter mean very fertile soils, and these, too, can mean problems of excess vigour in the vine leading to yield and ripening challenges. But the vine must have some organic matter in the soil. Very poor, infertile soils can indeed encourage the vine to put down very deep roots in search of nutrients, but if it doesn't find those nutrients anywhere it won't flourish.

IN THE VINEYARD

Great wine is made in the vineyard. That is an article of faith, and is quoted even by winemakers who, judging by the results, must do a great deal of their making in the cellar. It's not that winemakers are becoming, on the whole, less technical: every year there's a new bit of kit that is a must-have for anyone with the budget. Many of these new acquisitions, however, are aimed at enabling winemakers to identify the very best grapes as the crop arrives at the winery, and then to ease their transformation into wine with the least possible trauma. Technically this allows you to have grapes as perfect as possible. But might you not be losing a bit of personality along the way? And what does perfect mean?

Manual sorting tables in the winery were upstaged by optical sorting tables; now, if you pick by machine, you can sort ripe grapes from unripe by using the Tribaie system, which was invented by Bordeaux grower Philippe Bardet. First a vibrating belt excludes small, green berries. Then the berries pass into a rotating drum. Damaged berries stick to the walls, while sound ones fall into a container of fresh grape must. Ripe grapes sink, unripe grapes float: it's the same principle as ducking for witches, really.

But we're jumping the gun. We're supposed to be talking about vineyards, not wineries; and it's been interesting to see the way that biodynamism (see page 26 for a full description) has moved closer and closer to the mainstream. In Bordeaux, Châteaux Latour and Margaux, among many others, have a hectare or two under biodynamic trial; Château Palmer makes its own compost from its vine prunings; Château Climens in Sauternes and Château Pontet-Canet in the Médoc are fully biodynamic, and the latter is moving from tractors to horses for vineyard work. These châteaux are not run by wannabe hippies. They are run by businesspeople and scientists who want to make better and better wine. If it works, they'll do it, and only afterwards try and work out why it works. What we have now is a spectrum of practices with industrial winemaking at one end, spraying fertilizers, herbicides and anti-rot sprays with a merry heart, supplying the brands that want total reliability and consistency. Then there is *lutte raisonnée*, which generally means using as few chemicals as possible but as many as necessary; then there is organic viticulture, the definition of which varies according to which organic certification organization you choose to belong; then there is biodynamism; then there are natural wines.

Plastic grow tubes in a new plantation of vines in Walla Walla, Washington State, USA. The tubes resemble a mini greenhouse speeding up growth. The first crop of grapes can be as much as a year earlier when tubes are used. And if there are any little critters in the vineyard who'd like to chew on the shoots, they can't get at them.

Natural wines are also undefined, though biodynamism is usually a given. But they also embrace wines made in clay amphorae, white wines macerated on the skins, known as 'orange wines' (these may well be amphora vines) and zero or minimal sulphur wines. It's a holistic approach that brings cellar and vineyard into the same embrace; and while many drinkers may find intentionally oxidized wines and bottle variation an embrace too far, the effect of natural wines is being felt throughout the mainstream – whatever that is. So we feel justified in beginning this chapter with a tangential discussion about cellar practices: cellar and vineyard are starting to merge into a continuum. Which, one could argue, is what they always used to be, before chemicals effected a revolution in the vineyards, and technology a revolution in the cellar. The problems that made chemicals sprays so popular in the first place do not, however, go away.

Phylloxera: the coming of the plague

Phylloxera still necessitates grafting every vine in most parts of the world. It is the only protection there is against *Phylloxera vastatrix*, the tiny aphid that at one stage of its complicated life cycle feeds on vine roots and kills the vine. Phylloxera invaded from the east coast of the United States in the second half of the 19th century, and vines throughout Europe were wiped out. It found its way to Australia in 1877 (though parts of Australia are still phylloxera-free), California in 1873 and New Zealand and South Africa by 1885.

A familiar sight in 1990s California wine country: burning phylloxera-affected vines that had been planted on insufficiently resistant rootstocks. This bonfire is in St Helena, Napa Valley.

The emergence of Biotype B in California attacked vines which had been grafted on to AXR1 rootstocks that were only semi-resistant to phylloxera, and widespread replanting took place in the 1990s.

A question of graft

The vines used for phylloxera-resistant rootstocks are bred from native American species like *Vitis riparia*, which have natural immunity. There are dozens of different rootstocks now in common use, but it takes some years to determine by means of field trials just which rootstocks are most suitable in which circumstances and on which site. Some, like 333 EM and 41 B, are particularly tolerant to lime in soil, and are thus suitable for chalky soils like those found in Champagne. Others, like 110 R, are tolerant of drought; SO 4, Ramsey, Dog Ridge, 1613 C and others are resistant to nematodes, but SO 4 gives high yields and roots only shallowly; Rupestris St George, 99 R and 110 R are very vigorous, and produce huge canopies in their scions, which can be a problem; 101-14 and Riparia Gloire, by contrast, have very low vigour and so can be used to curb the vigour of vines in very fertile soils. Ramsey is good at seeking out water, and is resistant to salinity, which can make it useful in Australia. 101-14 is useful for deficit irrigation.

Nematodes are tiny worm-like organisms that live in the soil and spread viruses – and they in themselves can be a good enough reason for grafting vines on to resistant rootstock, even if there's no phylloxera problem in your region. Even in relatively disease-free Chile, where nematodes are not generally a big problem, they can flourish in the sandy parts of Casablanca, and nematode-resistant rootstocks are being found to be necessary if vines are to flourish there.

It was only really in the latter part of the 20th century that producers began to realize the importance of having the correct rootstock. At Château Angélus in St-Émilion, for example, the current generation say that when they took over in the 1980s, only about 30 per cent of the vines were on the right rootstocks. The choice of rootstock can affect the berry size, the leaf area and the leaf-to-berry ratio, all of which have quality implications for the wine.

To clone or not to clone

While clones of some varieties like Pinot Noir, as we've seen on pages 11–12, can vary widely in quality, in vine varieties where there is less variation the choice of clone probably only accounts for between 1 and 5 per cent of the quality of the wine. So your choice of clone if you are growing Riesling is less important than if you are growing the genetically more variable Pinot Noir. And in both cases good viticulture is more important. To make a decision not to plant clones is almost impossible in many parts of the world: if your entire region was planted with scratch from clones, as is the case in much of the New World, then clones are what

OLD VINES

In the case of vines, age and beauty generally (though not always) go hand in hand. European vines are not permitted to produce *appellation contrôlée* (or the equivalent) wine until their fourth year, but that is still infancy in vine terms. After about 20 years vines become less vigorous and produce smaller crops of more concentrated fruit, and at a certain age – usually about 30 years – the producer has to decide if he values the increased quality more than the drop in yield. Most vines are uprooted at this point, and replaced.

Wine from old vines thus has some extra cachet. The only problem is defining how old is old. Some wines which blazon the words 'old vines' or '*vieilles vignes*' on the label may be made from vines of only about 35 years, which is not even middle-aged, far less old.

So where can you find really old vines? Australia has plenty of Shiraz and Grenache, especially in the Barossa, which was planted more than a century ago: Turkey Flat can boast some of 170 years old. Chile has many vines of 70 years or more, and California has old Zinfandel and Carignane vines that are more than 100 years old. Madera County in particular has large plantings of old Carignane left over from a previous fashion that nobody ever bothered to uproot.

In Piedmont, Italy, Marcarini has some centenarian Dolcetto on its own roots in the cru of La Morra, and in Campania there is some Syrah and Aglianico, around the same age, which have grown to the size of small apple trees. In Germany's Mosel Valley, Ernst Loosen has some Riesling vines of 100 years old in the Ürziger Würzgarten vineyard.

Old vines certainly produce wine of greater intensity, but don't forget that many of the great 1961 Pomerols and St-Émilions came from vines that had been replanted after the devastating winter of 1956.

Pre-phylloxera vines

Technically these are vines that date from before phylloxera took hold at the end of the 19th century – which would make them very old indeed. But the phrase is sometimes used to refer to ungrafted vines – and these are less rare than one might suppose. In Europe some of the most famous are at Quinta do Noval in the Douro Valley in Portugal, which produces its rare Nacional port from half-a-dozen small terraces of ungrafted vines. The vines all around are grafted, and attempts to extend ungrafted vines to other parts of the vineyard have always failed. For some reason, phylloxera never affected those particular parts of the vineyard – nobody knows why.

Elsewhere in Europe the Middle Mosel is still free of phylloxera and there are many vines planted on their own roots.

In the New World ungrafted vines are remarkably common. They are the rule in Chile, where phylloxera has never struck; there are also many in Argentina, in Australia, especially in South Australia, and some in New Zealand.

Gnarled Shiraz vines more than a century old growing in Clare Valley, South Australia. Vines like these survived because fashion passed them by, yet now they make some of the world's most sought-after red wines. Old-vine Grenache and Mourvèdre are in almost as much demand now as old-vine Shiraz.

Spraying the vines in early spring in Marlborough, New Zealand. Chemicals are used in a far more restrained way nowadays, but are still essential for a healthy crop in most vineyards.

you have. Even in Austria, Toni Bodenstein of Prager has to label his massal selection vineyards as 'experimental' in order to make them legal; and he spent three years searching for good Riesling and Grüner Veltliner vines from which to take cuttings, eventually settling on 25 Riesling selections and 110 Grüner Veltliners. He prefers them because their yields are naturally lower and he has to spend less time green-harvesting; 'it enables you to work better with nature,' he says.

Some top producers are dubious of the value of clonally selected vines. Jacques Seysses of leading Burgundy estate Domaine Dujac believes that in the wrong hands they are a bad thing: because they give higher yields it is necessary to put far more work into the vineyard, and thin the crop rigorously. It is an irony familiar to all producers in Europe that clonally selected vines give yields that are higher than the legal *appellation contrôlée* maxima (see page 28).

Does having a variety of clones in a vineyard give greater complexity to the wine? Yes, say some growers; no, say others. In any case, many New World countries have only a limited range of clones officially available. The Mendoza clone of Chardonnay, for example, which is still found quite widely in the New World – actually it is a massal selection rather than a laboratory clone – can produce high yields of loosely structured wines, but some producers with old plantations produce outstanding wine from Mendoza due to its propensity for what is called 'chicken and hen' – small and big berries in the same bunch. Some of the original Pinot Noir clones were really only suitable for making sparkling wine – high-yielding, pale in colour, thin in texture – but where plantations are now 30–40 years – for instance, in parts of New Zealand, South Africa and the USA – these old vines can produce delicious, deep, scented wine.

This paucity of clones has given rise to the most elusive sort of clone of all, the so-called Samsonite clone: vine cuttings brought in illegally, tucked into luggage, to avoid the lengthy quarantine imposed by many countries in an effort to curb the spread of vine diseases. Selfish? Irresponsible? Offenders – and many top producers are offenders – say they get their cuttings checked out by laboratories, and much faster than would be the case if they brought them in openly. The debate over the quality implications of clonal versus massal selection will continue. To quote Aubert de Vilaine of the Domaine de la Romanée-Conti, Burgundy: 'There are advantages to clonal selection and advantages to massal selection. A great wine needs a population: it is a mistake to have only athletes. But they must still be the sort of people you want.'

Mustard being grown as a cover crop between rows of 100-year-old Zinfandel vines in Sonoma County, California. Mustard can help to control virus-spreading nematodes in the soil.

In sickness and in health

The vine is subject to so many ailments and pests that sometimes one wonders how it manages to survive at all. One Oregon vineyard lists its particular local enemies thus: 'Crown gall, weevils, thrips, birds, bears, other critters.' Another vineyard, this time in California, knows when the grapes are ripe because whole tribes of raccoons come and eat them. In Germany wild boar can be a problem; in Australia it's kangaroos; in South Africa, baboons. Deer, rabbits and birds are universal predators on defenceless grapes. That is, of course, assuming that the grapes succeed in surviving downy mildew (which likes warm, wet weather), powdery mildew (which prefers dry weather), early bunch stem necrosis and grey rot, and that the vine manages to avoid black goo, Pierce's disease, fan leaf virus, leaf roll virus and eutypa or 'dead arm' (after which d'Arenberg's Dead Arm Shiraz is named), to name only a few.

Vine diseases can be bacterial, fungal, viral or phytoplasma; some are controllable, some preventable, some deadly. Some of the fungal diseases, including downy mildew, powdery mildew, anthracnose and grey rot, can be controlled with the use of fungicide sprays, of which Bordeaux mixture, a solution of lime, copper sulphate and water, is the oldest.

Virus diseases like fan leaf and leaf roll are spread either by cuttings or by nematodes in the soil. Even vines that are virus-free when planted tend to become infected – though they certainly stay healthier longer than young vines that have no pretensions to being virus-free. (As one Burgundian puts it, 'You'll stay in better shape if you go to Africa healthy than if you go there unhealthy.') But then even the term virus-free is only relative: of about 20 identified viruses only six are really dangerous, and it is those of which virus-free vines are free. When it comes to the others, vines must take their chance.

Pierce's disease, on the other hand, is a bacterial disease which kills vines fast. It is one of the main targets of national quarantine regulations, and is spread by an insect called a sharpshooter which has sadly little respect for national boundaries; the disease is so far confined to certain parts of the USA – including Napa, although it seems to be under control – and Central and South America, but while the blue-green sharpshooter is relatively manageable and home-loving and seldom strays more than 15m (50ft) from streams, the larger and more mobile glassy-winged sharpshooter, which originates in the Southern states of the USA, is beginning to encroach into Oregon, Napa and Sonoma, and is capable of spreading Pierce's disease very fast and very efficiently. Sharpshooters are found in Europe as well. European growers are not looking forward to an encounter with Pierce's disease.

Other insects, this time leaf hoppers, spread flavescence dorée, one of a group of phytoplasma diseases known as grapevine yellows. This will kill young vines and weaken old ones, and is potentially even more destructive than phylloxera, partly because of the speed with which it spreads, and partly because there is no treatment. Northern Italy has suffered greatly from it; it is found in France, Germany, Switzerland, Australia, New York State and elsewhere.

The only route open to growers is to control the leaf hopper population; this can be done by spraying insecticide. But here one comes up against one of the main tenets of modern viticulture: that biodiversity is a good thing, that sprays of all kinds should be reduced as much as possible, and that it is desirable, where possible, to move in the direction of organic viticulture. Try telling that to the authorities in Burgundy who took a very quality-conscious biodynamic grower to court for refusing to spray against leaf hoppers because it would wreck his biodynamic status. Neighbouring non-biodynamic growers would be a lot more concerned about his pristine vines being a haven for the leaf-hopper, and thus a source of grapevine yellows infection. A cause célèbre that will recur.

It should also be noted that the role of all insects may not be fully understood. For example, we now know that yeasts are spread by wasps, which harbour *Saccharomyces cerevisiae* in their gut, passing it to the grapes when they bite into them. Birds and other insects can also spread yeasts; wasps also carry other microorganisms which they transfer to the grapes, and which can add flavours to the wine. And then there are ladybirds. Lovely things, ladybirds: they eat aphids and mealybugs, and may be encouraged by

A YEAR IN THE VINEYARD

Work in the vineyard follows a basic pattern the world over, but the timing of each operation is dependent on both the prevailing climate and the weather conditions of the year.

Winter: the leaves fall, the sap descends and the vine becomes dormant. Pruning can be done at any time during the winter.

Spring: the sap starts to rise, and the first signs of growth appear: the young vine must be protected against frost from now until early summer. Sprays may be necessary to protect the buds against pest and disease, though organic and biodynamic growers use other methods. Ploughing and hoeing aerates the soil and clears weeds; fertilizer may be applied. As the ground warms up, new vines can be planted. Once the vines begin to shoot, the new growth needs to be tied to the wires, otherwise the foliage would shade the fruit and prevent it from ripening; the final trellising (pages 28–29) takes shape.

Summer: eight weeks after budbreak, the vine flowers for about 10 days and then the fruit sets; cold or wet weather at this time can cause poor fruit set and thus reduce the quantity of the harvest. Summer pruning, or leaf plucking, may be necessary to allow more sunlight to reach the fruit. At *véraison* (the point at which the fruit changes colour) a green harvest may be done to reduce the size of the crop, and superfluous clusters removed. Netting or bird scarers may be used to protect against bird damage.

Autumn: picking usually begins in September/October in the northern hemisphere, or February/March in the southern, but climate change is bringing these dates forward.

Budbreak on a Cabernet Sauvignon vine; this normally occurs 20–30 days after the sap starts to rise in the spring.

Young Pinot Noir clusters begin to appear a few weeks after flowering. Rain and cold may yet prevent the young berries from developing evenly.

Cabernet Sauvignon changing colour at véraison*; this marks the beginning of the final ripening stage and occurs about 50–70 days after flowering.*

BIODYNAMISM: NEW AGE HIPPIEDOM OR SOUND SCIENCE?

Biodynamism is the most extreme position taken by growers who follow the path from conventional viticulture to integrated management to organic. It is based on the theories of Rudolf Steiner (1861–1925), and emphasizes the importance of working with the movements of the planets and cosmic forces to achieve health and balance in the soil and in the vine.

It is different in kind as well as in degree from organic viticulture. Biodynamism uses natural herbal, mineral and organic preparations in homeopathic quantities: key applications are of horn dung, horn silica and dung compost. The first is made by burying a cow's horn filled with dung over the winter; it benefits the roots. Horn silica again involves a cow's horn, but this time filled with powdered silicum and buried over the summer; it aids photosynthesis. Dung compost is self-explanatory but not simple, since ideally every vineyard should have its own recipe, to reflect its unique needs; it helps the soil.

These preparations are mixed with water and 'dynamized' by stirring them first one way and then the other: horn manure should be stirred into warm water if possible, and stirred briskly for precisely one hour. This should be done by hand, not by machine: one of the leading exponents of biodynamism, Loire producer Nicolas Joly, compares the patterns created in the water to the patterns of Celtic art. 'The air is unceasingly taken into the spirals and intensifies all sorts of exchanges,' he says. The solution should then be applied at sunrise to the leaves of the vine, in quantities of just 40–50 litres of water per hectare.

The notion that briskly stirring a liquid introduces air is not strange to the many winemakers who stir the lees of maturing wine: they know that creating a spiral gives a different result to moving a stirrer backwards and forwards. The suggestion that this somehow relates to some lost Celtic wisdom might, however, upset those who have problems accepting the odder elements of biodynamism. Joly also stresses that manure must be buried in a cow's horn rather than a bull's horn because the former accentuates the 'primordial feminine', and when choosing manure, you should take into account not just the diet of the animal (for choice, one third leaves, one third roots and one third hay) but also its temperament: horses, he says, are dominated by heat.

What is also controversial is the stress on the influence of the stars and planets on the growth and well-being of the vine. Different times of the day and different phases of the moon are held to be suitable for different activities and different remedies: growers turning to biodynamism must be prepared to be out in their vines at three or four in the morning if required. It is an extremely dedicated, expensive and time-consuming way of growing grapes.

Perhaps the oddest thing of all is that it seems to work. Biodynamism seems to produce wines with purer flavours and better balance. Indeed, the argument has moved from whether biodynamism works to why it works, and many growers are now taking a more scientific approach – which in the long run can only be to the benefit of biodynamism.

Is the success of biodynamism due to the intensely detailed attention biodynamic growers pay to their vines and soil – in other words, would that degree of care pay off whether you were biodynamic or not? What you put on the soil must make a difference: Dominique Lafon of Domaine des Comtes Lafon in Burgundy found that after he went biodynamic the acid balance returned to his wines after just three years. 'You can feel in the wine that the vine is healthy,' he says. Soil expert Claude Bourguignon finds that microbial life in the top 30cm (12in) of soil is much the same in an organic vineyard as in a biodynamic one. But, he says, if you go deeper than this, the microbial life in a biodynamic one is far greater. Reactions between the roots of the vine, which can descend 30m (100ft) into the subsoil, and the microfauna around them, generate the oxygen and minerals that are vital for a strong, healthy plant. Deep-rooting vines give thicker, stronger grapeskins, and thus more resistance to disease, more flavour, and longer maturing wine.

For many biodynamic growers in France, one of the attractions of the system is that it enables their wines to reflect their *terroir*: the application of standard commercial preparations, they fear, has a 'dumbing-down' effect on the individuality of each vineyard.

growers because of this. Trouble is, if they get into the grapes at harvest, and into the presses, they produce IPMP (2-isopropyl-3-methoxypyrazine) which taints the wine with a bitter, herbaceous flavour detectable at about one part per thousand, and which is ineradicable. Ontario, in Canada, lost a lot of wine to ladybird taint in the early part of the century. Potassium metabisulphite seems to be an effective deterrent to the insects; otherwise it's back to insecticides, which rather defeats the object.

Going organic

How feasible is it for all growers in all wine regions to adapt either organics or biodynamism? To be realistic, not very. The happy situation we described at the beginning of this chapter, with vineyard and cellar viewed as a whole, is still found in only a small minority of estates. But lip service is paid to it in very many more, and most producers, apart from die-hard industrial bulk producers, move in the direction of sustainability as and when they can.

Most producers, one has to remember, are constrained by the necessity of having a given amount of wine to sell every year. Losing a third of the crop or more to rot or mildew is not financial good sense. Few of us non-wine producers, either, would willingly risk a third of our income for the sake of an ideal. Warm, dry climates are best suited to thoroughgoing organic production, and regions like the South of France, Chile and California have a built-in advantage in not suffering from the humidity that can lead to mildew, rot and other ailments. In cooler, wetter regions like much of Germany, for a grower to go organic may mean losing an estimated 30 per cent of his crop in many years. Total loss is not unknown. Sustainability, *lutte raisonnée*, integrated management, are preferable for most. Sustainability might involve using pheromone capsules against insect pests rather than pesticides; you might plough and hoe your vineyards; you might encourage wildlife corridors and biodiversity; you might plough cover crops between the rows and then plough them in. Mustard can help to control nematodes; other crops, like clover, can unfortunately encourage leaf hoppers. The increase in ground cover in vineyards and the resulting increase in the leaf hopper population is certainly a factor in the spread of grapevine yellows. Biodiversity works both ways.

When you start to interfere in an eco-system, even though you act with the best possible intentions, you cannot be sure of the effects of your actions. If, when planting a vineyard in virgin land, you spray against a pest, you may find you have destroyed a predator of that pest. Organic and semi-organic producers prefer to introduce predators to control pests – in Chile, a coleopter called Ambrysellus can be used against red spider; ladybirds can be used against aphids (see above). (This is, of course, just another form of interference, but so far the results seem to be more beneficial to the vine.) Sexual confusion (the use of pheromones to control populations of certain insects) can be used against grapeworm and other pests.

All this is far more work than simply loading up a tractor with a container of chemical spray. It requires far closer study of your vineyards, and far more time spent on seeking solutions, but more and more serious growers seem to be prepared to make the commitment. No doubt many more would go fully organic but for the devastating reality of rot and mildew. Bordeaux mixture is permitted under organic rules, but presumably only for pragmatic and historic reasons, since it eventually builds up in the soil and can cause copper toxicity.

Full-blown organic viticulture forbids the use of any industrially synthesized compound, though the details vary from organization to organization. Fertilizers must therefore be natural – compost and manure – and the addition of these is also normal in integrated management.

Do organic methods produce better wines? One feels they should; yet too many wines with organic accreditation are poor quality. The range of quality is in fact exactly the same as is found in non-organic wines: from disappointing to very good. Clearly you can be a careful vine grower without being a good winemaker as well; whatever the reason, an organic logo on a label is not in itself a guarantee of a good wine.

The direction of the rows of vines can make a big difference to how your grapes ripen: you might want to protect the vines from the prevailing wind; you might want to let the wind in, for greater ventilation; you might want more or less sun exposure. These vines are in Marlborough, New Zealand and face north–south. Most regions show little consensus, but the most common directions are variations of north–south.

The optimum crop

Optimum, you notice, not biggest, smallest or even best. Frankly, some grape varieties give better results with higher yields, some don't. And there's no doubt you can cut the yields too low and quality can actually suffer. Winemakers, like everyone else, have to cut their coats according to their cloth, and it is sad but true that some possible improvements in quality are not initiated, particularly in less fashionable appellations, because the improved wine would not fetch a substantially higher price, even though the investment might be substantial. On the other hand, among fashionable appellations and fashionable producers the sky seems to be the limit, be it in the lengths they will go to to improve their wines, the amount of money they will invest, or the prices they can charge. Life is no fairer to wine producers than to anyone else.

Yields

The rule used to be so simple: it was that in wine, quantity and quality did not go together. Higher yields equalled lower quality, and vice versa. We have, in recent decades, explored both extremes. Work on canopy management, emanating originally from New Zealand, demonstrated that yields could rise more than anyone had thought and still give ripe grapes; *garage* wines, originally from the Right Bank in Bordeaux, took lesser *terroirs*, paid extravagant attention to the vines, and made super-concentrated wines from tiny yields, aged them in extravagant amounts of new oak and charged extravagant prices for them.

Where are we now? Growers of New Zealand Sauvignon Blanc have been urged, in recent years, to restrain their yields to avoid a damaging glut. Many want to do so anyway, because they feel that very high yields contribute to banality of flavour.

And *garage* wines? Where are they? Either in the mainstream (Valandraud on Bordeaux's Right Bank equipped itself with better *terroir* and is now a Grand Cru Classé) or out of the picture and regarded almost as a recent historical oddity.

It all comes down to ripeness and concentration. It all depends on your *terroir* and what sort of yield you can ripen; and also on what sort of wine you want to make. Frederic Panaiotis of Champagne Dom Ruinart, for example, chooses less ripe Pinot Noir for his prestigious Dom Ruinart Rosé than he does for the more fruit-forward Ruinart Rosé, because it sits better with the Chardonnay in the blend. Yields can be higher in Champagne than further south on the Côte d'Or because the desired end result is different; likewise yields are lower (or should be) for a Grand Cru Burgundy than for a Village one.

What has become clear is that you don't make high-quality wine merely as a result of having low yields. In practice, most (though not all) of the world's great wines come from low-yielding, low-vigour vines – but while low yields are associated with high quality, they do not in themselves confer it.

It's a question of getting the right amount of sunshine on to the leaves and grapes, and thus encouraging optimum ripening. Low-yielding, low-vigour vines have small, open, leaf canopies, and thus good leaf and fruit exposure to the sun. (This may be one reason why old vines, with their less vigorous leaf growth, give such good grapes.) Shade means less sugar, less flavour, often a streak of unripeness and, in red wines, less colour.

If you halve the yield, you may double the price of the wine, but not necessarily the quality: there is an optimum yield, which is the point at which the vineyard is in balance. That point is different for every grape variety in every vineyard in every vintage; not surprisingly, relatively few vineyards are in perfect balance.

Above and below that optimum yield the law of diminishing returns sets in. Reduce your yields in the vineyard too harshly and there will not be the increase in quality to justify the reduction in quantity. Increase your yields too generously and eventually quality will fall faster than quantity rises as the wine becomes increasingly dilute. To take the example of Riesling grown on steep slopes in the Mosel Valley: there the turning point for quality is said to be between 120 and 150 hectolitres per hectare; after that quality drops. Even the lower of those figures is far more than a good Mosel grower will admit to; but then asking a wine grower about his yields is a little like a doctor asking a patient how much he drinks. The doctor makes a mental adjustment of the figure he is given, and perhaps we should do the same.

Yield is expressed in various ways around the world. Some countries favour tonnes of fruit per hectare; others prefer tons per acre. Many European countries express yields as hectolitres of juice per hectare of vines. It is difficult to convert hectolitres per hectare to tonnes per hectare with precision because, of course, much depends on how much juice is extracted from the grapes; in fact, more efficient extraction of juice is one reason (though by no means the only one, or even the main one) why vineyard yields have seemed to be rising in recent years. Roughly speaking, one metric tonne of grapes will produce between 550 and 750 litres of juice, depending on how heavily they are pressed. White grapes may yield less than red, because very gentle pressing is essential for many varieties if harsh flavours are to be avoided. Riesling is an example: its thick skins, if pressed hard, will yield coarse-tasting phenolic compounds. But as a rough guide, one tonne per hectare is more or less equal to seven hectolitres per hectare. One ton per acre is more or less equal to 17.5hl/ha.

European wine laws, though not usually those outside Europe, impose limits on yields for appellation wines. France, however, also has the *plafond limite de classement*, or PLC which allows the yield in any given year to be raised by as much as 20 per cent or even more over the basic yield. This is why yields in Bordeaux, even for leading appellations, may be as high as 55 or 60hl/ha, when the official base yield is just 45hl/ha. Many, many vineyards around the world give yields that are too high for optimum quality. In Alsace, for example, between 1945, when the appellation was set up, and 1975, when the first Grand Cru vineyards were established, the basic permitted yield was increased two-and-a-half times. In Germany, whole lakes of poor quality Müller-Thurgau are produced at 300hl/ha.

Geneva double curtain trellising in action in Argentina. Systems like this are being tried around the world, but they're not suitable for all situations.

Density of planting

What is crucial, however, is not the yield per hectare or acre but the yield per vine. This is based on the number of bunches per vine, and the weight of the bunches, both of which are a direct result of viticulture – planting density, pruning, pest control and other vineyard management policies – and, of course, the weather.

Planting density can vary hugely, even within countries and regions. In Spain's Penedès, it varies from 800 to 2000 vines per hectare; in Australia and New Zealand it can be as low as 1000 per hectare in some vineyards or as high as about 4000/ha in others, with the odd experimental planting of 9000/ha; in California the traditional figure is about 1125/ha. Chile's usual figure is about 3000/ha, though one company is trying 25,000/ha in the South. In Burgundy's Côte d'Or, there are usually 10,000/ha, though the Domaine de la Romanée-Conti is experimenting with 16,000/ha; in Bordeaux's Médoc, there are usually 10,000/ha while in Entre-Deux-Mers this can go down to 2700/ha.

Generally growers are moving towards higher densities as part of the search for quality, but the choice of density ideally depends on the soil. In fertile, irrigated soils where the vines are vigorous, each vine will need enough room to spread its canopy out: a vigorous canopy cramped into a small space means shaded leaves and fruit. On less fertile soils, quality can improve if the vines are more crowded and yet produce no more wine per hectare. In Chile, Bordelais winemakers have often attempted Médoc-style higher densities; and, interestingly, these have not always proved successful. Better quality has

Gobelet A head-trained, spur-pruned system traditional to Mediterranean countries. The vine forms a low bush, which shades the fruit from excessive heat and encourages the grapes to ripen more slowly.

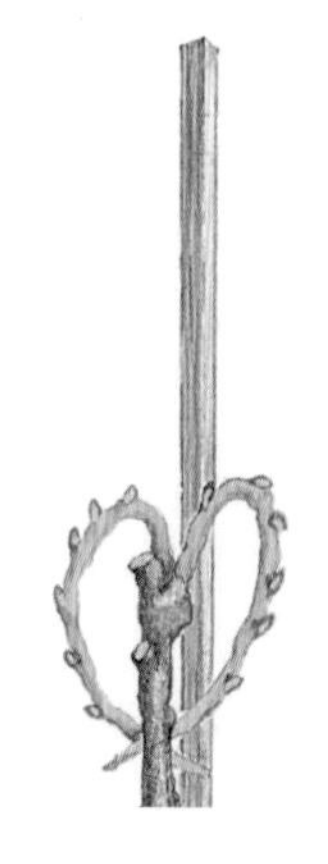

Mosel arch
A head-trained, cane-pruned system used in Germany's Mosel region. Each vine has its own supporting post, with wires to which the foliage is tied.

often been obtained from lower densities. But densities have changed over time in Europe as well: Ancient Roman viticulture used to have some 50,000 vines per hectare of vineyard.

It is often the method of vineyard management that, rightly or wrongly, determines the density of planting: 19th-century European viticulture, with a vine density of 10,000–15,000 per hectare of vineyard, was designed around the horse, and how it was used to work the land. In Chianti, until quite recently, the usual density was 2700 vines per hectare, and was designed around the tractor.

Co-planting

All vineyards, once upon a time, were co-planted. Varieties were mixed up any old how; sorting them out and planting them block by block was one of the great advances of the second half of the 20th century. It meant that each variety could be picked at optimum ripeness and treated individually.

But a few old, mixed vineyards survived: in Vienna, where the resulting wine, Gemischter Satz, was a local tradition; and in the Douro, in odd corners where the seductive replanting subsidies of the World Bank did not reach.

An alternative approach to canopy management is minimal pruning, seen here in Napa, but now largely gone out of fashion: all manipulation of the canopy is jettisoned, and pruning is stopped. The vines look alarmingly shaggy, but the idea is – or was – that over several years the vine finds its own balance. Questionable.

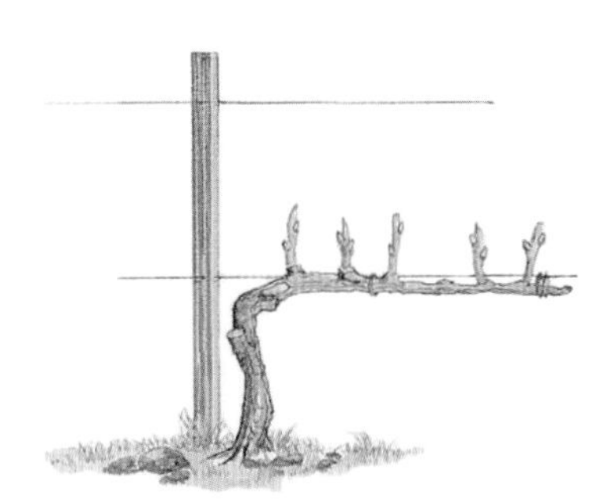

Cordon de royat A cordon-trained, spur-pruned vine, this is the system used in Champagne. The spurs are the short pieces of cane left by the pruner.

Lyre A cordon-trained system in which the canopy is horizontally divided.

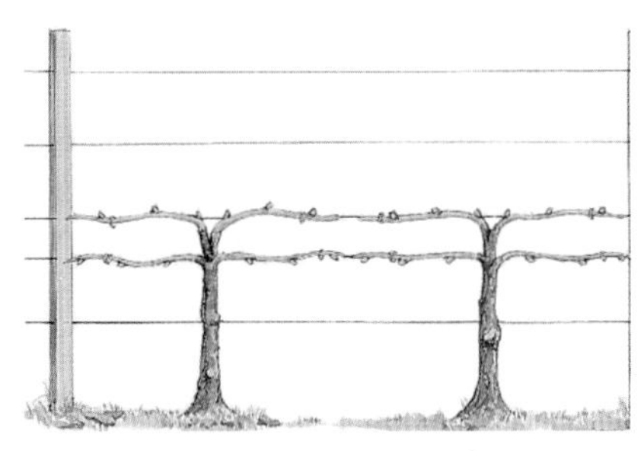

Scott Henry In this system half the shoots are trained up and half are trained down. In high-vigour sites a larger crop can be ripened, with higher sugar levels and lower acidity.

Double Guyot An example of a head-trained, cane-pruned vine. At the end of the season the canes will be cut off and new ones selected for the next year.

Geneva double curtain A cordon-trained system with the canopy horizontally divided and all the shoots hanging down. In high-vigour sites vines can ripen more fruit this way, though grapes may get too much sun, leading to phenolic, oily-tasting wine.

Now mixed vineyards are the new big thing. Gemischter Satz is a fashion rather than a fading memory, and its proponents insist that while you lose optimum ripeness for each variety, you gain harmony and cohesion in the wine. In the Douro, wines from the old vines that escaped uprooting sell for enormous prices, and if you happen to own such a plot, sharp-eyed winemakers will be knocking on your door three times a day.

Where else might mixed vineyards survive? Probably in many places. The constant trickle of rediscovered old varieties in Italy and Spain suggests the presence of the sort of mixed plantings in which such discoveries might be made. In Alsace, Marcel Deiss is a long-time believer in field blends. And some of California's more respected reds are from old field-blend vines.

Drip irrigation of young vines in New South Wales, Australia. Each vine has its own controlled water supply, and soil humidity can be regulated. Nutrients can also be delivered to the vine this way.

Canopy management: letting in the sun

Canopy management embraces pretty well everything one might do to a shoot or a leaf, from tying it in a particular place on a trellis to positioning it in the sun to cutting it off. It thus covers all aspects of pruning and training vines. The purpose of it is to get vines into that coveted balance in which they will produce the optimum yield of optimum quality grapes.

It takes about eight leaves to mature a cluster of grapes. But those eight leaves must have sunlight; if they are shaded by other leaves, the vine will produce too little sugar and the crop won't ripen. (Dense canopies also encourage humidity and therefore rot and other ailments.) The answer is not necessarily to prune the vine hard: if you do that to a vigorous vine it will simply produce an even denser canopy, with more shading effect and consequent green flavour. Leaf plucking, or removal of excess leaves in summer, can help this. If a vine is in good balance, the weight of prunings cut off in winter will be between one-tenth and one-fifth the weight of the crop of grapes. But growers have to bear in mind that changes in viticultural methods won't show any benefits for a couple of years: with vines you're always working for the next vintage but one.

Different trellising systems are being tried out for many varieties, all over the world. For example, in New Zealand a combination of adequate sunshine, yet cool climate and soils of often high potential vigour can give problems of underripeness. In such conditions the vines grow fast and furiously and the canopy ends up shading the fruit: the result is green, harsh, under-ripe flavours. Irrigation and the application of fertilizers only exacerbate the problem. Elaborate systems like Scott Henry have been successful in producing riper grapes.

With warmer summers and higher alcohol levels, there is an increasing need to reduce the amount of sun reaching the grapes. Leafplucking may now only be done on the less sunny side of the row, and more shading encouraged. At the very least, this can prevent sunburn, which bakes the aroma out of the grapes and gives harsh flavours in the wine.

Flood irrigation in Argentina. This is cheaper and far less controllable: it cannot be regulated to induce the sort of mild water stress that is desirable for top quality, but can control phylloxera.

Pruning and training

The traditional forms of trellising used in France, Germany and some other parts of Europe have the general name of Vertical Shoot Position, or VSP. This term embraces everything from gobelet to the pergolas used in parts of Italy, and includes the guyot and cordon methods.

The shoots are trained upwards as they grow, and the fruit positions itself towards the bottom of the canopy. Europe's low potential vigour soils produce relatively open canopies with this system, and leaf plucking and correct pruning to keep the vine in balance are generally enough to produce canopies of the required openness. In high vigour sites, on the other hand, VSP systems may lead to excess shading.

Newer systems are masterpieces of architecture. Canopies are divided vertically or horizontally, or both: some shoots may be trained upwards while others are, with difficulty, trained downwards.

Most systems are either head or cordon trained, and either spur or cane pruned. In head training the spurs or canes that produce the fruit-bearing shoots are positioned close together at the top of the trunk. In cordon training there are one or two long arms from which the fruit-bearing shoots will grow. These are part of the vine's permanent structure. In spur pruning, all the canes are cut back to one or two buds each spread along the cane. In cane pruning, one or more canes from a previous year's growth are left behind at pruning, though cut to the required length. These canes bear the buds that will shoot to produce fruit.

Green harvest

If your crop of grapes is too heavy – and many are – you can go through the vineyard during the summer and cut off the excess clusters. The idea is that the remaining clusters will have more concentrated flavours. It is not an ideal solution, and does nothing to get the vine into balance. The ideal is to have the vine producing the right quantity in the first place, but this end is extremely hard to achieve – and since frost, disease and hail can all reduce the size of the crop, many growers like to have a few extra clusters on each vine as an insurance policy. A green harvest just after *véraison*, to remove those green grapes that are clearly lagging behind the rest, is also a way of ensuring a more homogeneously ripe final crop. This is particularly useful if poor weather dragged out flowering over a long period.

Irrigation

It used to be said that vines had to struggle to produce great wine. The trouble is, vines that have to struggle too much produce wine that is very far from great; and the main cause of stress in vines is lack of water. Drought simply causes the vine to shut up shop: photosynthesis, and thus the ripening process, ceases. A little water stress can be a good thing, but not too much.

Managing irrigation properly is one of the greatest skills a grower must learn. Traditionally it was banned in European countries, but it is creeping in through the back door, and there is a case for it if hotter, drier summers are to become more common: it can nudge grapes to full physiological ripeness and prevent the vine shutting down and giving grapes with unripe, green flavours.

In much of the New World rainfall patterns are different from those in Europe. If you get no rain at all between spring and autumn, as is often the case, then obviously you have to irrigate to be able to grow vines. Burgundy, for example, gets 60 per cent of its annual rainfall during the growing season: Napa gets just 15 per cent. But trying to use irrigation to replicate European rainfall patterns in the New

BOTRYTIS CINEREA: A BETTER KIND OF ROT

Not all rot on grapes is undesirable. *Botrytis cinerea*, when it affects ripe white grapes, produces luscious sweet wines such as Sauternes, Tokaji or Beerenauslese. (It destroys the colour of red grapes although very rare red Beerenauslese is made.) Ironically it is the same fungus that, in unripe grapes, causes grey rot or botrytis bunch rot, which is the biggest headache for any grower coping with a damp climate.

Grapes affected by noble rot turn golden, then pinky purple, then go brown as they dehydrate. Eventually they shrivel and become covered in a grey mould.The fungus metabolizes both sugar and acids, generally reducing the sugar content of the grape by one-third, tartaric acid by five-sixths and malic acid by one-third. But the loss of water, which is partly the effect of the fungus and partly the effect of evaporation as the skins are punctured by the fungus, results in great concentration of what is left.

Heavily botrytized Sémillon grapes at Ch. Rieussec in Sauternes. These grapes will make dense, lush golden sweet, and expensive, wines. .

But that is not the whole story. The chemical make-up of the grape juice is changed by the fungus, and glycerol, acetic acid and enzymes such as laccase are formed. The result is that wines made from botrytis-affected grapes have a flavour utterly different to that of other sweet wines, and combine it with such longevity that sometimes they seem almost immortal.

Machine harvesting of Chardonnay in Narbonne, Languedoc, France just before dawn. Picking at night like this means that the grapes are cooler when they arrive at the winery, oxidation can be kept to a minimum, acidity and freshness are maximized, and no extra cooling is needed at the winery.

World doesn't necessarily work either: that's what leading Bordeaux winemakers Bruno Prats, formerly of Château Cos d'Estournel, and Paul Pontallier, winemaker at Château Margaux, originally tried to do at Domaine Paul Bruno, their property in Chile. They had thought that imitating a wet spring, followed by water stress up to the harvest, would produce better quality; it didn't. 'It was a huge lesson in modesty,' they say.

Growers see irrigation as one of the more effective tools for adjusting what nature has given them: 'a way of imitating great *terroir*', as one puts it. But there are no easy answers about how best to do it. Overgenerous irrigation, for example, simply increases the amount of wine you make, by effectively diluting it with water. Producers of bulk wine in places like California's Central Valley or Australia's Riverland may irrigate relentlessly to try to keep the soil full of water because quantity is their overriding consideration, not quality. What may be cleverer from the quality point of view is to introduce deficit watering just after flowering. This means giving the plant only enough water to grow sufficient leaves and shoots to ripen a crop; it causes mild stress which causes vegetative growth to slow and then stop, but not enough to stop photosynthesis and the ripening process. It also seems to affect cell division inside the grapes, so you get smaller berries. This means a higher proportion of skins and pips to flesh, and thus more colour, flavour and intensity.

The usual method of irrigation is drip irrigation. Thin rubber or plastic tubes run along the rows of vines, delivering carefully controlled quantities of water, drop by drop, to each vine. Ideally, constant monitoring of the humidity of the soil – and even of the pressure in the leaf cells, and the daily expansion and shrinkage of the vine trunk – enables the quantity of water delivered to be correctly regulated. In Chile and Argentina, however, some vineyards still rely on the more old-fashioned ditch irrigation, in which ditches dug at intervals through the rows are periodically flooded. Obviously this is far less controllable, and it is less common now that plantings are more likely to be on slopes. Certainly one of the problems of drip irrigation is that it encourages root growth near the surface of the soil, when what all growers want is deep roots that search out nutrients at lower levels.

With water in increasingly short supply in some places, efforts are being put into reducing irrigation to a minimum. Partial root drying is one way: you irrigate on one side of the row, then the other. Another method is to bury the irrigation pipes, reducing evaporation. Controlling irrigation to the last drop not only saves scarce water, but can improve the pH and thus the balance of the wine.

Heavily botrytis-affected Sémillon grapes being bought in from the vineyard at Château d'Yquem in Sauternes, probably the world's most famous sweet wine producer.

Finally, picking

The grape harvest may conjure up an image of jolly pickers tucking into rustic banquets washed down with last year's wine, but the reality is usually sunburn and backache – or cold fingers and rheumatism, if you happen to be picking in November or December at the northernmost reaches of the northern hemisphere. (In Germany, Canada and the East Coast of the USA, picking may even take place on a freezing day in January, if you are aiming to make Icewine. For this, the grapes are picked and pressed while they are still frozen; very concentrated, sweet juice runs from the press, and the water stays behind in the form of ice.)

And in many regions, the grape harvest takes the form of one man on a mechanical harvester, perhaps picking in the cool of the night before the morning sun warms the grapes too much for good winemaking.

Some producers distrust mechanical harvesters but there is no universally best method of picking, and how you pick your grapes must be governed by the dictates of ripeness, the weather and the type of wine you want to make. Plus, of course, your budget.

Picking at optimum ripeness is, whatever the climate, the most difficult to achieve. A cool spell in late summer may hold the grapes back, only for a hot week to accelerate them to the brink of overripeness. In those circumstances you need to act fast, since overripe grapes will produce soft, flabby wines which you will have to acidify (if you're allowed to). Or the grapes may be ripening so dreadfully slowly that you fear they will never get there at all, and perhaps rain is forecast. Do you then pick them before they are properly ripe, hoping to chaptalize – add sugar – (if you're allowed to) to compensate for the lack of sugar, or do you hold your breath and wait? One thing you should not do is pick in the rain: it is equivalent to simply pouring water into the juice.

When you have to pick fast, mechanical harvesters come into their own. They can work round the clock, and in hot climates picking at cool nighttime temperatures is a quality plus. They work by repeatedly hitting the vines so that the fruit, either in the form of loose berries or whole clusters, falls on to conveyor belts. Producers who use machines say that quality is not compromised provided that the vines have been properly prepared first, which usually means removing unsuitable fruit and clearing away any vegetation that might get picked as well. Inevitably there is some risk of bruising the fruit, and if the grapes have to travel a long distance to the winery there may be a risk of excessive skin contact with juice from broken grapes, but the effects of oxidation can be minimized with sulphur dioxide. (See In the Winery, pages 33–37.)

Machines are crucial where labour is scarce or expensive, but they cannot select the best grapes in the vineyard the way that well-trained pickers can. In Sauternes, and for Germany's Beerenauslesen and Trockenbeerenauslesen, where it is essential to pick selected grapes at particular levels of noble rot, you have to have pickers making several passages or *tries* through the vineyard. Like Sauternes, Monbazillac also relies on rigorous selection in the vineyard if the grapes are to achieve the super-ripeness needed for sweet wines. Nevertheless, mechanical harvesters were permitted for a period here until they were phased out from 1994. This is just one example of how a region's laws can favour the lazy until quality producers demand a change.

Other things being equal, however, there are no quality differences implied in machine versus hand picking. Machines are far gentler than they used to be, and not all human pickers are good at following instructions.

IN THE WINERY

Winemakers have got into the habit of saying 'wine is made in the vineyard'. Does this mean that as soon as the grapes arrive at the winery the die is cast? If the style and quality of the wine is inherent in the grape, is there nothing the winemaker can do to improve upon the raw material? Can he only destroy, not create? Well, obviously not. The winemaker still has choices, even if he chooses, for quality reasons, to do as little as possible. He can stress flavours or suppress them; he can coax a wine to precocious maturity or hold it in a state of arrested development. But if he does nothing but sit on his butt reading the newspaper, the best grapes in the world will swiftly turn into vinegar. The huge amounts of money being invested in new wineries all over the world suggests that something important is going on in those vats.

The first thing a decent winemaker must have is a vision of the style and flavour he wants to achieve in his wine. Indeed, the vines should have been cultivated with that vision in mind. In the winery he may favour minimal intervention, or else every technique under the sun, including the use of selected yeasts, must concentration, oak chips or new oak barrels, pre-fermentation cold maceration and goodness knows what. There is no New World–Old World pattern to winemaking any more. Everywhere, the most refined technology goes hand in hand with a new awareness of the past, as traditional techniques are found, after all, to have something to say for themselves.

Yeast: wild or cultured?

The winemaker may allow the native yeasts, present in the cellar and on the grape skins, to ferment the must to wine, or he may prefer the greater predictability of cultured, selected strains of yeast. If he chooses the latter, he may further choose a yeast that emphasizes particular aromas and flavours or one that is neutral in its effect on aroma.

The native yeast population in any cellar or vineyard will contain many different strains, and the balance of strains will vary from year to year. Winemakers who rely on wild yeasts welcome these differences, and the subtle year-to-year variation they give to the wines. The wine may also gain greater complexity from this varied population, though the relative unpredictability of wild yeasts means that there is a small risk of the fermentation sticking before it is complete, and refusing to restart. This can result in bacterial spoilage, and at best means that the wine will be sweet

HOW WINE IS MADE

Rosé wine

Rosé wine is made either by fermenting the juice of red grapes with the skins for a brief period, until the desired degree of colour is obtained, or by blending red and white wine. With the major exception of Champagne, this method is illegal in most of the EU, but with the advent of Pink Pinot Grigio, pink Prosecco, and the libertarianism of Vin de France, I suspect from now on, anything goes.

Red wine

Red grapes are usually crushed and may be wholly or partly destemmed. Grapes which are not destemmed undergo what is called whole bunch or whole berry fermentation (see Carbonic maceration, page 35).

The juice is fermented with the skins and (if wanted) the stems in vats of wood, cement, fibreglass or stainless steel, sometimes in small barrels, sometimes in cement 'eggs' or amphorae. Chaptalization or acidification may be done at this stage.

During fermentation the skins and stems rise to form a layer, or cap, on top of the fermenting must or juice. This layer must be kept broken up and in contact with the must if colour and tannin are to be extracted. The usual method is to pump juice over the cap (*remontage*). Alternatively, the cap may be punched down (*pigeage*) either manually or mechanically. In some cases it is trodden, by human foot or mechanical foot.
The skins and juice, and the stems if present, are left to macerate after the end of fermentation, to extract more colour and tannin. The wine is run off the skins into wooden barrels or into vats made of stainless steel, cement or fibreglass.

The skins are pressed to extract all the liquid; this is called press wine. It will be matured separately and blended in later, if required.

White wine

White grapes are destemmed and crushed. Occasionally skins and juice are left in contact for a short period to extract flavour from the skins, but overdoing this can release harshness. So usually the destemmed grapes go straight to the press, which separates juice and skins.

The juice is fermented in wood, stainless steel, cement or fibreglass. Chaptalization or acidification may be done at this stage.

The new wine is run off into wooden barrels or into vats made of stainless steel, cement or fibreglass.

Sweet wine

Sweet wine may be made by stopping fermentation at the desired balance of alcohol and residual sugar, either by adding sulphur to kill the yeasts, or by centrifuging the wine to remove the yeasts. If the must is very high in sugar, the fermentation may stop by itself before all the sugar has been turned into alcohol because the yeasts can't work at high alcohol levels, usually around 15% alcohol. Great sweet wines are made like this.

The wine may undergo malolactic fermentation. This changes the sharp-tasting malic acid to the rounder-tasting, sometimes creamy lactic acid.
While the wine is maturing in vat or barrel, the fine lees (the dead yeasts left from fermentation) may be stirred periodically (*bâtonnage*).
Alternatively, the wine may be racked off its lees by being run from one barrel or vat to another.
All the vats or barrels are tasted, and the final blend put together.
The wine may be fined and/or filtered and/or cold stabilized to remove impurities.

The wine is bottled.

with unfermented grape sugar (residual sugar) where none was intended. If cultured yeast is used, the wild yeasts must be killed by the addition of sulphur to the must: sulphur is used throughout the winemaking process as an all-purpose antioxidant and antiseptic.

Wild yeasts, of course, always come from somewhere. Research is being done on how far the action of wasps might spread different families of yeast from one vineyard to another; at the moment it looks as though the same families of yeasts will be found in the same vineyard over a number of years – which equates to millions of yeast generations. Yeasts can also be brought in on new oak barrels, and research indicates that this is how some families of yeast have arrived in some New World vineyards.

Fermentation temperature

Controlling the temperature of fermentation was the 20th century's single biggest advance in winemaking. Wines fermented at cool temperatures retain their aromas, freshness and varietal character; those fermented too hot lose these desirable attributes and taste tired and stewed.

However, just because cool is good, colder is not always better. The temperature has to be above 10°C (50°F) for yeasts to work effectively, althought there are rare strains of yeast which can operate at lower temperatures, sometimes barely above freezing point. Between 12° and 20°C (54° and 68°F) is normal for white wine, and temperatures at the lower end of this scale favour the development of floral aromas, leafy perfumes and assertive fruit flavours; reds are usually fermented hotter, at between 25° and 30°C (77° and 86°F), to extract colour and tannin. The vast majority of fermentation takes place in large vats of stainless steel, cement or wood, though a few properties ferment reds in oak barrels, a quite common practice for whites. Amphorae or concrete eggs (see below) can also be used.

Chaptalization and acidification

It takes approximately 18 grams of sugar in 1 litre of must to produce one degree of alcohol. If the must is insufficiently high in sugar, more can be added in the form of beet or cane sugar or must concentrate, either before or during fermentation. The process is known as chaptalization, and is legal (and normal) in northern Europe and cool climates elsewhere, though illegal in the warmer climates of southern Europe. In southern Europe, however, 'enrichment' by the addition of concentrated grape must is often permitted, and has the same effect of raising alcohol levels. There are legal limits on how much the alcohol level may be raised by chaptalization or enrichment.

A sorting table, like this one at Robert Mondavi's Carneros winery in California, is common at leading estates. The first selection of grapes is done by the pickers in the vineyard, but when the grapes arrive at the winery they must be checked again, and any unhealthy or underripe clusters removed. Some estates also use vibration, laser beams and goodness knows what else to remove every imperfect berry. Which makes you ask: What is a perfect berry?

Acidification is the warm climate counterpart of chaptalization: adding acidity to compensate for lack of it in the must. It may be added before fermentation or afterwards, but if you want it to blend subtly, adding before fermentation is better. Tartaric is the best, least intrusive acid though malic and citric can also be used.

Must concentration

This can be thought of as a way of fattening wines for market. It involves removing water from must, by evaporating water at a low temperature in a vacuum. The effect is to concentrate all the flavours in the wine. A process called reverse osmosis can also be used to remove water from wine, thus concentrating it. However, the jury is still out on the long-term development of such wines. It may be that it makes absolutely no difference to flavour and development in the long run, in which case it is useful only as a way of producing the sort of lush flavours that win press plaudits at the all-important moment when the young wines go on sale, in the spring following the vintage. Or it may be that must concentration handled badly can unbalance the wines because, remember, when you concentrate the must, you concentrate all its component parts, good and bad. More fruit might be fine. But more tannin? More acid?

Maceration of red wines

Red wines are usually left to macerate with the skins after the fermentation is finished. The aim is to extract colour, tannins and other phenolic compounds, all of which are necessary to red wine, though just how much colour and tannin is desirable or possible depends on the style of wine required and the grape variety being used.

The approach to tannin, in particular, has changed greatly in recent years. Nobody these days wants harsh tannins in wine. Red wines are drunk much younger than they used to be, and having to cellar a wine merely because it is undrinkable young is not an attractive option. Yet of course fine reds do improve and develop with age, and so they still need tannin just as much as ever. But the tannins must be ripe, and not as obtrusive as of old. Techniques for handling tannins are one of the most crucial parts of modern red winemaking, and they start in the vineyard, with growers paying great attention to the need for ripe skins and pips, not just high sugar levels, in the grapes.

Different compounds are extracted at different temperatures and at different concentrations of alcohol. Tannin, for example, is extracted faster if there is alcohol present – so if you want more tannin, you macerate the wine with the skins after fermentation. A long post-fermentation is supposed to extract more tannins, but smoother, more integrated ones.

If you want colour but not excessive tannin, then you may want to do a 'cold soak': macerate the juice with the skins, at a low temperature, before fermentation; a warm soak, at over 30°C (86°F), for 24 hours before fermentation is also used by some winemakers.

Amphorae, eggs and macerated white wines

Some skin contact may be used to increase aromas and flavours, though too much can give a harsh, phenolic taste to the wine.

What is newer, however, is orange wine. This is made with extended skin maceration, often in amphorae: you put the grapes in the (beeswax-lined) amphora, put the lid on and come back six months later. In Georgia they never stopped making wine like that, and winemakers in Slovenia, Croatia, Friuli-Venezia Giulia picked it up; from there it has spread across the world. Oxidized, textured, earthy and the polar opposite of the modern paradigm, they range from extraordinarily good and complex to plain grubby and tasting of tarpaulin.

Amphorae are catching on for reds, too; the results are equally unpredictable. At the moment they are mostly used experimentally. Concrete eggs, however, have rapidly become mainstream. These egg-shaped vessels, intended for fermentation and/or maturation, are usually made of concrete, though oak ones can also be found, and they are said to give finer balance and a better structure to the wine made in them. The concrete is unlined, and the rough surface is said to harbour tiny bubbles of air, which naturally micro-oxygenate the wine; similar to the effect of oak. The temperature difference between the top and bottom of the egg, about 1°C, helps to keep the wine in constant, slow circulation, aided by the shape: shape and material are of equal importance.

Composition of the grape

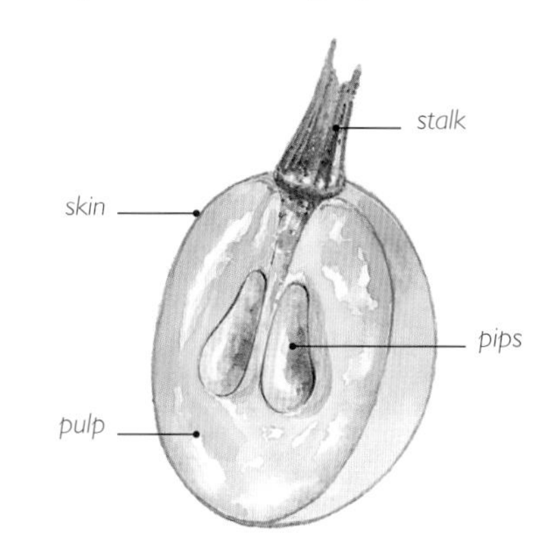

The juice, or squeezed pulp, of the grape contributes water, sugar, acids and flavour compounds, but it is not the only ingredient of wine. Colour, tannins and flavour compounds also come from the skins, and, depending on the type of wine, some tannin may also come from the stalks. The pips contain bitter oils, and modern winemakers avoid crushing them.

Other shapes

While the standard stainless-steel vat and oak barrel are not about to disappear from wineries, other shapes are finding favour, usually because they can help in producing ever-more silky tannins. Conical fermentation tanks for reds are popular, in wood or steel; upside-down cones can also be found; again, it is said that they give very gentle extraction. Concrete is enjoying a major return to favour: the new winery at Château Cheval Blanc consists of 52 curvy concrete tanks (for 40 parcels of vines: that's luxury).

The red wine fermentation chai at Domaine de Chevalier in Pessac-Léognan, Bordeaux. The post-fermentation maceration also takes place in steel vats like these at most estates; the wine is often run into barriques for the malolactic fermentation.

Carbonic maceration

If you want light fruity reds, bright in colour and low in tannin, this is the process to use. It is most usually associated with simple Beaujolais – a wine which fits the above description perfectly – but the process is used in many parts of the world, on a wide variety of grapes. Serious cru Beaujolais is not normally made this way.

It involves putting whole clusters of grapes, uncrushed and unbroken, into a closed fermentation vat, and smothering the lot in carbon dioxide. With oxygen excluded, an intracellular fermentation takes place within each berry which, apart from producing some alcohol, also produces aromatic compounds and glycerol, and reduces malic acid levels. This sort of maceration may continue for between one and three weeks; usually the grapes are then pressed and a normal fermentation completes the transformation of sugar into alcohol.

In practice, events are never quite this clear-cut. The weight of all those grapes tends to crush the ones at the bottom, so that the juice runs; it is only the grapes at the top of the vat that undergo a pure intracellular fermentation. Nevertheless, the wines have a characteristic aroma and flavour of bananas or bubblegum that can be very attractive. Such wines are intended for drinking young.

Even if it's not actual carbonic maceration, similar principles apply to reds made by whole cluster fermentation anywhere in the world. Even if the vat isn't doused in carbon dioxide, this gas will be given off by the fermentation of the juice at the bottom of the vat, and will have the effect of excluding some or all oxygen at the top. If any winemaker uses whole cluster fermentation, he's doing it to get bright, forward fruit flavours into the wine. It can be very successful with varieties like Tempranillo and Carignan. Some serious Burgundy producers, and Pinot Noir producers elsewhere, as well as some Syrah producers, use a proportion of whole bunch fermentation, for reasons of texture, and also for a particular sappy, almost green, herb-scented flavour from the stems which adds a fascinating complexity to the wine.

Oaked wine has not necessarily seen the inside of a barrel. Other and cheaper forms of oak (above) include granular virgin oak, oak beans or cubes, and oak chips, with different toast levels. These can be dunked into fermenting wine to give some new oak flavour. The traditional way is by aging the wine in new oak barrels, which are shaped and toasted over a fire, like these (left) at the coopers Tonnellerie Lasserre in the Bordeaux region. The precise degree of toast will have been specified in advance by the buyer of the barrel.

Carbonic maceration is not used for white wines, which acquire pretty strange flavours if treated this way.

Malolactic fermentation

This secondary, bacterial fermentation changes the sharp-tasting malic acid in wine to the riper-tasting lactic acid and is almost universal in red wine. It does not reduce the total acidity in the wine, but makes it taste less aggressive, and adds to the wine's apparent weight in the mouth. In white wines it may give a creamy, buttery flavour, but will certainly soften the wine and reduce the up-front fruitiness. A winemaker can decide to put all, part, or none of his wine through malolactic fermentation, depending on the flavour he wants.

The effect of oak

Wine may be stored in a variety of materials, including stainless steel, concrete, fibreglass and wood. (Wood usually, but not always, means oak; in Savennières in the Loire, for example, chestnut and acacia are traditional. Italy also uses some chestnut. Austria and Croatia use some acacia.) In most of these, except in wood, the wine, if it is kept at a low temperature and not allowed to oxidize, will not change or develop, and will retain its youth and freshness. Indeed, one Sauternes estate, Château Gilette, keeps its wine in concrete vats for up to a couple of decades before bottling it. At bottling the wine tastes almost as young as it did when first made.

But aging in oak barrels is different. First, there is a small exchange of gases through the pores of the oak; this slow oxygenation softens the astringency of the young wine and reduces the fresh primary aromas. Old oak barrels, which impart no flavour of their own to wine, are sometimes used to achieve this.

New oak barrels also give certain flavours to wine, most obviously vanillin. This has a delicious affinity with the flavour of some grapes, notably Cabernet Sauvignon and Chardonnay, and these are often aged in small new oak barrels (usually 225-litre *barriques*, often known as *barriques Bordelais* simply because they are the size traditionally used in Bordeaux) for maximum oak flavour. If overused, however, new oak can dominate the flavour of the wine and make all wines taste alike.

BLENDS VS. VARIETALS

It is a fact of modern wine that while consumers like varietal wines (those made from a single grape variety), winemakers often find blends more interesting – the reason being that a good blend is more than the sum of its parts, while only a few grape varieties are interesting enough on their own to produce complex, profound wines. Many authorities believe that no matter how good a varietal wine is, 5 per cent of another grape will always make it better. Blends are a useful way of hedging your bets in regions like Bordeaux, where not every variety can be relied upon to ripen perfectly every year.

De-alcoholization

Winemakers are struggling with rising alcohol levels caused by warmer summers, vigorous vines and long hang times. Too much alcohol means less aroma, and heavy, hot-tasting wines. There are two basic techniques for removing alcohol from finished wine: one is the spinning cone, which uses centrifugal force; the other is reverse osmosis. Wine can be balanced at more than one level of alcohol; but in removing alcohol you are also in danger of removing the esters, aldehydes and organic acids that give aroma and flavour. To solve the problem at source, in the vineyard, is the better long-term solution.

French oak generally imparts less strident flavours than American oak, though American oak treated in the same way as French (air dried rather than kiln dried; split rather than sawn) can be reasonably subtle, and the handling of American oak has improved. (French oak barrels are also approximately twice the price of their American equivalents.) Oak from Minnesota and Wisconsin is much used, though some winemakers find it too tannic; Oregon oak is now beginning to be used, though only on a small scale so far. French oak may come from a variety of different regions: Limousin oak is wide-grained and tannic and often used for brandy; Tronçais, Allier and Nevers is tight-grained and much used for wine; Vosges, too, is tight-grained. Oak from Slavonia in Croatia has long been popular with Italian producers, who favour it for their large old casks. German oak, which can give a spicy note, may also be used, and Russian and Hungarian oaks are making a reappearance.

But it is not just the oak that imparts different characters to the wine: the choice of cooper is equally important. Winemakers experiment with different coopers, and settle on those most suited to their wines, just as much as with different oaks.

And then there is the question of toast. All barrels are bent into shape over a fire, and the fire inevitably toasts the inside of the barrel. Barrels given a heavy toast impart a spicy, toasty, roasted coffee flavour to wines, but give less oak flavour because the toast acts as a barrier between wine and wood tannins. Barrels with a medium toast will create wine with more tannin and vanilla; barrels with a light toast will impart more tannin again.

Obviously, the smaller the barrel used the greater the effect of the wood on the final wine. The most common size is the 225-litre Bordeaux *barrique* but this is beginning to lose popularity in favour of the 500-litre barrel as consumer tastes move away from profoundly oaky wines. The Burgundy *pièce* is slightly bigger, at 228 litres, and while most regions have their speciality sizes, few producers in Chablis still use the traditional *feuillette* of 132 litres.

Traditional German barrels are the Mosel's 1000-litre Fuder and the Rhine's larger 1200-litre Stück. Italian *botti* come in various sizes, as do the pipes for aging port in Portugal. In Australia hogsheads are commonly 300 litres, but nobody is bound by tradition these days and winemakers choose their barrel sizes according to the effect they want the wood to have on their wine style.

Botti *and* barriques *– the old and the new at the Giordano winery in Valle Talloria d'Alba in Piedmont, Italy. Piedmont's powerful reds used to be made solely in big* botti *casks, then there was a big swing to* barriques*, which threatened to swamp the individuality of the wines, and many of the best estates now use both.*

New oak *barriques* have another effect as well: they help to fix the colour of red wines, and polymerize the tannins. (In other words the molecular chains of the tannins are lengthened so that the tannins taste softer.) The sooner the wine goes into *barrique*, the better: the malolactic fermentation is often done in *barrique* for this reason. Oak chips or oak staves have the same effect of fixing colour and polymerizing tannins; in addition, if chips are added during racking they reduce the necessity of adding sulphur dioxide, since they actively protect the wine against oxidation. They also give a touch of oak flavour, but they can't imitate the oxidative effects of proper barrel aging. Oak chips are generally illegal in Europe for *appellation contrôlée* wines or the equivalent – which is not the same thing as saying they are not used.

Co-fermentation

This is an old Rhône habit being enthusiastically adopted in the New World. Also called co-pigmentation, the practice involves fermenting a few white grapes with the red. It still happens in Côte-Rôtie, in the Rhône, where a tiny per cent of Viognier may be added to Syrah – but it is not the same as blending white wine with red. In Côte-Rôtie the red and white grapes must be fermented together, and this could be crucial. It seems that if a red fermentation has what are called 'co-factors' added before fermentation, then these co-factors may produce extra colour. Paradoxically, these cofactors may be found in certain white grapes. Yalumba of Australia, which is researching co-pigmentation, thinks that quercetin, flavonoids and procyanidins from the skins and pips of Viognier are just such cofactors – and that if you add Viognier to a Shiraz ferment you get more colour than you would from Shiraz on its own. The colour also seems to be more stable over time, mouthfeel and texture are enhanced, and flavours and aromas are more vibrant. In South Africa, Radford Dale has co-fermented Cabernet Sauvignon with Chardonnay, half and half, for an impressively dark, silky wine of great finesse. There are also South African experiments with co-fermenting Pinotage or Malbec with Viognier.

A–Z OF GRAPE VARIETIES

The A–Z section from page 40 to page 309 covers entries on grape varieties from all over the world.

The symbols after each heading indicate whether the grape is dark-skinned, usually for making red or rosé wine, or light-skinned, usually for making white wine. If you cannot find the grape variety you want in the A–Z, it may be described under another synonym, so try the Index of Grape Names and Their Synonyms on page 308.

The A–Z section contains special features on 17 of the world's great classic grape varieties, along with information on geography, including a world map locating major plantings, history, viticulture and vinification, finding the grape around the world and enjoying the grape. A Consumer Information box with Best producers and Recommended wines, and Maturity charts with advice on when to drink various wines, complete each special feature.

In addition to the classic grapes, the A–Z contains two-page features on a further 15 major varieties. Best producers have been given for all the classic and major grapes and, where appropriate, also for other minor A–Z entries.

Immaculate-looking vines, here in the Quintessa Vineyard, near Rutherford in the Napa Valley, California with Mount St Helena looming in the distance. Most of the classic French grape varieties are grown in California and recent replantings have done much to match varieties to the most suitable locations. This vineyard grows the red Bordeaux varieties of Cabernet Sauvignon, Cabernet Franc and Merlot and the end result is a rich, complex and concentrated wine.

ABOURIOU

Tannic, low-acid grape of southwest France, found in a few lesser wines. It's on the way out, and won't be much missed.

AGIORGITIKO

Greece's most widely planted red variety makes everything from rosés to dense, ageable reds, and even occasionally sweet wines from sun-dried grapes. Its tendency to overcrop needs to be controlled, it is susceptible to disease and it needs a long growing season to ripen its tannins. And, as one might guess, it provides a ready home for viruses. Its cultivation is centred on the Peloponnese, where it is the sole grape in the gutsy, spicy, ripe, plum-tasting wines of Nemea. Here it is cultivated at between 250 and 800m (820 and 2600ft), with the longest-living wines coming from the slopes of the high plateau of Asprokambos; elsewhere the grape's low acidity can limit its wines' longevity. But it has good fruit and colour, blends well with Cabernet Sauvignon, and on its own can also make attractive rosés. It is also known as Mavro Nemeas. Best producers: (Greece) Antonopoulos, Boutari, Gaia, Skouras.

AGLIANICO

This grape now seems to have originated in Italy, where it is first mentioned around 1520. It is found mainly in Campania (especially in the provinces of Benevento and Avellino) and Basilicata (especially in Potenza and Matera). Calabria, Puglia and the island of Procida, off the Neapolitan coast, also grow some.

Its most prestigious wine is Aglianico del Vulture in Basilicata. The production zone is centred on Mount Vulture, and vines are grown at between 450 to 600m (1500 and 2000ft); soils are volcanic, though more recently planted areas have more clay. Quality is generally good, and rising, and Taurasi in Campania also produces examples worth buying.

The weighty, concentrated, berried, sometimes smoky flavours of the wine, and its good acidity in warm climates, have attracted interest from the warmer parts of Australia, as well as in McLaren Vale and the cooler Margaret River. There is also some in California. It is early budding and rather late ripening, and favours a dry, sunny climate. Best producers: (Italy) Basilium, Antonio Caggiano, D'Angelo, D'Antiche Terre, De Conciliis, Di Majo Norante, Feudi di San Gregorio, Galardi, Le Querce, Mastroberardino, Montevetrano, Cantina del Notaio, Paternoster, Rivera, Sasso, Giovanni Struzziero, Terredora, Villa Matilde; (USA) Seghesio.

Aglianico vines on the slopes of Monte Vulture, Basilicata in southern Italy. Aglianico is suddenly one of the most fashionable grapes of a newly fashionable region: look for it on a label near you soon. Alternatively, look for it on an Australian label.

AÏDANI

This Greek grape grows on Santorini and other islands: it is attractively floral-scented and generally used for blending. There is also a black Aidani Mavro, some of which is used for the local sweet dried-grape wines. Best producers (white): Koutsoyiannopoulos.

AIRÉN

Airén is the major white grape of the vast La Mancha region in central Spain, and the low density of planting there – usually between 1200–1600 vines per hectare – means that the region has long held the title of largest area planted to one variety in the world. But that is changing as fashion turns towards red grapes.

Prices for red grapes are three times or more those for Airén, and that differential is especially painful given the huge investments made in state-of-the-art equipment for Airén. (As one consultant winemaker puts it, 'the tartaric acid bills are very high in La Mancha'.) Reds can earn it back; Airén can't. A lot has been uprooted, and a lot more will be.

Ironically, the wine has never been so good. The grapes' thick skins and reasonably generous yields (even with irrigation yields seldom reach the legal limit of 85 hectolitres per hectare, but at such low density that is still quite a lot of wine per vine) make it ideal for this region of extremes of heat and cold. The full weight of modern technology gets thrown at it, and the wines are faultlessly made, utterly clean and fresh – but usually have no character whatever. That's why they have been unable to command any price premium. You can't dislike them, but there's no real reason to buy them unless they're cheap.

A large part of La Mancha's production of Airén is distilled, and much eventually finds its way to Jerez for brandy, and into port as fortifying spirit. Some is blended with red wine to make light-coloured reds at low prices – La Mancha is one of the few DO (or equivalent) regions able to do this under EU law. And aficionados of traditional methods will be glad to know that some is still made in the old yellow, oxidized way for the local market. Best producers: (Spain) Ayuso, Vinícola de Castilla, Rodriguez y Berger.

ALBANA

An all-too-often unexciting Italian grape, given ideas way above its station by the granting of DOCG status to Albana di Romagna in 1987. The reasons usually given for this promotion are political and pragmatic: the authorities wanted a white DOCG, and the other possibilities were out of the question. The best Albana is exotically aromatic, with honeyed, soft fruit; too many examples, though, lack aroma and settle for being merely correctly made. Albana can be dry, off-dry, sweet or *passito*, and the latter two, though accounting for a very small proportion of production, can be the most interesting. Raisined grapes or, sometimes, nobly rotten ones are used, and there is some experimentation with *barrique* aging.

Plantings are concentrated in Emilia-Romagna in central Italy. The vine yields generously and is quite fussy: it needs good rainfall but is susceptible to grey rot in damp conditions. The most commonly planted clone, Albana Gentile di Bertinoro, has thicker skins which offer some protection against rot, and give quite deep-coloured wines: some authorities consider it a separate variety because of its smaller bunches and lower propensity to produce masses of vegetation. It is either the parent or the offspring of Garganega. Best producers: (Italy) Celli, Umberto Cesari, Leone Conti, Stefano Ferrucci, Fattoria Paradiso, Tre Monti, Uccellina, Zerbina.

ALBARIÑO

See pages 42–43.

ALBAROLA

Italian grape found in Liguria and used both for somewhat neutral wine and for the table. As a wine grape in Cinqueterre and La Spezia, it is blended with other grapes.

ALBILLO

There are several Albillo vines in Spain, and DO regulations do not differentiate between them. Albillo de Albacete, aka Albilla de Manchuela, is from Manchuela; Albillo Mayor is from Ribera del Duero and makes aromatic, full-bodied wines that are low in acidity and usually blended, often with Macabeo. Albillo Real is from Valladolid and likewise makes aromatic, low-acidity wines. Other Albillo vines may be different vines again, or may be Chasselas or Verdejo or something else entirely under an alias. Best producers: (Spain) Dehesa de los Canonigos.

ALEATICO

Sweet, scented red Aleatico is, at its best, an appealing dessert wine. It is either a parent or offspring of Muscat Blanc à Petits Grains, and shares the same heady aroma of roses. It is found in Italy in Lazio, Puglia, southern Tuscany, including the island of Elba, and on the French island of Corsica. It is also found in Kazakhstan and Uzbekistan, and to a very small extent in Chile. It has a rare white form, Aleatico Bianco, which is less fertile than standard white Muscat. Best producers: (Italy) Avignonesi, Francesco Candido.

ALFROCHEIRO PRETO

Interesting Portuguese grape found in Dão and Bairrada, and further south in the Alentejo and Terras do Sado. It is thought to have originated in the Dão or Alentejo. It is susceptible to rot, but gives wines with good colour and blackberry and spice flavours and soft tannins. Its name in the Dão is Pé de Rato, or 'mouse paw' – either because of the Portuguese passion for naming grapes for animal references (see Esgaña Cão, Periquita, Rabo de Ovelha) or perhaps because of unusual local food and wine matches. Best producers: (Portugal) Caves Aliança, Quinta dos Roques, Sogrape.

ALICANTE BOUSCHET

Actually more than one grape, but all are known by this name, and all were bred by Henri Bouschet from Petit Bouschet and Garnacha. Petit Bouschet was itself a crossing of Teinturier du Cher with Aramon (see page 44), produced earlier by Henri's father. One of Henri's crosses is Alicante Henri Bouschet; another is Alicante Bouschet no. 2; they are almost indistinguishable. The idea was to have a deep-coloured grape to blend with the prolific but light-coloured Aramon, but one that was of better quality than Teinturier du Cher. Alicante Bouschet (of which there were umpteen versions, crossed at different times and some worse than others) spread rapidly through the south of France in the years after 1885, and was planted, too, in the Southwest (including Bordeaux), Burgundy and the Loire Valley. In Spain a version is known as Garnacha Tintorera (see Garnacha, pages 102–111).

It is now in decline everywhere in France, and extinct in many places. (Its blending partner Aramon, which has one-15th of the colour of Alicante, is also in decline.) It is early ripening and quite high-yielding, though at very high yields of 200 hectolitres per hectare or more, which it can attain on flat, fertile land, the alcohol level drops to 10 per cent or less.

In Portugal's Alentejo region it produces its most successful wine, with good colour, tannin and fruit, and is also grown with varying results in southern and central Italy, Israel, North Africa, Croatia and Bosnia. In Chile it is usually blended with varieties like Cabernet Sauvignon, though it can make good concentrated, rough-fruited varietal wines. In California it supplied home winemakers during Prohibition, but its acreage has fallen drastically since. A few serious winemakers in Napa and Sonoma produce some, and it's a beefy, brawny, foursquare mouthful. Best producers: (California) Papagni, St Francis, Topolos at Russian River; (Portugal) Quinta da Abrigada, Quinta do Carmo, D F J Vinhos, Esporão, Herdade do Mouchão, J P Ramos.

ALIGOTÉ

Burgundy's second white grape probably originated in the region and is the offspring of Gouais Blanc and Pinot, which makes it a sibling of Chardonnay, Gamay and others. It comes a long way after Chardonnay both in reputation and in area planted (1700ha as opposed to 12,765ha for Chardonnay). Plantings are scattered throughout the region, mainly in the Côte d'Or and Saône-et-Loire departments, and on sites not suited to Pinot Noir and Chardonnay. On the Côte d'Or, where it was once interplanted with Chardonnay for the sake of its acidity, it is now confined to the hilltop vineyards and to the plain. Bourgogne Aligoté is traditionally mixed with cassis to make kir, but in a good year, and particularly from Bouzeron in the Côte Chalonnaise, where it now has its own appellation, the wine is well worth drinking on its own for

A. et P. de Villaine
Aubert de Villaine's day job is running Domaine de la Romanée-Conti, but he lives in Bouzeron and makes a sublimely pure, fresh, mineral-streaked Aligoté from his own grapes.

the sake of its fresh, mineral and buttermilk flavour. Some examples can age well, but generally it is for drinking young. It does not take well to oak aging. Aligoté can also be included in the blend for Crémant de Bourgogne. There is also a patch of vines at Die east of the Rhône Valley.

In Eastern Europe, where Romania, Bulgaria, Russia and the CIS countries grow it in large quantities, it is grown on flat land and its Burgundian hillside yields of 50–70hl/ha can triple, with a consequent drop in quality. There is also some in Chile and in California. Best producers: (Burgundy) d'Auvenay, M Bouzereau, Coche-Dury, A Ente, J-H Goisot, Jayer-Gilles, Denis Mortet, Tollot-Beaut, A et P de Villaine.

ALTESSE

An old variety indigenous to Savoie, this is also known as Roussette in Savoie. It's a very high-quality grape, lateish ripening, with high acidity and an unusual, pungent, mineral, mountain herbs and lemon pith aroma, and it ages extremely well. It is low-yielding but resistant to rot.

In Savoie its wine is called Roussette de Savoie, with several appellations (Frangy, Marestel, Monterminod and Monthoux). If the name Rousette is not followed by a specific appellation name, then up to half the blend may be Chardonnay, making for a good but less distinctive wine. It may also be blended into the local sparkling wines to add aroma.

Altesse de Bugey, which comes from scattered vineyards between Savoie and Lyon, is usually a blend of Altesse with a little Chardonnay. Best producers: (France) Maison de Chautagne, Maison Mollex, de Monterminod, Prieuré St-Christophe, Varichon & Clerc.

ALBARIÑO

This is the grape that brings flavour in bucketfuls to the white wines of northwestern Iberia. It has all the genetic diversity that points to great age (Terras Gauda in Rías Baixas has collected 115 different clones, with great differences between them), though it's not clear which side of the Portuguese–Spanish border it originated. In terms of aroma, there is always a fresh sea breeze to match the grapefruit and apple blossom scent, and an acid minerality to temper its yeasty ripeness of texture. Its fruit is white-fleshed, attractively crunchy even when ripe and the briney crash of the Atlantic waves on mossy rock is quite discernable even in its riper form.

Albariño is by far Spain's most fashionable white variety, particularly when grown in Rías Baixas, and rumours that far more is drunk in Galicia than is actually grown there probably have some truth in them. In Portugal it becomes Alvarinho, and is a key component (and often bottled on its own) of Vinho Verde. This is the grape in its lightest form and high yields are the reason. While better producers train their vines on wires and control yields for greater flavour, some Vinho Verde is still produced on pergolas, where yields are high and flavours low. You might even see some of the vines trained up poplar trees, as in Roman times. Even when trained on wires, canopies have to be large to allow for the vine's vigour in a wet, humid climate. Thirty to 40 buds per vine is normal and even at that level of cropping the wines can still reach 12–12.5 per cent alcohol.

There is some experimentation with *barrique* fermentation and aging and in Portugal, in Dão and Estremadura, the vines are being pruned and trained with a much smaller canopy, and lower crops. Despite being native to the rainy Northwest, Albariño withstands heat well, and the Australians thought it would do well there until they discovered that all their plantings were Savagnin – and pretty good, too. New Zealand, California, Oregon, Washington, New York and Uruguay have some tasty, though softer examples.

The taste of Albariño

When it's ripe, you can find golden gage plum skins and white peach flesh, but usually Albariño is light and zesty, merely hinting at exotic fruit, and more dominated by grapefruit blossom, lemon zest and an attractive ability to express a rocky, briney minerality. The grape has thick skins and a high proportion of skins and pips to flesh, which means lots of aroma, but a slight tendency to bitterness. Greater ripeness might eradicate the latter, but would reduce the wine's ethereal lightness. The standard of Albariños greatly improved in the 2010s.

Fefiñanes from Rías Baixas in Spain's Galicia region is delicious to drink young but has the rare ability to age brilliantly – I've had wonderful examples more than 10 years old. Even so, new Albariños are released very young – often before Christmas of the year of harvest.

Pazo Señorans

A classic modern Albariño, full-flavoured and remarkably citrussy, and intended for early drinking. Vintages vary greatly from year to year in this part of Spain, but advances in the vineyard and winery are producing super, mouthwatering wines every year.

Grand'Arte

Most Portuguese Alvarinho is grown in the country's warm, wet north and it is the main quality component of Vinho Verde. Yet Alvarinho is also useful further south – this lighter, fresh style comes from near Lisbon.

Left: Lagar de Fornelos vineyards in the Rías Baixas DO, Galicia in northwest Spain. Albariño's thick skins help protect it against rot in the wet climates of Galicia and of Vinho Verde, across the border in northern Portugal. All that rain, plus the vine's natural vigour, means that whether it is trained on pergolas or wires it must be allowed to have a big, spreading canopy. If the canopy is cramped and dense, rot and underripeness become big problems. Above: Theories abound about what is or is not Albariño. Science disproves many of them, but these are some. Albariño is related to Riesling and southwest France's Petit Manseng. But it isn't. And Madeira's Alvarinho Liláz is not Albariño either.

CONSUMER INFORMATION

Synonyms & local names

Albariño is the grape's Spanish name and Alvarinho the Portuguese name, though it is often called Cainho Branco. The Spanish grape called Albarín Blanco is probably not the same.

Best producers

SPAIN/Galicia/Rías Baixas Agro de Bazán, Castro Martín, Martin Códax, Condes de Albarei, Quinta de Couselo, DO Ferreiro, Fillaboa, Forjas del Salnés, Adegas Galegas, Lagar de Besada, Lagar de Fornelos, La Val, Marqués de Vizhoja, Gerardo Méndez, Viña Nora, Palacio de Fefiñanes, Pazo de Barrantes, Pazo San Mauro, Pazo de Señorans, Pazos de Lusco, Raúl Pérez, Santiago Ruiz, Terras Gauda, Tricó, La Val, Valmiñor, Zárate; **Ribeira Sacra** Dominio do Bibei, Guimaro, Adegas Moure; **other** Raimat.
PORTUGAL/Vinho Verde Afros, Quinta do Ameal, Quinta da Aveleda, Quinta de Azevedo/Sogrape, Quinta da Baguinha, Quinta de Gomariz, Quinta da Lixa, Quinta de Lourosa, Quintas de Melgaço, Anselmo Mendes, Palácio de Brejoeira, Provam, Quinta de Simães, Quinta de Soalheiro, Quinta do Tamariz, Caso do Valle; **rest of Portugal** D F J Vinhos, Quinta do Feital, José Maria da Fonseca; **USA/California** Bonny Doon, Cambiata, Hendry, Marimar Estate, Tangent; **Oregon** Abacela; **New York** Bedell, Palmer; **Australia** First Drop, Gemtree; **New Zealand** Coopers Creek, Stanley.

RECOMMENDED WINES TO TRY

Fifteen Spanish Albariño wines

Agro de Bazán Granbazán Ambar Rías Baixas
Castro Martin Rías Baixas
DO Ferreiro Rías Baixas
Fillaboa Rías Baixas
Forjas del Salnés Rías Baixas
Gerardo Méndez Rías Baixas do Ferreiro Cepas Vellas
Adegas Moure Moure Abadia da Cova Ribeira Sacra Albariño
Palacio de Fefinañes Rías Baixas
Pazo de Barrantes Rías Baixas
Pazo Señoráns Rías Baixas Selección de Añada
Pazos de Lusco Lusco Rías Baixas
Raúl Pérez Rías Baixas
Tricó Rías Baixas
La Val Rías Baixas
Zárate Rías Baixas

Twelve Portuguese Alvarinho wines

Quinta da Aveleda Alvarinho
Quinta do Azevedo Vinho Verde
Quinta do Feital Minho
Quinta do Gomariz Vinho Verde
Quinta da Lixa Pouco Comum
Quintas de Melgaço Vinho Verde Alvarinho
Doña Paterna Vinho Verde Alvarinho
Grand' Arte Alvarinho
Manoel Salvador Pereira Dom Salvador Vinho Verde Alvarinho
Anselmo Mendes Alvarinho
Sogrape Morgadio da Torre Vinho Verde Alvarinho
Quinta do Soalheiro 'Allo' Vinho Verde Alvarinho

ALVARINHO

See Albariño, pages 42–43.

AMARAL

Aka Azal Tinto: it's a low-yielding grape that forms part of the Vinho Verde tinto blend and usually lacks the character to be bottled alone.

AMIGNE

There are only 20ha (50 acres) of this vine in the world, and 16 of them are around the Swiss town of Vétroz in the Valais. Its wines are often slightly off-dry and may be late-harvested: at their best they have concentration, length and individuality. Their flavour has been described as resembling that of brown bread, but they tend to lack acidity – as does brown bread, of course. Best producers: (Switzerland) Bonvin, Fontannaz, J-R Germanier, Caves Imesch.

ANCELLOTTA

A lesser part of the Lambrusco blend (see page 127), Ancellotta is found also in many other parts of Italy and there is some in Argentina, Brazil and Switzerland. It is generally blended for its deep colour, not for its rather cloddish flavour.

ANSONICA

See Inzolia, page 126.

ANTÃO VAZ

A white grape much grown in the hot Alentejo, Estremadura and Terras do Sado regions of southern Portugal. It withstands heat, keeps its balance and in recent years has been making attractive, tropically fruited wines.

ARAGONEZ

See Tempranillo, pages 270–279.

ARAMON

This highly productive variety covered the plains of the south of France from the mid-19th century, when it was planted for its resistance to oidium, until the 1960s when the rather better-quality Carignan began to take its place. At low yields and in good sites it can give attractive, earthy, spicy, somewhat rustic wines; at high yields it always needed the extra colour – and flavour, however rough – provided by its traditional blending partner, Alicante Bouschet. Today it is rarely taken seriously except by the Mas de Daumas Gassac estate north of Montpellier who produce a herby, lush but light wine at a local cooperative. Its main claim to fame is that through its parent, Gouais Blanc, it's a half-sibling of 80 or more other varieties.

ARBANE

A Champagne grape which is almost extinct, but included here because Bollinger has some experimental plantings, and you never know. Champagne Moutard-Diligent bottles a varietal Vieilles Vignes Arbane.

ARBOIS

See Menu Pineau page 135.

ARINTO DE BUCELAS

Any grape that can retain its acidity in the baking summer temperatures of southern Portugal is bound to be popular; and acidity is Arinto's *raison d'être*. It can age well in bottle and when made well has considerable finesse and appealing lemony, peachy fruit. Some is now being fermented and aged in new French or Portuguese oak, and provided the oak is handled with a light touch this seems to add complexity to the wine. It forms 75 per cent of the blend in Bucelas, the rest being Esgana Cão, and is on the increase in Bairrada, Alentejo and Ribatejo. In Vinho Verde it is called Pedernã: here again high acidity is its hallmark. Other Arintos are usually aliases for something else: Arinto de Dão is Malvasia Fina, Arinto dos Açores is Sercial. Best producers: (Portugal) Alcântara Agricola, Quinta do Avelar, Quinta da Boavista, Campolargo, J M da Fonseca, Quinta da Murta, Luís Pato, Quinta dos Pesos, Quinta da Romeira, Quinta do Valdoeiro, Caves Velhas.

ARNEIS

This elegantly and exotically perfumed Piedmontese grape has only found popularity as a varietal wine relatively recently. Its traditional role was as a softener for Nebbiolo in Barolo and elsewhere, and a few rows would be planted alongside the Nebbiolo for this purpose. Few dedicated vineyards existed until a couple of producers began to take a serious interest in the vine in the 1970s and 1980s. The first example I had was from the great Barbaresco producer, Bruno Giacosa.

Without that change in fortune it might well have become extinct. Its problems include low acidity, a tendency to oxidation, susceptibility to powdery mildew, low yields and a temperamental nature, but better viticultural practices, and planting on the chalky, sandy soils of the Roero gives good acidity and structure, and improved clones can help with the mildew problem. Blending in some wine grown on sandy clay soil will add perfume. Typically the wine has a quite powerful aroma of almonds, pears and peaches and occasionally of hops. Plantings are on the rise in the Roero, encouraged by high prices; the DOC wine is called Roero Arneis. Some producers vinify and/or age the wine in oak; there are also *passito* versions. It is also supposed to be added in small quantities to Nebbiolo-based red Roero, though this practice is on the decline. It is producing some tasty whites in Australia. Best producers: (Italy) Almondo, Araldica, Brovia, Carretta, Ceretto, Cascina Chicco, Correggia, Deltetto, Giacosa, Malabaila, Malvirà, Angelo Negro, Prunotto, Vietti, Gianni Voerzio; (Australia) Dromana Estate, Kingston, Pizzini; (New Zealand) Trinity Hill; (USA) Seghesio.

ARNSBURGER

This German crossing of two Riesling clones is grown on Madeira for table wines. It produces generous yields of slightly floral wines.

ARRUFIAC

A variety grown in the deep southwest of France for alcoholic, perfumed, somewhat heavy wine. It is found in the Pacherenc du Vic-Bilh appellation in northeast Béarn. It is usually blended with Gros and Petit Manseng, and Courbu, and is beginning to receive a little more attention from producers.

ASSARIO BRANCO

See Malvasia Fina pages 132–133.

ASSYRTICO

Steely, minerally fruit, high acidity and good length are the keynotes of this high quality Greek grape. It is the main vine on the island of Santorini, where the vineyards are ungrafted and many vines are 70 years old. It is also grown in Attica, Halkidiki and Drama on the mainland; outside Santorini, flavours tend to be less minerally, but broader and fruitier. Surprisingly, it oxidizes easily and is sometimes oak-aged, though, in the case of dry wines, seldom successfully. It may be blended with the less acidic Savatiano and the fatter, more scented Malagousia and also with Aidani in sweet dried-grape wines. Best producers: (Greece) Domaine Carras, Gaia, Koutsoyiannopoulos.

ATHIRI

A Greek grape grown both for wine and for the table. It makes decent wine with a delicate aroma, though cool conditions and a long growing season can produce better wines. It is often blended with other grapes, particularly Assyrtico. Athiri Mavro is a dark-skinned version. Best producer: (Greece) Domaine Carras.

AUBUN

This vine of Mediterranean France is now in retreat, replaced by grapes of better colour and better quality. Some cuttings went to Australia in the early 1830s, and a few plantings still exist there, and also in California.

AUXERROIS

An old Alsace-Lorraine variety, still widely planted there, and sometimes called Pinot Auxerrois. It makes fairly neutral wine, quite broad, fat and honeyed at its best, but is seldom bottled on its own; instead it is blended with Pinot Blanc, perfectly legally, and sold under that name, or, in a wider blend, as Edelzwicker. It's well connected: through its parents, Pinot and Gouais Blanc, it has a vast number of relatives.

Auxerrois is also the Cahors name for (black) Cot, or Malbec. See Malbec, pages 130–131.

AVESSO

Portuguese grape grown mostly in the south-east of the Vinho Verde region close to the Douro river. Yields are high and the grapes are large; alcohol levels are quite high, and acidity is lower than with most Vinho Verde grapes, so the wines feel relatively full and weighty. Best producers: (Portugal) Caves do Casalinho, Quinta de Covela, Paço de Teixeró.

AZAL BRANCO

Late-ripening, high- to very high-acid Portuguese grape grown for Vinho Verde in northern Portugal. Best producers: (Portugal) Quinta do Outeiro de Baixo, Casa do Valle.

BACCHUS

This heavily scented German vine is the result of crossing a Silvaner x Riesling with Müller-Thurgau. It reaches high sugar levels but lacks acidity, and only when it is fully ripe does its exotic character really emerge. But at low yields or at Beerenauslesen or Trockenbeeren-auslesen levels, it can be overpowering: with Bacchus, more is not necessarily better.

High yields, though, are the norm in the Rheinhessen, its main field of operations, where it is blended into everyday QbA wines. There is also some planted in Mosel-Saar-Ruwer and Franken, but it achieves star status only in England, where it can be England's best white table wine grape, hedgerow-scented and tasting of elderflower and pears. This is about the closest that England can come to the herbal pungency of Sauvignon Blanc. Not to be confused with the American *Riparia x labrusca* hybrid, more commonly called Clinton, which, in any case, is black. Best producers: (England) Barkham Manor, Camel Valley, Chapel Down, Sharpham, Three Choirs; (Germany) Juliusspital, Markgraf von Baden/Schloss Salem, Klaus Zimmerling.

Harvesting Baga grapes from 70-year-old vines in Bairrada, its principal home in Portugal. Baga can be an austere, tannic grape and needs careful handling in the winery to bring out its fruit. A proportion of Touriga Nacional in the blend can often work wonders and add extra sugar and depth.

BACO

Baco Noir is a French hybrid widely planted in the eastern USA and in Canada for soft, fruity, smoky-tasting red wines. It is a crossing of Folle Blanche with a *Vitis riparia* vine, and was produced in 1902; in the years that followed it was planted quite widely in many parts of France but has since been largely uprooted from French vineyards, as has its stablemate Baco 22A or Baco Blanc. Best producers: (Canada) Stonechurch Vineyards.

Baco Blanc, also a French hybrid, was bred in 1898 by the same eponymous nurseryman, and was a crossing of Folle Blanche with Noah, an American hybrid. Until the late 1970s Baco Blanc was the main grape for Armagnac in southwest France but is in the process of being phased out in favour of Ugni Blanc, though traditionalists still favour it. It could also be found at one time in New Zealand.

Pub quiz special: which is the only hybrid allowed in a French *appellation controlée*? Baco Blanc, for Armagnac.

BAGA

Baga's bugbears are tannins of the most aggressive sort, coupled with high acidity. This Portuguese vine forms most or all of the blend in Bairrada, on the Atlantic coast, and is also found in nearby Dão and further south in Ribatejo. Traditionally the tannin problem was exacerbated by fermenting the stalks with the must, and softened (up to a point) by leaving the wines in bottle for some years – as many as 10, 15 or even 20.

These days vinification techniques are aimed at controlling the astringency and producing rounder, softer wines from the start, in which Baga's rich, berried fruit is to the fore. Other varieties, especially Touriga Nacional (see pages 282–283), may be blended in.

Baga lasts very well in bottle, even up to 20 years. It attains greater depth with age, but it is never a wine of enormous finesse. It is fairly high yielding, giving up to 8 tonnes per hectare in Dão, and more than 12 tonnes per hectare for Vinho Regional wines.

To make a serious wine it needs to be grown on very well exposed slopes, but even so, one of the most go-ahead producers in the region, Sogrape, attributes the improvement in quality in Dão in the 1990s to the reduction and eventual exclusion of Baga and the inclusion of other grapes in the blend. That's not very kind since Baga is from the Dão region, but its role has now been taken by Touriga Nacional – also a Dão native, but one that's easier to love. Baga, though still common in the Dão region, is now allowed only in the regional Beiras IGP. Elsewhere varietal Baga, which manages to balance the slightly dry tannins with good fruit, can be fairly attractive and occasionally shows quite piercing blackcurrant fruit.

Its synonyms include Bago de Louro, Poerininha, Tinta Bairrada and Tinta de Baga. Best producers: (Portugal) Caves Aliança, Quinta das Bágeiras, Bussaco Palace Hotel, Quinta do Carvalhinho, D F J Vinhos, Gonçalves Faria, Quinta de Foz de Arouce, Caves Messias, Luís Pato, Caves Primavera, Quinta da Rigodeira, Casa de Saima, Caves São João, Sogrape.

BARBAROSSA

An obscure Italian variety found in Emilia-Romagna and on the French island of Corsica. In Provence it is known as Barbaroux and is occasionally used in the blend for Côtes de Provence. It is not the same grape as the equally obscure Barbarossa found across the Italian border in Liguria. Best producer: (Italy) Fattoria Paradiso.

BARBERA

Barbera is best known outside Italy for being Piedmont's second-best red grape, after Nebbiolo, but inside Italy it has a wider market. It grows virtually all over the country, popping up in the most unlikely blends. The wine can be young and fruity, or dark, serious (sometimes a little self-consciously serious) and *barrique*-aged: indeed, new oak barriques seem to have a more natural affinity for Barbera's straightforward cherryish, sappy flavours than for Nebbiolo's exotic perfumes. But while Barbera gains complexity with oak it also loses some varietal character; and there are many very good examples, rich and sumptuous, which have never seen the inside of a *barrique*.

It is believed to have originated in the Monferrato hills, in central Piedmont, and here it is still planted on the best sites; further west, in Barolo and Barbaresco, it has to concede these to Nebbiolo. Overall about half the Piedmontese vineyard is given over to Barbera, and it is from Piedmont that the finest, most concentrated examples come. Barbera d'Alba has the most complexity and power, along with a deep colour; Barbera d'Asti is brighter in colour, with elegance and finesse. It is in Piedmont, too, that there has been a traditionalist-versus-modernist debate similar to that over Nebbiolo, although Barbera has more conclusively swung towards modernism and use of the *barrique*.

The grape's high acidity makes it ideal for warm climates, and its low tannin and good colour are very much in the currently fashionable mould. Indeed, the New World is now showing quite a bit of interest in all things Italian, and Barbera is an obvious candidate for anywhere warm. But in fact it has been more than just a bit player in the Americas for some time. In the pre-Cabernet days in California it was an important variety – especially since many wineries were of Italian origin – and it is making a bit of a comeback now. The same is true in Argentina, where it came over with the tides of Italian immigration, and, despite being largely used for everyday wines, has shown that its fruity, fresh acid style can be a success in the modern world. In both countries yields will have to be kept down if we are to see how good New World Barbera could be. Certainly it is finding this transcontinental leap easier to make than Nebbiolo, and Australia, especially the Barossa, is starting to produce some decent wines. Elsewhere in Europe, there is a little in Slovenia, Greece, Romania and Israel. But watch this space.

THE TASTE OF BARBERA

Barbera's most evident characteristics are high acidity, low tannin and ripe, sometimes almost dark fruit. It can be young and cherry-fresh; or weighty, moreish and with a sour-cherry twist at the end; or again *barrique*-aged, plummy and rounder, with a touch of spice. This is the most serious style, with vibrant aromas and lots of body; but Barbera never has quite such exuberance as Cabernet Sauvignon, for example. It has a danger of becoming a little raisiny if overripe, but the acidity rarely fails, whatever the style, and we should see more and more of this adaptable grape in the future.

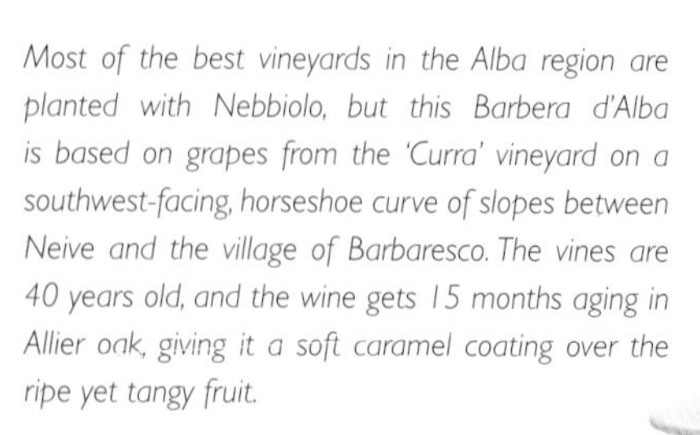

Most of the best vineyards in the Alba region are planted with Nebbiolo, but this Barbera d'Alba is based on grapes from the 'Curra' vineyard on a southwest-facing, horseshoe curve of slopes between Neive and the village of Barbaresco. The vines are 40 years old, and the wine gets 15 months aging in Allier oak, giving it a soft caramel coating over the ripe yet tangy fruit.

Coriole
Coriole were one of the first producers in the McLaren Valley to begin experimenting with Italian varieties – long before we began talking about global warming.

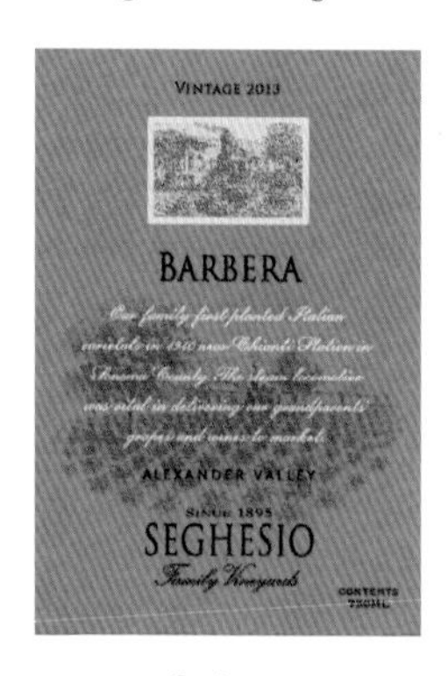

Seghesio
Based in Sonoma County, California, Seghesio have been successful grape-growers of Italian varieties, both red and white, for over a century.

Left: New oak barriques in the cellar of Angelo Gaja in Barbaresco, Piedmont. The wine takes up polysaccharides from the oak, and these increase its richness and reduce its astringency. New wood also gives particular sorts of tannins – hydrolysable – to the wine which act as antioxidants. Above: Barbera grapes. All the best Piedmontese Barberas are made from yields of less than 45hl/ha; from yields above that level the wines can be very attractive, but they will not be great. As for where Barbera comes from: there is no mention of it in Piedmont before the end of the 18th century, and it may be the result of an unknown vineyard crossing.

CONSUMER INFORMATION

Synonyms & local names

Local names include Barbera d'Asti, Barbera Dolce, Barbera Fine, Barbera Forte, Barbera Grossa, Barbera Riccia and Barbera Vera.

Best producers

ITALY/Piedmont/Barbera d'Alba Gianfranco Alessandria, Elio Altare, Azelia, Enzo Boglietti, Brovia, Burlotto, Cascina Fontana, Cascina Luisin, Ceretto, Cigliuti, Domenico Clerico, Elvio Cogno, Aldo Conterno, Giacomo Conterno, Conterno-Fantino, Corino, Giacosa, Elio Grasso, Bartolo Mascarello, Giuseppe Mascarello, Mauro Molino, Monfalletto-Cordero di Montezemolo, Andrea Oberto, Armando Parusso, Pelissero, Ferdinando Principiano, Giuseppe Rinaldi, Albino Rocca, Bruno Rocca, Giovanni Rosso, Luciano Sandrone, Vajra, Mauro Veglio, Vietti, Gianni Voerzio, Roberto Voerzio; **Barbera d'Asti** Araldica/Alasia, La Barbatella, Pietro Barbero, Bava, Bertelli, Braida, Cascina Castlèt, Coppo, Hastae, Franco M Martinetti, Il Mongetto, Perrone, Pietro, Laiolo Reginin, La Spinetta, Vietti, Vinchio e Vaglio Serra; **Langhe** Elio Altare, Boglietti, Silvano Bolmida, Bongiovanni, Cascina Fontana, Clerico, Aldo Conterno, Giacomo Conterno, Conterno-Fantino, Gaja, Rocche dei Manzoni, Marchesi di Gresy, Bartolo Mascarello, Giuseppe Mascarello, F Nada, Vigneti Luigi Oddero, Parusso, Punset, G Rinaldi, Roagna, Giovanni Rosso, Trediberri, Vajra, Roberto Voerzio; **Emilia-Romagna** Montesissa, La Stoppa, La Tosa.
AUSTRALIA Chain of Ponds, First Drop.
USA/California Palmina, Preston, Renwood, Jeff Rundquist, Seghesio.
ARGENTINA Norton, Santiago Graffiner.
SOUTH AFRICA Fairview.

RECOMMENDED WINES TO TRY

Ten Barbera d'Alba wines

G Alesssandria Barbera d'Alba
Burlotto Barbera d'Alba
Cascina Luisin Barbera d'Alba
Aldo Conterno Barbera d'Alba Conca Tre Pile
Ferdinando Principiano Barbera d'Alba Pian Romualdo
Giuseppe Rinaldi Barbera d'Alba
Albino Rocca Barbera d'Alba Gèpin
Giovanni Rosso Barbera d'Alba
G D Vajra Barbera d'Alba
Roberto Voerzio Barbera d'Alba Riserva Pozzo dell'Annunciata

Ten Barbera d'Asti wines

La Barbatella Barbera d'Asti Superiore Vigna dell'Angelo
Bava Barbera d'Asti Superiore Stradivario
Bertelli Barbera d'Asti San Antonio Vieilles Vignes
Braida Barbera d'Asti Bricco dell'Uccellone and Barbera d'Asti Ai Suma
Coppo Barbera d'Asti Pomorosso
Franco M Martinetti Barbera d'Asti Montruc
Perrone Barbera d'Asti Tasmorcan
Pietro Barbero Barbera d'Asti La Vignassa
Laiolo Reginin Barbera d'Asti
Vietti Barbera d'Asti

Five New World Barbera wines

Chain of Ponds Barbera The Stopover (Australia)
First Drop Under the Gun Barbera (Australia)
Norton Barbera (Argentina)
Renwood Amador County Barbera (California)
Jeff Rundquist Barbera, Dick Cooper Vineyards (California)

BAROQUE

A grape of southwest France, which had the great good luck to be picked up by a local Michelin three-star chef – Michel Guérard of Les Prés d'Eugénie – when he decided to turn winemaker, saving it from extinction. It's not bad in a rather hearty, alcoholic style with a slightly Sauvignon Blanc-type apples and pears flavour. Possibly a crossing of Folle Blanche and Sauvignon Blanc, it is part of the blend in the Tursan *appellation*.

BASTARDO

Portuguese name for the grape known in France's Jura region as Trousseau (see page 285). In Portugal it is one of the permitted grapes for Port, though it is not one of the recommended five varieties. It is also found further south, especially in Dão. It is valued as a blending grape – it gives good alcohol and substance to a blend – but it yields poorly and so is being grown less and less. There is an old Portuguese saying that to plant Bastardo is an excellent way for a grower to become poor. It does, however, have potential for fair quality if grown and made well. It is grown to a small extent over the border in Galicia under the names of Merenzao or María Ordona; some Australian 'Touriga' is in fact Bastardo. There is a white grape called Bastardo grown in Portugal and the Canaries, which may just be a white version of red Bastardo. In fact, there are quite a few Bastardos, Bastardinhos, Spanish Bastardos, Little Spanish Bastardinhos – and the rest – in Portugal, which can be construed either as a comment on the difficulty of getting a crop or some long-standing grudge against their Spanish neighbours. Best producers: (Portugal) Caves Aliança, Quinta do Giesta, Caves São João.

BICAL

A quite robust, ageworthy grape used for sparkling and still wines in the Portuguese regions of Bairrada and Dão; in the latter it is known as Borrado das Moscas, or 'Fly Droppings', because of its speckled skin (not its flavour). The wine combines high acidity with high alcohol and can become honeyed after some years in bottle. Quality is improving, and many bright, aromatic examples now reflect the intrinsic quality of the grape. Skin contact before fermentation seems, however, to produce a slightly soapy, flowery aroma. Its high acidity means that it is also used for sparkling wine in Bairrada. Best producers: (Portugal) Caves Aliança, D F J Vinhos, Luís Pato, Caves São João, Sogrape.

BLACK MUSCAT

See Muscat of Hamburg page 155.

BLANC DE MORGEX

See Prié page 202.

BLATINA

Late-ripening, very dark-coloured grape found in Bosnia and Herzegovina, and often made as a varietal, but sometimes as a field blend because it needs the presence of other vines for its flowers to be pollinated. Medium weight; keeps its acidity well.

BLAUBURGER

An Austrian variety produced in the 1920s by crossing Portugieser and Blaufränkisch. The wines are straightforward, low in acidity, and light, though a few growers make more concentrated, blackberry-scented wines. Burgenland producer Willi Opitz makes a sweet wine. Best producer: (Austria) Willi Opitz.

BLAUBURGUNDER

The German and Austrian synonym for Pinot Noir. Spätburgunder is a more common synonym in Germany. Best producers: (Austria) Albert Gesellmann, Fritz Wieninger.

BLAUER PORTUGIESER

A lightweight, high-yielding grape used for both wine and the table and found in Austria and Germany, where it's known as Portugieser (see page 202). Most authorities think it originated in Austria, though there is a view that it arrived there from Portugal, unsurprisingly. At best it makes everyday wines, easily giving crops of 160hl/ha. The colour is pale, and acidity is low. In Austria its synonyms are Vöslauer and Badener. Grown widely in Niederösterreich, it is the country's third most planted black grape but is gradually declining.

BLAUER SPÄTBURGUNDER

A German synonym for Pinot Noir (see pages 188–199). Plain Spätburgunder is a more common name.

BLAUER WILDBACHER

An Austrian grape found in the region of Weststeiermark, where it is almost all made into the local pink speciality, Schilcher. It may also be called Schilcher. Schilcher is notable both for its startlingly high acidity, and for the speed with which it sells out, to locals and visitors alike. It has a light redcurrant flavour, a lot of redcurrant acidity, and is best drunk young and at a local café. There are some pretty tart examples of red wine from the grape, too. Wildbacher Spätblau is a different grape, and anyway almost extinct.

BLAUFRÄNKISCH

A potentially very good-quality grape found mainly in Austria, Germany (where it is called both Lemberger and Limberger, see page 128) and points east.

The best Austrian examples are intense and zesty, often unoaked, with flavours of blueberries, red cherries and redcurrants. Oak is coming more into balance these days, and the wines have an appetizing, savoury freshness. The vine produces 100hl/ha or more with ease, but the wines get thin and weedy if made from overcropped grapes. In a blend, low-yield Blaufränkisch brings structure and acidity. The vine needs warmth, and flourishes in the Neusiedlersee and southern Burgenland regions of Austria. In Hungary it is called Kékfrankos; in the Czech Republic, Croatia and Vojvodina (Serbia), Frankovka; in north-east Italy, Franconia; in Canada and the USA usually Lemberger. Best producers: (Austria) Feiler-Artinger, Albert Gesellmann, Gernot Heinrich, Kollwentz, Krutzler, Nittnaus, Ernst Triebaumer; (USA) Channing Daughters.

BOAL

The name of Boal or Bual has been given to many Portuguese vines: there is Boal Bagudo, Boal Cachudo, Boal Carrasquenho, Boal Branco, Boal de Alicante, Boal Espinho and Boal Bonifacio, or Vital, and a raft of others. Of these, Boal Cachudo and Boal Branco are the same and in mainland Portugal more often called Malvasia Fina (see pages 132–133).

Boal Branco is recommended for Madeira and gives the name Boal to a fairly sweet type of the wine. Any wine labelled Boal should be made from this grape; but minute plantings have not increased; the shipping houses prefer to buy the cheaper Tinta Negra Mole, the island's workhorse grape, and there is a limit to what the small Madeira market will bear. Also growers are unwilling to plant what have become unfamiliar grapes. Boal, in any case, is a poor yielder, being subject to poor fruit set if there are high winds at flowering time. Best producers: (Portugal) Barros e Sousa, Blandy, Cossart Gordon, Henriques & Henriques, Leacock, Rutherford & Miles.

BOBAL

A dark-coloured, robust grape used for bulk wine and grape concentrate in most of south-east Spain, but showing real class in Manchuela.

In Utiel-Requena replanting with Tempranillo is recommended, but Bobal still accounts for 84 per cent of the vineyard there. It can have attractive black cherry fruit and it gives deep, juicy rosé. Best producers: (Spain) Gandía.

BOĞAZKERE

Turkish grape giving unusual, tannic but delightfully scented red wines of real personality. Its wine can be on the aggressive side but it is increasing in popularity and is often blended with Öküzgözü, with the latter's sleeker tannins and somewhat higher acidity. It ages well either alone or in a blend. As Turkey attempts to develop an export market, it is to be hoped they resolutely stick with the real individuality of their native grapes rather than internationalize their wines.

BOMBINO BIANCO

High-yielding Bombino Bianco is found in Emilia-Romagna, the Marche, Lazio and in southern Italy. It also goes under the name Pagadebit due to its propensity to pay a grape grower's debts by virtue of its reliable cropping. Another synonym is Straccia Cambiale. Trebbiano d'Abruzzo may be the same grape.

A great deal has long been shipped from southern Italy northwards to Germany for blending with strongly aromatic German varieties to produce EU table wine or cheap Sekt. It also makes decent raisins. But it deserves better than that. In the north of Italy its wines can be distinctly tasty, though seldom very aromatic; as Trebbiano d'Abruzzo (if it is) it can display much more depth than true Trebbiano. But a lot depends on how it is cultivated and handled. Best producers: (Italy) Giovanni d'Alfonso del Sordo, Rivera.

BONARDA

If you relish confusion, you'll love this one. Bonarda is not one but a whole clutch of Italian grapes, and all are different from the grape called Bonarda in Argentina.

Italy first: the Bonarda found in Piedmont is blended into Gattinara and Ghemme, along with Nebbiolo (see pages 166–175) and Croatina (see page 93). This is Bonarda Piemontese – proper Bonarda, if there is such a thing. The Bonarda of Oltrepò Pavese and Colli Piacentini is Croatina. The Bonarda of Novara, Vercelli and Pavia is Uva Rara. There are other Bonardas in Italy, too, but only Bonarda Piemontese is true Bonarda.

In the Novara and Vercelli hills in Piedmont there are said to be two clones: Bonarda di Gattinara, from Vercelli, and Bonarda Novarese, from Novara. The most important Italian Bonarda quantitively is the one more properly called Croatina from Oltrepò Pavese and Colli Piacentini, which makes soft and simple reds of deep colour and a certain plummy richness.

In Argentina (where there is a great deal of Bonarda) they may have finally pinned down what Bonarda is. Cheers went up when they decided it was the same as California's Charbono, which they also thought might be Dolcetto, until they discovered it was actually Douce Noir, a grape from France's Savoie region which used to be part of the Kingdom of Sardinia – along with Piedmont. Yes, I'd re-read this entry a couple of times if I were you. But in Argentina, call it what you will, it delivers. It is very late ripening, but in warm spots can outclass Malbec. It does, however, need to be allowed to ripen thoroughly, in which case it has a marvellous scent of wild strawberries and balsamic vinegar acidity. Best producers: (Italy) Mazzolino, Vercesi del Castellazzo; (Argentina) La Agricola, Altos Las Hormigas, Anubis, Catena, Zuccardi.

BORRAÇAL

Very dark-skinned Vinho Verde variety also grown in Galicia, where it is known as Caíño Tinto. It's high in alcohol and acidity, and in Vinho Verde is usually blended with Vinhão. Carbonic maceration can tame the tannins.

BORRADO DAS MOSCAS

See Bical, opposite.

BOSCO

Ligurian white grape found in Cinque Terre and blended with Albarolo and Vermentino in the local sweet wine, Cinque Terre Sciacchetrà, which is full of honey and nuts and dried apricot flavours.

BOUCHET

See Cabernet Franc (pages 50–51).

BOURBOULENC

One of five grapes used for white Châteauneuf-du-Pape, Bourboulenc has an incisive quality and a modicum of citrus perfume which makes it popular throughout the southern Rhône and Languedoc, as well as in much of Provence. Except in the La Clape sub-zone of Coteaux du Languedoc, it is always a minority partner in the blend. If picked too early it tastes lean and neutral, but when ripe (and it is a late ripener) it has good richness and depth, as well as citrus acidity and angelica freshness. In a warming world always on the lookout for white grapes that can hold acidity in hot conditions, Bourboulenc should be more widely planted. Best producers: (France) Caraguilhes, Lastours, la Négly, Pech-Redon, la Rivière Haute, la Rouquette-sur-Mer.

BOUVIER

A vine discovered in 1900 by Clotar Bouvier in Nether Styria in Austria (now Slovenia). It reaches high sugar levels but has low acidity, and is used for sweet wines of unremarkable quality in Austria's Burgenland region, though I've had one or two good fat examples. The best are blended with some other more acidic variety, often Welschriesling. Best producers: (Austria) Alois Kracher, Lenz Moser.

BRACHETTO

One of Italy's more unusual grapes, Brachetto makes every style from dry and still to its more usual type of sweet and sparkling. The colour is light red and the flavour reminiscent of wild strawberries of the most aromatic sort. It is found in Piedmont, and has its own DOC in Acqui. As a sweet red aromatic sparkler it is a delightful, refreshing oddity, but as a still *passito* wine it has more character and can age for many years. It is not the same as the French grape Braquet which is found in the wines of Bellet, near Nice. Best producers: (Italy) Viticoltori dell'Acquese, Banfi Strevi, Bertolotto, Braida, Contero, Matteo Correggia, Piero Gatti, Domenico Ivaldi, Giovanni Ivaldi, Giuseppe Marenco, Scarpa.

BROWN MUSCAT

The name given to the dark-skinned version of Muscat Blanc used in Rutherglen and Glenrowan in Northeast Victoria, Australia for fortified sweet wines. Best producers: (Australia) All Saints, Buller, Campbells, Chambers, McWilliams, Morris, Seppelt, Stanton & Killeen.

BRUNELLO

The name given to the Sangiovese grape by the producers of the Montalcino zone in Tuscany. It was long thought to be a separate clone. See Sangiovese, pages 222–231.

BUAL

See Boal (facing page) and Malvasia Fina (see pages 132–133).

BUKETTRAUBE

Produces light, acidic, ordinary wine in South Africa, and, supposedly, in Alsace.

CABERNET FRANC

Cabernet Franc is, in fact, the original Cabernet grape: the far more famous Cabernet Sauvignon is Franc's offspring. Yet nowadays Franc is the minor member of the family. This isn't fair. Sure, Cabernet Sauvignon is deeper, darker, richer, more tannic – but Franc has a delightfully mouthwatering perfume and a smooth, soothing texture that can tame the aggression and power of Cabernet Sauvignon. In Bordeaux it also ripens more easily and in difficult years produces much sweeter, more balanced fruit than Cabernet Sauvignon. Indeed, the cool soils of St-Émilion and Pomerol rarely ripen Cabernet Sauvignon, whereas Franc thrives there. It also thrives in the cooler soils of the Loire Valley, and, from the limestone or gravel vineyards of Chinon and Bourgueil, can be one of France's most lovely red wines.

The current view is that the vine probably came from the Spanish Basque country. But it may have originated in Bordeaux, being sent by Cardinal Richelieu to his abbey of St-Nicolas-de-Bourgueil in the Loire. The Abbé's name was Breton – Cabernet Franc's traditional Loire name is Breton. And there is some evidence that it actually originates in Brittany – Breton again. With these ancient vines, there are often various possibilities, no certainties. It is definitely treated with far more respect in the Loire than in Bordeaux. It rarely gets the warmest spots of soil in Bordeaux, but in the Loire it is regarded as highly soil sensitive, reacting quite differently to sand, gravel, or limestone soils. When not overoaked, the wine here displays thrilling texture and flavour as well as – from limestone – great longevity.

In the northern Italy region of Friuli Cabernet Franc used to be confused with Carmenère. Both these varieties produce delightful earthy, fruity Friuli reds. Elsewhere in Italy, and across the world, it is increasingly planted as a partner for Cabernet Sauvignon, often also with Merlot. In parts of Canada, New York State and Washington State it can be more successful than Cabernet Sauvignon, and good varietal wines are appearing from Australia, Chile, Argentina, South Africa, California and New Zealand. One of its most encouraging successes has been in warm, humid conditions in places like Brazil and Virginia, where it can resist the humidity and produce beautifully ripe fresh-flavoured reds. It also has the ability to deal with late season rains without suffering excessive dilution, making it popular in China.

The taste of Cabernet Franc

At its best, Cabernet Franc has an unmistakeable and ridiculously appetizing flavour of raspberries, pebbles washed clean by pure spring water and a refreshing tang of blackcurrant leaves. This is the kind of flavour that gets your taste buds going from Chinon and Bourgueil in France's Loire Valley. Northern Italy and Hungary can often achieve something similar, and New World examples, rare but good, generally emphasize the raspberry, sometimes getting quite rich but holding on to the ripe red teatime jam fruit and a dash of earthiness. Chile, Virginia, Brazil, California, South Africa and New Zealand have produced excellent examples in wildly differing conditions.

Bernard Baudrey is a leading grower in France's Chinon appellation in the Loire Valley. Son Matthieu makes the wine after stints in California and Tasmania, and Les Granges comes from 10–15-year-old vines mostly grown on gravel, sand and limestone soils. The wine sees no barrel-aging – spending a mere seven months in cement tanks and wooden vats – and is a fine example of the earthy, chewy, fresh fruit style Chinon excels at.

Château Cheval Blanc

There is some 60% Cabernet Franc in the blend of Cheval-Blanc, and that's massive for the St-Émilion appellation. But there is an unusual amount of extremely suitable gravelly soil in the vineyard.

Le Macchiole

Cabernet Franc is quietly making its mark round the world. Paleo is a fine example from the long-established estate of Le Macchiole in Tuscany's Bolgheri.

Left: Sorting machine-harvested Cabernet Franc to remove leaves and unhealthy grapes at Château de Targé in the Saumur-Champigny appellation in the Loire Valley. Cabernet Sauvignon never does as well here as its parent – the Loire is too far north and that bit cooler, which suits Cabernet Franc, but makes Cabernet Sauvignon struggle in all but the best years. ***Above:*** *Cabernet Franc grapes. The vine is very prone to mutation, and it lacks Cabernet Sauvignon's intensity and richness but is increasingly showing an ability to keep a deep, raspberry personality under all kinds of conditions.*

CONSUMER INFORMATION

Synonyms & local names

There are many French alternatives: the most important are Bouchet, sometimes found in St-Émilion, Pomerol and Fronsac on Bordeaux's Right Bank, and Breton in the Loire Valley. In Italy Cabernet Franc wine is often labelled simply as Cabernet. Bordo and Cabernet Frank are Italian synonyms.

Best producers

FRANCE/Bordeaux Ausone, Beauregard, Belair, Canon, Canon-la-Gaffelière, Cheval-Blanc, Clos l'Église, Clos des Jacobins, la Conseillante, Corbin-Michotte, Dassault, l'Évangile, Figeac, la Gaffelière, Lafleur, Larmande, Soutard, Tertre-Daugay, Tour-Figeac, Trottevielle, Vieux-Château-Certan; **Loire Valley** Philippe Alliet, Y Amirault, Audebert, Bernard Baudry, Baudry-Dutour, Blot/la Butte, la Bonnelière, T Boucard, P Breton, la Chevalerie, Clos de l'Abbaye, Clos Rougeard, Cognard-Taluau, de Coulaine, Couly-Dutheil, Daheuiller, Pierre-Jacques Druet, Filliatreau, S Guion, Hureau, Charles Joguet, Lamé-Delisle-Boucard, la Lande/Delaunay, Langlois-Château, R-N Legrand, A & J Lenoir, Logis de la Bouchardière, Frédéric Mabileau, Nau Frères, Nerleux, la Noblaie, Ogereau, Ouches, de Pallus, Ch. Pierre-Bise, Dom. de l'R, Olga Raffault, Raguenières, Richou, Rochelles, Roches Neuves, de Rochouard, la Sansonnière, P Sourdais/Logis de la Bouchadière, Joël Taluau, Targé, Villeneuve.

ITALY Ca' del Bosco, Duemani, Marco Felluga, Gasparini, Franz Haas, Le Macchiole, Pojer & Sandri, Quintarelli, Ronco del Gelso, Ronco dei Roseti, Russiz Superiore, San Leonardo, Schiopetto.

USA/California Bevan, Cadence, Cayuse, Delectus, Derenoncourt, Jeff Rundquist, Lang & Reed, St Francis, Viader, Villicana; **New York State** Schneider; **Washington State** Andrew Will, Cougar Crest.

CANADA Château des Charmes, Pelee Island, Thirty Bench.

AUSTRALIA Chatsfield, Clonakilla, Fox Creek, Frankland Estate, Grosset.

NEW ZEALAND Esk Valley, Providence.

CHILE Lomo Larga, Santa Rita, Valdivieso.

ARGENTINA Andeluna.

SOUTH AFRICA Antonij Rupert, Boekenhoutskloof, De Trafford, Druk my Niet, Mont du Toit, Raats, Warwick.

RECOMMENDED WINES TO TRY

Seven Loire Valley wines

Domaine Phillipe Alliet Chinon Vieilles Vignes
Yannick Amirault Bourgueil Les Quartiers
Pierre-Jacques Druet Bourgueil
Ch. du Hureau Saumur-Champigny
Domaine Charles Joguet Chinon Clos de la Dioterie
Domaine des Roches Neuves Saumur-Champigny Cuvée Marginale
Ch. de Villeneuve Saumur-Champigny

Five Italian wines containing Cabernet Franc

Ca' del Bosco Maurizio Zanella
Marco Felluga Carantan
Le Macchiole Paleo
Quintarelli Alzero
Russiz Superiore Collio Cabernet Franc

Five other Cabernet Franc wines

Barboursville Cabernet Franc (Virginia)
Lang & Reed Cabernet Franc (California)
Loma Larga Cabernet Franc (Chile)
Schneider Cabernet Franc (New York)
Viader Napa Valley Estate Wine (California)

CABERNET SAUVIGNON

Where would you start if you were determined to make a splash in the world of fine red wine? What grape varieties would you plant, especially if your country had no indigenous varieties, or, at least none that seemed likely to make a half-decent glass of grog? Would you plant Sangiovese, Nebbiolo, Tempranillo, Zinfandel, Touriga Nacional, Merlot even? No, you wouldn't, not if you were aiming for the top: you'd plant Cabernet Sauvignon, the world's most widely planted wine grape, and its top-quality red grape. King Cab, they call it; King Cab the colonizer, the conqueror. Cab the corrupter of other cultures, laying waste other grape varieties and other wine styles round the world with the brutal power of its broadsword, from Tuscany to Bulgaria, from Chile to Spain.

Yet at the same time Cabernet Sauvignon is the consumer's friend. It was the first grape to give such upfront flavours to red wine, flavours that were so easy to recognize and admire, that they turned on generations of drinkers who'd never come near a bottle of red wine before. Cabernet is both these things. It has been the most insidious of colonizers, infiltrating almost by stealth – and yet it is welcomed by consumers, to whom it offers a lifeline and a recognizable name. Virtually every winemaking country where red vines will ripen has some Cabernet Sauvignon planted somewhere. Love it or loathe it, it may be worth stopping for a second and asking, Why?

Aristocratic and magnificent, Cabernet Sauvignon is represented here by the sunburst, the emblem of France's king Louis XIV, also known as Le Roi Soleil or the Sun King. His brilliant court at the Palace of Versailles was filled with images of Louis' glory. The painting captures Cabernet Sauvignon's self-importance and regal position in the world of wine.

Partly because Cabernet Sauvignon tastes recognizably similar, wherever it grows. That's its appeal to consumers. Its appeal to producers has been different: they knew they could sell anything labelled Cabernet Sauvignon. But it's also a very obliging grape to grow and vinify. If you're a grower in an underrated region struggling to find the right way to grow your local vines but also trying to find a way to modernize your traditional wine styles, a judicious addition of Cabernet Sauvignon can be just what you need. Has Cabernet's presence improved the whole gamut of a region's wine? The answer is almost always yes.

Of course there's a danger here. Once Cabernet Sauvignon is in a region it tends to stay there – and its powerful personality means that it will hijack any wine to which it is added. Yet critics have to admit that it makes some of the most wonderful wines in the world. The blackcurrant and cigar box scented wines of Pauillac are, for many wine lovers, the greatest creations of Bordeaux. Napa Valley's memorable dense Cabernet Sauvignons literally define California wine.

The great wines of Pauillac, based on Cabernet Sauvignon, are an absolute delight, but their classic flavour is relatively simple, like so many great recipes in the kitchen. Blackcurrant fruit, seasoned by the closely related scents of cedar wood, pencil shavings and cigar box. That's the formula, as simple, as perfect as bacon and eggs or apple pie and cream.

And consequently, when the modern wine world was expanding like crazy in the 1970s and 1980s, Cabernet offered a classic style that seemed easy to understand, and, those pioneers thought, easy to replicate. As it happens, the great Bordeaux reds have proved very difficult to replicate, but all efforts to do so have brought forth many exciting interpretations of Cabernet Sauvignon from around the globe. And the similarities of fruit and texture – sturdiness of tannin, a dark ripeness of black cherry or blackcurrant fruit, and a distinct propensity to develop cedar and cigar box perfume with age – are ultimately of more importance than the differences.

Perhaps Cabernet Sauvignon does lack the perfumed subtlety of Pinot Noir, perhaps it doesn't possess the heady sensual onslaught of Shiraz or the easygoing plumpness of Merlot; certainly it doesn't demand the concentration and effort required by Nebbiolo or Sangiovese – but it is always itself. Wherever you plant it, however little money you have to invest in grand wineries and heaps of new oak barrels, you can still make a recognizable, enjoyable Cabernet. Prince or pauper, peasant or plutocrat, Cabernet Sauvignon will express itself reliably and recognizably for them all.

Cabernet Sauvignon: from Grape to Glass
Geography and History page 54; Viticulture and Vinification page 56; Cabernet Sauvignon around the World page 58; Enjoying Cabernet Sauvignon page 62

GEOGRAPHY AND HISTORY

Cabernet Sauvignon gets everywhere. Everywhere the sun shines, everywhere a grape will ripen. Everywhere someone decides they want to make a 'serious' red wine, they'll be planting Cab.That's why it's the most widely planted variety in the world. Even England has produced the odd – and I mean odd – bottle from vines grown in plastic tunnels. Germany has approved the vine for cultivation as far north as the Mosel Valley, where in order to have the faintest hope of ripening it would need the best and hottest sites – those currently allocated to Riesling. It's a mad world.

Cabernet Sauvignon's popularity, the fact that pretty well everybody who opens a bottle of wine for pleasure has heard of it, works to both its advantage and its disadvantage. For years it was the default red grape for any region trying to sell its wines

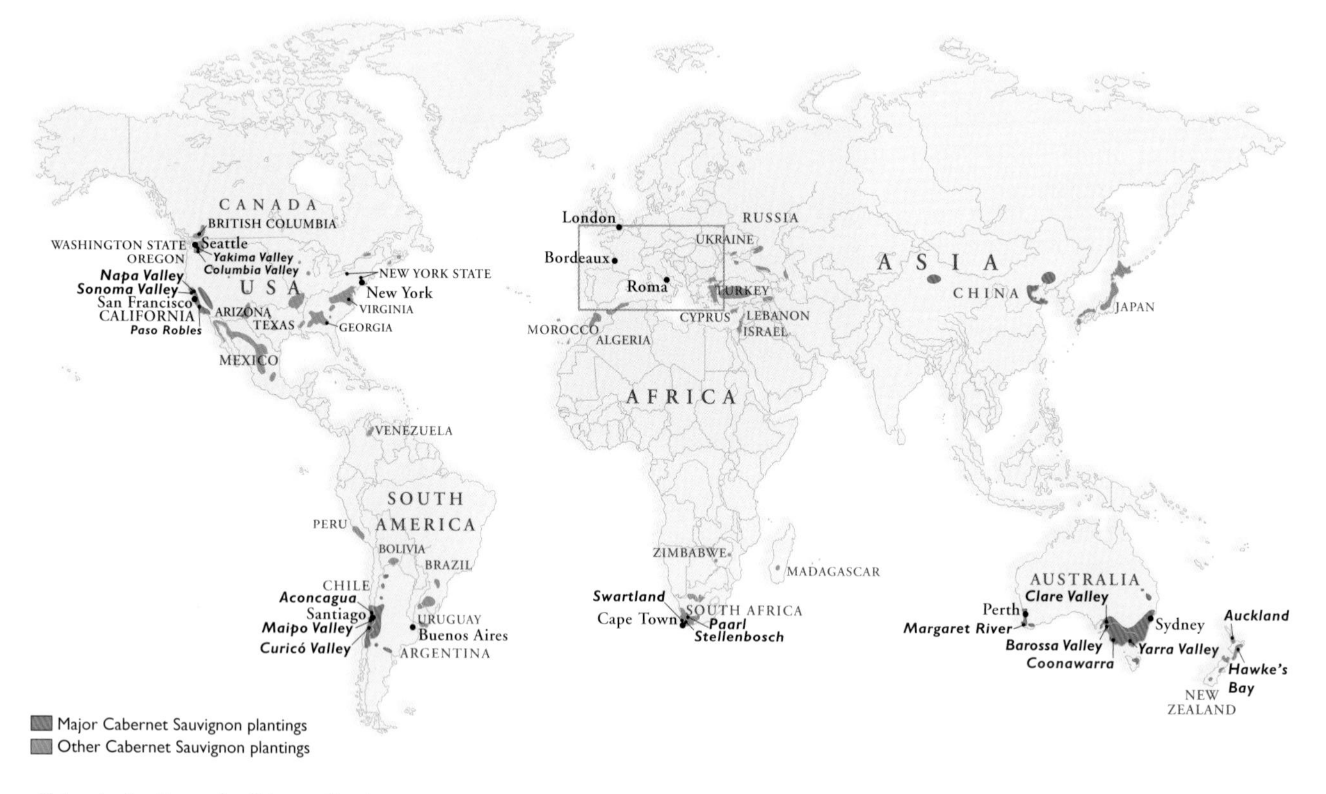

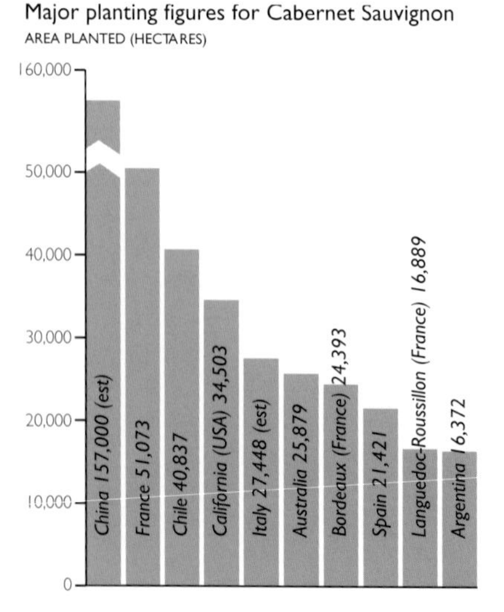

abroad and make a name for itself: now, there are as many Bordeaux-style blends in the world as any sane person could want. Probably more.

There had to be a backlash. The pendulum has swung towards indigenous varieties. Whether you're in Italy or Greece, Croatia or Turkey, the way you grab people's attention now is with some variety that only three people have heard of, and if it's good, people will try and buy it. International grapes like Cabernet Sauvignon have not exactly had their day, because in France, Italy, Australia and the USA they still produce some of the world's greatest wines. But Cabernet Sauvignon does tend to hijack any blend of which it's a part. It's good to see other grapes getting more of a look-in.

Is Cabernet driving out other varieties? Probably not. The biggest enemy of any wine region is not interloper grape varieties; it is the inability to sell wines internationally for a decent profit. Some countries like Portugal, Greece, Croatia and Italy are busily rejuvenating their industries with indigenous varieties, while Australia, South Africa, the USA, Chile, Argentina and China have all tended to call on Cab.

HISTORICAL BACKGROUND

The myths surrounding the origins of most vines are turning out to be a great deal less interesting than the reality. But the Cabernet Sauvignon myths were always less colourful than those surrounding, say, Syrah. For example, because the word Sauvignon is a bit like 'sauvage', perhaps it was originally a wild grape? But isn't that true of every other *Vitis vinifera* variety?

Or: in the 18th century Cabernet Sauvignon used to be known in Bordeaux as Petite Vidure; perhaps it took its name from the hardness of its wood (*vigne dure* = vidure)? Is there then a link with Carmenère, another historic Bordeaux variety, which used to be known as Grande Vidure?

The truth was revealed by DNA fingerprinting at the University of California at Davis in 1996: Cabernet Sauvignon is a chance crossing of Cabernet Franc and Sauvignon Blanc. The clue, ironically enough, was there in the name all along – and in the flavour. With the benefit of hindsight, how did we miss the simple fact that it tastes like both its parents? (Think of the green leafiness of unripe Cabernet Sauvignon, and the leafiness of Sauvignon Blanc.) We may never know when the crossing of the two occurred, but when the great wine estates of the Médoc were planted in the 18th century by Bordeaux's newly rich *noblesse de la robe*, Cabernet Sauvignon was established enough to form a major part of the blend for red Bordeaux and has remained so ever since.

Cabernet Sauvignon, incidentally, is a half-sibling of Merlot, which is in turn a half-sibling of Malbec. But then Sauvignon Blanc is related to Chenin and Savagnin – so Cabernet is also related to them. Which all makes Bordeaux a far more promiscuous place than anyone dared to imagine.

Entrance to the ultra-modern chai at Château Lafite-Rothschild in Pauillac. The 1980s saw a building boom in Bordeaux, with cranes and diggers moving into what seemed like every second property. It was funded by steeply rising prices and a series of good vintages, plus increased demand for fine wine all over the world. The 90s were more of a challenge, but prices for top Bordeaux in the 21st century have gone mad.

A glimpse of history – bottles dating back to 1848 in the cellar at Château Margaux. Red Bordeaux is one of the world's great classic wine styles, largely because of its phenomenal ability to age. Amazingly these bottles could still taste good.

Traditional fining of red wines uses egg whites – six per barrique here at Château Léoville-Barton. In Bordeaux, Spain and Portugal the yolks are used for sticky yellow cakes. In Bordeaux, these are called cannelés girondins. *Very fattening.*

VITICULTURE AND VINIFICATION

The fact that just about every wine country in the world grows Cabernet Sauvignon is inclined to make us take it for granted. If it grows everywhere, the logic goes, it must be easy to grow, making few demands on the knowledge or skill of the grower. It is true that it is less fussy about climate and soil than many varieties, is relatively disease-resistant and succeeds in producing wine that is recognizably Cabernet no matter where it is planted. But there are only a few places in the world where varietal Cabernet by itself is as good as or better than a blend. As one Australian winemaker memorably puts it: 'Bordeaux's greatest coup was in convincing the rest of the world that great red Bordeaux is pure Cabernet.' In other words, you guys plant pure Cabernet, then realize how much better our blends in Bordeaux are.

Climate

Distinguishing between the importance of climate and that of soil is always difficult, but seems more so with Cabernet Sauvignon precisely because at most quality levels it reflects its soil less than some grapes. In Bordeaux, soil type has traditionally determined what is planted where, but it is ultimately the temperature of the soil that is crucial; in Australia and California, more emphasis is given to climatic factors. But to quote Patrick Campbell, founder of Laurel Glen Winery in California's Sonoma Valley: 'Cabernet Sauvignon at the top level should speak of a site. That may not be possible or even necessary at lower levels, but at the top of the pyramid, Cabernet must be from somewhere' and must taste not just of its variety, but of its 'place' as well.

Cabernet Sauvignon needs warmth to ripen. It needs a warmer climate than Pinot Noir, or it will turn out green and sappy, with a flavour of green capsicum peppers; too much warmth, however, turns it soft and jammy, with a flavour of baked blackcurrants. Methoxypyrazines, the odour compounds that give Cabernet Sauvignon the green, herbaceous part of its flavour profile, are destroyed by sunlight as the grapes ripen; the detection threshold on the palate is 2ng/l. In warmer climates levels can fall from around 30ng/l at *véraison* to 1ng/l at picking.

Excessive green, vegetal flavours have been the bane of many regions which hoped their cool climates would give them Cabernets of Médoc-like elegance. In California these flavours used to be known as 'Monterey veggies'. The Monterey region is both cool and windy, and since vines shut down in high winds, eliminating excess vegetal flavours in Monterey Cabernet is very difficult, even with leaf removal. If you want to grow Cabernet in Monterey, you have to pick a warm spot. It is becoming clear that Russian River, too, is on the cool side for Cabernet.

The minty flavour sometimes found in Cabernet Sauvignon, particularly in Margaret River and Coonawarra in Australia, and in Washington State may be the result of a marginally cool climate. But soil could also be a factor. It crops up occasionally in Pauillac, for example, but not in Margaux. Personally, I love it.

Minty flavour, however, is known to derive from the proximity of eucalyptus trees. They're very insidious: the oils vaporize and get on to the grapes. In South Australia's Yarra Valley, vines within 50m (164ft) of eucalypts have been shown to have 15.5 micrograms per litre of eucalpytol (alias 1,8-cinceole) in the finished wine; a microgram is one-millionth of a gram. Vines between 50 and 175m (164–574ft) away had just 0.1 micrograms per litre. It's only found in red wines, because reds are fermented with the grape skins; and picking grapes by hand, and therefore keeping leaves out of the fermentation vat, significantly reduces the flavour. Producers in most countries can fell eucalypts if they're tainting the wine; in Australia it's illegal to fell native trees. If you don't like the taste, move the vineyard.

Cutting-edge winery architecture at Viña Almaviva, a joint venture between Concha y Toro and the late Baroness Philippine de Rothschild of Bordeaux in Chile's Maipo Valley. Outside investment and know-how – and a swanky name – equal premium prices. Luckily, in this instance, they make good wine as well – but quite a few such grandiose schemes deliver too much sizzle and not enough steak.

Soil

The fame of Cabernet Sauvignon was originally based on the gravel soils of Bordeaux's Médoc and Graves: they produced the wines that made the rest of the world want the vine. Cabernet Sauvignon likes gravel simply because it is warm. It drains well, warms up quickly in spring, and holds the heat well. All these factors suit this late budding, late ripening variety because they help to coax the grapes to ripeness in the marginal climate of somewhere like the Médoc. That is not to say that there are no spots in Bordeaux's St-Émilion, or even Pomerol, where Cabernet might thrive. But take gravel away from Bordeaux and you wouldn't have much Cabernet: the more clayey or limestone soils of St-Émilion and Pomerol are generally too cool. Nevertheless, in the 1960s until the mid-1970s the bureaucrats made it obligatory to plant Cabernet Sauvignon in St-Émilion – and, what was more, plant high-yielding clones on vigorous SO4 rootstocks. The results, says Stephan von Neipperg of Château Canon-la-Gaffelière, were 'fine if you wanted to make cola'. Rum and Cabernet, anyone?

Elsewhere, where the climate is warmer, gravel *per se* seems to be less crucial than soil that is well drained and of poor potential vigour. In Coonawarra, South Australia, there is terra rossa over limestone, but in the Rutherford and Oakville areas of California's Napa Valley the vine thrives on alluvial soil.

The Coonawarra in South Australia has proven to be particularly suitable to Cabernet Sauvignon. The vines are on a ridge of terra rossa soil over thick, free-draining limestone – perfect in Coonawara's cool, wet conditions.

Yields

The 1970s clones that were developed for high yields – and which were often responsible for giving green, thin, herbaceous flavours to the wine – are increasingly being uprooted and replaced with newer, better ones. However, these new virus-free clones also inevitably give higher yields.

In Bordeaux the Classed Growth châteaux may achieve 60hl/ha – the legal maximum is 50hl/ha, but there is also the *plafond limite de classement*, a legal dodge that enables *appellation contrôlée* regions to increase their yields in prolific years. Most years are now prolific enough to qualify. To keep yields down, green harvesting is usually necessary – chopping off excess clusters at *véraison* – and selection for the *grand vin* will be stringent or not according to the producer. Low yields, however, are not an end in themselves. Concentration is good, but too much becomes lack of balance. Bordeaux has received plenty of stick in recent years for copying Napa.

At the winery

Cabernet Sauvignon's high ratio of pip to pulp – almost 1:12, compared to Sémillon's 1:25 – and its high phenolic content mean that it can withstand both fairly high temperatures at fermentation, and long maceration. Fermentation temperatures of up to 30°C (86°F) are usual, and in Bordeaux, a maceration of three weeks was traditional because the cellar staff used to take the opportunity to shut the doors and go hunting. Where softer, earlier drinking wines are the aim, the maceration may only be a few days. In Australia and other New World countries, carbonic maceration has sometimes been used to produce soft, juicy wines.

Cabernet Sauvignon and oak

The wine has a startling affinity with new oak, blending its blackcurrant flavours brilliantly with the vanilla and toasty spice of the barrels. The success of Cabernet in new French oak made the 225-litre *barrique Bordelais* effectively the standard size wine barrel throughout the world.

In Australia, California and elsewhere, American oak, which gives a more assertively vanilla flavour, may also be used; but winemakers' sensitivity to the risk of over-oaking is increasingly leading them to use a mix of French and American, or a mix of new and used barrels. American oak can be processed like French oak, giving it subtler flavours, and different types of American oak are being identified: Oregon oak, for example, is more powerful in flavour than that from Missouri, Pennsylvania or Virginia.

THE BORDEAUX BLEND

Cabernet Sauvignon is almost never bottled as a varietal wine in Bordeaux: it usually lacks enough flesh in the middle palate, and needs its somewhat lean profile filled out with the fatter Merlot and the perfumed, fruity Cabernet Franc. This is the classic blend for red Bordeaux. Not that there's a standard recipe: each château has its own balance of vines, depending on its soil and climate, and its *grand vin* may or may not reflect that balance exactly.

With such a fickle climate, including the threats of both frosts and rains at vintage time, much depends in Bordeaux on the year. The reason that the region evolved its particular mix of grapes was that not every vine could be relied upon to ripen every year. Growing several varieties means that if one variety is hit by late frost, another may survive to give you a crop. If September storms arrive, at least some varieties may be ripe enough to pick early.

Such pragmatism may not be necessary in warmer sites like the northern part of the Napa Valley, or South Australia's Barossa or McLaren Vale. Here varietal Cabernet can be very successful – though even so it is not uncommon for a touch of Merlot to be added to tweak the final flavour. In cooler climates like New Zealand, Cabernet blends are usually more successful than varietal Cabernet Sauvignons.

In Bordeaux Petit Verdot and Malbec may also be added, though Carmenère, which was important in Bordeaux before phylloxera, is now hardly grown there. Malbec is grown patchily – there's some in Fronsac and Bourg – and the lighter soil of Margaux means that a small percentage of Petit Verdot is often added there for its dark colour and violet perfume. Petit Verdot is also grown in corners of Australia, California, Virginia, New Zealand and Spain for blending or as a varietal.

Further back in history the classic Bordeaux blend included Syrah, which might have been grown either in Bordeaux or in the Rhône Valley. This is echoed today in the classic Australian red blend of Cabernet and Shiraz; but Cabernet may be blended in other countries with almost every imaginable red grape. Tuscany, for instance, has made a notable success of blending it with Sangiovese, while the regions of Cataluña and Navarra in northern Spain make very good Cabernet-Tempranillo blends.

CABERNET SAUVIGNON AROUND THE WORLD

Now that Bordeaux can make wines of New World richness, and the New World is showing that its wines have structure and longevity, it is becoming increasingly difficult to ascribe particular flavour profiles to particular places. Styles can be defined as much by individual winemakers and individual sites as by regions. Christian Seely of Château Pichon Baron maintains that Pauillac is the only place in the world where Cabernet Sauvignon doesn't simply taste of itself. He may be biased – some leading Napa or Coonawarra producers might beg to differ – but he may also have a point.

Bordeaux

Cabernet Sauvignon is not the most widely planted grape in Bordeaux: that honour goes to Merlot. Back in the 19th century, Cabernet's popularity was increasing rapidly: growers loved its resistance to rot (its thick skins and loose clusters help here) as well as its tannic structure, acidity and good flavours. But the 1852 oidium epidemic in Bordeaux revealed its great susceptibility to that disease. Growers turned to Merlot instead, and confined Cabernet to the gravel outcrops of the Médoc and Graves. But where there is less gravel – in the northern part of the Médoc, for example – Cabernet can be too austere for fun, never mind for fashion, and needs plenty of Merlot to fatten it up. In St-Émilion it is a minority grape, with Cabernet Franc and Merlot taking over; and in Pomerol's clay it is hardly found at all.

Even in its most favoured spots, Cabernet Sauvignon will not produce sensational wines every year, though with global warming, the number of years it produces poor wine is now far outweighed by the number of good to excellent years. There's no doubt that most of the longest-lived red Bordeaux have a high proportion of Cabernet Sauvignon in their blend, though it's worth remembering that even in the Médoc and Graves, where Cabernet Sauvignon is likely to be the biggest single variety in the blend at the top châteaux, it may still be in a minority against the various other varieties in the blend.

Cabernet's style in the Médoc varies from the mineral austerity of St-Estèphe through violet-scented intensity in Margaux, classic lead pencils and blackcurrant in Pauillac, cedar and cigar boxes in St-Julien, softer and rich in Moulis and somewhat earthy in Listrac and the northern Médoc to minerally again in Pessac-Léognan. Lesser regions like the southern Graves produce good blackcurrant flavours, without the intensity of the best sites.

Cabernet Sauvignon revels in well-drained stony soil and all the best Médoc vineyards are planted on deep gravel beds. This is a Cabernet Sauvignon vine at Château Léoville-Barton in St-Julien.

Other French Cabernets

Cabernet Sauvignon is grown in the Southwest for lookalike red Bordeaux wines, and it produces its customary blackcurrant fruit, but in a lighter style than the best of Bordeaux can offer. These pruned Cabernet Sauvignon wines are at Ch. Léoville-Barton, in St-Julien, a Bordeaux village famed for the balance and ageability of its red wines due to its deep beds of gravelly soil sitting on the edge of the Gironde Estuary.

The super-Tuscan phenomenon

Cabernet Sauvignon in Italy is no longer the dangerous interloper it was once perceived to be. Winemakers have worked with it, learnt about it – and gone on to learn more about their native grape varieties. Cabernet on its own is no longer the first choice for anyone wanting to make serious wines. The grape was present in Italy long before Italy was a single country. It arrived in Piedmont in 1820 and is still grown there, even, it is said, being introduced into Barolo in significant proportions. Such a thing would of course be illegal for DOCG Barolo, but is believed to improve the colour and make the wine

Château Montrose

When tasters talk of the classic flavours of red Bordeaux, the blackcurrant, the cedar wood and pencil shavings and the tannnic grip, Château Montrose in St-Estèphe fits the bill perfectly.

Château Léoville-las-Cases

A super-expensive Second Growth St-Julien château, where the aim is to produce super-concentrated wine without losing the haunting blackcurrant and cedar wood beauty of top Bordeaux.

Isole e Olena

In 1988 Chianti Classico producer Paolo de Marchi at Isole e Olena was one of the first in Tuscany to experiment with planting classic French grapes like Cabernet Sauvignon and Syrah.

fruitier – both of which aims can be hard to achieve with Nebbiolo.

Legal blends of Cabernet and Nebbiolo take in the DOCs of Langhe or Monferrato; Cabernet and Barbera also blend well, with or without the addition of Nebbiolo, though the addition of two high tannin grapes to the high-acid Barbera can require the use of some new wood to add some sweet spice.

Piedmontese varietal Cabernets range from very good to excellent, but seem to need the best vineyard sites.

Cabernet's history in Tuscany has been still more controversial (apart from in Carmignano, where it has been part of the DOC blend since 1975), partly because of the so-called super-Tuscans – top-class wines which deliberately went outside the DOC system, often in order to add Cabernet, or by making Cabernet as a varietal. Cabernet in Tuscany has a beautiful deep blackcurrant and black cherry sweetness, and retains its acidity even when the alcohol reaches 14 per cent, as it can. The marriage of Cabernet and Sangiovese has also proved superbly fruitful; it is up to the winemaker to find a balance between the assertive Cabernet and the less dramatic Sangiovese.

Cabernet Sauvignon has long been conspicuous in Lombardy, where it is often blended with Merlot. Bordeaux-style blends can also be found in Emilia-Romagna, the Veneto, Friuli-Venezia Giulia and, to a lesser extent, the Alto Adige and Trentino. Teroldego and the Valpolicella grapes are other blending partners in their respective regions. Cabernet Sauvignon usually suffers from insufficient ripeness in the Alto Adige and Trentino – the dreaded green bean flavour again – particularly when it is trained on high pergolas and overproduces, as it usually is and does. Better clones, Guyot-trained, can help this problem, but so can growing the earlier-ripening Cabernet Franc instead of Cabernet Sauvignon.

In the South of Italy, Cabernet is blended with every conceivable red grape: Gaglioppo in Calabria; Merlot and Aglianico in Campania; Nero d'Avola in Sicily; and Cannonau and Carignano in Sardinia.

Spain

Nearly every region of Spain has some Cabernet Sauvignon planted, though often only as an experiment. But Cabernet experiments have a habit of turning out well. It has already shown it can produce good varietals in Penedès, and it was introduced to Rioja in the mid-19th century by the Marqués de Riscal. There are currently some 70ha (173 acres) planted there, and there are moves to get it added to the list of approved varieties for Rioja. It is already being used at many bodegas, but is it a cuckoo in the nest? It seems to have less pronounced tannins in Rioja than in

Leap of faith

Back in 1976 there was a tasting in Paris of top French wines against the best of their counterparts from California. It was organized by wine merchant Steven Spurrier, and its effect was electric. Known as the Judgment of Paris tasting, it awarded top place among the reds to Stag's Leap Wine Cellars 1973 Cabernet Sauvignon. This Paris triumph made the cover of Time magazine; there is now a bottle of Stag's Leap 1973 in the Smithsonian Institute.

Warren Winiarski (left), founder and until recently the owner of Stag's Leap, planted his Cabernet vines in 1970, after years of exploring the Napa Valley and noting where the vegetation changed, where there was frost damage, and what the growing conditions were – and this was an unfashionable attitude at the time in California. The emphasis then was on grape variety, not place. Winiarski's first career was teaching political theory at the University of Chicago. His fascination with wine began when a friend brought a bottle of wine to lunch – it came from the East Coast and was made from hybrid grapes. Eventually Winiarski and his wife decided to make wine themselves, and drove across the desert to California..

Stag's Leap Wine Cellars

S.L.V. stands for Stag's Leap Vineyards – the original estate vineyard planted in 1970. The 1973 Cabernet which won the Judgment of Paris tasting (see above) was from this vineyard.

most places, and in cooler spots in the region, like Haro, it doesn't ripen well, and still only has experimental status there, so I don't think we need to worry too much. It has had more of an effect in Navarra and Ribera del Duero.

Other Europe

The bargain red wine of the 1980s, Bulgarian Cabernet Sauvignon has lost popularity to fruitier, softer versions from Australia and Chile. Cabernet is widely grown in Hungary, Moldova, Romania, Croatia, Slovenia and throughout the former Eastern Bloc, and individual examples can be attractive, particularly where Western winemaking techniques are available. It is grown on a small scale in Austria, but seldom ripens well there, and some growers are replacing it with Merlot. It is successful in Greece, Turkey and Israel, and forms part of the blend at several Lebanese wineries.

USA: California

The appearance of a new biotype of phylloxera in California in the 1980s, and the subsequent replanting of many vineyards, did not bring about the reduction in the amount of Cabernet Sauvignon, and the increase in other varieties that some people hoped for. Quite the reverse, in fact: Cabernet's acreage more than doubled between 1988 and 1998. In the Napa Valley north of Yountville the vineyards are almost solidly Cabernet Sauvignon now, with some Cabernet Franc and Merlot: there is very little Chardonnay or Sauvignon Blanc still grown in this part. The smaller hillside regions like Mount Veeder, Howell Mountain, Diamond Mountain and Spring Mountain, with their slow-aging, tightly structured styles of wine, are more Cabernet-dominated than before.

Sonoma's leading Cabernet region, Alexander Valley, has now been replanted with better clones that are less likely to give green, herbaceous flavours, and more Merlot and Cabernet Franc have been planted. But for every producer experimenting with blends of Merlot, Cabernet Franc, Malbec or Petit Verdot, there is likely to be another increasing the proportion of Cabernet Sauvignon to emphasize the strong character it has here.

Growers have generally learnt to make better balanced wines, with the fashion for extremely long hang times now mercifully waning – for some producers, anyway. Dry Creek Valley, Sonoma Mountain and Sonoma Valley are up-and-coming regions, and Mendocino County is showing great promise. Further south, Paso Robles in San Luis Obispo County, Santa Cruz Mountain and Monterey's Carmel Valley have only small amounts of Cabernet, but are making some stylish wines. Of the more established regions, Stags Leap District makes supple, well-structured, black cherry wines, and warmer Oakville and Rutherford are more blackcurrants and plums, richer and with firmer, dusty tannins.

The main stylistic difference in California is between hillside and valley floor wines. Hillsides, with their thinner soil, give lower yields (1–2 tons per acre, compared to 4–8 on the valley floors). The berries are smaller, and flavours are more austere and intense, chewier and less opulent, in a slow-maturing, Bordeaux style. But there are exceptions, notably the valley-floor-grown Opus One, with its Bordeaux-style restraint.

Californian Cabernets, especially from Napa, where the influence of Robert Parker has focused growers on making super-ripe, super-rich, super-polished reds, can reach over 15 per cent and are seldom below 14 per cent. They argue that Bordeaux regularly hits 14 per cent nowadays, too: true, but the greater acidity of Bordeaux makes for more balanced, fresher wines. Napa growers have a horror of anything that could be construed as 'green', and if that means making wines that taste of prunes and raisins, then so be it. Their wines may be textbook-perfect, but they often reflect their hang-time and their additions in the winery – routinely, acid and extra tannins – more than their *terroir*.

USA: Washington State

Cabernet Sauvignon could have greater long-term potential here in Washington State even than Merlot – particularly now that better vineyard management is beginning to reduce its unripe green flavours. It has now overtaken Merlot as the state's Number One red variety, even though Merlot has proved faster at showing its worth. To thrive Cabernet needs the hottest sites: the Yakima Valley is generally on the cool side, and warmer parts of the Columbia Valley are more suitable. Its great advantage, as far as the growers are concerned, is its resistance to winter cold.

The owners of Screaming Eagle in the Napa Valley (above) aim to make California's greatest wine, bar none. With just 200 cases of intensely concentrated Cabernet made each year, few people will ever have the chance to judge if they're succeeding. Not surprisingly, the wine has cult status.

Cabernet's trademark in Washington State is its forthright fruit, with more freshness than Californian examples, and it can make early-drinking styles – though there is an increasing number of growers who are producing impressively dark, brooding reds.

Rest of North America

There are small quantities of Cabernet Sauvignon planted in Oregon, mostly in the Umpqua and Rogue Valleys; other states, including Texas and Arizona, also grow it. Its toughness in the face of cold winters makes it attractive to Canadian growers with Southern British Columbia having the most success.

Australia

Coonawarra led the way in the 1970s with blackcurrant and mint fruit and fine structure, though Cabernet's reputation is now based as much on the black-fruited, dustily herby and tightly structured wines of Margaret River, the balance and elegance of Yarra Valley, the sweet, focused fruit of Clare Valley and the rich, heavy wines of Barossa and McLaren Vale.

All these areas, and various other subzones too, make individual and at times exceptional Cabernets that truly reflect regional characteristics, and this has given rise to the latest wave – trying to express actual vineyard character; much more enjoyable than the unlamented slavish chase after a dense, choking, 100-pointer style.

At the less expensive end there are abundant, ripe-fruited examples blended from different regions and even between states. Plantings are still rising, although at about one-sixth of the rate of Shiraz.

New Zealand

Hawkes Bay is the key region here. The ripest Cabernet Sauvignons come from this climatically diverse North Island region, though often still retain a green flavour as a reminder of its relatively cool climate. This is exacerbated by high yields: the fertile alluvial soil of much of the flat land means vigorous vines. Better-adapted rootstocks and lower yields can help produce riper fruit that emphasizes cassis flavours; canopy management has already made big improvements to ripeness; and development of warm gravel beds in areas like Gimblett Gravels has produced exciting Cabernet results. Waiheke Island, in Auckland, also has some impressive wines. Even so, blends with Merlot are nearly always more interesting than pure Cabernet Sauvignon, particularly when Merlot dominates.

Central and South America

Not surprisingly, Cabernet Sauvignon is found in just about every wine country in Central and South America. In Mexico, the wines can be earthy and four-square; in Uruguay, they can have nicely balanced blackberry fruit. Brazil has a few successes in the far south. In Argentina Cabernet is typically blended with Malbec, especially at the top end but can excel on its own. These premium wines, full of tobacco and leathery fruit at their best, have considerable aging potential. Early-drinking wines tend to have sweeter, lighter fruit.

In Chile, climate is the main factor in determining what gets planted where, though soil is increasingly important and *terroir* is the new buzzword. Maipo, where a lot of the eucalyptus trees have been felled, is making much less minty Cabernet than before; there is still that piercing blackcurrant fruit to the wines but with an increasingly long-lived structure. Aconcagua wines are fairly solid, closed in but still sweet and ripe at heart. Warmer Curico gives richer, softer Cabernets, and Colchagua generally gives fast-developing wines with soft tannins, sweet fruit and less acidity. As in Argentina high yields and over-oaking can be a problem, but Chilean Cabernet has the potential to be a world-beater.

South Africa

Cabernet Sauvignon from new clones, coming on stream in the vineyards in the mid-1990s, has ripe, sweet fruit in place of the high acidity and unripe herbaceous notes of the old, virused clones. Later picking, and better winemaking which avoids volatility, are also crucial. Location is now a major factor in style: Constantia Cabernet has minty, herbal flavours to Stellenbosch's structure and ripe fruit weight. A blend of the two might be about right. The west coast north of Cape Town, cooled by coastal breezes, is an interesting new region.

Rest of the World

China is the big story. She is the fastest-growing wine producer in the world, with Cabernet Sauvignon regarded as the star variety. According to latest available statistics, China already has the world's largest Cabernet Sauvignon plantings. Inland regions like Ningxia look more exciting than coastal Shandong and each vintage brings an increased flow of tasty, characterful wines.

Japan and India make occasional decent examples, but the climatic conditions in these countries don't generally favour making big, structured reds.

Shafer
Napa is one of the world's classic Cabernet Sauvignon regions, and Stags Leap District produces its supplest wines. This is Shafer's top Cabernet wine, with the capacity to age for 20 years or more.

Petaluma
Petaluma makes a classic Coonawarra wine, elegant and restrained, but with ageability firmly to the fore. The wine is predominantly Cabernet Sauvignon , with some softening Merlot blended in.

Te Mata
One of New Zealand's top reds, this Hawkes Bay blend adds Merlot and Cabernet Franc to a backbone of Cabernet Sauvignon in true Bordeaux style and greatly improves with age.

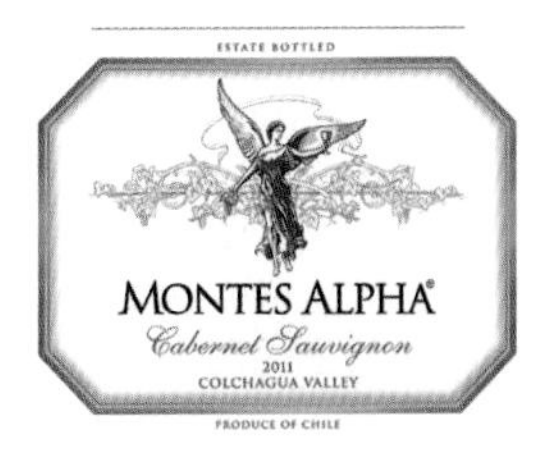

Montes
Montes have long championed Cabernet Sauvignon in Chile's Colchagua Valley, in particular planting the challenging Apalta slopes. This is 80% Cabernet Sauvignon, blended with Merlot and Cabernet Franc.

ENJOYING CABERNET SAUVIGNON

Cabernet Sauvignon makes wines that can shine at a century old, yet it can also make delightful wine that is unbeatable a mere six months after vintage. Bordeaux is where Cabernet Sauvignon first showed what it could do, in terms of both flavour and longevity. Questions are sometimes raised about the aging ability of the currently fashionable richer, fleshier style of red Bordeaux. The 1982 vintage, the first in this style, has aged unpredictably, as have the 1989 and 1990, with some leading wines fading surprisingly quickly but then regaining richness and structure for no apparent reason except that Cabernet is a grape that is determined to age well if it can. The top 1982s are still inspiring wines. It still seems fair to say that top red Bordeaux needs at least 10 years to come round, and in a good vintage should last two or three decades longer. French practice is to drink them at a few years old. Although top Italian and Spanish Cabernets could easily take 10 years of aging, most will also be consumed at only a few years old.

Cabernet changes so much in bottle that it would be a shame to forget the pleasures of mature wine. But it would also be a shame to forgo the pleasures of young Cabernet. The top Cabernets of Australia need around 10 years, but most Aussie Cabs are excellent at five years old, many peak at only two to three years, some last for twenty. Top Californian vintages like 2006 or 2012 can last for two, perhaps three decades, but since so much Californian Cab is drunk on release without any further aging, it isn't surprising to find that, below the top level, most are ready at two to three years old. South American Cabernets – especially Chilean examples – are bursting with flavour at only a couple of years old though they definitely do age. South African examples, though softer and riper than they used to be, still often need six to eight years. New Zealand Cabernets are usually ready quite young but do age well, with sweet blackcurrant fruit, even if they rarely lose their streak of leafy greenness.

The taste of Cabernet Sauvignon

There is no mistaking the blackcurrant scent of Cabernet. Young wines taste of black cherry and blackcurrant; mature wines add the classic nose of pencil shavings, cedar and cigar boxes.

At lower ripeness levels Cabernet exhibits a telltale greenness, a green bell pepper nose that at its worst is raw and vegetal. It is a flavour winemakers try to avoid. More pleasant tastes and smells are those of tobacco, mint and eucalyptus, and fruits like blackberry and black cherry; blackcurrant is generally present, even in less than super-ripe examples, though less so in Bordeaux and Napa. Overripe Cabernet goes jammy-tasting – stewed prunes and dates at worst. Some growers like to pick a mixture of slightly underripe, perfectly ripe and slightly overripe grapes, believing that the combination of all these flavours gives extra complexity. Others say no, the ideal is to pick everything at optimum ripeness: that way the grape tastes most like itself.

New World examples, particularly at the less expensive end, show sweeter fruit than Bordeaux of equivalent quality: they are juicier and more forward and, at lower price levels, more attractive. Basic Bordeaux Rouge, regardless of how much Cabernet Sauvignon it contains, is more likely to emphasize the austerity of the style over the fruit.

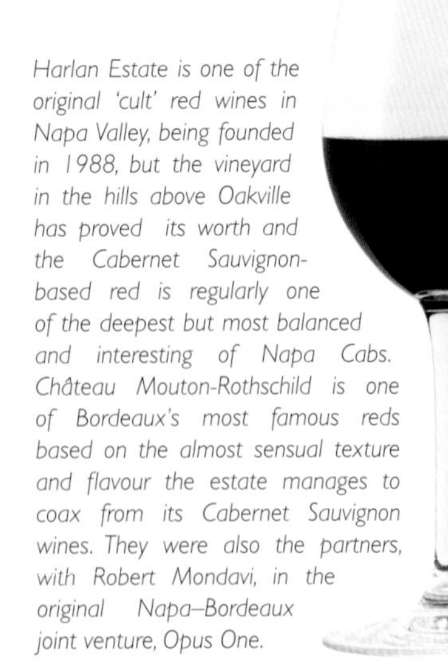

Harlan Estate is one of the original 'cult' red wines in Napa Valley, being founded in 1988, but the vineyard in the hills above Oakville has proved its worth and the Cabernet Sauvignon-based red is regularly one of the deepest but most balanced and interesting of Napa Cabs. Château Mouton-Rothschild is one of Bordeaux's most famous reds based on the almost sensual texture and flavour the estate manages to coax from its Cabernet Sauvignon wines. They were also the partners, with Robert Mondavi, in the original Napa–Bordeaux joint venture, Opus One.

MATCHING CABERNET SAUVIGNON AND FOOD

All over the world Cabernet Sauvignon makes full-flavoured, reliable reds: the ideal food wine. Classic combinations include *cru classé* Pauillac with roast milk-fed lamb; super-Tuscans with *bistecca alla fiorentina*; softer, riper New World Cabernet Sauvignons with roast turkey or goose. Cabernet Sauvignon seems to have a particular affinity for lamb, but it partners all plain roast or grilled meats and game well and would be an excellent choice for many sauced meat dishes such as steak and kidney pie, beef stews, rabbit stew and any substantial dishes made with mushrooms.

CONSUMER INFORMATION

Synonyms & local names

France's many historical synonyms include Petite Vidure and Bidure.

Best producers

FRANCE/Bordeaux Calon-Ségur, Dom. de Chevalier, Cos d'Estournel, Ducru-Beaucaillou, Grand-Puy-Lacoste, Gruaud-Larose, Haut-Brion, Lafite-Rothschild, Lagrange, Latour, Léoville-Barton, Léoville-Las-Cases, Léoville-Poyferré, Lynch-Bages, Ch. Margaux, Montrose, Mouton-Rothschild, Pichon-Longueville, Pichon-Longueville-Comtesse de Lalande, Pontet-Canet, Rauzan-Ségla; **Other** La Grange des Pères, Tour des Gendres, Trévallon, Verdots.
ITALY Antinori, Gaja, Isole e Olena, Lageder, Le Macchiole, Ornellaia, San Leonardo, Sassicaia, Tasca d'Almerita, Tua Rita.
SPAIN Abadía Retuerta, Blecua, Enate, Jané Ventura, Marqués de Griñon, Torres.
PORTUGAL Esporão.
USA/California Araujo, Beringer, Bryant Family, Buccella, Cakebread, Caymus, Chimney Rock, Corison, Dalla Valle, Diamond Creek, Dominus, Dunn, Grace Family, Harlan, Hartwell, La Jota, Ladera, Laurel Glen, Long Meadow Ranch, Peter Michael, Miner Family, Mondavi, Newton, Oakville Ranch, Joseph Phelps, Ridge, St Supery, Screaming Eagle, Shafer, Silver Oak, Spottswoode, Stag's Leap Wine Cellars, Terra Valentine, Titus, Viader; **Washington State** Andrew Will, Cadence, Corliss, DeLille, Dunham, Fidelitas, Hedges, Januik, Leonetti, Quilceda Creek, Three Rivers, Woodward Canyon.
SOUTH AMERICA Almaviva, Altair, Aristos, Atamisque, Carmen, Catena Zapata, Cobos, Concha y Toro, Errázuriz, Haras de Pirque, Kaiken, Mendel, Santa Rita, Terrazas de Los Andes, Miguel Torres.
AUSTRALIA Tim Adams, Balnaves, Cape Mentelle, Cullen, Forest Hill, Fraser Gallop, Giaconda, Grosset, Hardys, Henschke, Houghton, Howard Park, Katnook, Leasingham, Leeuwin, Majella, Moss Wood, Mount Mary, Parker, Penfolds (Bin 707, Bin 169), Penley, Vasse Felix, Voyager, Wendouree, The Willows, Wirra Wirra, Woodlands, Wynns, Xanadu, Zema.
NEW ZEALAND Craggy Range, Esk Valley, Man O'War, Stonyridge, Te Mata, Trinity Hill, Vidal, Villa Maria.
SOUTH AFRICA Beyerskloof, Boekenhoutskloof, Buitenverwachtung, de Toren, de Trafford, Neil Ellis, Grangehurst, Jordan, Kanonkop, Le Riche, Meerlust, Rustenberg, Saxenburg, Thelema, Vergelegen, Waterford.

RECOMMENDED WINES TO TRY

Ten Bordeaux Classed Growths

Domaine de Chevalier Pessac-Léognan
Ch. Ferrière Margaux
Ch. Haut-Bailly Pessac-Léognan
Ch. Lagrange St-Julien
Ch. La Lagune Haut-Médoc
Ch. Langoa-Barton St-Julien
Ch. Léoville-Poyferré St-Julien
La Mission-Haut-Brion Pessac-Léognan
Ch. Montrose St-Estèphe
Ch Palmer Margaux

Ten other good Bordeaux wines

Ch. d'Angludet Margaux
Ch. Batailley Pauillac
Ch. Chasse-Spleen Moulis
Ch. La Gurgue Margaux
Ch. Labégorce-Zédé Margaux
Ch. Monbrison Margaux
Ch. Pibran Pauillac
Ch. Potensac Médoc
Ch. Poujeaux Moulis
Ch. Sociando-Mallet Haut-Médoc

Twenty New World Cabernets

Andrew Will Klipsun Cabernet Sauvignon (USA)
Balnaves Coonawarra Cabernet Sauvignon (Australia)
Catena Alta Cabernet Sauvignon (Argentina)
Chateau Ste Michelle Col Solare (Washington State)
Corison Cabernet Sauvignon (California)
Diamond Creek Cabernet Sauvignon(California)
Errazuriz Viña Chadwick (Chile)
Grosset Gaia (Australia)
Hedges Family Estate Red Mountain (Washington State)
Jordan Cabernet Sauvignon (South Africa)
Le Riche Auction Reserve (South Africa)
Ridge Monte Bello (California)
Sandalford Cabernet Sauvignon (Australia)
Santa Rita Casa Real (Chile)
Tim Adams Clare Valley Cabernet (Australia)
Valdivieso Reserve Cabernet Sauvignon (Chile)
Vasse Felix Cabernet Sauvignon (Australia)
Vergelegen Vergelegen Red (South Africa)
Vistalba Tomero (Argentina)
Wynns Coonawarra Cabernet Sauvignon (Australia)

Cabernet Sauvignon's small berries can make reds soft and fruity enough to enjoy at one to two years old, as well as the majority of the world's genuine long-distance wines.

Maturity charts

Cabernet Sauvignon is potentially one of the longest lasting of red grapes, but much depends on the producer.

2009 Médoc Second Growth Cru Classé

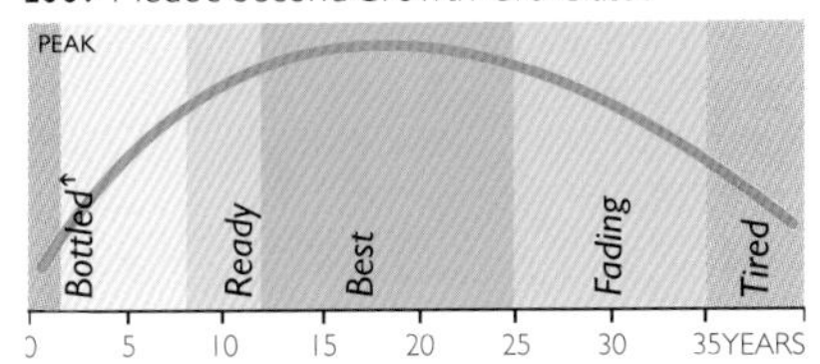

2009 was an excellent vintage throughout the Médoc. It was particularly strong in St-Julien, Pauillac and St-Estèphe.

2009 Top Napa Cabernet Sauvignon

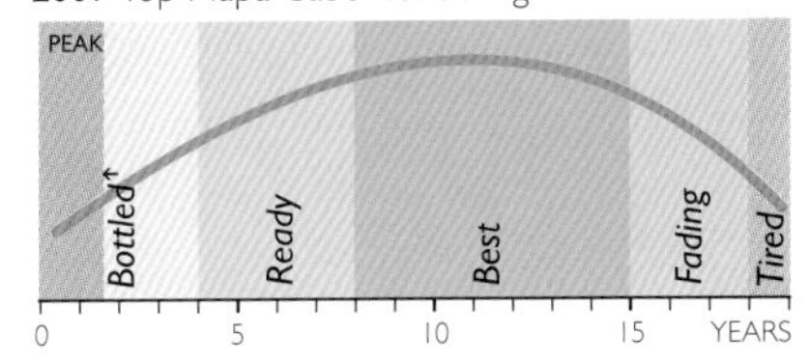

2009 was a very good year. A cool summer produced wines of balance, concentration and a fresh acidity that will ensure long life.

2010 Coonawarra Cabernet Sauvignon

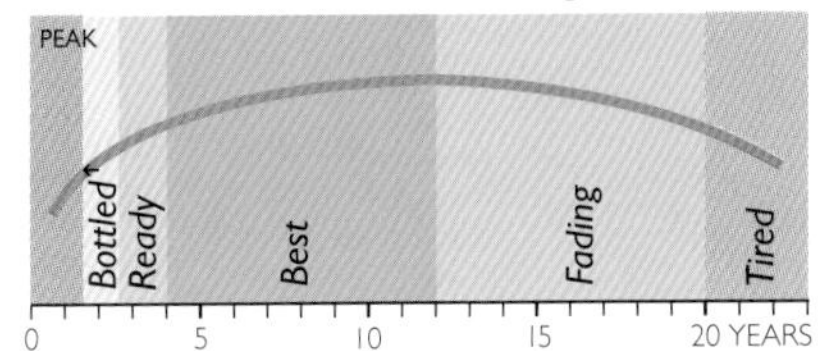

A relatively cool year for Coonawarra, one of the world's classic Cabernet regions. The wines have great intensity and dark, black fruit.

CAÍÑO BLANCO

Found in the Rosal valley in Rías Baixas, northwest Spain, and to a much lesser extent in Portugal, this has largely been abandoned by growers in favour of Albariño, which is easier to grow and gratifyingly fashionable. But Caíño Blanco has its fans: its tropical aromas, density of flavour, creamy texture and good structure mean that some think it's more interesting than Albariño. It transmits its *terroir* more faithfully than Albariño, but gives less juice, which immediately makes it less rewarding, and is susceptible to mildew and botrytis; it also ripens late. Best producer: Terras Gauda.

CALADOC

A recent French crossing of Grenache and Malbec with good colour and body, enticing loganberry fruit but a slight tendency to volatility. It is attracting attention in South America and Portugal, among others.

CALABRESE NERO

A synonym for Nero d'Avola, see page 176.

CALLET

The name, in Mallorcan dialect, means 'Black'; sadly unimaginative, but clear. Its wines are often not that dark in colour, however, and it is often blended with other local varieties, including Manto Negro for fruit, or Fogoneu, with which it is often co-planted. Acidity and alcohol are seldom high, but the wines can be appealingly perfumed at their best.

CAMARATE

Portuguese grape with umpteen synonyms, but not much variety. It's mostly found in Bairrada, where it makes quite soft, silky reds. Castelão de Nosso, Castelão Nacional, Vide Prete and Negro Mouro are all Camarate.

CANAIOLO

A perfumed red grape that seems to have been the main constituent of Chianti until the late 19th century, and thereafter was used for softening the astringency of Sangiovese. It is no longer a required part of the Chianti blend, and has been in decline since the onset of phylloxera, when it proved a tricky vine to graft. The available clones have also been generally poor, though there have been moves to remedy this. Some Tuscan growers still treasure their Canaiolo, and blend it with Sangiovese. There is some in Lazio, the Marche and Sardinia, and a white version, Canaiolo Bianco, in Umbria, alias Drupeggio, is a different vine.

CANNONAU

The Sardinian name for Garnacha Tinta (see pages 102–111). In Sardinia it gives powerful, often slightly earthy, toffeeish table wines and some exciting, rich fortified wines.

CARIGNAN

A hot climate vine that probably did more than another other grape to fill Europe's wine lake in the late 20th century, Carignan is now in decline. It still covers large tracts of Languedoc-Roussillon, producing yields of up to 200hl/ha with ease. The wine has quite dark colour, loads of tannin and acidity, plenty of astringency, and as a bulk wine gives very little pleasure. But it has shown itself to be eminently suitable for vinifying by the Beaujolais method of carbonic maceration when the colour deepens, the astringency softens and a rustic but attractive fruit and perfume appear out of nowhere. Blend this with Grenache or Syrah and good wine can result.

Only in exceptional sites, with first-class exposure and good drainage, and with very good winemaking can it produce fine wine on its own. However, such wines do exist, both in the south of France, especially Corbières, and elsewhere. Indeed, some Languedoc-Roussillon producers call Carignan 'our Pinot'. When Carignan makes a serious wine, yields are low and the vines are usually old – certainly 50 years or more. It can also make attractive herb-streaked rosés.

Outside Languedoc-Roussillon it is found in the southern Rhône, but it cannot travel too far because, being both late budding and late ripening, it needs a warm climate to ripen. It is also susceptible to rot and both kinds of mildew. There is also a white version, Carignan Blanc, found in Languedoc-Roussillon.

In Italy it is found as Carignano, especially in Sardinia, but also in Lazio. In Spain it is called Cariñena, and in fact it originated in Aragon. Today, however, it plays only a small part in Cariñena, the wine named after it. Instead it is mostly found in Cataluña, especially in Ampurdán-Costa Brava, Priorat and Tarragona. Under its alternative name of Mazuelo it may be a small part of the Rioja blend, and is valued for the very acidity that is not relished in the Languedoc. Its colour and tannin are also useful in Rioja.

California can produce some very good Carignane, as it is called here, but it can also produce high-yielding wine quite as dreadful as any other. Nearly all the vines are old, simply because no one has got round to replanting with something else. Paul Draper of Ridge

A healthy crop of Carignan from 70-year-old vines in Chile's Colchagua Valley. This vineyard, belonging to Villalobos Winery, was never trained, and thus grew wild – in clumps along the valley floor or, in this case, up trees. The vines aren't tended during the year (horses help with the pruning by nibbling the shoots), and the forest has grown up around the vines. The harvest requires gauntlets, stepladders and a good sense of balance.

Vineyards makes some excellent Carignan from vines planted in 1880. Here, as almost always, it is blended with other varieties. Chile has small amounts of old Carignan which are now greatly in demand and produce beautifully deep, minerally wines; so much so that more is being planted. There are some old vines in South Africa. Best producers: (France) Aupilhac, Clos Centeilles, la Dournie, Mont Tauch co-op, Pech-Redon, Rabiéga, Roc des Anges, la Voulte-Gasparets; (Italy/ Sardinia) Argiolas, Mauritania, Santadi co-op; (California) Bonny Doon, Cline Cellars, Fife; (Chile) De Martino, Odfjell, Viña Segú, Miguel Torres, Valdivieso; (Spain) Vall-Llach.

CARIGNANO

See Mazuelo page 135.

CARIÑENA

The Spanish name for Carignan/Mazuelo.

CARMENÈRE

See pages 66–67.

CARNELIAN

This US cross, bred in 1949 from Mazuelo and Cabernet Sauvignon, was planted in some spots in Western Australia because it was thought to be Sangiovese, and it's actually done quite well. It has plenty of tannin, as one might expect from that parentage, and good acidity. It's in decline in the USA and is mostly found in warmer parts of California.

CARRICANTE

A late-ripening Sicilian white grape, which is often trained as a bush vine and is a staple of Etna Bianco blend. It does well on its own, though: aromatic, fresh, orange-zesty and complex.

CASETTA

One of many old Italian varieties being rescued from extinction, this is found in the far north, where it was abandoned because of its susceptibility to fungal diseases. It makes very dark, tannic acidic wine redolent of plums and tobacco. Albina Armani is the main producer.

CASTELÃO

This is the principal name for the widely planted southern Portuguese grape better known by its nickname of Periquita, or 'little parrot', a name it acquired from a vineyard, Cova de Periquita, where it was planted by José Maria da Fonseca in the early 1850s. Its other names include João de Santarém, Bastardo Espanhol (in Madeira) and Trincadeira, though this is also a separate variety sometimes known in the Douro as Tinta Amarela.

It mostly makes appealing, upfront wines, quite low in acidity and high in alcohol, generally on the light side and with good raspberryish fruit. The Setúbal peninsula seems to be one of its best regions, though a little added acidity may be necessary for balance and here, particularly in Palmela, ripe, rich reds appear. It is prone to over-ripeness in Alentejo but can still produce some big juicy reds. Much is blended with other grapes, though it also appears as a varietal under several of its names.

Best producers: (Portugal) Quinta da Abrigada, Bacalhôa, Quinta do Casal Branco, J M da Fonseca, Quinta de Pegos Claros, Casa Santos Lima, Sogrape.

CATARRATTO BIANCO

Widely planted Sicilian white grape that can attain good quality if yields are controlled, but only sometimes does so. Catarrato Bianco Comune and Cararratto Bianco Lucido are the same grape. It used to be grown for Marsala, but now is mostly either distilled or turned into grape concentrate. It features in several DOCs in Sicily, and if well made the wine can be crisp and vaguely interesting. Best producers: (Italy) Calatrasi, Rapitalà, Spadafora.

CATAWBA

On the eastern US seaboard New York State grows Catawba for pink wines in various styles. Its skins are dark pink in colour, and it needs help from thermovinification to produce anything that could be described as red. The wine is decidedly 'foxy'-tasting. It was first spotted growing beside the river Catawba in North Carolina in 1801, and is perhaps a *labruscana*, a crossing of *Vitis labrusca* and *Vitis vinifera*. Its synonyms include Mammoth Catawba and Francher Kello White. Best producers: (New York State) Lakewood.

CAYETANA BLANCA

An Iberian white variety with more synonyms than you could shake a stick at. In Spain it can be found as Jaén Blanco or Pardina, or as Mourisco Branco in Portugal, as well as Doradillo in Australia – though there is also a Doradillo in Málaga which is different to the Australian Doradillo. To have more than 50 synonyms suggests a high degree of popularity, but it's not in all honesty very thrilling: neutral, oxidizes easily. The Australians fortify it, distill it or uproot it.

CENCIBEL

The name given to Tempranillo in central and southern Spain. See pages 270–279.

CERCEAL

A Portuguese grape found in Bairrada, Dão and the Douro, but distinct from the Sercial of Madeira. It adds good acidity to a blend – always an advantage with Portuguese whites.

CESANESE

Old, interesting but relatively rare vine found in Lazio, near Rome. Best producers: (Italy) Casale della Ioria, Casale Marchese, Villa Simone.

CÉSAR

An ancient grape variety that originated either in northern France or Germany. Now it is practically extinct on the Yonne, and not permitted for AC wine in the Côte d'Or. The César in Chile may or may not be the same variety, and has mostly been uprooted. It gives dark-coloured, rather brutal wines.

CHAMBOURCIN

A French hybrid, one of the best in existence, that produces wines of an intensely purple colour and a pronounced flavour of black cherries and blackberries, sometimes with a touch of spice or game. The wine is best drunk when young and fresh. It has been planted only since the 1960s and is found to a small extent in France's Pays Nantais at the western end of the Loire Valley, and in southwest France, though only for table wine. In Australia Chambourcin is sometimes blended with Shiraz for the sake of its colour, but varietal versions (occasionally sparkling) are full of meaty, black cherry fruit. Sightings have been reported in Vietnam. Best producers: (Australia) D'Arenberg (Peppermint Paddock); (USA) Naylor.

CHARBONO

A now rare Californian vine that was thought to be the same as Italy's Dolcetto (see pages 94–95) but is the same as Argentina's Bonarda, which isn't the same as the Italian Bonarda (there are six of these) but is the same as Douce Noire from France's Savoie . . . zzzz – hey, wake up! There's not much Charbono in California, but the wine is good, strong and smokily rich, and as Bonarda in Argentina – oh, just look back to page 49! Best producers: (California) Duxoup, Etude.

CHARDONNAY

See pages 68–79.

CARMENÈRE

What did you get if, in the 19th century, you took cuttings of Merlot from Bordeaux and planted them in Chile? A field mix of Merlot and Carmenère, that's what. And (according to most estimates) between 60 and 90 per cent Carmenère, which perhaps says something about the relative unimportance of Merlot in pre-phylloxera red Bordeaux.

The two vines look fairly similar, and are in fact closely related, the main difference in appearance being that Carmenère's young leaves are red underneath, while Merlot's are white, and that the central lobe of the Merlot leaf is longer. Chileans just assumed that Carmenère was a slightly weird clone of Merlot – well, you would. In Bordeaux Carmenère used to be considered to be just as good as Cabernet Sauvignon, but unlike the latter it proved an irregular yielder when grafted and it was phased out in the 20th century. It ripens some three weeks after Merlot, which makes a field blend problematic. The difference between the two varieties was officially recognized in Chile in 1996, and it has been possible to label wines as Carmenère since 1998, though some growers still call Carmenère 'Merlot' and 'Merlot' 'Merlot Merlot'. All new plantings are of either Merlot or Carmenère, and Chilean growers are learning how to grow Carmenère. It dislikes irrigation or rain between winter and harvest time: water at this time exacerbates the green pepper flavour, as do poor soils which cause the vine to need more water. Because it gets high sugar levels before the tannins are ripe, it needs a long growing season, but in too hot a site the alcohol goes too high and the balance disappears. Even so, it is rapidly proving itself to be a really interesting grape with an unusual savoury quality to its taste. It also blends superbly with Cabernet, adding perfume and bright juiciness to the mix, rather as Cabernet Franc does in Bordeaux. In the winery it needs careful handling: its low acidity can leave it open to problems of bacterial infection and oxidation or reduction.

There are a few vines – literally a handful – in Bordeaux. Northern Italy has some, often labelled as Cabernet Franc. There are a few in California and quite a few more in Washington State, but the biggest plantings outside Chile are in China, where it is known as Cabernet Gernischt.

THE TASTE OF CARMENÈRE

Carmenère's low acidity gives it really sweet-tasting fruit, which makes it even more important to keep the green peppers under control: when ripe it has blackberry, black plum and spice flavours, rich, round tannins, and a marvellous savoury array of flavours – coffee, grilled meat, celery and soy sauce. This sweet/savoury flavour plus a full mouth-massaging texture make Carmenère a real original whose character often improves the palate of both Cabernet Sauvignon and Merlot.

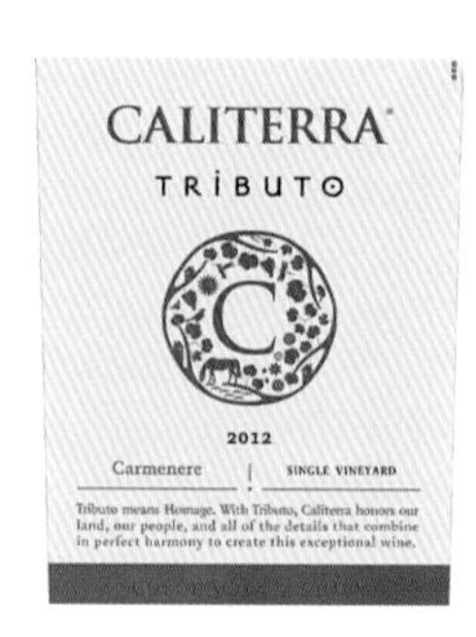

Caliterra
Caliterra planted their Colchagua vineyard 15 years ago, in a west-facing horseshoe of slopes looking out towards the Pacific Ocean. The wines are made in a fresh, modern style with good varietal definition.

Purple Angel is one of Chile's leading Carmenères, having a dark sensuous fruit richness and lush perfume that is tempered and made just a bit more serious by the addition of 8% of the tough, chewy Petit Verdot to the blend. The wine is a mixture of Carmenère from the daunting, powerful granitic Apalta vineyards and the more easy-going, clay-based El Arcángel vineyard in Marchigüe.

Falernia
Elqui is the most northerly of Chile's main wine regions, way up towards the Atacama Desert. Originally just a source of grapes for distillation, the Elqui Valley produces remarkable focused, cool climate Carmenère and Syrah.

Left: *New World extremes in Chile. The broad, flat, easy-to-cultivate expanses of Colchagua's valley floor, contrasted with Carmenère on the steep, rugged slopes of Clos Apalta owned by Casa Lapostolle. No one's going to mechanize these slopes.* ***Above:*** *Carmenère is late-ripening, dislikes water during the growing season and has a tendency to give green flavours in cool years. In addition, the vines seem to need to be mature to give good flavours: many winemakers agree that vines of under eight years old are likely to give vegetal-tasting wines without the admixture of fruit richness that marks out the grape's eventual personality. So far all plantings are massal selections: research to identify the best clones started in late 2007, led by Talca University's Dr Yerko Moreno.*

CONSUMER INFORMATION

Synonyms & local names

Grande Vidure is the best known of several historic Bordeaux synonyms. This name was occasionally used in Chile, when the grape was first identified there. Cabernet Gernischt is its Chinese name. It is sometimes labelled Cabernet Franc in northeastern Italy.

Best producers

CHILE Almaviva, Apaltagua, Arboleda, Bisquertt, Caliterra, Carmen, Casa Donoso, Casa Lapostolle, Casa Rivas, Casa Silva, Concha y Toro, De Martino, Luis Felipe Edwards, Errázuriz, Geo, Gracia, Los Robles, Montes, MontGras, Odfjell, Santa Rita, Terra Andina, Terranoble, Veramonte, Viu Manent.

RECOMMENDED WINES TO TRY

Ten Chilean Carmenère wines

Caliterra Arboleda Carmenère
Carmen Grande Vidure Reserve
Casa Silva Carmenère Reserva
Concha y Toro Terrunyo Carmenère
De Martino Reserva de Familia Carmenère
Luis Felipe Edwards Carmenère
Falernia Carmenère
Gracia Carmenère Reserva Especial Callejero
Montes Purple Angel
MontGras Carmenère Reserva

Ten top Chilean reds containing Carmenère

Apaltagua Envero Carmenère
Arboleda Carmenère
Casa Lapostolle Clos Apalta
Casa Rivas Reserva Carmenère
Casa Silva Gran Reserva (tinto)
Luis Felipe Edwards Reserva Carmenère
Santa Rita Triple C
Terranoble Carmenère Gran Reserva
Veramonte Primus
Viu Manent Secreto

De Martino
Until a few years ago, Carmenère was always blended, usually with Merlot. This wine, from an old-established winery and one of the Maipo Valley's rising stars, is a sign of how seriously the grape is now regarded in Chile.

E·R

CHARDONNAY

You can have too much of a good thing. Don't let anyone tell you otherwise. You can be the most adored and fawned upon, the most popular, the most respected and craved for. You can be the blonde-haired prince, the golden Queen, the vinous Almighty. And it can all go terribly wrong. How quickly can respect and reverence turn to sneering and abuse. How casually can a great reputation be mangled and past glories trodden underfoot in the rush to find something new to gawk at and proclaim. Just ask a Chardonnay producer.

If one grape variety led the New World wine revolution it had to be Chardonnay, the grape that could produce such lush, golden fruit, that could suck up the vanilla richness of oak barrels with such aplomb and create a wine bursting with ripe allure, oozing warmth and exotic promise, and yet somehow still be regarded as a dry white wine. Australia and California excelled at this, but where did they get the idea from? Well, those trailblazers back in the 1970s and 1980s chose Chardonnay to plant in their vineyards because one of the few European white wines with any reputation was white Burgundy – and the grape they used for white Burgundy was Chardonnay. They couldn't take the Burgundy vineyards home, but they could try to study how the great wines were made – and they could plant the same grape – Chardonnay. Which they did, with awesome determination.

There seem to be more flavours associated with Chardonnay than with any other grape and it also has a wonderful affinity with new oak barrels. So here carved in the fresh new oak are many of these flavours, including cloves, hazelnuts, warm brioche and a host of different fruits.

The rest of the world was watching, and it saw the rapidity with which Chardonnay took off. It saw how the drinking public swooned over this new experience – white wine that was soft, dryish, definitely fruity, often with a whiff of vanilla. And it sold like hot cakes to an entirely new generation of wine drinkers. Every single wine producer in every country who was just beginning to wonder if he or she could possibly attract the world's wine buyers to even consider taking a slurp of their grog was hit by the same blinding explosion of light at the same time. And that starburst of light was a glittering sign in the brains of the winemakers of the world that flashed the single word – CHARDONNAY – with all the brilliance of a Broadway billboard full of 1000-watt bulbs.

So now, every time I try to discover a new region or country – what's the wine I can't avoid? Chardonnay. The south of France saw Chardonnay as their great white hope. Spain caught on, Italy even more so, Bulgaria, Hungary, Romania, Greece, India – yes, India, and it's not too bad. Did I mention Moldova, Slovenia, Israel, China, Belgium, England? I should have done. They've all got Chardonnay, too.

But I haven't even started on the New World. California, Australia, Chile, Argentina, New Zealand, South Africa – everywhere in the New World anyone has ever thought of planting a vine, Chardonnay is pretty well top of the list. Whether there are suitable conditions or not – it doesn't matter. Chardonnay can make serviceable wine where it's too cold and where it's too hot, too dry, too windy, too wet.

So what went wrong? Chardonnay is simply too amenable. It can make pretty much any style and wine you want, from lean, crisp and minerally in northern Europe, through an astonishingly elegant savoury beauty in France's Burgundy but also in the best examples from Australia, New Zealand and the USA, right through to tropical, voluptuous mouth-fillers. If it had stopped there, Chardonnay's reputation would still be high, but sadly its very versatility and popularity meant that the industrial-scale wine producers, especially in California, Australia and Southern France, flogged its good nature to within an inch of its life, drowning its reputation in a floodtide of sugary, often cloddish and frankly unrefreshing liquid – or sometimes, in France and Italy, a liquid so pale and watery that you'd be excused for thinking they'd connected the filling machine to the cold water tap near the wine vat. I suppose it proves Chardonnay's stunning versatility, but it's done tremendous damage to its reputation. Even so, the good guys are fighting back, and along with the dross, there are more exciting Chardonnays in the world today than ever before.

Chardonnay: from Grape to Glass

Geography and History page 70; Viticulture and Vinification page 72; Chardonnay around the World page 74; Enjoying Chardonnay page 78

GEOGRAPHY AND HISTORY

Chardonnay is a global brand – has been for ages. But its brand manager has had to be pretty flexible in recent years. First there was boom, then – not exactly bust, not that; but certainly a move to ABC (Anything But Chardonnay). The grape became too successful, too ubiquitous. For a while it was seen as the answer to every struggling producer's problems, and French AC rules were reviled because they forbade the uprooting of old local varieties and the planting of Chardonnay for yet another oaky. tropical-flavoured me-too. Then came the rise and rise of New Zealand Sauvignon Blanc, and more recently, a fascination with just those old, indigenous varieties that would have been grubbed up in favour of Chardonnay.

So Chardonnay has had to reinvent itself, and it has. Australian Chardonnay has done a 180° turn, from fat and

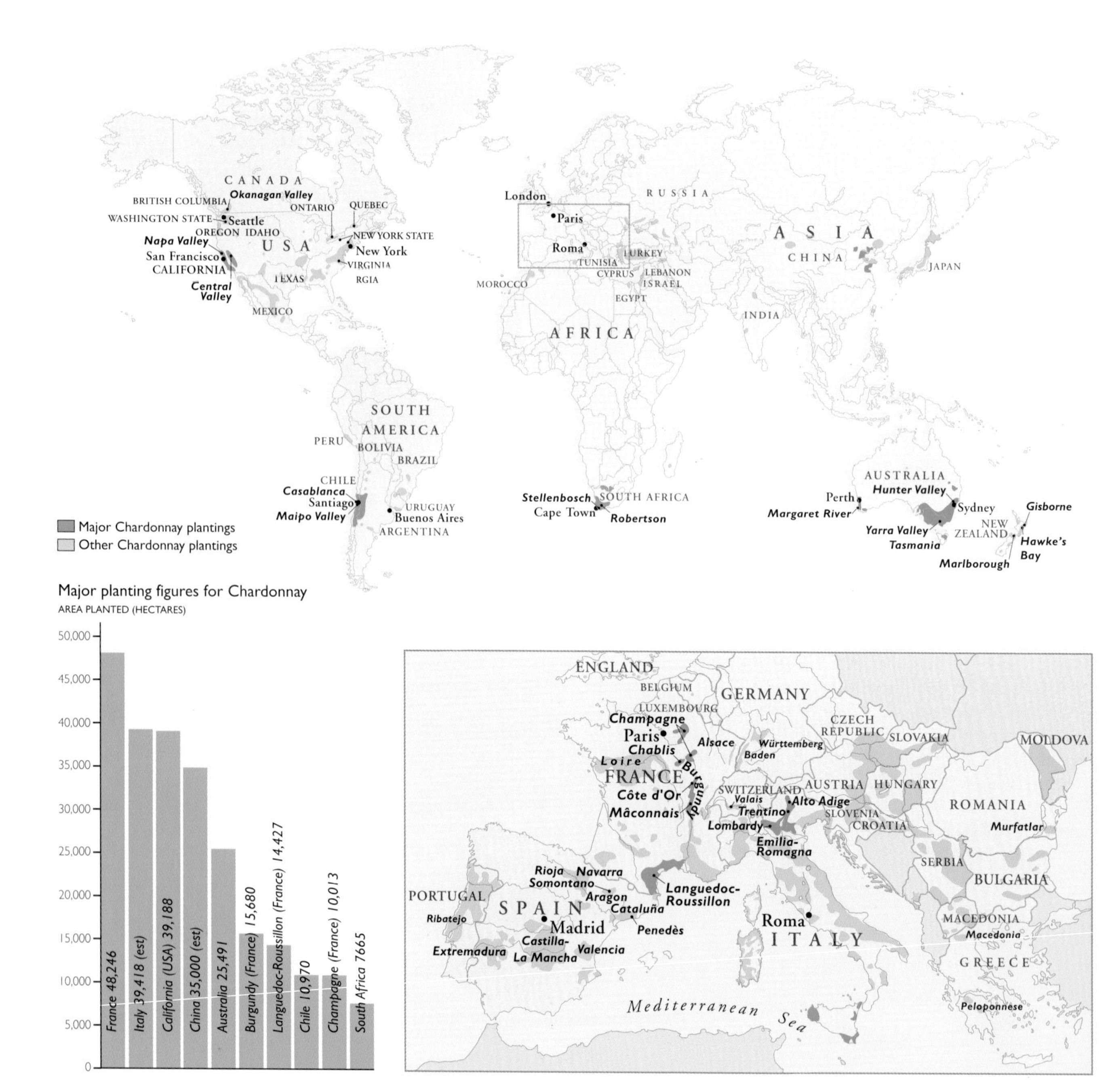

buttery to lean and mineral. The rest of the Chardonnay world has watched, and done the same. Chardonnay across the world is like the 'after' in a TV makeover programme – elegant, tailored, slender.

Chardonnay, you see, can do pretty well anything you want. Its natural self is relatively neutral, but it willingly accepts almost any climate or winemaking technique. In cool, limestoney Chablis it produces wines of piercing minerality that no one in the world has been able to copy. In Le Montrachet and other top Burgundy vineyards it gives wines of such towering complexity that you can hardly believe they come from mere grapes. In warm climates across the globe it produces lush, creamy, peachy wines; in cooler ones, nuttier, stonier ones. It's grown for sparkling wines everywhere, from Chile to Hampshire. It can be a cheap commodity wine, or it can improve in bottle for decades. A world without Chardonnay is impossible to imagine.

HISTORICAL BACKGROUND

Chardonnay appears to have originated in Burgundy, though precisely when is more difficult to say. It is the offspring of Pinot and Gouais Blanc – the latter being a mediocre white variety that is now thought to have originated somewhere between northeastern France and southwestern Germany. Gouais Blanc was widely planted in the Middle Ages, though is now almost extinct (there's a tiny bit in Haute-Savoie), but it became a sort of Queen Victoria figure – mother or grandmother to half Europe, with at least 81 offspring to its credit. Aligoté is a full sibling of Chardonnay, as is Gamay, so the vineyards of Burgundy and Beaujolais are every bit as crammed with close relatives as those of Bordeaux.

Chardonnay is first reliably identified in the late 17th century, in the village of St-Sorlin, now called La Roche-Vineuse, in the Mâconnais. It takes its name from the village of Chardonnay, also in the Mâconnais.

Pinot Blanc, which is a white mutation of Pinot Noir, looks extremely similar to Chardonnay, and confusion between the two has been common. Indeed, in places such as Australia and California, Chardonnay was sometimes labelled as 'Pinot Chardonnay' until late in the 20th century. The leaf shape is almost the same, but where Chardonnay has naked veins on top of the leaf, Pinot Blanc does not. The hairs on the shoot tips are slightly different.

There is a third vine that looks confusingly like both Chardonnay and Pinot Blanc: Auxerrois. Some French nurseries can be a little cavalier about correct identification – at least, it seems, when they are selling to foreigners: the 'Chardonnay' vines that were sold to Baden in the mid-1980s turned out to be a mixture of Chardonnay and Auxerrois; the same happened when South Africa bought 'Chardonnay' in the 1980s. One German vine expert even recalls visiting Chablis a few years ago and realizing that by no means all the vines he saw there were Chardonnay.

The seven Grands Crus of Chablis are situated on the appellation's warmest, southwest-facing slopes just above the little town of Chablis. The differences between the Grands Crus are subtle, but real, and they can all too easily be hidden by an excess of new oak in the winery. Les Clos (above) produces the nuttiest, most honeyed wine of all, without any need for extra new oak.

Olivier Leflaive standing proudly in front of row after row of Burgundy's creamy new oak barrels which he will use to age wines from both the Côte Châlonaise and Côte d'Or. Bâtard-Montrachet is a particular favourite of his.

Le Montrachet: for most people the wine from this vineyard is the pinnacle of white Burgundy. The slope is gentle – just 3% – and the soil thin, marly limestone, but the vines manage to catch more sunlight than those of any neighbouring vineyard.

VITICULTURE AND VINIFICATION

There could not have been a white grape better suited than Chardonnay to launching the New World into the big league of winemaking. It cheerfully adapts to most soils and most climates, with its only problem in warm spots being a tendency to overripen rapidly and lose acidity.

In the decades when the winemaker was king, and the technical jiggery-pokery of the winery the focus of all attention, Chardonnay responded brilliantly, strutting its stuff with all the aplomb of a supermodel parading Versace in the morning and Armani after lunch. It is all too easy to make all Chardonnays taste alike – but the top producers in Burgundy, Australia and California are all now producing wine of vibrant, memorable individuality. Indeed, Chardonnay can express *terroir* and react to sensitive winemaking so well that the one boon offered by an excess of Chardonnay in a red wine-obsessed world has been that the best producers have had to make great efforts to search out and then maximize their grapes' *terroir* personality to make them stand out from the golden crowd.

Climate

With Chardonnay, *terroir* does make a difference even though it settles down happily in a far greater range of *terroirs* than most other varieties; it reflects its *terroir* in the wine.

In cool climates it produces leaner, steelier flavours and reasonably keen acidity, though no other region so far has been able to mimic the mineral steeliness of Chablis. Chablis, Champagne and Tasmania are among the few regions where Chardonnay can taste green and unripe, though good viticulture, global warming and the grape's early-ripening nature can combine to defeat this in all but cold and wet years. But it buds early, as well, so is prone to frost damage in the spring: frost is the bane of most Burgundians' lives. Even if your vineyards are frost-free, cool wet weather during flowering can produce the uneven fruit set to which Chardonnay is also particularly prone. Late pruning, which delays flowering by up to a fortnight, can push flowering into a period of warmer, drier weather.

At the higher end of the temperature scale, Chardonnay can turn broad, heavy and flabby, with insufficient acidity to balance its melony fruit. Ideally, Chardonnay likes a long, slow ripening period in which it can develop flavour. In warm climates the temptation can be to pick before the acidity plummets; but more acidity in such cases can mean less flavour.

Soil

Limestone, chalk and clay are important for Chardonnay, though it will produce good flavours on a huge variety of soils. In France, key Chardonnay regions juggle these three soil types in different proportions: solid chalk in Champagne; limestone with clay on the Côte d'Or. The soil of the heart of Chablis is Kimmeridgian chalky marl and marly limestone. The scattered areas of Petit Chablis, the lowliest Chablis appellation, spill on to Portlandian limestone which gives less finesse.

On shallow, limy soils Chardonnay produces its tightest, most minerally wines; clay gives more weight and depth. In Meursault Perrières, for example, where the topsoil is barely 30cm (12in) deep over the limestone, the wine is restrained, powerful and slow maturing; in Meursault Charmes, where the topsoil is nearly 2m (6ft) deep, the wine is rounder, richer and more seductive.

In Walker Bay, South Africa, Anthony Hamilton Russell of Hamilton Russell Vineyards believes that his low-vigour, stony, shale-derived soils, with their high proportion of clay, compensates for a climate that is warmer than Burgundy to produce relatively tight, restrained, minerally Chardonnay. Fruit grown on sandstone in the same area produces far richer and broader wine.

Dominique Lafon looks like a moody French film star but is, in fact, one of the most serious and talented grape growers and winemakers in Burgundy, specializing in outstanding Meursault.

In the warm Hunter Valley in New South Wales, by contrast, Chardonnay is grown on the same light sandy soil that is favoured for Semillon: drainage is all important in a region where rain can make heavy, poorly drained soil impassable.

Yields

Chardonnay's ability to ripen and attain good sugar levels even in unpromising sites makes it less yield-sensitive than, say, Pinot Noir. Go over about 80hl/ha in the low-vigour soils of France, however, and you're risking serious loss of quality – and watery, dilute wine. On the hills of the Midi 50–60hl/ha is probably about right for *vin de pays*. For really top Burgundian Chardonnay yields of about 30–45hl/ha are probably necessary. In Champagne yields may well exceed 100hl/ha – finesse, not concentration, is the aim here. On high-vigour soils, yields of 60–100hl/ha may well be necessary for vine balance. New Zealand manages to combine high yields with some of the most intense Chardonnay flavours in the New World; so it is possible

Clones and sub-varieties

In Burgundy, there is little quality difference between massal selection and clones. The clones, of course, start off virus-free, which massal selected cuttings do not; many growers prefer clones for this reason. Each has its disadvantages: plant virus-free clones and your yields may be too high; plant virused cuttings and your vines run a serious risk of dying in 10 or 15 years.

Commercially available clones worldwide have improved hugely in the past 15 years, and choosing the right clone can make all the difference between producing a wine that sells for a low price, and one that attracts a much higher one. New Dijon clones, better adapted to cooler climates, are helping produce richer flavours in Oregon. New Zealand produces remarkably powerful flavours despite generally nondescript clones, and Chile is also short of decent clones. Chardonnay Musqué, with its surprisingly Muscat-like aroma, is found in the Mâconnais, particularly in Clessé; there is also a Chardonnay Rose, with intriguing pink berries. Old selections, like Wente in California and Mendoza in Australia and New Zealand, with mature vines, are now producing exceptional full-bodied wines.

At the winery

This somewhat neutral grape variety, once it gets into the winery, can be subjected to as many different treatments as the jawline of an aging socialite. All are directed towards influencing, refining or even adding flavour. Techniques usually thought of as Burgundian – barrel fermentation, malolactic fermentation and *bâtonnage*, or lees stirring – add, or bring out, nutty, toasty, creamy flavours in the wine; New World techniques – maceration with the skins, cold fermentation, ultra-hygiene – give tropical, fruit-forward flavours. From the 1980s onwards each camp has increasingly adopted the techniques of the other, which has led to cleaner, fresher white Burgundy and more elegant, restrained and subtle New World Chardonnays. Present a winemaker blind with a line-up of top Chardonnays from around the world, and it is increasingly likely that a Burgundian, an Italian, an Australian and a Californian will all mistake each other's wines for their own.

In Burgundy, and increasingly elsewhere for top wines, wild yeasts are allowed to have their way with the must, though producers might resort to cultured yeast if the fermentation is too slow in starting. The more mature the grapes, the longer the fermentation may last – though few continue as long as one of Domaine des Comtes Lafon's 1963 Chardonnays, which persisted for four or five years; the wine was eventually bottled in 1968.

Fermentation in barrel requires the right proportion of sediment to juice and, of course, clean sediment. Too many solids in the juice, says Dominique Lafon, give 'really weird wines' with green, bitter flavours. Fining the juice before fermentation, which might be necessary with machine-picked grapes, can, however, make the juice difficult to ferment.

In Champagne chaptalization is still normal, though climate change is giving more and more vintages where it's not necessary, and Champagne producers have been heard wondering whether acidity levels in the region might be falling below what they need. But as the best vintages in Champagne have nearly always been the ripest ones, there is probably no need for alarm yet.

Indeed, riper base wines (and there are more good Chardonnay vintages in Champagne than there are good Pinot vintages), more precise viticulture (which normally means parcels divided so that each parcel has homogenous vigour and ripeness) and picking at the optimum moment, plus ever more gentle winemaking, mean that zero-dosage Champagnes are becoming more common. Even where producers don't want to go that far, dosage levels generally have been falling: 9 grams per litre of dosage would be about normal for a non-vintage blend now, and less for a vintage. A generation ago this might have been nearer 15 grams per litre. Vintage wines tend to be riper anyway, and the longer times on the lees means that less dosage is needed.

Lees aging has an extraordinary effect on flavour. The breakdown of yeast cells releases many different substances into the wine – amino acids, peptides, proteins, fatty acids and others. Some of these give the wine extra aromas, others act as flavour enhancers. Amino acids give flavours of toffee, hazelnuts, honey, and cocoa. There's a protein called thaumatin, which is 2500 times sweeter than sugar.

Non-vintage Champagne must by law spend 12 months on the lees before disgorgement, which is just about long enough for some yeast autolysis character to start showing in the wine. Sparkling wines from other parts of the world, though often made from a similar blend of Chardonnay, Pinot Noir and perhaps Pinot Meunier, may decide against any lees aging at all. In France the preference is for drinking very young Champagne – brisk, refreshing, occasionally tart. Britain and the USA generally prefer Champagne with a little more age. Champagne kept in cool conditions can easily improve for 10–20 years. Even non-vintage blends can benefit greatly from 5–10 years' maturity. As the wine ages, the colour deepens, the flavour becomes much richer, toastier, nuttier, and the bubbles become less insistent and softer. Some of the most heavenly old Champagnes I've luxuriated in have been Blanc de Blancs wines based on Chardonnay grapes from the Côte des Blancs.

THE PREMOX PLAGUE

Or, if you prefer, the pox. It doesn't sound a lot better, but it's not meant to. It started with the 1995 and 1996 vintages of white Burgundy. Both were good years; the wines were supposed to last 10 or 15 years with no problem, before reaching their peak of complexity and resonance. But early tastings rang alarm bells: first one wine, then another, tasted very odd. Instead of tasting tight and closed, all structure and freshness, they tasted oxidized. People were quick to blame the corks, and cork probably was a factor – but this wasn't random, with a bottle or two per case being affected. It gradually became clear that whole wines, whole vintages, were oxidizing years before they should.

Now, a few years on, it's possible to identify what happened; and there were multiple causes. This was not a quick-fix problem. One factor was riper years giving wines with lower acidity: acidity helps to protect wines against oxidation. Another was the entirely laudable desire of growers to make purer wines, with lower levels of sulphur dioxide, the all-purpose winery antioxidant. Another was more reductive winemaking, in which the juice was protected against oxidation; it's counter-intuitive, but if the juice oxidizes early the wine is protected against oxidation later. Another was the fermentation of purer juice, with less solid matter in it and lower levels of phenolics – phenolics are also an antioxidant. Another was oxygen entering at bottling stage. And cork might have been part of the problem: the massive improvement in cork quality has been in this century, not last.

So what are Burgundian producers doing now? Vintages since 2002 have been less affected, but the problem hasn't gone away. They're using more sulphur, which they don't like, and they're allowing more solids in the juice. They're doing much less lees-stirring, that technique used during a wine's aging to keep the lees in suspension: it adds fatness to a wine, but could have been a factor in premox. Some have looked hard at their corks. What they want to do now is bring sulphur levels back down, but without having their wines die early. They're getting there, but it's best not to think of white burgundy as being the reliably long-lived wine it once was.

CHARDONNAY AROUND THE WORLD

The difference between Chardonnay styles is increasingly not one of regions, but one of climates and techniques – although it is uncertain that anybody outside Burgundy has yet produced a convincing imitation of the sublime flavours of Montrachet. But watch this space. The wine world is full of ambitious producers busting a gut to do so.

Burgundy

Great white Burgundy is the epitome of Chardonnay. It is the wine that persuaded the rest of the world to plant Chardonnay, and to try and copy the flavours of the Côte d'Or. Yet the absence of the grape's name from most Burgundy labels means that it seemed to appear from nowhere and take over the world in one mighty bound.

It is planted widely on the Côte de Beaune; and there is relatively little on the Côte de Nuits. Montrachet is smoky and immensely concentrated; Puligny-Montrachet is structured, savoury and tight; Chassagne-Montrachet oatmealier and softer; Meursault creamy and oatmealy; and Corton-Charlemagne rich yet minerally. Further north in Chablis, where the geology is quite different, the wine takes on a flinty, austere mineral character, flecked with honey, especially if it is made and aged without oak.

Further south in Burgundy a bit of a quality revolution is quickening. Côte Chalonnaise wines are lean but nutty. The best Mâconnais are fleshy, rich, yet balanced. All are for drinking young. But it's not for nothing that so many Côte d'Or growers have been seeking land down here.

Champagne

Chardonnay does not reach full ripeness here, even on the best, east-facing exposures of the Côte des Blancs. It's not the hours of sunshine that hold Chardonnay back – Champagne has as many of those as Alsace – it's the temperature. The west wind sweeps in across these low hills, keeping the annual mean temperature to around 10.5°C (51°F) – just half a degree above the absolute minimum needed to ripen grapes. Winemakers here look for creamy, nutty, flowery characters in the Chardonnay; for vinosity, elegance and aroma. Mention the word 'fruit' to them and they say, 'Oh yes, of course'; but one feels that it comes a good way down the list of priorities.

Because Champagne is usually a blend of regions and more often than not a blend of grapes, the different villages and districts are valued for the qualities they bring to the party. The Côte des Blancs villages of Cramant, Oger, Mesnil and Vertus all add the desired attributes of elegance and aroma, while Chardonnay from the eastern end of the Montagne de Reims is leaner and zesty; that from the Côte de Sézanne to the south is creamier, more lush.

Other French Chardonnays

Chardonnay has spread relentlessly from its bases in Burgundy and Champagne into the Jura, Savoie, the Ardèche, the Loire and even Alsace. These wines are usually light, and generally fairly mineral and lean but give Chardonnay another chance to express a different face. In the Midi, where the grape used to be given the Australian treatment, styles have calmed down to a better balance. In Languedoc it is best grown on the hills, and produces better quality than in Roussillon, where it is 3–4°C hotter on average; the higher altitudes of the Limoux or Pic St-Loup vineyards seem to be capable of producing better quality still. Indeed, Chardonnay from Limoux's limestone hills is developing a good reputation for its restrained and age-worthy style. Most Chardonnay from the Midi is pleasant rather than special – the soils are a bit fertile and the yields a bit high. Some of the best wines have a little scented Viognier blended in.

Rest of Europe

In Italy, Tuscan Chardonnay is becoming increasingly refined as the vines grow older. Over-oaking can still be a problem, as can over-generous yields in the north of the country. There may well be some confusion, either intentional or unintentional, with Pinot Blanc: Nicolas Belfrage, in *Barolo to Valpolicella*, quotes an unnamed Alto Adige producer as saying 'the best Chardonnay in Alto Adige is that made with Pinot Bianco'.
In Lombardy much goes to the sparkling wine industry; Franciacorta is particularly high quality. Some of the best still wines come from Piedmont, where the cooler climate gives the wines greater elegance. All over Italy it is blended with every conceivable white grape: Cortese, Favorita, Erbaluce, Ribolla, Albana, Trebbiano, Vermentino, Procanico, Incrocio Manzoni, Verdeca, Grecanico, Catarratto, Nuragus, Viognier – and even Nebbiolo vinified off the skins.

Spain, for the most part, is a bit hot for Chardonnay and its star has waned in the

Vincent Dauvissat
Dauvissat is one of Chablis' great traditionalist stars, using old oak barrels to produce thrillingly pure and characterful Chablis.

Domaine Leflaive
Domaine Leflaive is probably the most famous producer in Puligny and its Montrachet, produced in tiny quantities, is rare and fabled.

Billecart-Salmon
This family-owned Champagne house makes wines of great elegance and delicacy. The name dates from 1818, when a M. Billecart married a Mlle Salmon.

Vineyards owned by the Chalone winery in the remote Gavilan Mountains above Soledad in Monterey County, California. In the 1970s Chalone was one of the first Californian producers to seek out limestone soil and the white wines have always possessed wonderful depth and balance.

face of an explosion of interest in Spain's own native varieties like Verdejo, Godello and Albariño. Penedès, Navarra, Somontano and Costers del Segre are its strongholds; it is permitted in Rioja. In Penedès it may be blended into sparkling Cava, generally with good results.

There is only a little Chardonnay in Portugal, but it is made in a generally international style. Germany has some fairly beefy examples. Austria's is somewhat leaner, and Austria as a whole is getting over its over-oaking habit. It is found throughout Eastern Europe, with light, attractive examples coming from Hungary, often under Australian tutelage. Bulgarian Chardonnay is generally attractive, without any particular character of its own. Slovenian examples can be more elegant and Croatia does it well unoaked; Swiss ones are light and attractive. There are isolated but good examples from Turkey, Israel and Greece. And England, Belgium and Denmark have all planted it, with surprisingly pleasing results. In England, especially, it plays a crucial role in the burgeoning sparkling wine business, especially since many soils in southern England are remarkably similar to those of Champagne, both being part of the chalk and limestone system called the Paris Basin.

USA: California

Despite winemakers showing a keen interest in less familiar white varieties such as Viognier and Pinot Gris, Chardonnay is still California's most important variety, accounting for almost 20 per cent of the harvest in 2013. Much of it is in Napa, Sonoma and Santa Barbara counties; some of the best is on Sonoma Coast, but an awful lot is also in the torrid Central Valley. Replanting after the Biotype B phylloxera infestation has concentrated it where it should be, in the cooler spots: Sonoma Coast, the new favourite area, where Chardonnay has real elegance and freshness; Carneros, with nutty fruit and brisk acidity, and Russian River offering a bit more substance and flinty fruit. Monterey Chardonnay is reminiscent of mango or guava; while wines from Santa Maria and Santa Barbara are richer but less tropical. Alexander Valley is creamy and silky. Carneros, Russian River and Anderson Valley are all key areas for sparkling wine. Basic Chardonnay from the Central Valley is generally off-dry and dull.

Styles are evolving away from super-ripeness and sweetness towards much more elegance and freshness, though it's taking a while for some to find poise at this new level. Growers seem to favour a bit less alcohol, even in Napa, which leaves Napa Cabernet Sauvignon rather out on a limb as the only wine there still clinging to the lifebelt of size and weight and the mantra 'more is more'.

How far can you go in modifying styles in the Napa, where the intrinsic style is big and alcoholic? It is possible to remove some alcohol by using reverse osmosis, osmotic distillation or with a contraption called a Spinning Cone. Experiments suggest that reverse osmosis is the best method, and that if you gradually reduce the alcohol from 14 per cent down to 12 per cent you will hit two or three different levels of alcohol where the wine is in balance. Choosing one of these allows a winemaker to fine-tune a wine in the cellar. Others prefer to do it in the vineyard, with a combination of clever choice of site, canopy management and control of yields.

Rest of North America

Washington Chardonnay is not dissimilar to Californian Chardonnay, generally with an

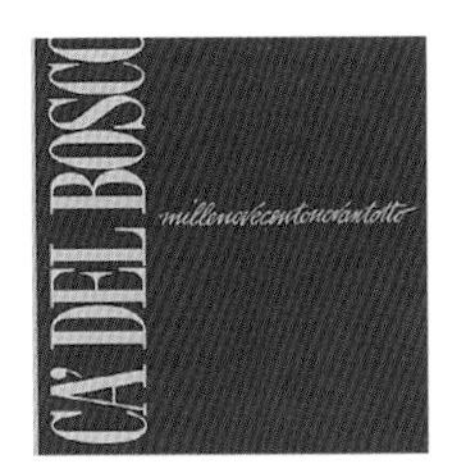

Ca' del Bosco
Winemaker Maurizio Zanella's aim with this Chardonnay is to equal Burgundy. So it's barrel-fermented, and is a deep, smoky, buttery delight.

Newton
Newton introduced the concept of unfiltered Chardonnay to California. It's an intense, refined wine of tremendous complexity.

Roederer Estate
Roederer uses grapes from the cool, damp Anderson Valley in northern California to produce a fine, sparkling wine in the style of Champagne.

emphasis on fruit flavour rather than creamy texture, but the top producers are beginning to produce good, savoury, nutty, Burgundian styles. Oregon's wines are becoming richer as it plants better, Dijon, clones. In New York State Chardonnay flourishes especially in Long Island and Finger Lakes – helped by its resistance to winter cold, and Virginia makes some serious examples. In Canada, too, it seems at home, producing rich wines in Niagara and oatmealy classics in Prince Edward County, Ontario and lighter ones in British Columbia.

Australia

Australian Chardonnay is here to stay. Styles are constantly evolving, with more refined, cooler climate wines at the top end, and more complexity lower down the scale, derived from Burgundian techniques like lees stirring, as well as from new and better Dijon clones. Over-oaking is much less of an affliction than it was, and there is less charring on the oak. The quest now is for minerality and flintiness. Today's Chardonnays are nothing like those of the 1980s, when deep golden-coloured, broad, fat, oily wines full of rich butterscotch flavours were the rule. Now colours are paler, structure is better and the fruit flavours are those of nectarine or white peach instead of melon and pineapple. Partly it's a question of picking earlier, perhaps before the stems are fully lignified; partly it's a question of gentler pressing and whole-bunch pressing, for the right phenolics, but not bitterness. Warmer regions routinely add acidity in the winery, and this is apt to show on the palate.

Chardonnay displays regional variations less than some grapes, but Hunter Chardonnay is buttery, viscous and opulent: the style of Chardonnay most often thought of as Australian, in other words. More elegant wines of varying styles but increasingly high quality come from the Yarra Valley with some outstanding oatmealy, minerally wines from Mornington Peninsula. There is considerable complexity in the Eden Valley and Adelaide Hills wines. Margaret River produces wines of outstanding concentration and complexity, and Tasmanian wines go from delicate and citrous to complex and resonant. The Riverland makes everything from inexpensive bulk wines to surprisingly high-quality examples. The coolest regions, including Tasmania, Geelong and the Macedon Ranges, have some conditions comparable to that of Champagne, and produce sparkling wine of great finesse and style.

Basic Australian Chardonnay, however, displays none of these regional characteristics. It is a commodity wine, made to a price, manipulated in the winery, and you get what you pay for.

New Zealand

New Zealand makes some of the most intense, powerful and balanced Chardonnays in the New World, marrying ripeness and depth with good acidity. Indeed, as some producers appear to have lost their confidence in making the really snappy Sauvignons that catapulted New Zealand to global recognition, they seem to have rediscovered how to make deep, savoury Chardonnay in its place. The most powerful come from Hawkes Bay, though Marlborough and Canterbury can produce impressive results and Gisborne wines are lush and delicious. Wairarapa, Nelson and even Auckland also do well. Chardonnay suffers from fashion swings here, but the best are world class.

Australian Chardonnay: a new classic

Chardonnay vineyards at Leeuwin Estate in Margaret River, Western Australia (right). Leeuwin Estate makes complex Chardonnays that approach top Burgundies in their structure and ageability – and few people could have dreamt, 30 years ago, that Australia was capable of such quality. The grape is no newcomer to Australia. It was first planted in the 19th century, but winemakers didn't fall in love with it until after Murray Tyrrell produced a commercial version in the Hunter Valley in 1971, having acquired his vines by hopping over the fence into a neighbouring Penfolds vineyard to remove a few thousand prunings. The first reactions to the wine were, ironically, that Australians would never drink white wine with oak. Wine show judges gave it six marks out of 20, and as Murray's son Bruce Tyrrell says: 'even the spit bucket gets eight'.

Grosset

This is the direction of top Australian Chardonnay today: cool climate fruit from regions like Adelaide Hills in South Australia, restrained alcohol levels and subtle aging in oak barrels.

Chile and South America

As winemaker Ignacio Recabarren puts it, 'in terms of quality, winery improvements took Chardonnay from under the table to on top of the table. To get to the roof we must work on viticulture. We are currently about halfway to the roof.' One of the battles is getting better clones into the vineyards; the other major factors here are climate and yields. The latter are generally high to very high; the former, in this long thin country, are infinitely adjustable. Let's take a look at the cool Casablanca Valley: plantings started here in about 1990. There's a frost risk in spring here; Recabarren characterizes it as cool as Mâcon in Burgundy, but warmer than Marlborough with harvest usually mid to late March. (Chile's Central Valley would be early February.) However, Casablanca is closer to the Equator than comparable regions elsewhere, which means that the grapes can be left on the vine for longer and still benefit from the sun's warmth. The other side of the coin is that leaving the grapes on the vine longer means lower acidity.

High yields do not in themselves imply lower quality: the balance of the vine is the crucial point. Casablanca's yields are high for two reasons, one of which is specific to the valley and tied to its frost risk; the other is common to the whole of Chile, and indeed to other Southern Hemisphere countries.

Casablanca's frost risk is very real. Usually it would be combated by air propellers, or by water sprinklers. Sprinklers, however, use four times as much water as drip irrigation, and water is scarce here. Propellers need an inversion layer, which is not always there. So Casablanca growers often use a system of leaving extra shoots on the vines at pruning to allow for frost damage. However, these shoots must be thinned after the danger of frost has passed, and before flowering if the vines are spur pruned; before the berries are pea-sized if the vines are cane pruned. Not to do so will mean too high a crop, and lower quality, in that year. Bit of a temptation for the greedy.

The second reason for high yields is that, more generally, Chile's level of sunlight induction on the vine's half-formed buds is higher than that in Europe. Dijon clones therefore naturally give higher crops here than they would back home in Burgundy.

But Casablanca isn't Chile's only cool climate paradise. Limarí, further north, is cooler yet and may well supplant Casablanca in quality terms. Coastal Leyda and San Antonio are also big new names for Chardonnay. All this is in keeping with the trend away from too much fat and too much oak: Chilean Chardonnays have moved fast to keep up with consumer tastes. In general, Chilean Chardonnays, even at the lowest price points, are fruity, balanced and probably the wine world's best value for money Chardonnay. So far Argentina has lagged behind Chile, partly because of overproduction, and partly because the vines grown there were usually brought in to make sparkling wines, and grapes with ripeness and concentration of flavour were not priorities. But better clones, more ambitious producers and the development of cool areas like Gualtallary and Uco are bringing exciting results, sometimes in a relatively oaky style, but increasingly tasting savoury, balanced and refined.

The cacti in the foreground make you think: dry, desert – and yes, you'd be right. The Limari Valley in northern Chile is dry, but it isn't that hot due to strong cold coastal breezes, and Limari Chardonnay is some of Chile's best.

South Africa

Climates are seldom very cool here, and even the coolest parts, like Walker Bay, are warmer than Burgundy, though perhaps cooler than Adelaide Hills. But within that framework, exciting Chardonnays with a Cape individuality are being produced. Hemel en Aarde on the far south of the country can produce exceptional full but minerally styles. Elgin is a little more delicate. Stellenbosch, though warm, does benefit from the coastal influences and makes a fresh, nutty style, as does Franschhoek. The biggest plantings are in Robertson – where unoaked styles can be very good off limestone soils – Stellenbosch and Paarl.

Glenora
This barrel-fermented Chardonnay hails from New York State's Finger Lakes region. Steely acidity marries brilliantly with rich, toasty oak flavours.

Neudorf
Nelson, on the northern tip of New Zealand's South Island has a reputation for being so laidback no one ever wants to leave. In which case, Neudorf's outstanding, waxy, nutty Chardonnay should make it even easier to stay.

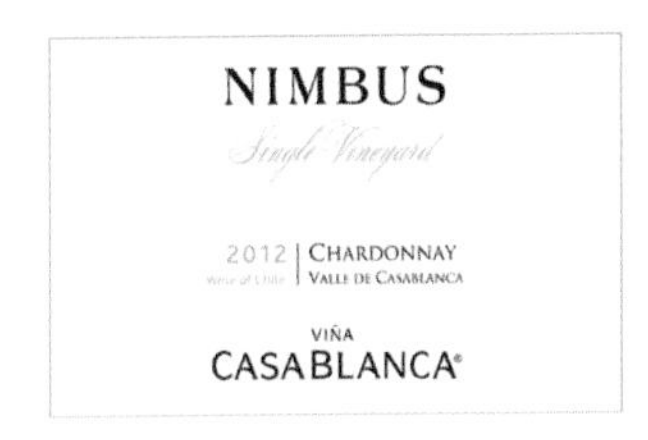

Viña Casablanca
Chile is only just developing the concept of single estate wines, but as vines mature in places like the Casablanca Valley, some vineyards, such as Nimbus, are starting to stand out.

Hamilton Russell Vineyards
Tim Hamilton Russell used to lament the fact that South Africa didn't stretch 300km (200 miles) further south so that he could find really cool vineyard conditions. Even so he made South Africa's first cool climate Burgundian Chardonnay in the 1970s.

ENJOYING CHARDONNAY

To get the most out of a great Chardonnay – great, not merely good – you have to give it bottle age. True, Corton-Charlemagne can be utterly seductive from the barrel, but such wines are so rare and so expensive that drinking them straightaway can seem wickedly frivolous. Keeping wine like that for several years somehow both amortizes the expense and prolongs the pleasure. Am I showing my puritan streak? No. I'm showing that I'm such a hedonist that I can enthusiastically embrace the waiting period because I know that the eventual pleasure will be massive.

In a top year, from a leading producer, such Grand Cru white Burgundies can last up to 30 years, and really should not be drunk before they are eight or 10. But much depends on the style of the producer and how concentrated the wine is to begin with. Premiers Crus are less long-lived – 20 years might be the upper limit – and village wines should be drunk earlier again, within eight to 10 years. Bourgogne Blanc and Côte Chalonnaise are best drunk within five years; the best Pouilly-Fuissé can be allowed up to eight. At least, those are the traditional guidelines. Bear in mind that the premature oxidation problem (see box, page 73) should make anyone with potentially long-lived white burgundies in the cellar check on their progress regularly from a relatively young age.

There are as many rules for Chardonnays elsewhere as there are producers. Concentrated, balanced, cool climate wines from areas as diverse as Carneros, Russian River, Uco, Hawkes Bay, Mornington Peninsula and Margaret River may easily last a decade or more. Others may tire after only a few years. Much depends on the philosophy and skill of the producer. But simple, warm climate Chardonnays do conform to an unbreakable rule: drink them early for their brief, bright-eyed burst of fragrance and fruit.

The taste of Chardonnay

The taste of Chardonnay is hard to pin down: unoaked wines from high-yielding vineyards may taste of not very much, while a minerally, concentrated bottle from the Côte d'Or or Margaret River may display a greater range of flavours than almost any other white grape. Some of these flavours may derive at least in part from vinification techniques; some come from the *terroir*, some from just the climate, some from the clone. Barely ripe Chardonnay tastes of green apples; riper, cool climate examples where the grapes had a long hang time have flavours of pears and acacia, lemons and grapefruit, nuts and biscuits, butter, honey and popcorn. There may also be a minerally, flinty or smoky streak to the fruit, some toast from the oak and, of course, balancing acidity which may seem tight and piercing in youth. White Burgundy and Burgundy lookalikes can boast oatmeal and a hint of savoury sulphur like a struck match.

Warmer climate Chardonnays get tropical, with mango, cream, banana, pineapple, melon and peach, plus butterscotch and more butter, honey and toast. There may be some spice, too, or boiled sweets. Everything but the kitchen sink, really.

Champagne is often just crisp and creamy or flowery in youth, and develops fresh breadcrust flavours and winey depth with age.

Marimar Torres' objective in establishing her vineyard in the chilly Russian River Valley was to make lean, pure experiences of the Burgundian grapes Chardonnay and Pinot Noir. This Acero Chardonnay is made without oak and the flavour of the variety shines through. Bouzereau uses oak for his Meursault, but in the subtle, understated way that is the hallmark of all the best Burgundy producers. Meursault-Charmes is a Premier Cru, but interestingly, Bouzereau also includes the name of the parcel of vines inside the Premier Cru – Les Charmes Dessus – as an extra point of difference on the label.

MATCHING CHARDONNAY AND FOOD

With its broad spectrum of flavours and styles – from steely, cool-climate austerity to tropical lusciousness – there is a Chardonnay for almost every occasion, and most are superb with roast chicken or other white meat. The relatively lean end of the spectrum is one of the best choices for simple fish dishes. Top Burgundy and the really full, rich New World blockbusters need rich fish and seafood dishes. Oaky Chardonnays are good with tricky-to-match smoked fish, and work well with garlicky dips such as guacamole; unoaked styles are better with spicy Southeast Asian or north Indian food; they are also good all-rounders with, say, festive turkey.

CONSUMER INFORMATION

Synonyms & local names

The world's favourite white wine grape variety is rarely called by any other name and most of the old European synonyms have fallen into disuse; Austria, especially in Steiermark, still calls it Morillon.

Best producers

FRANCE/Chablis Billaud-Simon, R & V Dauvissat, Droin, Fèvre, Louis Michel, C Moreau, Raveneau; **Côte d'Or** H Boillot, J-M Boillot, Bonneau du Martray, Bouchard, Carillon, Coche-Dury, Marc Colin, Colin-Morey, Drouhin, Arnaud Ente, J-N Gagnard, Jadot, A Jobard, Lafon, H Lamy, Dom. Leflaive, Bernard Morey, Niellon, Ramonet, Roulot, Sauzet, Verget; **Mâconnais** D & M Barraud, Bret, Ferret, Merlin, Saumaize-Michelin, J Thévenet.
AUSTRIA Bründlmayer, Tement, Velich, Wieninger.
GERMANY Huber, Johner, Knipser, Rebholz, Wittmann.
ITALY Bellavista, Ca' del Bosco, Gaja, Isole e Olena, Lageder, Lis Neris, Castello della Sala, Tiefenbrunner, Vie di Romans.
SPAIN Chivite, Enate, Manuel Manzaneque, Muñoz, Torres.
USA/California Au Bon Climat, Calera, Chateau St Jean, Dutton Goldfield, Gary Farrell, Flowers, Hanzell, HdV, Iron Horse, Kistler, Littorai, Marcassin, Merryvale, Peter Michael, Newton, David Ramey, Ridge, Saintsbury, Sanford, Shafer, Stony Hill, Talbott; **Oregon** Domaine Drouhin, Domaine Serene, Evening Land.
CANADA Le Clos Jordanne, Closson Chase, Flat Rock, Norman Hardie, Joie Farm, Pearl Morissette, Quails' Gate, Tawse.
AUSTRALIA Bannockburn, Bindi, Brookland Valley, Cape Mentelle, Coldstream Hills, Cullen, Curly Flat, Diamond Valley, Giaconda, Grosset, Howard Park, Leeuwin, Moorooduc, Oakridge, Penfolds, Pierro, Savaterre, Shaw & Smith, Tapanappa, Tarrawarra, Tyrrell's, Vasse Felix, Voyager, Woodlands, Yabby Lake.
NEW ZEALAND Ata Rangi, Babich, Bell Hill, Brancott, Church Road, Cloudy Bay, Craggy Range, Dog Point, Dry River, Escarpment, Felton Road, Fromm, Kumeu River, Matua Valley, Neudorf, Ngatarawa, Palliser, Peregrine, Saint Clair, Seresin, Te Mata, TerraVin, Trinity Hill, Vavasour, Vidal.
SOUTH AFRICA Ataraxia, Bouchard Finlayson, Chamonix, Paul Cluver, Crystallum, Hamilton Russell, Jordan, Mulderbosch, Thelema, Vergelegen.
SOUTH AMERICA Aquitania, Catena, Cobos, Concha y Toro, De Martino, Errázuriz, Leyda, Maycas del Limarí, Tabalí, Tapiz.

RECOMMENDED WINES TO TRY

Ten classic white Burgundies

Bonneau du Martray Corton-Charlemagne
Brocard Chablis Montmains
Coche-Dury Meursault les Perrières
René et Vincent Dauvissat Chablis la Forêt
Joseph Drouhin Beaune Clos des Mouches
Girardin Chassagne-Montrachet les Caillerets
Louis Jadot Chevalier-Montrachet les Demoiselles
Domaine Leflaive Le Montrachet
Raveneau Chablis les Clos
Robert-Denogent Pouilly-Fuissé les Carrons

Thirteen New World Chardonnays

Ataraxia Chardonnay (South Africa)
Bergstrom Chardonnay (Oregon)
Crystallum Clay Shales Chardonnay (South Africa)
Dog Point Chardonnay (New Zealand)
Grosset Piccadilly Chardonnay (Australia)
Hamilton Russell Chardonnay (South Africa)
Leeuwin Estate Art Series Chardonnay (Australia)
Marimar Estate Chardonnay (California)
Maycas del Limarí Chardonnay (Chile)
Neudorf Chardonnay (New Zealand)
Norman Hardie Chardonnay (Canada)
Penfolds Yattarna Chardonnay (Australia)
Pierro Chardonnay (Australia)

Five Champagnes (Blanc de Blancs)

Billecart-Salmon Blanc de Blancs Vintage
Deutz Blanc de Blancs Vintage
Jacquesson Blanc de Blancs Grand Cru Vintage
Krug Clos de Mesnil Vintage
Ruinart Blanc de Blancs Vintage

Five sparkling wines

Ca' del Bosco Satein Franciacorta (Italy)
Deutz Marlborough Cuvée Blanc de Blancs Vintage (New Zealand)
Green Point/Domaine Chandon Blanc de Blancs Vintage (Australia)
Nyetimber Première Cuvée Blanc de Blancs Vintage (England)
Roederer Estate L'Ermitage Vintage (California)

Chardonnay looks so similar to Pinot Blanc that in Italy they were only officially differentiated in 1978, though France managed it in 1872. Small quantities of Pinot Blanc can still be found on the otherwise Chardonnay-dominated white vineyards of the Côte d'Or.

Maturity charts

Most Chardonnay should be drunk early: only top wines are designed to improve with years of bottle age.

2010 Chablis Premier Cru

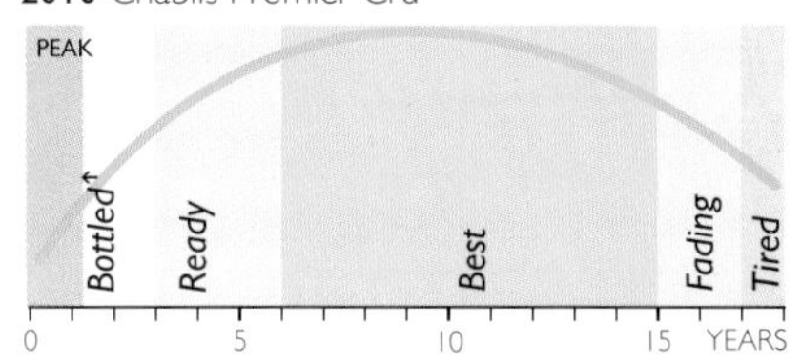

2010 was a good year for white Burgundy generally, but excellent for Chablis. The wines have fabulous purity, minerality and racy, ripe acidity.

2011 Côte de Beaune White (Premier Cru)

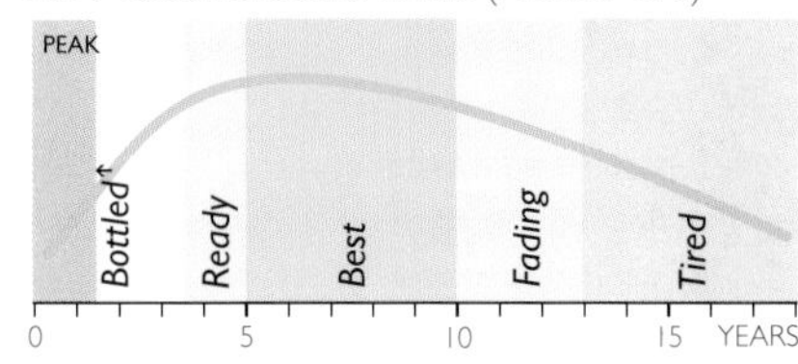

Burgundy's conditions were a bit erratic in 2011, and a cool summer ended with rain bringing good, medium-weight wines.

2011 Adelaide Hills Chardonnay

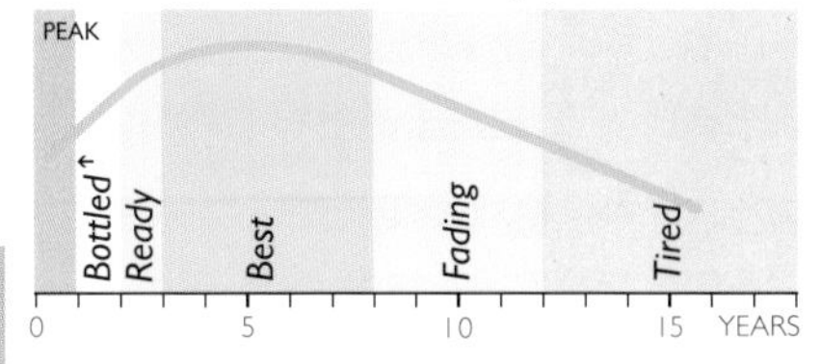

A fine year for Chardonnay in the cooler climes of Adelaide Hills, now emerging as one of Australia's most attractive wine regions.

CHASAN

A recent crossing of Palomino (see page 178) and Chardonnay (see pages 68–79) which, in mixing the neutrality of the former with the flavour of the latter, manages to produce a lightweight, neutral imitation of Chardonnay. It is grown in the south of France, where the local name for Palomino is Listán.

Sunset over snow-covered vineyards and Lac Léman at La Tour-de-Marsens, Lavaux in the Swiss canton of Vaud. Sometimes called Dorin in the Vaud, Chasselas produces 99 per cent of the canton's white wine and suffers from a certain folie de grandeur; elsewhere in Switzerland it may be called Fendant.

CHASSELAS

Switzerland's favourite grape variety reaches peaks of quality there that it attains nowhere else, although it is widely planted throughout the world, from Chile to Ukraine and goes under an awful lot of other names. We'll cover the main ones. As befits what seems to be a very old vine, it comes in many variations. It appears to have originated around Lake Geneva, and have taken the name of Chasselas from the Mâconnais village of the same name in southern Burgundy, where it was first grown in France.

In Switzerland its main interest for the drinker is its ability to reflect its terroir: on granite soil it tastes flowery with fair acidity, on chalk it is fruity and honeyed, on the deeper, more clay soils of Epesses it has more weight and character, and in Dézaley, also in the Vaud, it is minerally. But 'weight' is always a relative term with Chasselas: it is, at best, a lightweight, neutral wine, even though it has more acidity in Switzerland than elsewhere.

It is grown in Germany as Gutedel (see page 125) or Krachgutedel or Weisser Krachgutedel. In Croatia some people call it Queen Victoria. Don't even begin to ask. In Austria, where there is relatively little planted, it's called Junker or Moster or Gutedel. You can see that Chasselas follows the rule that the more neutral the grape variety is, the more vainglorious titles it acquires.

In France it is being replaced both in Alsace and in Pouilly-sur-Loire, where it was widely planted before phylloxera and produced table grapes for the markets of Paris. But the coming of the railways meant that the Midi could get its earlier ripening grapes to Paris faster, and Pouilly lost its market. It began then to make wine from its Chasselas, and the appellation of Pouilly-Fumé was so called to distinguish the superior Sauvignon Blanc wine from the Chasselas-based Pouilly-sur-Loire. The best French Chasselas comes from Crépy, in Savoie, where the wines are like the Swiss versions, only even lighter – if that's possible.

Not all its worldwide plantings go into the wine vat: it is much grown as a table grape. Most of its Romanian crop, for example, is destined for the table. Hungary, Moldova, Ukraine, north and south Italy and North Africa all have some Chasselas, as does Chile. But the grape known in California as Golden Chasselas is most likely to be Palomino (see page 178). Best producers: (France) Serge Dagueneau, Kientzler, Pfaffenheim co-op, de Ripaille, Guy Saget, Schoffit; (Switzerland) H Badoux, L Bovard, Dubois, E & L Fonjallaz, R Gilliard, Caves Imesch, J & P Testuz.

CHENIN BLANC

See pages 82–91.

CHIAVENNESCA

The name for Nebbiolo in the Valtellina region of Lombardy, northern Italy. See pages 166–175.

CIENNA

A red crossing of Cabernet Sauvignon and Sumoli produced in Australia but really only being championed by Brown Brothers in their warmer vineyards because it's a slow ripener and clings to its tough tannins even when fully ripe. Yalumba tried it in cooler Wrattonbully – it didn't work. But it does have lavish black fruit flavours, and Brown Brothers have had some success turning it into a fruity, low alcohol, sweet, fizzy red.

CILIEGIOLO

The 'little cherry' grape has a flavour, not surprisingly, of cherries and spice, and its lowish acidity makes it a good blending partner for Sangiovese (of which it seems to be either a parent or offspring) in Chianti. It is sometimes bottled as a varietal, in keeping with the current passion for rediscovering little-known varieties, and it can be found as far south as Sicily and as far north as the Val d'Aosta. Best producers: Antonio Camillo, Rascioni & Cecconello, Sassotondo.

CINSAUT

This vine's reputation for high quantity and poor to middling quality is only partly deserved. It can certainly give high yields, but at low yields it can produce characterful, lush, sweetly rich wines which play a part in such star names as Lebanon's Chateau Musar. Here yields are as low as 25hl/ha.

France's Languedoc, too, can produce Cinsaut with finesse, providing that yields are kept down and the wine is given a long maceration with the skins. It is aromatic in youth, with soft, supple fruit, though with a nip of tannin, and is often used to calm down the tougher Carignan. But if allowed to yield heavily its quality falls rapidly. It is classed as a *cépage améliorateur* or improving variety, but lacks the prestige of other grapes used similarly such as Syrah or Mourvèdre. On its own, it can make attractive rosé. In France it is often spelt as Cinsault. It buds relatively late and is susceptible to mildew and oidium, though in Algeria it has proved more resistant than Aramon to drought and drying winds like the Sirocco. It has long been a popular grape in North Africa, and is Corsica's main variety. In Spain, the old variety Samsó is none other than Cinsault. Saying it aloud gives the clue . . .

In South Africa its popularity has been eclipsed by its offspring, Pinotage (see pages 200–201), the result of crossing Cinsaut (known locally as Hermitage) with Pinot Noir (see pages 188–199), but there's been a revival of interest since it has been realized that the astonishing longevity of old Cape Cabernet Sauvignon is due to a healthy dollop of Cinsaut in the blend. Now people are seeking out ancient plots of Cinsaut and making delicious reds full of raspberry coulis fruit. As a table grape, it can be found in France under the name of Oeillade: the berries are apparently judged too small to be of interest to export markets. Best producers: (France) l'Amarine, Caraguilhes, Clos Centeilles, Le Clos du Serres, Mas de Daumas Gassac, Mas Jullien, Pech-Redon, Val d'Orbieu, Vignerons Catalans; (Chile) De Martino; (USA) Morrison Lane.

CLAIRETTE

Once the great standby of southern French white blends, Clairette seems to have had its day. It doesn't fit the modern idiom: it is too unstructured, too quick to oxidize, too high in alcohol and too low in acidity.

It still plays a part, albeit a shrinking one, in many wines. On its own it makes Clairette de Bellegarde and Clairette du Languedoc, the second of which can be dry, sweet or *rancio*. It appears with Muscat Blanc à Petits Grains in the Clairette de Die appellation, and pops up in Châteauneuf-du-Pape, Côtes du Rhône, Côtes de Provence, Cassis, Bellet, Palette and many other southern appellations, often along with higher-alcohol grapes like Ugni Blanc. In the right *terroir* and at yields below 50hl/ha it can make interesting wine: fat and perhaps a little heavy, and with 14 per cent alcohol, but with a certain musky attraction. Modern winemaking, too, can hold back its tendency to oxidize. But at high yields of 100hl/ha or more it is the sort of wine that gave the Midi a poor reputation.

In South Africa it is regarded as a low-alcohol variety because it can achieve full ripeness at as little alcohol as 12% and though usually blended is now appearing successfully on its own from a few old plantations. In Australia's Hunter Valley it used to be called Blanquette, and should not be confused with the Blanquette variety found in the Languedoc region of Limoux, which is Mauzac. Clairette is also a synonym of several other grapes in the south of France, including Ugni Blanc (known as Clairette Ronde in the Languedoc) and Bourboulenc. There is also a Clairette Gris, which has a pink tinge to its skins. Best producers: (France) Achard-Vincent, Clairette de Die co-op, D Cornillon, Faure, J-C Raspail.

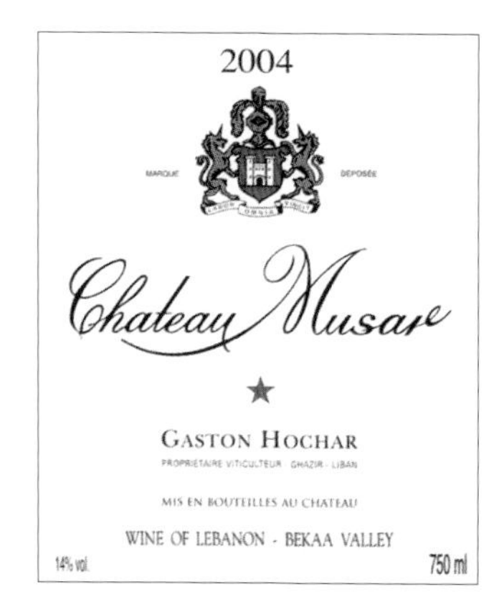

Chateau Musar
This blend of Cabernet Sauvignon, Cinsaut and Syrah is not only Lebanon's most famous wine, but a world- class classic.

CLEVNER

Clevner (or Klevner) is the Alsace name for members of the Pinot family, usually Pinot Blanc. In Switzerland's Zurich region Klevner is the local name for Pinot Noir (see pages 188–199).

COCOCCIOLA

A southern Italian white grape, recently rediscovered as a varietal. The wine has good acidity and a herbal streak.

CODA DI VOLPE BIANCA

The white fox-tail grape comes from Campania in southern Italy and used to be found only in blends, but it's an old grape and can produce high quality wines, mineral- and apricot-tinged. Best producers: D'Antiche Terre, Crogliano, Vini Antico Palazzo.

COLOMBARD

The French variety Colombard seems to have been the result of a crossing of Gouais Blanc and Chenin Blanc, and to have originated in the Charente region north of Bordeaux, where it was long made into white wine, and later distilled into Cognac. Many of the grapes are now wanted for still wine – to which Colombard is in fact much better suited than brandy, since it is higher in alcohol and lower in acidity than the usual brandy grape, Ugni Blanc.

Colombard is particularly effective in contributing peach and nectarine fruit and citrus lemon perfume to Vin de Pays des Côtes de Gascogne (from the Armagnac region) and, to a lesser extent, Vin de Pays Charentais (from Cognac). It is also planted in other parts of southwest France, and pops up in many white blends – in lesser Bordeaux regions such as Blaye, Colombard actively improves the perfume and acidity of whites.

Colombard is a useful, warm-climate grape, too, but few producers have maximized its potential. Even so, it's on the rise in South Africa. It maintains its acidity well and gives good crops of fresh, grapefruit and peach-flavoured whites for early drinking. Several good examples have also appeared in Australia.

California's Colombard, known as French Colombard, used to be prized for its good acidity in hot conditions, but it is in decline there. Until 1991 it was the state's most widely planted white vine. Most but not all plantings are in the Central Valley. Yields are very high – up to 12 tons per acre (210hl/ha) – and the wine is usually blended into everyday jug wines. Best producers: (France) Jean Aubineau, Brumont/Montus, Joy, Meste-Duran, Producteurs Plaimont, Tariquet; (South Africa) Graham Beck, Bon Courage, Botha, Longridge, Robertson Winery, Swartland Wine Cellar; (Australia) Primo Estate.

COMPLETER

Switzerland has a number of grapes which are very rare but extraordinarily good, and this is one of them. It makes a weighty, structured white wine with lovely honeyed flavours and appley, quince fruit, concentrated and quite high in alcohol. Best producers: Completer-Kellerei, Donatsch, Peter & Rosi Hermann, Schloss Reichenau, Hermann Schwarzenbach, Thomas Studach.

COMPLEXA

A red-fleshed grape found on the Portuguese island of Madeira, where it was planted in the 1960s. Used for table wine and fortified, it gives wine that is lighter in colour than the more widely planted Tinta Negra Mole, and less astringent, but it is very susceptible to rot.

CHENIN BLANC

Well, at least the Chenin Blanc is versatile. It can make just about any style you could dream of. In the Loire Valley in France, its wines go from scaringly dry, to dry, to fairly dry to vaguely off-dry, to off-sweet, sweet, very, very sweet – and there's good Chenin fizz too. In South Africa, Chenin wines go from mild and innocuous to snappy, dry and fruity, to full-bodied, to dense, rich, nutty, oaked styles with a wild flash in their eyes, to seriously sweet. Oh, and Chenin makes an awful lot of brandy there, too.

So. Versatility like that should make Chenin a pretty popular grape. But it doesn't. If we remove the Loire Valley and South Africa, we've removed all the memorable examples of Chenin except one doughty example from New Zealand. Shouldn't it be more popular? Well, only up to a point. Lots of other countries have tried to make exciting Chenin – California and Argentina still have well over 2000 ha (4500 acres) each, yet they just can't coax any character out of it. In most places, it's regarded as a high yielding variety, with useful acidity in warm conditions, and spends most of its time getting lost in blends of more expensive varieties like Chardonnay. And as Chardonnay loses its allure, Chenin's usefulness fades. All round the world, it's being ripped out, except in the centre of the Loire Valley and in South Africa. Because in each of these places, Chenin can produce thrilling wines, and there's a new generation of young growers and winemakers determined to trumpet its talents.

Floating on air and water, the château of Chenonceau stretches across the river Cher in Touraine, with the river Loire beyond. Chenin Blanc was first planted here in the heart of the Loire Valley in the 15th century and Anjou and Touraine are still where it produces its most exciting wines, whether sweet or dry, sparkling or still.

A new generation. Crucial. A new generation who don't have to shoulder their parents' baggage of mediocrity. A new generation not wealthy, merely with the passion to excel. They couldn't afford to plant new vineyards or bid up the price of the sexy varieties. They had to use what was there. And in the Loire Valley and South Africa it was Chenin. Yet these two areas took very different routes to success.

The Loire Valley had Sauternes in Bordeaux to thank. During the 1980s Sauternes had a series of splendid vintages for rich dessert wines, so by the 1990s prices had risen dramatically and there was an enthusiastic band of well-heeled consumers who'd got the sweet wine bug and who would pay. The only other French region with a classic sweet wine tradition was the Loire Valley, and during the 1990s, the Loire was luckier than Bordeaux with its weather. Spurred on by promises of the mighty dollar and with a brood of young winemakers much more aware of the modern wine world than their parents, the misty side valleys of the Loire, and in particular Bonnezeaux, Quarts de Chaume and Coteaux du Layon south of Angers, produced a series of gorgeous, rich, and startlingly original sweet wines from noble-rotted Chenin Blanc grapes. Suddenly Chenin had a standard-bearer for quality. But the noble rot which Chenin enthusiastically welcomes occurs in just these few rare side valley sites. The rest of the Loire has traditionally struggled to make even half-decent dry wine from the slow-ripening Chenin. Yet success breeds success. New World winery attitudes of cleanliness, discipline, attention to detail, and in the vineyards lower yields, better growing methods and that old-fashioned virtue – the courage to wait out poor autumn weather until your grapes are ripe – these have transformed dry Chenin in the Loire. Bone dry, yes, but streaked with minerals and softened with angelica and honey.

In South Africa, as much Chenin gets ripped out as is re-planted, but the story here is the old vines, in Stellenbosch, Paarl, up to Malmesbury and into the Swartland, where hardy bush vines sprawl across the dry hillsides and produce grapes of quince and peach syrup wildness that add nut intensity and sour cream richness by being fermented in oak. The final result is wines of shocking, unnerving brilliance, to drink or to age for a decade, and are possibly the best white wines in South Africa.

Why doesn't anyone else do it? Chenin isn't famous, isn't popular and so is an absolute bastard to sell, unlike Chardonnay or Sauvignon. It would be nice to say: Chenin is on the march, the world is getting the message. But it isn't. And it isn't.

Chenin Blanc: from Grape to Glass

Geography and History page 84; Viticulture and Vinification page 86; Chenin Blanc around the World page 88; Enjoying Chenin Blanc page 90

GEOGRAPHY AND HISTORY

A glance at this map might make one think that Chenin Blanc is one of the world's favourite grapes. It covers nearly a quarter of the vineyard in South Africa; it flourishes in some of California's warmest spots; Argentina produces it in abundance. It can be found in Canada, New York, New Zealand, Australia, Brazil, Uruguay and Mexico. Yet, with a few exceptions, none of these plantings produce wines that are interesting enough to warrant international attention.

The reason Chenin Blanc can be regarded as a classic white grape variety is one small region of France and a few patches of old vines in South Africa. Anjou-Touraine, where the vineyards cluster on the river Loire and its tributaries, is the source of startlingly intense, concentrated, long-lived wines, sweet, medium, dry and sparkling. Stellenbosch and Swartland produce small crops from bush vines that make South Africa's most individual dry whites.

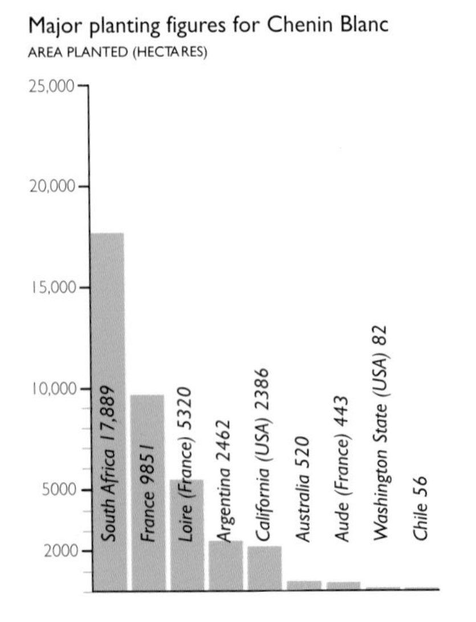

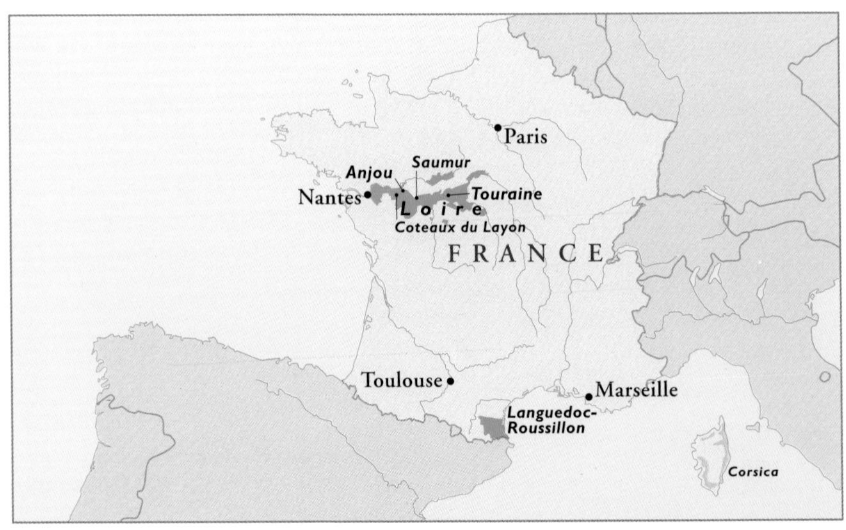

Why even here? Especially when Anjou-Touraine is so cool, and Chenin is so ponderously late-ripening. Growers there measure other grape varieties' ripening cycles by how much sooner than Chenin their fruit is ready to pick. Yet this is where the human element enters in. Chenin is a vigorous vine and will produce almost as many grapes as you want it to. I've seen vines in South Africa so laden with fruit the vine needed propping with wooden posts. But keep the yields right down, in Savennières, or Saumur, or Vouvray, and the result is unique, steely wine with a flavour of greengages and angelica, which ages sublimely for a decade or more. In South Africa, the result is wild, funky, almost syrupy, nutty whites, rich yet dry. As for sweet wine, Chenin is prone to noble rot, and in the side valleys of the Loire's tributaries, autumn mists and warm sunny days produce small crops of noble-rotted Chenin that is both piercingly fresh yet succulent when young and which will age for generations to a deep amber brilliance of honey and barley sugar and quince.

HISTORICAL BACKGROUND

Chenin Blanc has been growing in the Loire Valley for hundreds of years, and probably originated there: in 1496 it was planted near Chenonceaux by Thomas Bohier, and some 30 years later was taken by his brother-in-law, Denis Briçonnet, Abbot of Cormény to the monastery of Mont-Chenin in Touraine, where he planted a number of different varieties, and it was probably from here that it took its name. It is a full sibling of Sauvignon Blanc, with Savagnin for one parent and an unknown grape for the other.

South Africa now has the world's greatest plantings of Chenin – it first travelled there in 1652 with Jan van Riebeeck, the Dutch East India Company's first Commander at the Cape and the country's first winemaker: no doubt Chenin's ability to retain acidity even in warm climates, and its obligingly high yields, endeared it to the early settlers.

It seems to be genetically fairly stable and less given to mutation than other very old *Vitis vinifera* vines: genetic variation is not regarded as a problem in the Loire Valley, and most growers approve of the quality of the commercially available clones, even if they disapprove of the higher yields they usually produce. So there are a few complaints.

One vine, sometimes referred to by Loire growers as a sub-variety of Chenin, is not Chenin at all, though it is a half-sibling and shares Savagnin as a parent. It's actually Verdelho. A little can sometimes be found interplanted in the Chenin vineyards of the Loire: it is strictly forbidden by the appellation rules, but it's been there a long, long time and because it ripens two weeks before Chenin, it can be useful in a cool year. Its association with these vineyards is hardly new: back in 1928, in his book *Le Vigneron Angevin*, Dr Maisonneuve referred to Verdelho as 'an interesting *cépage*...which at first glance has a certain resemblance to Chenin Blanc'.

Château de la Roche-aux-Moines viewed from above the Coulée-de-Serrant vineyard at Savennières, with the river Loire on the left. The terroir of Savennières is far from homogeneous, but all the vineyards share a steep slope down to the Loire, and hot, dry soil.

The Moulin de la Montagne amid Chenin Blanc vines at Bonnezeaux. It's windier here than in the Coteaux du Layon vineyards nearer the river but, in spite of this, noble rot arrives virtually every year.

Chenin Blanc grapes being harvested in Vouvray. The comparative rarity of noble rot in this appellation means more dry wines, and medium-sweet wines that keep the varietal flavours of Chenin.

VITICULTURE AND VINIFICATION

Loire growers of Chenin Blanc both love the vine and despair of it. 'It's an ungrateful vine,' says Florent Baumard of Domaine Baumard in Coteaux du Layon: you do everything for it, you control its yield, you cosset it and fuss over it, 'and then at the moment of harvest, it rots, or stops ripening'. Perhaps a warning should be sent to South Africa, telling Chenin aficionados just what they are in for as they fall in love with the vine. When it is ripe, and properly balanced, it is wonderful, but unripe Chenin is nasty, and high-yielding Chenin is dull and bland. It is by no means an all-purpose vine, although that is what it was long used for in the Cape, where an increasing number of low-yield examples have emerged. Indeed, its main task was to produce vast harvests of thin wine for distilling into brandy. Now its ambition is to make South Africa's finest whites.

Climate

France's Loire Valley is marginal for Chenin Blanc, which of course is one reason why it produces wines of such great finesse there. Go even a little way further north and the wines become not so much refined as acidic: Jasnières is the most northerly appellation for Chenin in France, and the wines here are distinctly thinner, or *pointu*, as the French say. Further south, it is grown in the Languedoc, primarily at Limoux.

There are suggestions that the characteristic mineral flavour of Loire Chenin derives from a degree of unripeness. In cool, wet years, it has a flavour of green apples and hard, unyielding acidity, raw and unpleasant. But times have changed: growers point to global warming. They pick 10 days or so earlier than in the past and still have ripe grapes.

It is climate, too, far more than soil, that determines whether Loire Chenin is to be sweet or dry. Vouvray and Montlouis have much the same climate and make excellent fizz as well as bone dry, medium and sweet wines. Coteaux du Layon is less continental and more influenced by the Atlantic Ocean, and so gets more botrytis than Vouvray and Montlouis. Savennières is windy and less foggy, and thus less subject to botrytis; nevertheless, according to Mme de Jessey of Domaine du Closel, while at the beginning of August there is more sugar in the grapes in Coteaux du Layon than in Savennières, by 1 October the situation is reversed, and Savennières has the higher sugar. The picture is complicated by the abundance of rivers, large and small, in the region, which means that humidity varies from one vineyard to the next. The climate also varies hugely from year to year.

Soil

If it is climate that determines whether Loire Chenin is to be dry or sweet, it is the soil that gives the wine its style. Chenin reflects its soil as clearly as Riesling or Pinot Noir. Sandy soils give light wines which are attractive young, and which mature relatively early. Clay gives weightier, richer wines, and is conducive to the development of botrytis: Coteaux du Layon is rich in clay, with a chalk subsoil. Limestone gives wines with fine, scything acidity, wines from silex are vivacious, tingling on the tongue. Calcareous clay – or *argilo-calcaire* – produces perhaps the most rounded wines, with both acidity and weight: Vouvray has plenty of this soil, as well as some silex in the best sites. Savennières produces one of the world's driest wines, and has dark bluish schist, fairly friable on top but very hard underneath, hot and well drained; there is schist and quartz, too, at Bonnezeaux.

In South Africa the two major soils are granitic ones, and Table Mountain sandstone. On granite soils Chenin is lean and acidic, with citrus-lime flavours; on sandstone it is heavier, with more pineapple, yellow apple, pear and quince notes; it has structure, but less acidity.

Yields

Chenin Blanc is vigorous enough to produce wine like water, if that is what you require of it. Even on the poorest soils of the Loire, it will give 80hl/ha unless checked; on richer soils it can do twice that. In South Africa Chenin is

Nicolas Joly in front of his Coulée-de-Serrant vineyard in Savennières. He is a passionate, almost mystical believer in biodynamism, the system based on the early 20th-century teachings of Rudolf Steiner. His wines are some of the most individual Chenins in the world.

often left to overproduce to its heart's delight. Though good producers claim to keep their crop down to 50hl/ha, greedy producers will produce at least three times that amount – and counting. Chenin at this level is simply not interesting: most of the problems of South African Chenin can be related to overcropping. The rest are probably caused by planting Chenin in regions too warm for it to give interesting wines.

For dry wines, 40–50hl/ha is a generally accepted yield if quality is the aim. (The legal maximum in Savennières, for example, is 50hl/ha.) For sweet wines, Bonnezeaux's legal maximum is 25hl/ha, and most growers get 20–25hl/ha; the same applies in Coteaux du Layon. At Domaine Font Caude at Montpeyroux in the Languedoc, Alain Chabanon grows it for sweet wine on poor *argilo-calcaire* soil, with yields of between 5 and 11hl/ha.

To keep yields to these levels, a green harvest – i.e. a thinning of the crop in midsummer – may be necessary, as well as a choice of *riparia* or *rupestris* rootstocks to reduce vigour. Many estates also sow grass between the rows to reduce vigour.

Old vines, being less vigorous, also play their part. The average age of the vines at Château de Fesles in Bonnezeaux is 35 years, and at Domaine du Closel in Savennières there are some vines approaching their centenary. (Old vines give more complex botrytis flavours, too: botrytis on young vines gives floral flavours, but not much depth. With old vines, it's the opposite.) Curiously, phylloxera seems no longer to be a problem at Closel. Here, Mme de Jessey has some vines still on their own roots, propagated by the age-old technique of layering; Château d'Épiré, also in Savennières, practises the same method on some vines. So far neither domaine has found any sign of the dreaded louse returning, but nor has anyone been able to come up with a convincing explanation. Phylloxera moves in mysterious ways.

At the winery

One major factor distinguishing New World Chenin wines from Loire versions is the fermentation temperature. In the Loire temperatures of between 16° and 20°C (61° and 68°F) are normal: that's quite warm for a white wine these days. But what the Loire growers actively want to avoid is the tropical fruit aromas and flavours produced by fermentation at colder temperatures: it is these flavours that distinguish New World Chenin wines.

South Africa has a fair amount of old Chenin vines, but it needed young tigers like Eben Sadie, pictured here with his egg-shaped concrete fermenters, to make people realize what a precious resource they were. Sadie and his peers are now producing an array of fascinating, powerful, rather wild at heart Chenin wines.

In the Loire the flavour of new oak is unwanted. Usually a proportion of the barrels (one-fifth to one-third) is renewed each year. In Savennières chestnut or acacia barrels are traditional: neither gives a vanilla flavour to the wine, though chestnut gives a slightly buttery taste, and the flavour of acacia is stronger. In the New World the flavour of new oak is considered more welcome, adding toastiness and spice and considerable richness.

However, Loire winemaking has changed greatly. The producers acknowledge that what consumers want is less acidity and more fruit flavour in their wines. Clean Chenin grapes have shown themselves to be very amenable to skin contact – leaving the lightly crushed grapes to macerate in their own juice before fermentation starts. This brings out the greengage and angelica flavours. To reduce acidity some producers let the malolactic fermentation take place, thereby fattening the texture of the wine, and many age the wine on the fine lees, stirring the lees and wine together by *bâtonnage*. Even dry wines may be made with a few grams of residual sugar.

Bottling times vary: Philippe Foreau in Vouvray seeks wines (both sweet and dry) with high acidity and freshness; he therefore avoids the malolactic fermentation and bottles in the spring following the vintage; at Château de Fesles in Bonnezeaux the wine normally spends up to 24 months in wood.

TRI, TRI **AND** TRI **AGAIN**

One of the keys to making great Chenin in the Loire Valley is successive selected pickings, or *tries*. Over a period of a month to six weeks in the late autumn pickers must be sent through the vineyards picking perhaps whole clusters, but more likely individual grapes, at precisely the right degree of ripeness. If grey rot, rather than the desired noble rot, is invading the vineyards, then affected grapes must be removed in separate pickings. And the decision on whether to pick today or wait for greater ripeness, whether to make a dry wine, a demi-sec or a sweet one, is made on a day-to-day basis and may change as the vintage progresses.

The Loire Valley's convoluted map of soil and climate means that Chenin Blanc never ripens all at the same time. Grapes from warm schist soils ripen before cooler clay ones; noble rot appears more regularly where the river mists creep up the hills at night, to be dispelled by sunshine during the day. In addition, most producers have the option of making wines at different levels of sweetness. Every year, in other words, they are trying to hit several moving targets.

The degree of the grapes' ripeness and the lateness of the season between them will make the grower decide how much dry wine to make. In a cool year in Vouvray a lot will go to make sparkling wine; in great years almost no dry wine at all is made. Philippe Foreau of Clos Naudin says that if your grapes aren't ripe enough for triage, then they're not ripe enough to make sweet wine. But here again years vary: in some years grapes of 12.5° potential alcohol will be ripe, in other years not.

Four to six pickings are normal for Vouvray, perhaps three for Savennières. The purpose of each picking depends on the year: growers taste their grapes every day. The first picking might be for dry wine; others might be for sweet. If, in a hot, dry year there is no botrytis, the grapes may be picked *passerillé*, or shrivelled; botrytis may arrive later, or it may not.

Château de Fesles in Bonnezeaux reckons to pick one or two grapes per cluster on the first picking, the majority of the grapes in the second and third pickings, and the remaining one or two per cluster thereafter. The third and fourth pickings, they say, are generally the finest.

CHENIN BLANC AROUND THE WORLD

Wine styles in the Loire have moved to greater ripeness; elsewhere winemakers are mostly still feeling their way towards higher quality. We don't know yet quite how versatile Chenin will prove to be, but examples from South Africa especially and also Australia, New Zealand and the Americas suggest we could have a rising star on our hands.

France: Loire Valley

The year that saw the beginning of today's revival of Chenin quality in the Loire Valley was 1985. In that year, when botrytis was plentiful, growers rediscovered the art of selection in the vineyard. France was in the middle of a golden decade, the sun shone on a regular basis, money was being made and a whole battalion of growers suddenly decided to put quality first. Thank goodness they did. Loire Chenin had been on the slide since the 1960s, and we were in real danger of losing one of France's unique wine styles.

Good growers aim for the greatest possible richness in their grapes; climate, along with vintage variations, soil and the exposure of the slope will determine whether the result will be sweet or dry, or in between. In Anjou and Saumur there is more chance of botrytis: in Bonnezeaux and Quarts de Chaume, the tiny subregions of Coteaux du Layon, botrytis can be relied upon to appear to some degree practically every year. Further up the river in Touraine, Vouvray and Montlouis get botrytis only about four years in 10. In non-botrytis years these sweet wines are made from over-ripe and shrivelled – *passerillé* – grapes, and this goes part of the way to explaining the difference in flavour between sweet Vouvray and Bonnezeaux or Quarts de Chaume.

Sweet Vouvray is much more likely to have the characteristic varietal flavour of Chenin: a flavour of apples, greengage and minerals, which matures to flavours of honey and acacia and quince. The sweet, botrytized wines of Anjou-Saumur are more likely to have more typically botrytized flavours of peach, barley sugar and marzipan. (*Botrytis cinerea* tends to destroy varietal character: see page 31.)

Soil makes a difference, too: Vouvray's calcareous clay and silex gives different styles to the schist of Savennières and Bonnezeaux. Early bottling, in the spring after the vintage, and the avoidance of the malolactic fermentation (both more likely in Vouvray) preserve freshness; sweet wines given long aging in wood before bottling are likely to have more rounded, Sauternes-like characters. The sweet wines are seldom high in alcohol: 12–13 per cent is normal for Vouvray Moelleux, with the grapes being picked at a potential alcohol of 18–20 per cent. If the alcohol level is higher, the wine can become unbalanced and heady.

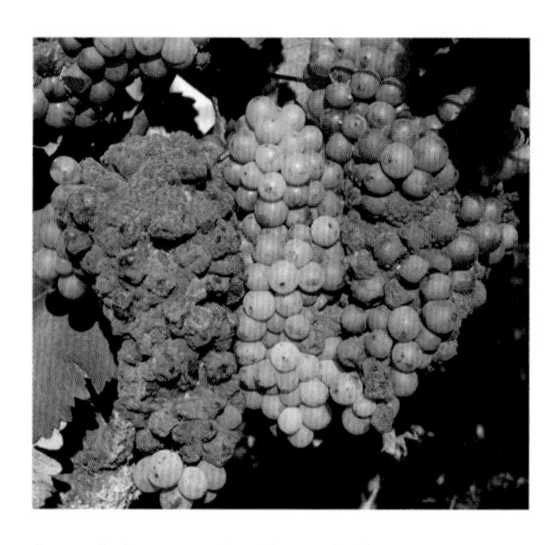

As a rule in sweet wine vineyards, the more repellant the grape, the better the wine will be, as noble rot infects the bunch. These Chenin grapes show typical patchy Vouvray botrytis – the bunch on the left heavily infected, on the right lightly infected and still clean and green in the middle.

Among the dry wines, Anjou Blanc Sec and Saumur Blanc are the simplest and earliest maturing (and up to 20 per cent of Chardonnay or Sauvignon may be added to the blend for Saumur); Savennières and Vouvray Sec are the most complex and longest-lasting. The maximum alcohol permitted in Savennières is 13.5 per cent: if in a very ripe year the wine has more than that, a special dispensation must be sought from the *appellation contrôlée* authorities. Expect this to happen more often.

Other French Chenins

There is a little Chenin Blanc planted in Limoux in the western Languedoc, where it is blended with Mauzac and Chardonnay for sparkling wine, and elsewhere in the region – for example, further east at Montpeyroux in the Coteaux du Languedoc appellation.

South Africa

With 18 per cent of total plantings, Chenin Blanc is still by far the most widely grown variety in South Africa, even though it is in steady decline (it was 32 per cent in 1990). Its traditional local name, Steen, has gone out of fashion as interested producers, led by the Chenin Blanc Producers' Association, set up in the late 1990s, begin to take the grape more seriously. Most is planted in Paarl and Worcester and in Swartland, around

Domaine Ogereau
A sweet, honeyed, botrytis-affected wine from top producer Ogereau using the Coteaux du Layon Saint Lambert appellation.

Domaine Huet
Huet is Vouvray's most famous producer and Clos du Bourg was the first vineyard to become biodynamic in the 1980s.

François Chidaine
Montlouis is on the south bank of the Loire producing lighter wines than nearby Vouvray. This sweet wine is from biodynamic vineyards.

There is far more Chenin Blanc grown in South Africa (above) than in the Loire Valley, but it is only very recently that growers have started to take it seriously. Research into the best sites is only just beginning, but in all the Cape's 18,000ha (45,000 acres) of Chenin there will be some exciting spots. These vines are at Fairview Estate in Paarl, looking out towards the Simonsberg Peak.

Malmesbury, and 90 per cent or so is crushed by the cooperatives. In the past much of the harvest ended up being distilled, but more recently has found its way into bottle, yet a lot of it still proves that low-quality Chenin is really not worth the trouble. There is still a lot of work to be done on identifying the best vineyards, many of which are very old, and on trying to discover which soils and climates suit the vine best.

South African growers do not necessarily admire Chenin. What the French see as finesse and subtlety, they see as raw fruit and acidity. While South Africans are now putting some emphasis on bottle aging, they are still more likely to seek flavours of guava and banana, pear and pineapple in their Chenin than the sort of mineral tightness that takes 10 years in bottle to open out. Most wines have high alcohol and some sweetness, actual or perceived, and growers are finding that the riper Chenin is, the more flavour it has. Prices have shot up for top Stellenbosch and Swartland examples.

California

The majority of California's Chenin is planted in the hot Central Valley, and is treated as a bulk wine suitable for inexpensive blends. It lacks the tight acid structure of Loire Chenin, and when made with residual sugar, as it often is, the result is a loose-knit and bland wine. A great deal has been uprooted since Chenin's high days of the mid-1980s when plantings totalled over 16,188ha (40,000 acres); and in 2013 it accounted for just over 1 per cent of California's harvest. However, occasional good Chenin wines do appear. The Clarksburg region in the Sacramento Delta produces some stylish fruit, and there are odd good Chenins from elsewhere in the state.

Rest of North America

In Oregon, Chenin's 1990 figure of 18ha (4 acres) has declined until it has vanished off the statistics. There is still some in Washington State, though here, too, it is in decline. Terminal? More than likely.

Australia

Plantings of Chenin Blanc are slowly declining. The current 700ha (1730 acres) of vines are widely scattered across the country. The wines have soft fruit-salady flavours; critic James Halliday describes the taste as 'tutti-frutti'. However, the Swan Valley and Margaret River in Western Australia have produced some tauter, more impressive examples.

New Zealand

Every vintage in New Zealand seems to produce one or two Chenins of such balance and beauty – all greengage and angelica, flecked with honey, sharpened up with lemon acidity – that you cry: why doesn't New Zealand produce more? Simple reason. You can sell Chardonnay and Sauvignon Blanc for a lot more money. So Chenin remains a marginal variety despite being very suitable for New Zealand's climate. In 2014 there were less than 30ha (74 acres), mostly in the North Island. There is an occasional delectable sweetie made, too.

Rest of the world

Chenin is quite widely planted in Argentina, mostly used to pad out Chardonnay. Mexico has some and there is a bit in Chile, Brazil and Uruguay, but it hasn't caught on. Israel has some and India manages to make some quite typically greengage-y Chenin. But in most places, it is usually vinified carelessly, with cavalier use of sulphur, so it's a bit difficult to know whether it could be any good or not.

Dry Creek Vineyard
Dry Creek is a Sonoma winery, but they grow their Chenin in Clarksburg in the Sacramento Delta, where river breezes give surprisingly cool conditions..

Millton Vineyard
Milton makes exceptional, off-dry Chenin in warm, humid Gisborne, using biodynamic methods that manage to keep rot and vine disease at bay.

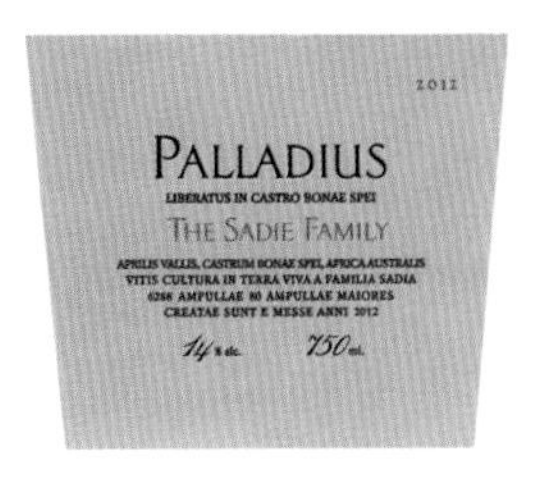

The Sadie Family
Eben Sadie is a true original. Palladius is a majestic, challenging Chenin, made even more so by the label being in Latin.

ENJOYING CHENIN BLANC

Sweet Loire Chenin Blanc is one of the most long-lived of all wines. In Vouvray, the richly botrytized years of 1921, 1945, 1947, 1955, 1959 and 1976 are all still drinking well now; a sweet Vouvray reaches maturity at around 12 years old, and will last for a century. Coteaux du Layon from Moulin Touchais, a family company that makes a speciality of older wines, seems to be at its best with 20 to 30 years in bottle, though a decade in bottle would be a more usual rule of thumb for those who want to drink top-quality Coteaux du Layon at the start of its maturity. As so often happens, it is often the years with the highest acidity that last the longest; you can measure their progress in generations rather than decades.

Demi-sec wines also need about 10 years to really shine; again, this applies to wines from very good producers. Lighter, less concentrated wines can be drunk earlier. Good demi-sec Chenins will stay at their peak for 30 years or so. The finest dry wines are hardly faster to mature. Good dry Vouvray needs a decade or so; even sparkling Vouvray can last a few years, though sparkling Saumur (like the fresh, fruity, but bone dry still Saumur) should be drunk young.

Savennières reaches its peak after 10 years or so, and lasts for many, many years. When the producers of Savennières formed themselves into an association in the 1950s, one of their members volunteered to supply the wine for their first formal dinner, saying that 'I have enough of the 1851 available to go with the main course'.

However, Chenin does go through a closed phase. It is delicious for a few months after bottling, but then may retire into itself for seven or eight years before it begins to emerge with the beguiling flavours of maturity.

The taste of Chenin Blanc

No other grape can imitate the flavours of Loire Chenin Blanc. In extreme youth the taste of crisp green apples is mixed with greengage, angelica and something earthy, chalky, minerally; the dry wine can taste pretty hard in youth, particularly once it begins to close up in bottle. But even in this least generous phase you can peer through the acid carapace and get a brief glimpse of complexities to come. And trust your instincts. They do come.

In most cases acidity is a little less aggressive than it used to be, but that's still comparative. It may be softer acidity, made less assertive by the malolactic fermentation; it may just be the riper acidity of more mature grapes from lower-yielding vines. *Bâtonnage* or lees stirring also helps to clothe the acidity in more creamy weight. So young Chenin is less unapproachable than it was. But this mineral thing is still there, and it doesn't go, in non-botrytized wines, even when maturity brings a flood of acacia and honey, brioche, quince and greengage.

Botrytized Chenin wines have less of the Chenin green apples, and more of the peach and pineapple, barley sugar, marzipan, quince and cream that comes with botrytis. They are generally less weighty and alcoholic than Sauternes, and they never seem to lose that piercing Chenin acidity, however sweet they are.

South African Chenin can be rich, nutty and toasty. Other New World Chenins are more tropical – bananas, guava and pineapple.

Domaine Huet has been Vouvray's leading producer for generations, making every style that Vouvray can deliver, from sparkling to bone dry to medium and luscious and sweet, depending on the quality of the vintage. However, their Demi-Sec – semi-dry – almost always makes an appearance, not quite dry, but definitely not sweet, and full of wild and challenging flavours. Mullineux are much more recently established in South Africa's Swartland but are taking advantage of South Africa's Chenin bush vines to make stunningly original wines, both sweet and dry.

MATCHING CHENIN BLANC AND FOOD

Chenin Blanc makes wines ranging from averagely quaffable dry whites to the great sweet whites of the Loire Valley. The lighter wines can be good as apéritifs or with salads as well as with light fish or chicken dishes. The medium-sweet versions usually retain enough of their acidity to counteract the richness of pâté and creamy chicken and meat dishes such as the Loire speciality of pork with prunes. The sweet wines are good with most puddings and superb with those made with slightly tart fruit. They are also marvellous with fresh fruit, foie gras or blue cheese.

CONSUMER INFORMATION

Synonyms & local names

Traditionally called Pineau or Pineau de la Loire in the Loire; Steen has long been its traditional name in South Africa but Chenin Blanc is becoming more usual as the grape achieves higher status; called Pinot Blanco in parts of South America and in Mexico.

Best producers

LOIRE VALLEY/Bonnezeaux M Angeli/la Sansonnière, Fesles, les Grandes Vignes, Petit Val, la Petite Croix, Petits Quarts, Terrebrune, la Varière; **Coteaux de l'Aubance** Bablut/Daviau, Dittière, Giraudières, Haute Perche, Montgilet, Princé, Richou, Rochelles/J-Y Lebreton; **Coteaux du Layon** P Aguilas, P Baudoin, Baumard, la Bergerie, Cady, P Delesvaux, Dom. F L, Forges, de Juchepie, Ogereau, Passavant, Pierre-Bise, Pithon-Paillé, Quarres, J Renou, la Roulerie, Sablonettes, Sauveroy, Soucherie; **Montlouis** Alex-Mathur, L Chanson, L Chatenay, F Chidaine, L & B Jousset, des Liards/Berger, Le Rocher des Violettes, F Saumon, Taille aux Loups/Blot; **Quarts de Chaume** P Baudoin, Baumard, Bellerive, Bergerie, Forges, Laffourcade, Pierre-Bise, Pithon-Paillé, Plaisance, J Renou, Suronde, la Varière; **Saumur** Clos Rougeard, Hureau, F Mabileau, Roches Neuves, Villeneuve, Yvonne; **Savennières** Baumard, Closel, Coulée-de-Serrant, Épiré, Dom. F L, Forges, Laffourcade, Laureau, aux Moines, Morgat, Pierre-Bise, Taillandier; **Vouvray** Aubuisières, Bourillon-Dorléans, Brunet, Champalou, F Chidaine, Clos Naudin/Foreau, Gautier, Huet, F Pinon, Taille aux Loups/Blot.

SPAIN Can Ràfols dels Caus, Escoda Sanahuja.

USA/California Chalone, Chappellet, Dry Creek Vineyard, Graziano, Husch, Pine Ridge; **Washington State** Chateau Ste Michelle, Covey Run, Hogue Cellars, Kiona, L'Ecole No 41, Paul Thomas.

NEW ZEALAND Astrolobe, Esk Valley, Forrest, Millton Vineyard.

SOUTH AFRICA Alheit, Beaumont, Botanica, Jean Daneel, DeMorgenzon, De Trafford, Flagstone, Ken Forrester, Kanu, Kleine Zalze, Mulderbosch, Mullineux, Nederburg, Sadie Family, Spice Route, The Winery of Good Hope.

RECOMMENDED WINES TO TRY

Ten sweet Loire wines

Patrick Baudouin Coteaux du Layon Sélection des Grains Nobles
Clos Naudin Vouvray Moelleux Réserve
Dom. Philippe Delesvaux Coteaux du Layon Cuvée Anthologie
Ch. de Fesles Bonnezeaux
Jo Pithon Coteaux du Layon St-Aubin Clos des Bois
René Renou Bonnezeaux Les Melleresses
Dom. Richou Coteaux de l'Aubance Cuvée les Trois Demoiselles
Dom. de la Sansonnière Bonnezeaux
Ch. Soucherie Coteaux du Layon la Tour
Dom. de la Taille aux Loups Montlouis Cuvée des Loups

Ten medium-sweet Loire wines

Dom. des Aubuisières Vouvray les Girardières Demi-sec
Didier Champalou Vouvray
Clos Naudin Vouvray Demi-sec
Ch. Gaudrelle Vouvray Réserve Spéciale
Huet Vouvray Clos du Bourg Demi-sec
Dom. des Liards Montlouis Vieilles Vignes
Gaston Pavy Touraine Azay-le-Rideau
Ch. Pierre-Bise Anjou le Haut de la Garde
Dom. Richou Coteaux de l'Aubance Sélection
Taille aux Loups Montlouis Demi-sec

Five classic dry Loire wines

Bourillon-Dorléans Vouvray Coulée d'Argent Sec
Clos de la Coulée-de-Serrant Savennières Coulée-de-Serrant
Clos Rougeard Saumur Blanc
Dom. du Closel Savennières Clos du Papillon
Ch. du Hureau Saumur Blanc

Five sparkling wines

Dom. de l'Aigle Crémant de Limoux
Clos Naudin Vouvray Pétillant Vintage
Gratien & Meyer Crémant de Loire Brut
Huet Vouvray Mousseux Vintage
Langlois-Château Crémant de Loire Quadrille Brut

Six New World wines

Astrolobe Chenin Blanc (New Zealand)
Chappellet Napa Valley Old Vine (California)
Ken Forrester Chenin Blanc (South Africa)
Millton Te Arai Chenin Blanc (New Zealand)
Mullineaux Chenin Blanc (South Africa)
Sadie Family Chenin Blanc (South Africa)

When Botrytis cinerea attacks white grapes like this Chenin Blanc, the berries first go mauve-brown in colour and then begin to shrivel. At that point good growers will send the pickers through the vineyards for a first tri, with instructions to pick only the shrivelled berries.

Maturity charts

Chenin Blanc from the Loire Valley can be immensely long-lived. It can also be somewhat unfriendly in youth.

2011 Savennières (dry)

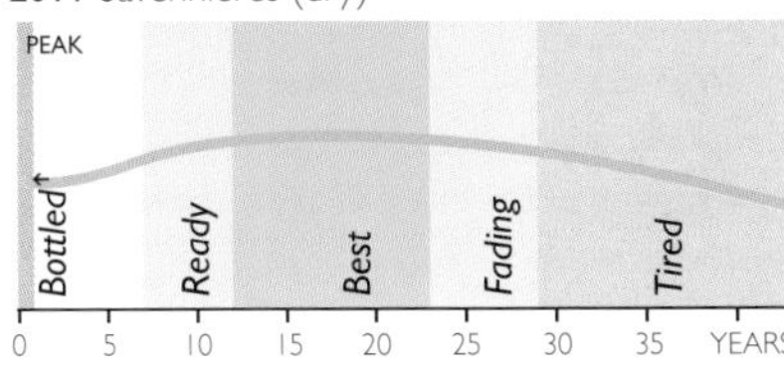

A good but not great year in Savennières. Many wines are now being made in an earlier-maturing style than previously; this chart applies to traditional styles.

2010 Quarts de Chaume

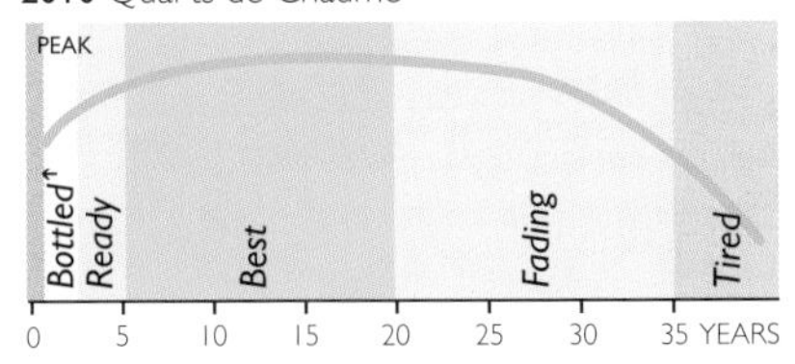

Nowadays as rich as Bordeaux's Sauternes, succulent, with a thrilling fruity acid that keeps them fresh and exciting for two decades at least.

2014 South Africa old vine Chenin

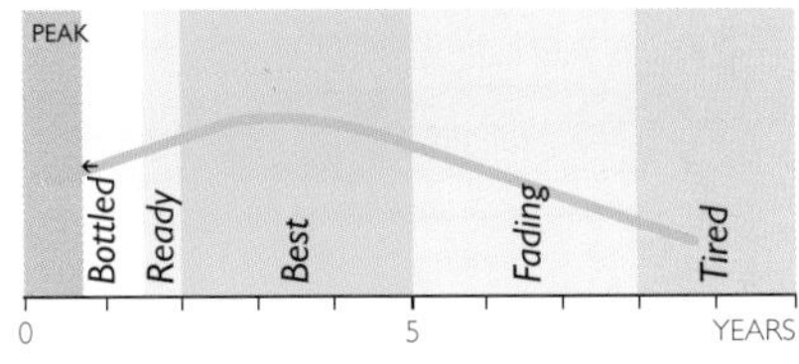

Chenin dominates South Africa's vineyards and the wine is increasingly good. Old bush vine Chenin fermented in barrel can be wild, toasty stuff.

CONCORD

This *Vitis labrusca* grape was the one that coaxed many North Americans into drinking wine, until the California wine revolution introduced them to Chardonnay in the 1980s. It is grown widely in the northeastern states, particularly New York State, in Canada and to some extent in Brazil. It is better than most *Vitis vinifera* varieties at withstanding harsh winters (conditions in Brazil are unfavourable in the opposite direction), but has the strangely aromatic flavour typical of *labrusca* vines. This flavour is usually referred to as 'foxy', but is in fact closer to the smell of mayblossom, wild strawberries and nail varnish, and is a long way from any wine aroma familiar to drinkers of *vinifera* wines.

The vine got its name from Concord, Massachusetts, where one Ephraim W Bull planted the seeds of a wild vine in 1843. It is successfully used for grape juice and jelly as well as for wine, which is often very sweet.

CORNALIN

This is the Humagne Rouge of Switzerland, but its Italian name is Cornalin, and Italy is where it originated, in the Valle d'Aosta. The Swiss Humagne Blanc is unrelated. The grape known as Cornalin in the Valais is Rouge du Pays. There's far more Humagne Rouge in the Valais than there is Cornalin in Italy, however. It makes richly tannic, even chewy reds redolent of blackberries and smoke.

CORNIFESTO

Portuguese red variety grown occasionally in the Douro. There's the odd patch in Australia and South Africa. It makes attractive light wines, but is quite susceptible to disease.

CORTESE

Despite the high prices charged for Cortese's most famous wine, Gavi, this Italian grape is traditionally seldom more than pleasant. It is now showing that it can muster an attractively full texture to go with its typical apple peel and lemon zest austerity. It has the virtue of retaining its acidity even in hot summers, which makes it a good bet in its homeland, northwest Italy. Plantings are concentrated in Piedmont, and spill over the border into Lombardy. Gavi from the town of Gavi, now called Gavi del commune di Gavi, is usually better than plain Gavi. In other parts of Piedmont and Lombardy it produces DOC wines including Cortese del Alto Monferrato and Colli Tortonesi; it is also part of the blend in Bianco di Custoza.

At its best it has good body and a nose of limes and greengages, but yields must be kept down if the wine is to have sufficient body to balance the acidity. In cool summers the acidity dominates and the wine can be unattractively lean. Some producers use *barrique* fermentation or put some of the wine through the malolactic fermentation in an attempt to rectify this. Best producers: (Italy) Battistina, Bergaglio, Gian Piero Broglia, La Chiara, Chiarlo, La Giustiniana, F Martinetti, San Pietro, La Scolca, Castello di Tassarola, Villa Sparina.

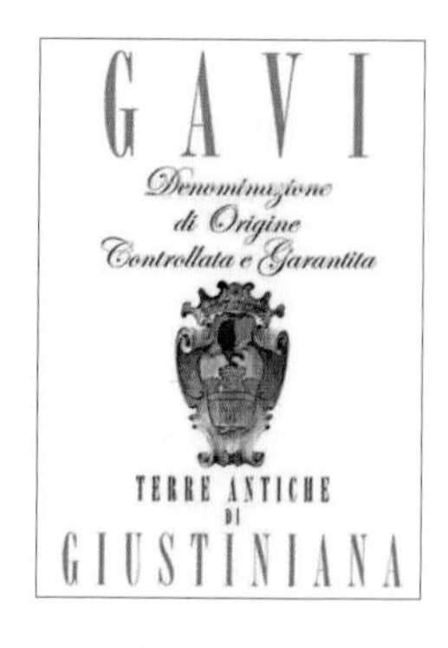

La Giustiniana
Gavi in Piedmont is the main area for the Cortese grape and it is usually bottled as a lean, lemony style. This is a more substantial single-vineyard style.

CORVINA

Corvina is the mainstay of Valpolicella and Bardolino, both of them light red wines from northeastern Italy. It will therefore come as no surprise to learn that Corvina seldom has much colour or tannin, but what it does bring to the party are aroma and acidity. When it is overcropped, as it all too often is, its wines are insubstantial and poor quality, as any drinker of inexpensive Valpolicella will testify. But on good hillside sites, and on the right soil – volcanic *toar*, which gives the most perfumed wines, or chalk, or alluvial – it can produce wines of considerable charm, lush texture and floral, cherryish perfume. Controlling yields can be a problem, because the first few buds on the cane don't fruit; it therefore needs a long cane, and pergola or *spalliera* training, to produce a crop at all.

Corvina really comes into its own as a grape for drying to make recioto and amarone wines. It has small berries with thick skins, the latter helping to protect the grapes from rot while they shrivel. Best producers: (Italy) Accordini, Allegrini, Antolini, Bertani, Boscaini, Brunelli, Tommaso Bussola, Castellani Viviani, Corte Sant'Alda, Dal Forno, Gamba, Guerrieri-Rizzardi, Masi, Quintarelli, Le Ragose, Rizzardi, Le Salette, Speri, Tedeschi, Tommasi, Valentini Cubi, Zenato.

Corvina's main job is to make light, juicy reds in Bardolino and Valpolicella, but it has a nobler role to play in being the chief grape used in the creation of Italy's most famous 'dried grape' wines, Amarone and Recioto. Leading producer Masi uses trays to dry and shrivel the grapes in their winery loft before turning them into Amarone.

CORVINONE

DNA fingerprinting has identified Corvinone as a separate variety rather than a variation of Corvina, and it is increasingly being planted separately, though most vineyards are still a mix of the two. It can replace Corvina as part of the Valpolicella blend. Corvinone has larger berries (as the name suggests), more colour and tannin and reaches higher sugar levels, and it does not have Corvina's problem of infertile buds at the base of the cane. It also makes wines far more suitable for aging.

COT

Cot is the proper French name for Malbec. See pages 130–131.

COUNOISE

A southern French grape that plays a small but valued part in some reds of the southern Rhône, Provence and Languedoc. Its close resemblance to Aubun makes it difficult to identify in the vineyard, though the wine is of better quality. It has a peppery, spicy character, damson fruit and good acidity, though not great colour or tannin, and it is tricky to grow, being low-yielding and late-ripening. But it is disease-resistant and adds body and fruit to blends. Best producers: (France) de Beaucastel, Mas Blanc, Romanin.

CRIOLLA GRANDE

There are a number of Criolla grapes in the Americas; the name means 'Creole' and it implies European descent, indicating, usually, that they were grown from seed or cuttings by settlers. Criolla Grande is a low-quality, pink-skinned grape grown in Argentina for the most basic white wine. It still covers huge areas, especially in Mendoza. There is also a darker-skinned vine called Criolla Chica in Argentina; this is more properly called Listán Prieto, and is the same as the País of Chile, and the Mission of California.

CROATINA

This northwest-Italian grape plays a part in Gattinara and Ghemme, alongside Nebbiolo (alias Spanna) and Bonarda. But Croatina itself has the name of Bonarda (see page 49) in Oltrepò Pavese (Lombardy) and Colli Piacentini (Emilia). The wine is frequently juicy, perfumed and delightful. Croatia, predictably, makes a Croatina, but from an entirely different variety.

CROUCHEN

A neutral French grape now abandoned by France, but still found in decreasing amounts in South Africa. It is concentrated in the Paarl and Stellenbosch regions and is mainly used in blends. On its own the wine is fairly steely. It is known there as Cape Riesling, South African Riesling or Paarl Riesling. (Proper Riesling is called Weisser or Rhine Riesling in South Africa.) Crouchen has almost vanished from Australia, where it was called Clare Riesling until 1976 when its identity was established by ampelographer Paul Truel. Best producers: (Australia) Brown Brothers; (South Africa) Bon Courage, Fairview.

CSÓKASZOLO

This dark Hungarian grape is being rediscovered. It's low-yielding and gives good quality, spicy, vibrant wines with a touch of Pinot Noir to the flavour and a touch of Blaufränkisch. Best producers: Bussay, Attila Gere, Kaló Imre, József Szentesi.

CYGNE BLANC

Cygne Blanc is a white seedling of Cabernet Sauvignon, discovered by grower and winemaker Dorham Mann in his garden in Western Australia's Swan Valley region – hence, of course, the name, *cygne* being the French for 'swan'. First tastings indicate a soft, ripe wine slightly reminiscent of Sauvignon Gris.

DEBIT

Croatian grape found widely in Dalmatia, making fresh white wines with good concentration.

DELAWARE

American hybrid named after the Ohio town where it was first propagated in 1849. Its flavour is less 'foxy' than that of Concord, and it ripens early, which makes it useful in New York State and Japan, where it is widely planted. It is part of the blend, along with Concord, for Austria's Uhudler – a supposedly aphrodisiac wine I've rather enjoyed on the odd summer's night.

DIMIAT

Or Dimyat. Bulgarian grape that covers large areas in the south and east of that country. Its wine is aromatic and simple, and usually off-dry to sweet. It is said to be named after a town in the Nile Delta, from where it was taken to Thrace at the time of the Crusades. In fact, it probably originated in Bulgaria.

Criolla was the workhorse grape of Argentina for centuries, used for eating, juicing and producing everyday wine and these ancient vines were planted in Salta in 1862. In fact, the term Criolla applies to a number of varieties in South America, either imported by the conquistadores, or conceived in South America by natural crossings.

DOLCETTO

'Little sweet one' is what Dolcetto means in Italian, but, as I try to find a Dolcetto which fits its description, I often feel like a mother shaking her head in long disapproval at a naughty son who is anything but a 'little sweet one'. Dolcetto wine can have all the sweetness and delightful winsomeness of a mischievous favourite son, but such examples are anything but common, and although it should produce the bright refreshing everyday reds of Piedmont, tannin, acidity and coarseness get in the way surprisingly often. Ideally Dolcetto should have moderate acidity, unintrusive tannin (at any rate, compared to Nebbiolo, Piedmont's most famous grape), a distinctive suggestion of orchard blossom perfume, and an appetizing bittersweet twist at the end. And it should be drunk at a year or two, when it is hopefully still brimful of fruit. The fashion, however, is for Dolcetto that is ever bigger, ever more alcoholic: producers in Piedmont seem determined to make a grand statement out of what should just be utterly drinkable and enjoyable. Dolcetto isn't really suited to the prestige treatment – it just isn't that sort of grape, and traditionally it has known its place. Because it is not as prestigious as Barbera or Nebbiolo – and also because it ripens a couple of weeks before Barbera, and up to four weeks before Nebbiolo – it gets planted in the cooler sites not suitable for the other two – OK in a good year, pretty iffy in a poor one. It is regarded as being easy to grow in Piedmont and consequently few growers lavish much attention on it. Dolcetto is traditionally best and most characterful in the Ovada and Alba zones and Alba is now the chief production area.

Decent Dolcetto is quite common. Exciting Dolcetto depends on talented and determined winemaking. When Ovada's greatest Dolcetto maker died, no one else was able to pick up the baton. There's a little Dolcetto elsewhere in Italy – Liguria, for instance, calls it Ormeasco and produces some interesting examples. Savoie produces something similar called Douce Noire, and everyone thought the deep, dark, chunky, chocolaty stuff called Charbono in California was Dolcetto, but it was Douce Noire.

Otherwise only Australia seems to have any – and we're talking just a few vines – but since the oldest ones go right back to the 1860s we should take them seriously, because they are probably the oldest Dolcetto plantings in existence.

THE TASTE OF DOLCETTO

Cherry flavours are typical of Dolcetto: ideally ripe black cherries on the nose and palate, and bitter cherries on the finish for that characteristic Italian twist. Less ripe versions – often 12.5 per cent abv – have a rougher cherry stone texture, cranberry fruit and an almost metallic chewiness. There can be flavours of prunes as well, and liquorice. If you're lucky you may also find a wine with intriguing perfume, but, in spite of its name, Dolcetto wines are dry. The grapes are not even notably high in sugar.

Vajra are special in Piedmont because not only do they make excellent Barolo, but they seem to put just as much effort into the supposedly lesser wines like Barbera and Dolcetto. Admittedly, Dolcetto doesn't get given the best vineyard sites, being early-ripening, but this example is full of power and spice with a rich rustic yet soft red fruit and buttery cream.

Araldica
An outstanding example of Dolcetto from the Dogliano zone of Piedmont. This single-vineyard wine comes from low-yielding, old vines.

Marcarini
Marcarini's punchy, powerful Dolcetto d'Alba comes from 100-year-old vines planted, unusually, on their own roots. The wine is unoaked – and is all the better for it.

Left: Harvesting Dolcetto in the vineyard of Aldo Conterno at Bussia near Monforte d'Alba in Piedmont. Dolcetto is generally regarded as the third-best red variety in Piedmont, after Nebbiolo and Barbera. And as it is earlier ripening than either, up to four weeks, it's often used on sites too cool for the top two. ***Above:*** *Dolcetto grapes are tremendously dark in colour, and need only a short maceration on the skins to produce equally dark wine. This short maceration is the reason why the wines are light in tannin: the grapes have as much tannin as any other, should a winemaker wish to extract it. Dolcetto is now being seriously studied for planting in Australia, and it will be fascinating to see what the Australians do with it if they go for it in a big way.*

CONSUMER INFORMATION

Synonyms & local names

Known as Ormeasco in the Riviera Ligure di Ponente zone in western Liguria. The French region of Savoie has the similar Douce Noire, as does California, under the name Charbono.

Best producers

ITALY/Piedmont/Dolcetto d'Acqui Viticoltori dell'Acquese, Villa Sparina; **Dolcetto d'Alba** Alario, Altare, Ascheri, Azelia, E Boglietti, Bongiovanni, Bricco Maiolica, Bricco Rosso, Brovia, Cà' Viola, Cigliuti, Domenico Clerico, E Cogno, Aldo Conterno, Conterno-Fantino, Corino, Gastaldi, E Germano, B Giacosa, Elio Grasso, Marcarini, Marchesi di Gresy, B Mascarello, G Mascarello, Moccagatta, Fiorenzo Nada, Oddero, Paitin, Armando Parusso, Pelissero, Ferdinando Principiano, Prunotto, Renato Ratti, G Rinaldi, Roagna, Albino Rocca, Bruno Rocca, Luciano Sandrone, Scavino, La Spinetta, Vajra, Eraldo Viberti, Vietti, Vigna Rionda, Gianni Voerzio, Roberto Voerzio; **Dolcetto d'Asti** Brema; **Dolcetto di Diano d'Alba** Alario, Bricco Maiolica; **Dolcetto di Dogliani** Abbona, Boschis, Quinto Chionetti, La Collina, Del Tufo, Devalle, Luzi Donadei, Luigi Einaudi, Gillardi, Marenco, Pecchenino, Carlo Romana, San Fereolo, San Romano, Schellino, Giovanni Uria; **Dolcetto di Ovada** La Guardia, Giuseppe Ratto/Cascina Scarsi, Annalysa Rossi Contini, Terre da Vino; **Liguria** Lupi, Lorenzo Ramò. **USA/California** Duxoup, Enotria, Palmina; **Washington State** Morrison Lane.

RECOMMENDED WINES TO TRY

Ten Dolcetto d'Alba/di Diano d'Alba wines

Alario Dolcetto di Diano d'Alba Costa Fiore
Boglietti Dolcetto d'Alba
Brovia Dolcetto d'Alba Solatio Brovia
Conterno-Fantino Dolcetto d'Alba Bricco Bastia
Marcarini Dolcetto d'Alba Fontanazza
Pelissero Dolcetto d'Alba Augenta
G Rinaldi Dolcetto d'Alba
Roagna Dolcetto d'Alba
Albino Rocca Dolcetto d'Alba Vignalunga
G D Vajra Dolcetto d'Alba Coste e Fossati

Ten Dolcetto di Dogliani wines

Marziano & Enrico Abbona Dolcetto di Dogliani Papa Celso
Luzi Donadei Dolcetto di Dogliani
Quinto Chionetti Dolcetto di Dogliani Briccolero
Antonio Del Tufo Dolcetto di Dogliani Vigna Spina
Luigi Einaudi Dolcetto di Dogliani Vigna Tecc
Gillardi Dolcetto di Dogliani Vigna Maestra
Pecchenino Dolcetto di Dogliani Sirì d'Jermu
Carlo Romana Dolcetto di Dogliani Bric dij Nor
San Fereolo Dolcetto di Dogliani San Fereolo
San Romano Dolcetto di Dogliani Vigna del Pilone

Five other Dolcetto wines

Duxoup Napa Valley Charbono (California)
Lupi Riviera Ligure di Ponente Ormeasco Superiore Le Braje (Italy)
Palmina Santa Barbara (California)
Lorenzo Ramò Riviera Ligure di Ponente Ormeasco (Italy)
Villa Sparina Dolcetto d'Acqui Bric Maiola (Italy)

DOMINA

A fresh, flavoursome but seldom elegant crossing of Blauer Portugieser and Pinot Noir, bred in Germany in 1927. It yields well, ripens well and is disease-resistant. Barrel aging seems to suit. It's mostly found in Franken.

DOÑA BRANCA

Iberian grape grown (as Doña Branca) in the north of Portugal, but not related to Doña Blanca over the border in Spain. Perish the thought. It forms part of the white port blend but is also grown for unfortified wines which can be strongly perfumed.

DORINTO

This is the new official name for Arinto no Douro and its synonyms Arinto do Douro, Arinto do Interior and Arinto de Trás-os-Montes. Arinto is a confusingly popular name for Portuguese white grapes, unfortunately. Arinto do Dão is Malvasia Fina, and Arinto de Bucelas is Bical. Dorinto is far less widely cultivated than Arinto de Bucelas, and appears in Transmontano, Duriense and occasionally in old mixed vineyards in the Douro. The wines are interesting, citrus and flavoursome.

DORNFELDER

Some of Germany's most attractive red wines are made wholly or in part from Dornfelder. True, they may not be great wines in the world-class league, and they can't really compete with the good-to-fantastic Pinot Noirs or Spätburgunders that Germany is now making. There's a certain honesty about Dornfelder: it doesn't pretend to be more than a well-coloured, juicily fruited grape for short- to medium-term drinking, and it fulfils that role very well. Plantings in Germany are concentrated in the Pfalz, Rheinhessen and Württemberg, where even at high yields of 120hl/ha it gives decent colour. Producers who want to age Dornfelder in oak and make a wine that will age for a few years will opt for lower yields.

Dornfelder is the 1956 offspring of Helfensteiner and Heroldrebe, which are themselves crossings bred in Germany in the 20th century. Dornfelder's pedigree is exceedingly complicated, and includes all of Germany's major red vines; it is also the parent of several new crossings, bred at Württemberg and introduced in 1999: Acolon is Lemberger x Dornfelder; Cabernet Dorsa is Dornfelder x Cabernet Sauvignon, and Cabernet Dorio is another Dornfelder-Cabernet Sauvignon crossing. Dornfelder is also found in England, where its wines are often blended with Pinot Noir and, as well as decent rosé, it gives some of England's few full-flavoured reds. There's a bit, too, in the Czech Republic, California, New York and elsewhere. Best producers: (England) Chapel Down, Denbies, Stanlake Park; (Germany) Graf Adelmann, Drautz-Able, Lingenfelder.

Dornfelder manages to balance high yields with deep colour and decent juicy flavours. This makes it popular in Germany's Pfalz and Rheinhessen regions, but these vines are at Denbies Vineyards in Surrey, England, where it still ripens well enough to give full-coloured reds and spicy rosés.

DOUCE NOIRE

This is not a name one's likely to see in big letters on a wine label, to be honest, but it's here because it's the correct name for both the Charbono grape grown in California (which is not the same as the Charbono that used to be found in Piedmont) and the Bonarda grape grown in Argentina – which, naturally, is not the same as the Bonarda (any of them) found in Italy. Douce Noire originates in the French region of Savoie, but there's little left there now. It makes pleasantly fruity wine.

DRUPEGGIO

This white grape is known as Canaiolo Bianco in Tuscany, but it's less confusing to call it Drupeggio because Canaiolo Bianco is a synonym for several varieties in Tuscany, including the better known Vernaccia di San Gimignano. Drupeggio is a minority part of the Orvieto blend and is grown in other parts of the Umbria and also in Lazio.

DUNKELFELDER

A dark-skinned, red-fleshed German crossing introduced to vineyards in the 20th century and found in the Pfalz, Rheinhessen and Baden regions. Its colour is its chief attraction; it's early-ripening, susceptible to drought and powdery mildew, and not big on flavour.

DURAS

Deep-coloured grape found in southwest France, where it is blended with other local grapes such as Négrette and Fer. The wine is peppery and structured. Plantings are slowly on the increase.

DURIF

A 19th-century crossing, probably of Peloursin and Syrah, which became popular in the south of France because of its resistance to downy mildew, though not for any other quality. Durif produces coarse, rustic red wine and plantings have virtually disappeared from French vineyards.

De Bortoli
Durif has shown itself to be particularly well suited to Australia's hot, irrigated Riverina region, and can produce dark reds of power and perfume.

In recent years it has been most famous as the alter ego of California's Petite Sirah (see page 181), although some of the latter turned out to be Peloursin; well, it's reasonable to look like your parent. In Australia Durif is grown under its own name, and produces dry, solid, four-square wines in warm climates that supposedly age for ever, and just occasionally, in Riverina and Rutherglen, wines of delightful perfume and depth. Best producers: (Australia) Campbells, De Bortoli, Morris, Rutherglen Estate, Westend.

Much of Cappadocia is a remarkable, volcanic landscape with its fairy chimney pinnacle rock formations and its frescoed rock churches. It's an ancient Christian area and the Hittites were making wine here thousands of years ago. Emir is the best local grape, grown in challenging continental conditions which usually require the vines to be earthed up in winter. Even so, a devastating winter-kill occurs every generation or so.

DUTCHESS

An American hybrid found in New York State, in Pennsylvania and in Brazil, where it's probably at its best. It makes rather 'foxy'-tasting wine.

EHRENFELSER

German crossing bred in 1929, definitely from Riesling, but no one seems quite sure what the other parent is. The usual suspect, Silvaner, has recently been ruled out. It was intended, like so many such crossings, to have all the advantages of Riesling (elegance, finesse, complexity, longevity) with a few more thrown in for luck, mainly earlier ripening and higher yields. The wine is actually fairly good, though acidity is low and quality is not nearly as high as that of Riesling. It made some impact in the Pfalz and Rheinhessen regions, but its success was never notable.

ELBLING

Elbling may well have been cultivated by the Romans, and it dominated the vineyards of medieval Germany and no doubt produced wines that were every bit as painfully acidic as they are today. Nowadays it is found mostly in the uppermost reaches of the Mosel Valley and over the border in Luxembourg, where both a red and a white version are known. Its main use is for sparkling wine. It yields generously but reaches only low sugar levels.

EMIR

One of the main Turkish white grapes, though in a country where most wine is red, that's not saying a lot and there are less than 100ha (247 acres). It likes warm summers and cold winters and well-drained soils and makes light wines with attractive freshness, faintly leafy in flavour.

ENCRUZADO

Potentially the best white grape of the Dão region in Portugal. Growers are experimenting with barrel fermentation and lees stirring to bring out its character and a little oak for aging seems to suit it rather well. At its best the wine is quite minerally, with some green leaf and floral aromas, and citrussy acid balance. Best producers: (Portugal) José Maria da Fonseca, Quinta dos Roques, Sogrape.

ERBALUCE

The earliest written record of Erbaluce in Piedmont is in 1606, and the vine seems to have originated in the alpine foothills here. Its piercing acidity makes it ideal for sweet wines, and Caluso Passito, made from dried grapes, is its finest incarnation. There is also some dry wine called Erbaluce di Caluso, but this has to be very ripe to combat the acidity. It can be a good, often very interesting grape, with attractive appley fruit. Best producers: (Italy) Antoniolo, Cieck, Luigi Ferrando, Orsolani.

ERMITAGE

Northern Rhône synonym for Marsanne (see pages 136–137).

Ferrando
Erbaluce is a high acid variety often at its best in 'dried grape' wines. Ferrando dry their grapes for five months in wooden cases before fermentation and aging in barrel.

ESGANA CÃO

The Sercial of Madeira is grown in many parts of Portugal, and on the mainland is generally known as Esgana Cão. Its unfortunate effect on canines – the name means 'Dog strangler' – is thought to be the result of its high acidity. It is found in Vinho Verde, Bucelas, where it is usually blended with Arinto de Bucelas, and the Douro Valley. See Sercial page 254.

ESPADEIRO

A source of light red Vinho Verde in northern Portugal. The Espadeiro grown further south around Lisbon is, in fact, Trincadeira, alias Tinta Amarela.

FABERREBE

This is the new official name for Faber, but the grape itself has not, alas, become any more interesting. It's a generally uninspiring crossing, either of Weissburgunder (Pinot Blanc) and Müller-Thurgau or of Silvaner and Pinot Blanc, bred in the 1920s and favoured in Germany's Rheinhessen region for its high sugar levels, high acidity and ability to ripen in cooler sites than those needed by Riesling. It nevertheless lacks the cardinal virtues of flavour and character, and is now on the decline.

FALANGHINA FLEAGREA

Falanghina is now one of Italy's new star white grapes; and to go with its new status it has a new and correct pedigree. Its new name differentiates it from Falanghina Beneventana, which also grows in Campania more locally, in the province of Benevento near Naples. The two varieties seem to be unrelated, and either way, you'll normally just see Falanghina on the label. Falanghina Fleagrea is an old grape, first mentioned in a poem in 1666. It has good acidity, an elegant apricot-tinged nose and palate and good weight and structure. It's like a more subtle yet scented version of Fiano, and while some producers like to emphasize the aroma, others, true to the Italian distrust of aromatic wines, keep it well controlled. Either way, it's a lovely wine.

FAVORITA

Favorita, with its large berries, can double as a table grape, though now its popularity as a wine grape seems to be gently increasing again, the table might have to do without. Its home is in Piedmont, where it grows in the Roero and Langhe zones.

Its wine is well structured and with good acidity, but without much aroma except on the rare occasions it evokes a fleeting memory of pears. At its weightiest and ripest it resembles good Vermentino – what a surprise, because latest research implies they're the same vine – but that doesn't mean that Vermentino grown down on the coast is going to taste the same as Favorita from inland Piedmont. It is late ripening and is sometimes blended with Nebbiolo to soften the latter. Best producer: (Italy) Deltetto, Gianni Gagliardo.

FENDANT

Chasselas is known by this name in Switzerland's Valais region. See page 80.

FER

Fer, or Fer Servadou, lends its perfumed, redcurrant fruit to a variety of blends in south-west France. It is a minority grape almost everywhere, though plays a more substantial part in the Marcillac, Entraygues et le Fel, and Estaing vineyards. It is also found in Madiran, though only to a small extent. It is actually quite a good grape, with concentration and character, and for once it's not on the decrease. In fact its suppleness means that it is useful in softening Tannat, and making the latter's wines approachable earlier.

There is a vine called Fer grown in Argentina, but it is believed to be a clone of Malbec. Fer is a grandparent of Carmenère, and originated either in southwest France or further south and west in the Spanish Basque country. Best producers: (France) du Cros, Laurens, Producteurs Plaimont, Marcillac-Vallon co-op, Jean-Luc-Matha.

FERNÃO PIRES

A fairly aromatic and very versatile variety found all over Portugal: in the Bairrada region it is known as Maria Gomes. It is probably Portugal's most planted white grape, and can make anything from sparkling wines to still dry ones to botrytized sweet ones, and can be successfully oak aged provided the oak isn't overdone. The wine is best drunk young and some examples tire within the year. Best producers: (Portugal) Quinta da Boavista, Quinta do Carmo, Quinta do Casal Branco, Quinta das Setencostas.

FETEASCĂ

The soft, vaguely Muscaty wine produced from this old eastern European grape is generally of reasonable, though not high, quality. It can be low in acidity in warm climates and ultimately lacks much character, though its gently peachy wine is agreeable enough. Romania boasts two Feteascăs: Alba, and Regală, which is a crossing of Fetească Alba and Grasa, and dates from the 1920s. Fetească Regală is the later ripening of the two, and by far the most planted. Its wine has more finesse and makes good rich late-harvest versions. Romania has large plantings of both Feteascăs, and there is some in Hungary, Bulgaria, the Ukraine and Moldova. Fetească Alba and Fetească Regală are similar to Leányka and Királyleányka in Hungary. The dark-skinned version, Fetească Neagra, is less widely grown but gives good spicy red in Romania and Moldova. Best producer: (Romania) Cotnari Cellars.

FIANO

Interesting, high-quality and very fashionable southern Italian grape responsible for Campania's aromatic Fiano di Avellino. At their best the wines are weighty and honeyed with notes of flowers and spice, and have the potential to improve in bottle. It's also proving a considerable success in Sicily and several growers in South Australia are giving it a go (Clare Valley's Jeffrey Grosset is one). Best producers: (Australia) Coriole, Fox Gordon, Jacob's Creek; (Italy) Colli di Lapio, Di Majo Norante, Ferrara, Feudi di San Gregorio, Mastroberardino, Paternoster, Planeta, Giovanni Struzziero, Terredora, Vadiaperti.

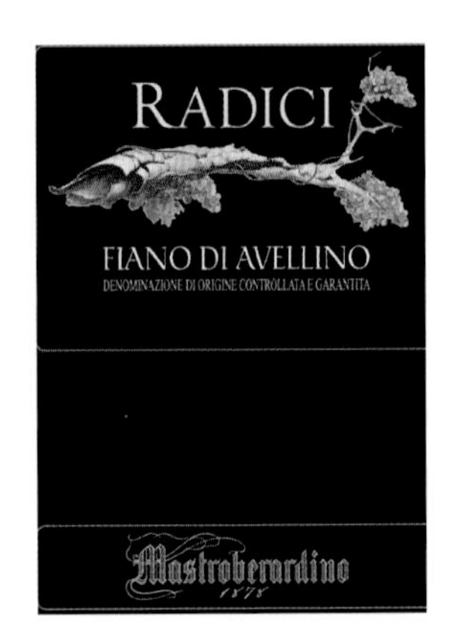

Mastroberardino
Mastroberardino took the lead in reviving Campania's historic varieties in the 1970s. Fiano may be directly descended from the varieties of Roman times.

FOGLIA TONDA

The name means 'Round Leaf' in Italian, and this Tuscan variety, which almost disappeared from the vineyards in the 1980s before being rediscovered, is a minority part of the Chianti Classico blend. It contributes colour, power and the ability to age.

FOLLE BLANCHE

Folle Blanche is rapidly becoming superfluous to requirements in western France, where once it flourished. In the Gers it is known as Piquepoul or Picpoul, but it is unrelated to true Picpoul (see page 182). It was widely grown for distillation into Cognac and Armagnac until the onset of phylloxera, but is susceptible to rot and has since been largely supplanted in these regions by other varieties, notably Ugni Blanc (although brandy from Folle Blanche is generally better than that from Ugni Blanc). As Gros Plant it is grown for the VDQS wine of the same name near Nantes in the western Loire, but here too demand is falling, and growers are replacing it with Chardonnay for Vin de Pays du Jardin de la France. It is late ripening, and must be picked two weeks after Melon de Bourgogne, alias Muscadet, the main grape of the area; its acidity is very high, and its flavour neutral. All of which implies that its wine is pretty horrible. Indeed, it is – until you plonk it down next to a plate piled high with *fruits de mer* – and then, just sometimes, it's the perfect wine.

Folle Noire is not related: this is a French synonym for several varieties, Négrette and Jurançon among them.

Bottles of Tokaji in the cellars of Oremus at Tolcsva, Hungary. The blend is Furmint, Hárslevelü and sometimes a little Muscat. The colour matches the wine's rich, golden hue, but sharp lemon peel acidity is also always present in good Tokaji.

FRANCONIA

The Austrian grape Blaufränkisch (see page 48) sometimes takes this name in Friuli, in northeastern Italy and there are moves to use the name in Croatia too.

FRAPPATO NERO

Soft, low-tannin Sicilian grape that makes wines of considerable charm and some aroma. It is usually blended with other red grapes, particularly Calabrese, Nerello and Nocera, but it has potential for light, fruity reds. It could have potential as a Merlot lookalike.

FREISA

A love-it-or-hate-it Piedmontese grape, high in strawberry and raspberry aromas and acidity and sometimes quite high in tannins. The problems arise with the bitterness detectable on the finish, and the residual sugar found in many examples which can be too much for some people, though others find it quite irresistible. I err on the side of the believers.

It is thought to have originated in the hills between Asti and Turin, and most authorities differentiate between Freisa Grossa, which has large berries and large clusters, and Freisa Piccola, which has small ones. Indeed, Freisa Grossa has now been shown to be a different variety, and identical to Neretta Cuneese. Freisa (Piccola) has more character and perfume. As Freisa de Chieri and d'Asti, it has its own DOCs. Styles of wine vary: Freisa can be *frizzante* (both sweet or dry, like super-Lambrusco), or dry, still and more serious, but not necessarily better. Best producers: (Italy) Caudrina, Poderi Colla, Coppo, Piero Gatti, Giuseppe Mascarello, Cantina del Pino, Giuseppe Rinaldi, Scarpa, Vajra, Rino Varaldo, Gianni Voerzio.

FRIULANO

See Sauvignonasse page 242.

FRONTIGNAC

Australian synonym for Muscat Blanc à Petits Grains (see pages 157–165).

FRÜHROTER VELTLINER

An Austrian red-skinned grape used for making white wine. It's unrelated to Grüner Veltliner, but with the same parents (Roter Veltliner and Silvaner) as Neuburger, and a half-sibling of Rotpgipfler and Zierfandler. Sadly for its future in the vineyards, it lacks the character of any of those, and is mostly being replaced by the infinitely more fashionable Grüner Veltliner. It can also be found to some degree in Hungary, Slovakia and the Czech Republic.

FURMINT

Very high-quality grape that, having survived the poor handling it suffered in Hungary's Tokaj region under Communism, is now at last beginning to come into its own.

It originated in Tokaj, and was grown to some extent for sweet Ausbruch wines in Austria's Burgenland region in the past, where it is enjoying a small revival now. Both green and yellow versions exist there, with the yellow being more highly prized. In Germany it is called Zapfner and in Steiermark, Mosler. It has been shown to be the same as Slovenia's Sipon, but different to Romania's Grasa (see page 123) and Croatia's Pošip. It has also been shown to be a parent of Hárslevelu, and is a parent of Zéta, formerly known as Oremus, and also used in Tokaji.

Its advantages are its complexity of flavour, its finesse, its longevity and its high acidity. Young dry Furmint has flavours of steely smoke, lime peel and pears. Sweeter wines, affected to a greater or lesser degree by botrytis, taste of apricots and marzipan, barley sugar and blood orange and become nutty, smoky and spicy, with flavours of tea, chocolate and tobacco and sometimes with a distinct note of cinnamon, as they age.

The problem with dry Furmint is in expressing these flavours. The Disznókö estate has found that it is necessary to pick non-botrytized grapes for dry wine after picking botrytized ones for sweet wine – the opposite of what one would normally expect. If done the other way round, the acidity in the dry wines is painfully high and flavour is lacking.

The other problem is that the joint venture companies, who came after the end of Communism, have been mostly interested in creaming off the 10 or 15 per cent of the crop that is affected by botrytis. The aszú wines for which these are used are of extremely high quality, and in terms of longevity seem to be immortal. The companies are only slowly becoming interested in making good dry Furmint. Best producers: (Austria) Wenzel; (Hungary) Château Megyer, Château Pajzos, Disznókö, Oremus, Royal Tokaji Wine Co.

GAGLIOPPO

An ancient variety which is a source of sturdy, red wine in Calabria, Abruzzo, the Marche and Umbria. Its most famous incarnation is as Cirò, on Italy's east coast – deep-coloured, alcoholic, weighty and often very good, especially if the producer has had only a light touch with the oak. Examples with no oak at all can be even better. Gaglioppo is currently attracting interest from winemakers, so we can expect to see more from this grape. Best producers: (Italy) Caparra & Siciliana, Librandi, Odoardi, San Francesco, Statti.

GAMARET

Gamaret won't be found far outside Switzerland. It is an unusual beast – a Swiss red variety with deep colour and fairly powerful structure. It is a cross between Gamay Noir (see right) and Reichensteiner created in Switzerland in 1970 and is becoming popular both as a blender providing colour and weight to Gamay and Pinot Noir, and as a varietal. There's a little bit in Italy and it is authorized for use in Beaujolais. I thought red Beaujolais had to be made solely from the Gamay grape, but there you go, what do I know?

GAMAY

The growers of Beaujolais have been enjoying both good times and bad times – good times because they've had several superb vintages which have shown off the results of a great deal of work in both vineyard and cellar. Bad times because it's still horribly hard to sell Beaujolais to much of the world. An enormous gap has widened between the good growers and the rest, between the crus (the villages of Brouilly, Chénas, Chiroubles, Côte de Brouilly, Fleurie, Juliénas, Morgon, Moulin-à-Vent, Regnié and St-Amour) and Beaujolais Nouveau. The latter still exists, and is still released on the third Thursday of each November, but it's difficult to know who drinks it. Good Beaujolais is delicious: mineral, focused, with fruit of raspberries, black pepper and cherries; it's never overstated or blockbusting, but it has character, balanced acidity, lightness and freshness. It's made with conventional red-wine vinification techniques, usually; carbonic maceration is for the less serious wines, the Nouveau, perhaps the basic Beaujolais.

Plain Beaujolais comes from the flatter southern part of the region, where the soil is sedimentary clay and limestone, while Villages wines (39 villages may call their wine Beaujolais-Villages) and the Crus come from the granite hills of the North. The vines are trained in *gobelet* form, which tends to restrain their natural vigour. Simple Beaujolais and Beaujolais Villages is meant to be drunk young, and that applies equally to Gamay from elsewhere. Cru wines will improve with a couple of years' aging, and the longest-lasting – Moulin-à-Vent, Morgon, Chénas and Juliénas – will improve for up to 10 years, gaining a character not unlike that of mature Pinot Noir.

Gamay needs very particular conditions for it to excel. The decomposed granite soils of Beaujolais' best villages seem to be ideal. There are 10 top villages, called 'crus'. These vines are growing at Régnié, the most recently created 'cru'.

North of Beaujolais in the Mâconnais and Côte Chalonnaise, the vine has been losing ground to Chardonnay. In the Ardèche and along the Loire Valley, in Touraine and in regions to the west, it flourishes on flinty silex soil, producing light, peppery wines with good aroma. It's also important in Savoie. In Switzerland it is often blended with Pinot Noir to the advantage of neither: the blends are called Dôle or Goron (this has slightly more Gamay) and often taste rather thick and dull.

Gamay was introduced to Italy in 1825, but there is only a little there now. It can be found, however, throughout eastern Europe, especially in the Balkans. It is often confused with Blaufränkisch (see page 48). Its early-budding, early-ripening nature makes it suitable for cool climates, though spring frosts can be a problem. There are occasional plantings in Canada and New Zealand, both countries where it could be interesting.

California boasts two grapes, one called Gamay Beaujolais and other called Napa Gamay. Gamay Beaujolais is not Gamay at all, but a poor clone of Pinot Noir. Plantings are gradually being replaced by Napa Gamay or better Pinot Noir, and the name Gamay Beaujolais was forbidden from 2007. Napa Gamay is more complicated: long thought to be true Gamay, it is now reckoned to be Valdiguié, a French grape so poor it has been pretty well kicked out of its homeland. Napa Gamay, too, is on the decline in California, but the occasional juicy, herby example makes you wonder first whether Gamay doesn't have a future there after all, and second, what the grape variety really is in California.

Touraine also has some small plantings of Teinturier Gamays, with deep red flesh and juice. These include Gamay de Chaudenay, Gamay de Bouze, Gamay de Castille, Gamay Mourot and Gamay Fréaux. Their wine is robust, solid and unaromatic – quite unlike Gamay Noir à Jus Blanc. Best producers: (France) Aucoeur, Aujoux, Berrod, Cellier des Samsons, Charvet, Duboeuf, Henry Fessy, Janodet, la Madone, Thivin, Thorin, Pelletier; (Switzerland) Caves Imesch, Caves Orsat.

Pieropan
Pieropan pioneered high quality Soave and were the first to bottle single vineyard Soave wine. La Rocca is a limestone outcrop (most of Soave is basalt) that allows the Garganega to get particularly ripe.

GAMZA

Hungary's Kadarka grape (see page 126) is known as Gamza in Bulgaria.

GARGANEGA

Garganega's great quality is often overlooked because of the general mediocrity of Soave wine. However, good growers with hillside vineyards can produce excellent dry, waxy, almondy examples of this wine.

This late-ripening, highly vigorous vine is the main grape behind Soave, Gambellara and other Veneto whites. It spills over into Friuli and Umbria as well, but Soave is the wine most associated with it, for good or bad.

Good Soave, and good Garganega, is exceedingly good. It has both delicacy and structure, finesse and just enough weight, and a flavour reminiscent of almonds, greengage plums and citrus fruit. When made as a sweet Recioto, from dried grapes, it is intensely sweet with good but not piercing acidity; Recioto wines will improve in bottle for a decade or more, and even good single-vineyard Soave from top producers in the Classico zone can sometimes improve for nearly as long. Garganega's problem is that its vigour has encouraged too many plantings on ultra-fertile soils in the flatlands outside the Classico zone, where it produces high yields. The wine then is at best thin, neutral and dull – a description which applies to too much Soave.

It is the same variety as Sicily's Grecanico Dorato. Best producers: (Italy) Anselmi, Ca' Rugate, Coffele, Filippi, Gini, Guerrieri-Rizzardi, Inama, Masi, Pieropan, Prà, Suavia, Tamellini, Tedeschi.

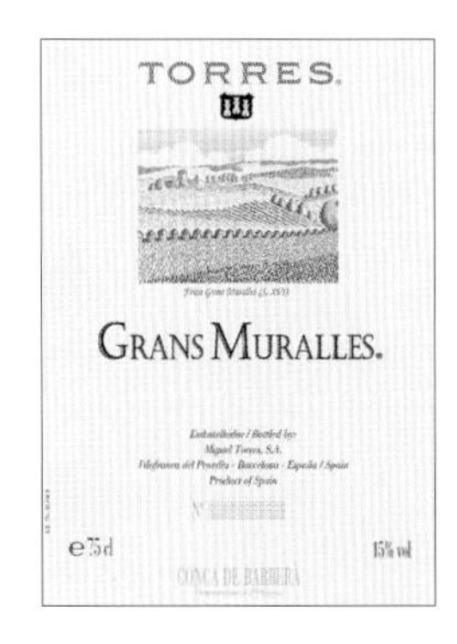

Torres
Torres use this label to highlight the ancient indigenous grapes they are saving from extinction. Most important so far is Garró..

GARNACHA BLANCA

The Spanish name for Grenache Blanc (see page 123) is the oldest and therefore the correct one, but there is a good deal more in southern France than there is in Spain. It is found in Spain's northeast, where it is the main grape in the Alella DO, in which it is surprisingly lighter and aromatic. It is permitted in Priorat, Tarragona and Rioja, where it makes much heavier wine, but only tiny plantings exist. Best producers: (Spain) Celler de Capçanes, Masía Barríl, De Muller, Bàrbara Fores, Scala Dei, Costers del Siurana; (France) l'Amarine, Casenove.

GARNACHA TINTA

See pages 102–111.

GARNACHA TINTORERA

A Teinturier grape that isn't actually Garnacha at all: instead it is another name for Alicante Henri Bouschet (see page 41).

GARRÓ

This old Catalan variety has been researched and cultivated by the Torres company, and is used in its Gran Muralles blend. It's a small-berried, dark-coloured, tannic variety, with fruit that acquires leathery flavours. It's good quality and the wine should age well.

GELBER MUSKATELLER

The German and Austrian synonym for Muscat Blanc à Petits Grains (see pages 156–165).

Priorat is known as one of Spain's most formidable red wines, but there is a small amount of white wine – dense, oily and powerful. Garnacha Blanca is the main grape and the Cellers de Scala Dei makes good use of them.

GARNACHA TINTA/GRENACHE NOIR

Grenache is one of the great wine experiences. It has a wonderful raw-boned power that sweeps you along in its intoxicating wake. It exudes a blithe bonhomie and a taste – all ruddy cheeks and flashing eyes, and fistfuls of strawberry fruit – that seduces you yet makes you think that surely it's all harmless fun. But it isn't. That juicy good nature usually hides a scarily high level of alcohol. And as your head spins from one glass too many – and it could be just your second – you realize woefully that you've been had again and when will you learn, but you hurl yourself anyway into the fandango of delight that is Grenache.

Grenache is for me the wild, wild woman of wine, the sex on wheels and devil take the hindmost, the don't say I didn't warn you. But many of us will have drunk more Grenache than we realize, because Grenache has spent the majority of its existence as a widely planted – indeed, until recently, it was globally the most widely planted of all red grape varieties – and alcohol-boosting blender throughout a large part of Spain and Southern France. If you drank a cheap Southern French or Spanish red, it was likely to have Grenache at its heart.

Grenache can easily ripen to 16 per cent alcohol all by itself – which meant that blenders loved it, and fortified winemakers liked it – but they were all a bit bashful about admitting they used it. 'Junk grape' some winemakers called it. But that didn't stop big plantings all round the Mediterranean basin, in Australia, in California and further south.

High on the skyline loom the gaunt ruins of the castle built by the popes during their sojourn at Avignon in the southern Rhône, instead of Rome. This was their new castle, their château neuf – and the vineyards spread around the castle walls are those of Châteauneuf-du-Pape. Gnarled old Grenache vines grow in a soil covered in large white pebbles, or galets roulés, that retain the heat of the southern sun long into the night. The pebbles make good homes for lizards, too.

It was the awakening of an interest in French Rhône wines during the 1990s that began to drag Grenache from out of the shadows. This Rhône fascination was led by the Syrah reds of the northern section, but attention swiftly continued south and discovered France's great warm weather classic – Châteauneuf-du-Pape. Though it doesn't say so on the label, this great, hulking gobful of pleasure is based on Grenache, and the wines can even be 100 per cent Grenache. As the modern age of instant gratification burst into life, Châteauneuf became a superstar, intoxicated with its own success.

Spain, too, had a slumbering legend that decided it was time to finish hibernating and start making a bit of a noise. Priorat, a dense, brooding red of enormous alcohol (as high as 18 per cent) made from tiny yields of primarily Garnacha grapes (as low as 5hl/ha) had been around for 800 years or so, but it took the rise of Catalan self-awareness and a bunch of ambitious young growers to revive its reputation during the 1990s and turn it into Spain's Garnacha superstar. Then things got better. Forgotten Spanish areas like Calatayud, Campo de Borja and Cariñena had large plantations of old bush vine Garnacha going to waste. New World 'flying winemakers' couldn't believe their eyes at this treasure trove when they turned up to try to resurrect the various fly-blown co-operatives that littered the regions – field upon field of the kind of grapes they would pawn their granny for back home. Nowadays, you want Grenache/Garnacha happy juice? This is where you find it.

You'd think that the warm New World would have embraced Grenache enthusiastically. Well, they did – but primarily because it has shockingly high alcohol levels and you could make a pretty decent fortified 'Port' style out of it. California even abuses it by producing an inferior version of White Zinfandel. Honestly, you guys should know better.

But it is the Aussies who have taken Grenache to their hearts. Once Shiraz got famous, they looked around and realized they had piles of Shiraz's Rhône stablemate Grenache in their vineyards – and what's more these were often enviably old vines, giving concentrated wines of great depth. Junk grape no more. And what's this? Mourvèdre? The third great Châteauneuf-du-Pape grape along with Grenache and Shiraz? Old vines of that too? We can make our own Aussie Châteauneuf. And we'll call it GSM – after the three grapes. A great big, irresistible, irrepressible party animal of a wine. From junk grape to New World classic in an uproarious, ribald cartwheel.

Garnacha Tinta/Grenache Noir: from Grape to Glass
Geography and History page 104; Viticulture and Vinification page 106; Garnacha Tinta/Grenache Noir around the World page 108; Enjoying Garnacha Tinta/Grenache Noir page 110

GEOGRAPHY AND HISTORY

Garnacha Tinta is a Mediterranean grape par excellence. It clings to those warm lands as tenaciously as a tourist from northern Europe on holiday, and that's just how it was long seen: as something undemanding, simple in its needs and somewhat raucous in its modes of expression. Like that tourist, it has only one thing on its mind: pleasure as undemanding yet warming as a few hours baking on the beach. What, sunburn too? No – but a hangover? Yes! Wherever Grenache grows, high alcohol is the objective, and those innocuous little rosés it makes in Spain, France and Italy are all far more potent than they seem – as anyone who's slumbered through the afternoon after a seemingly harmless few glasses of Provence rosé or Navarra rosado can tell you. Like a package tourist, it has not travelled far from the sun, although it changes its name regularly from

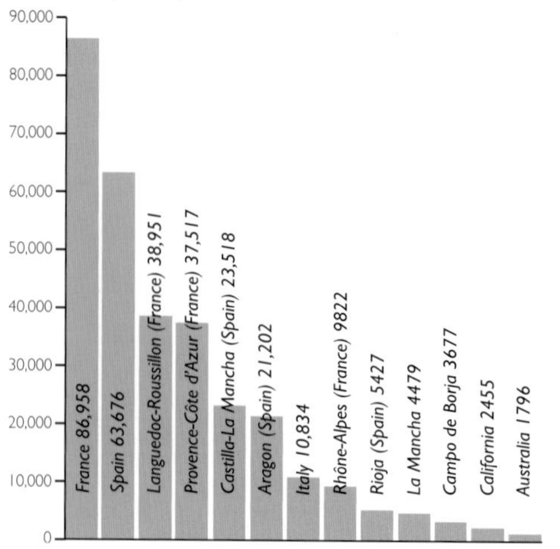

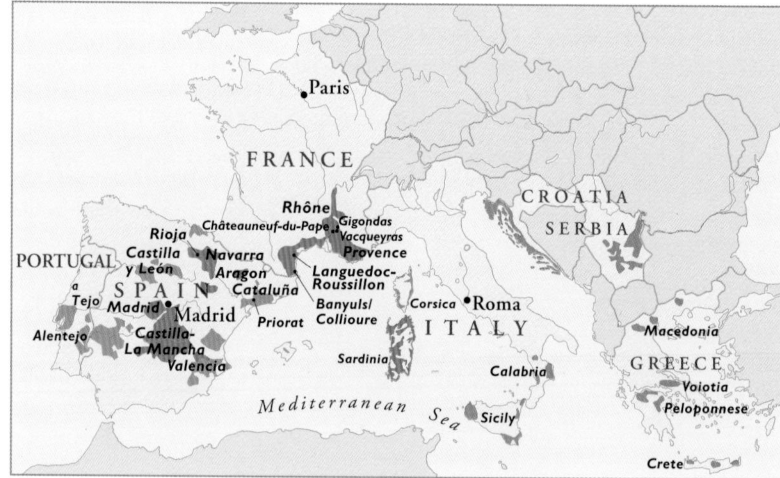

region to region: Grenache Noir in France and Cannonau in Sardinia. Even so, it has ventured right the way round the Mediterranean coast via Sardinia to Greece, Israel, Cyprus and North Africa in search of sun-drenched fields. But now that it's being taken more seriously by producers, it's showing that it can reflect its *terroir* remarkably precisely as well. Marc Parcé of La Préceptoire de Centernach in Maury calls Grenache 'the midwife of *terroir*'; it always tells the truth about its vineyard, he says. If Garnacha is rustic and raucous, it's because the producer's work in the vineyard has made it so; treat it properly and it can have great finesse.

Most of the time it's blended: with Tempranillo in Rioja, with Mazuelo (Carignan), Cabernet Sauvignon and others in Priorat, with Mourvèdre, Syrah, Carignan and others in southern France. In Australia it's the comeback kid: along with Shiraz and Mourvèdre it had dominated plantings when fortifieds ruled the market, but fallen right out of favour before steaming back with a shout of raucous joy in the late 1990s.

HISTORICAL BACKGROUND

Garnacha Tinta may be better known to English- and French-speaking wine lovers as Grenache Noir, but we give it its Spanish name here because all the evidence suggests that it is a Spanish grape that moved across the border to France and beyond. For once there seem to be no legends linking the vine with the Romans. Instead the Spanish are given full credit for cultivating the original Garnacha, probably on the east coast, in Aragon or Cataluña. One of its synonyms, indeed, is Tinto Aragonés – which should not be confused with Aragonez, a Portuguese name for Tempranillo. Aragon can, it seems, lay claim to having been the birthplace of both Spain's major red grape varieties.

From there it was but a short hop to France – and even shorter when one considers that until France annexed it in 1659, Roussillon was part of Spain and Garnacha was probably already established there in the Middle Ages, before changing its name to Grenache and marching off through the Languedoc to the Rhône Valley. It was one of the first varieties to be planted in Australia at the end of the 18th century.

Garnacha Blanca/Grenache Blanc is the white-berried form of the vine, discussed separately on page 101. There are also grey and pink versions (Grenache Gris and Grenache Rose), which are blended in the southern Rhône and the Midi with Grenache Noir to make fortified wines or *vins doux naturels* (see page 107). The downy-leaved red grape Garnacha Peluda is a variation on the same grape; as with all old grape varieties, Garnacha has undergone considerable mutation over the years.

Garnacha Tintorera, however, is not Garnacha at all. Instead, it is a synonym for Alicante, a Teinturier grape (one with red, rather than colourless pulp) which is found all over Spain and goes into many blends. Some DO regulations in Spain list the two names separately, others give only one name.

New Garnacha plantings at Clos l'Ermita in Gratallops, Priorat. This dramatically high, rugged region of Cataluña was once an inland sea. Now the vines on these mountains are planted up to 700m (3000ft)above sea level, and land prices are soaring. These are new vines, but the very best Garnacha grapes, from the oldest, centenarian vines, are in huge demand for wines that enjoy cult status with drinkers around the world.

Autumn at Gratallops. The crumbly schist soil of this area retains water remarkably well – an advantage in a region where there is very little rain – sometimes less than 400mm (16in) a year.

Torres' Priorat vineyards at Porrera are an example of the revitalization of the region which has made Priorat a powerhouse in Spanish wine. But the yields are minute – in some old vineyards as low as 5hl/ha.

VITICULTURE AND VINIFICATION

Garnacha is the most clubbable of grapes, fitting comfortably into many different blends. Indeed, it is usually happier in a blend than solo – which is not preventing more and more producers, in search of a new grape name to put on the label, from making it as a varietal.

In some ways it is a useful grape in the vineyard, particularly in dry climates, since it laughs at drought. But it is an irregular cropper, and in irrigated vineyards where drought is not part of the gameplan, Garnacha may respond by producing high yields of pale, dilute, forgettable wine – though the alcohol level may still be high.

To get serious wine from Garnacha, you must treat it seriously, and low-yielding, old bush vines on poor soil have produced all the best results so far.

Climate

Having said that Garnacha shrugs off drought, one should add that it needs a drought-resistant rootstock to do so. Otherwise its natural resistance to dry weather is somewhat undermined.

It is a late ripener and loves warmth. Even fierce, dry winds like the Mistral of southern France do not seem to worry it unduly. Some Australian producers like to plant it on hilltops, where it gets whatever harsh weather is going. It seems to produce far better wine when it is stressed, perhaps because it is naturally very vigorous.

But vigour does not necessarily go hand in hand with resistance to all disease. Garnacha suffers from *coulure*, or floral abortion, which can cut yields to unpredictable levels; and is susceptible to downy mildew, and to bunch rot, because of its tight clusters. Marginal climates increase the risk of *coulure* at one end of the growing season and of bunch rot at the other end, if the grape's late ripening pushes it into autumn rains.

Soil

Garnacha is irrevocably associated with hot, dry soils, preferably poor and well-drained. Apart from those basic requirements it is not over-fussy, though the best French Grenache often comes from schist or granite soil, or relatively high-altitude sites.

Châteauneuf-du-Pape has its famous, heat-retentive stones, its *galets roulés*, on the higher plateaux; opinions vary as to whether the *galets roulés* soils give the best wines. Schist is also important in Rioja, and especially in Priorat, where Garnacha produces some of its very best wines. Vineyards here are planted up to 700m (2300ft) above sea level.

Cultivation and yields

Gobelet or bush training, with four or more arms, and spur pruning, with two spurs per arm, seem well suited to this vigorous, upright vine, though guyot or royat training on wires is increasingly popular. Garnacha needs to be pruned hard and debudded if yields are to be kept within bounds: yields of under 35hl/ha will give very different wine even from yields of 50hl/ha (the base yield for Côtes du Rhône).

But training on wires and using irrigation sensibly can give higher quality at higher yields than bush training with no irrigation. In Priorat yields from very old vines may be as low as 5–6hl/ha while at Château Rayas in Châteauneuf-du-Pape yields average 15–20hl/ha. Charles Melton in the Barossa Valley recommends no more than 1.5 tons/acre (27hl/ha). By contrast, Garnacha/Grenache vineyards planted for high volume and low or lowish quality, like those in California's Central Valley, may give 100hl/ha, and maybe a good deal more; it is on such wines that the grape's reputation for uninteresting quality rests. Low yields give structure; at high yields the grape's tendency to low acidity is exacerbated, and colour and flavour disappear.

Anyone tasting a Grenache-based Priorat, or the Grenache-based Châteauneuf-du-Pape Château Rayas, would find it hard to believe that such powerful mouthfuls have anything to do with the lighter, lifeless jug wines of much of southern France, Spain and California.

Irrigation must therefore be treated with caution if quality is the aim. In Châteauneuf

The galets roulés *stones of Châteauneuf-du-Pape, shown here with bush-trained vines, absorb heat during the day and give it out during the night. Not surprisingly, the grapes reach super-ripeness and give wines with high levels of alcohol. But the subsoil, of red clay and ferruginous sands, is probably more vital to quality.*

irrigation is permitted, though officially limited; in practice, things being what they are, overwatering is not uncommon.

Garnacha/Grenache reaches high levels of ripeness easily in warm climates. Picking late, at about 15 per cent potential alcohol, seems to give the best balance and flavour; picking earlier for (supposedly) better acidity and elegance doesn't seem to work: you simply get green flavours and poor colour. The problem is, therefore, to get physiological ripeness without losing too much acidity: growing at higher altitudes can help, since cooler nights will help the retention of acidity. The other solution is to blend with some other grape for which acidity is less of an issue. It is not unknown for the alcohol content of Garnacha table wines to rise to 18 per cent without the aid of fortification, though a slightly more drinker-friendly 14.5 per cent is a much better mouthful.

There are some winemaking techniques that you simply can't bring up to date. Mas Amiel make a fortified wine from Grenache Noir and begin the aging process by leaving the new wine outside in the hot sunshine in 70-litre glass bonbonnes *for a year, encouraging the deliberately oxidized style called* rancio.

Clones

These vary widely in quality. Some clones are highly productive, others less so; some have better colour than others, or produce more or less irregularly. Galet (1998) lists clone 362 as being particularly good for *vin doux naturel* because of the high degree of ripeness it attains.

At the winery

Garnacha/Grenache must be handled gently: it oxidizes with extreme ease, and loses colour if care is not taken. The wine can have a tendency to green, herbaceous flavours, which can be made worse by the inclusion of too many stems in the fermentation vat. Over-harsh pressing or over-hot fermentation, both designed to extract more tannin than is natural to the wine, tend to give astringency. A long, slow fermentation, followed by a long maceration to extract tannin, is best, followed by as little racking as possible, to prevent oxidation.

In the southern Rhône and in Spain, old oak barrels are usual for aging. New oak, inevitably, is creeping in, and can help to prevent oxidation and fix the colour. Some think the flavour of new oak an aberration in Garnacha/Grenache; others welcome it. It is, perhaps, a matter of taste, but I don't think more than a small percentage of new oak adds anything to a good Grenache's unmistakably fruity style. If anything, it masks the fruit and strips it of individuality.

FORTIFIED WINES

Cast your mind back, if you will, to 1299. In that year the king of Mallorca granted a patent to a Catalan alchemist called Arnaldus de Villanova; the patent was for the process, which he had developed but not invented, of stopping the fermentation of grape juice by the addition of grape spirit. (One can imagine the conversation: 'Well, that's great, Arnaldus, but, like, what's it for?')

The king of Mallorca's realm included Roussillon, and Roussillon went on to become a centre for the production of what is now called *vin doux naturel* (VDN). It still is; and there is a long tradition both in Roussillon and in neighbouring Cataluña (the two regions were both under Spanish rule until 1659) of using Garnacha/Grenache for these sweet concoctions of partly fermented grape juice and spirit. These wines may be made in a *rancio* style by leaving them outside in glass demi-johns (or *bonbonnes*) or wooden barrels exposed to the air and to hot daytime temperatures for several years, until they acquire a maderized, sour *rancio* tang of nuts and raisins and cheese.

In Roussillon, Grenache-based VDNs are made in Maury and Banyuls. Banyuls is generally higher in quality than Maury, and, in addition, there is a rarely seen Banyuls Grand Cru appellation, for which 75 per cent of the blend must be Grenache Noir, compared to 50 per cent for normal Banyuls. Styles in Banyuls vary enormously: some wines are fruity, some dark and concentrated, some *rancio*. Some may be 20 or 30 years old.

In the southern Rhône, the village of Rasteau specializes in a *vin doux naturel* made from Grenache. Cataluña, in particular Tarragona, also makes sweet fortified Garnacha, allowing the juice to ferment for three days, then adding grape spirit to bring the strength up to 15–16 per cent alcohol. In Sardinia, too, where Garnacha is known as Cannonau, some fine fortifieds are made. This practice spread to the New World, in particular to Australia, where Grenache, sometimes blended with Shiraz or Mourvèdre, provided the backbone for the country's 'port' industry and produced many super examples.

Garnacha may be blended for these purposes with other grapes, often Cariñena and Tempranillo in Spain, and Syrah, Cinsaut, Carignan, Grenache Gris or Grenache Blanc in France.

GARNACHA TINTA/GRENACHE NOIR AROUND THE WORLD

Garnacha can be great, simple and charming, or dull and uninteresting. Only a very few unblended ones are great – but they should be enough to inspire the others. But are varietal examples necessarily in Garnacha's best interests? Not unless they're cleverly grown and made. The last thing the world needs is more overcropped, anonymous red.

Priorat

This ancient vineyard region in Cataluña leapt suddenly and dramatically to the forefront of Spanish red wines (after languishing in obscurity for centuries), making dark, heady wines that sell at crazy prices. Yields are extremely low and the wines very concentrated, though it is noticeable that in recent years they have becomes less oversized and more balanced.

The old way here, and it still just about exists, is to make powerfully alcoholic wines that are almost black in colour. The new way is to make wines which have huge blackberry fruit in youth, with perhaps a bit of Cabernet or Merlot added for aroma; these wines are drinkable pretty much on release.

Of the nine villages in the DO, Gratallops is the leader of the new wave. The pioneers of the new style arrived here in 1986, planted Syrah, Cabernet Sauvignon and Merlot alongside the existing Garnacha (which covers about 40 per cent of the Priorat vineyard) and Cariñena vines, installed drip irrigation and modern winery equipment, and started to make wines with more fruit and more new oak than was traditional. With time all Priorat seems to develop a flavour of tarry figs; whether you think it should be drunk young or left to age is very much a matter of taste.

Vines are planted at between 100 and 700m (330–2300ft) up, on terraces if the angle of the slope defies gravity too much. In Gratallops the soil is schist, but the typical Priorat soil is Llicorella, a rock composed of stripes of slate and quartzite that glitters black and gold in the sun.

Rioja

Curiously, Garnacha is less well regarded in Rioja than in many other parts of Spain, and it is the only red variety in decline in the region, having been planted heavily for volume in the past. Most of Rioja's 6700 ha (16,560 acres) of Garnacha are in Rioja Baja, where the greater heat and lower rainfall suits it. Soils are mostly sandy here, and many bodegas like to add a proportion of Garnacha – 15 to 20 per cent is common – to their Tempranillo for the body and alcohol it brings to the blend. However, it tends to oxidize quickly and yields must be kept low if it is to age well.

Navarra

Rosado used to be what Spaniards drank when they wanted something that wasn't red and/or oaky; real men didn't drink white wine, apparently. And Garnacha makes excellent pink wine: juicy, soft and only good young. Navarra makes a speciality of Garnacha rosado, particularly from the sandy-soiled, dry south of the region. Over half of Navarra's vineyards are planted with Garnacha, though the proportion is falling as more red and less rosado is produced, and the growers are encouraged to replant with Tempranillo. The official aim is to have 35 per cent Garnacha and 31 per cent Tempranillo.

Rest of Spain

Garnacha's stronghold is in the north and east of the country. In Calatayud it covers about 65 per cent of the vineyard, much of which makes rosado; in Campo de Borja it accounts for 75 per cent. It is more than half the vineyard in Cariñena and takes up a good chunk of most other DOs in this part of Spain.

Versatility is its key: it can turn out rosado and wonderfully juicy *joven*, or young reds if you ask it to, but if you put it into a barrel and blend in some backbone with some other grape, then you have a sturdy, oaky red. Quality is often best in the easy-drinking styles,

Southern Rhône and the Midi

The secret of good Châteauneuf-du-Pape lies partly in Grenache, which forms the bulk of the blend, but partly also in Mourvèdre, which adds tannin and earthy, savoury flavours to the wine, and Syrah, which brings structure and fantastic perfume. It is often said, particularly by growers of Châteauneuf and Gigondas, that the preponderance of Grenache in their vineyards is the legacy of domination by the Burgundian merchant houses, with their ceaseless demand (in the past, of course) for wine of high alcohol to beef up their

Mas Martinet
Cabernet Sauvignon, Merlot and Syrah are blended with Garnacha in this top-notch wine from Priorat. It needs at least five years' aging.

Gran Feudo
Navarra makes a speciality of Garnacha-based rosado wines. This example is strawberryish and fresh.

Domaine du Pégau
Paul Feraud and his winemaker daughter Laurence produce spicy, earthy, chocolatey and cherry-fruited Châteauneufs from a Grenache-based blend.

own pallid brews. AC yields here, at 35hl/ha, are low, and minimum alcohol, at 12.5 per cent with no chaptalization, is high. With Grenache, achieving the second is easy if the first is adhered to. But quality is mixed, and overproduction not uncommon. The most usual fault is high alcohol without the backbone to support it.

There is also some carbonic maceration done alongside more traditional vinification, which produces wines lighter and fruitier than the rich, spicy Châteauneuf of most drinkers' imagination; as a part of the blend, these fruitier wines can be very attractive.

The soil of Châteauneuf is famously stony in parts: the big round *galets roulés,* or pudding stones, do not cover the entire vineyard area, and they are not essential to good Châteauneuf. Certainly, they retain heat during the day and give it out at night, but lack of heat is not really an issue in this part of the southern Rhône.

Grenache is also the foundation of Gigondas, where it can form up to 80 per cent of the vineyard, and Vacqueyras. Officially, the maximum yield in both is 35hl/ha, and wines with a fair amount of Syrah and Mourvèdre in the blend will have better structure and more substance, but if Grenache gets up a head of steam out in those torrid, rocky vineyards, it can make as burly, as chewy, as intoxicating a red as any in the Rhône Valley. Grenache is traditionally aged in large, old wooden barrels. In Lirac Grenache must take up at least 40 per cent of the vineyard, and makes reds and rosés; in Tavel it makes only rosés. Grenache is also the staple grape of Côtes du Rhône (apart from those sections of the appellation in the Syrah-only North) and Côtes du Rhône-Villages, and plays varying parts in the wines of Provence, Languedoc, Minervois, Corbières, Fitou and Roussillon, but it is here that we have seen the biggest changes in how Grenache is regarded. Many growers here are in love with Grenache, and it shows.

The most famous old vines in Australia's Barossa Valley are Shiraz, but there were fellow-travellers from France back in the 1840s – Grenache and Mourvèdre, the two main grapes in the southern Rhône Valley. These were initially disregarded as 'junk grapes', but old examples of these two are now highly prized. Langmeil has some of the Barossa's oldest vines and these beauties date from 1848 and are probably the oldest in the world.

Australia

Time was if you saw a vine in Australia, particularly in South Australia, it was likely to be Grenache. But that was before the fashion for Cabernet and Chardonnay. The vine-pull scheme of the 1970s ensured that many bush-trained, low-yielding vineyards of old Grenache were uprooted along with Shiraz vines of equal quality; but luckily some survived. They are now much in demand. Plantings are rising slowly but steadily again, and while blends of Grenache, Shiraz and Mourvèdre (known as GSM) are classic, there are also an increasing number of varietals. Some of these are excellent, though some lack backbone – the usual Grenache problem. Barossa Valley versions are intense, even jammy; McLaren Vale Grenache is spicy and lusciously rich and Clare Valley herby and memorable.

USA

In California Grenache is used for both cheap jug wines, often pink, and high-quality old-vine wines. The vogue for Rhône varieties has helped it, and producers like Alban and Bonny Doon, show what can be done if the grape is planted in the right place. Washington's tiny amount is good and juicy.

Rest of the world

In Sardinia, as Cannonau, it produces some earthy reds, often of very interesting quality; there is also a sweet, porty, high alcohol version under the name of Anghelu Ruju from Sella & Mosca. There is a little in both Greece and South Africa, and it may yet follow Syrah into Chile. In North Africa, in common with every other planted variety, it awaits a revival of interest in winemaking.

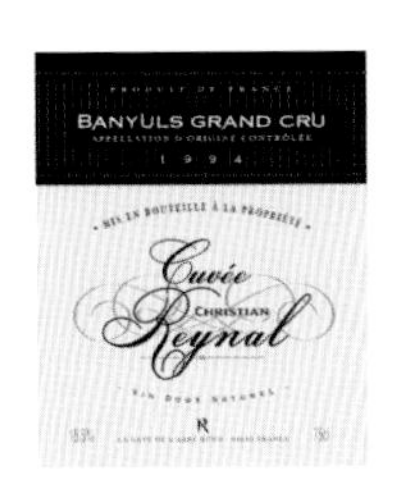

Terres des Templiers
Astonishing sweet fortified Grenache that was aged for one year in glass jars out of doors, then for 38 years in 350hl oak casks before bottling.

Charles Melton
A deep, dark rosé with a touch of tannin which should be drunk young. Charles Melton is one of the leaders of the Barossa's Grenache revival.

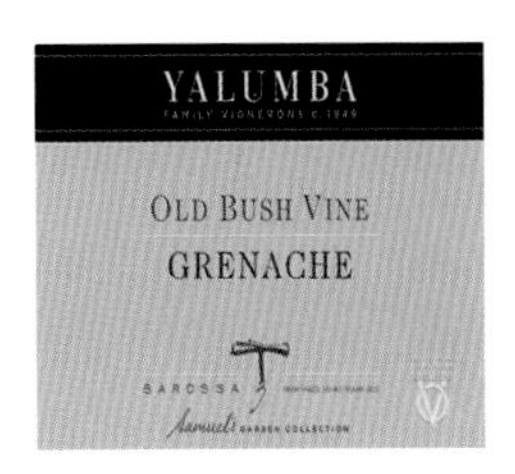

Yalumba
Yalumba are great preservers of Barossa Valley traditions. This Grenache comes from vines between 35 and 80 years old.

ENJOYING GARNACHA TINTA/GRENACHE NOIR

It is too easy to answer the question of how Garnacha/Grenache ages with the simple answer that it doesn't. But some Garnacha/Grenache ages brilliantly, though it has to be admitted that the great majority is best drunk young. Most Garnacha/Grenache wines oxidize fast, and so ought to be drunk in the first ruddy bloom of youth.

The exceptions are the best wines of Châteauneuf-du-Pape and of Priorat, and good examples of these can both happily age for a decade or much more. Of course, loads of Riojas age well, but these are usually mostly made from Tempranillo: in general Garnacha/Grenache ages much better if it has another grape to wrap its fat juiciness around.

A good Châteauneuf, made from low-yielding vines so that the Grenache has structure and concentration, can last from five or six years up to easily 20: nearly as long as a good Côte-Rôtie. A low-yield Châteauneuf superstar like Château Rayas can last a good deal longer than that, but commercial blends of Châteauneuf and Gigondas won't improve much beyond five or six years.

Many modern Priorats, despite their powerful flavours, are made to be drinkable early: they don't, on the whole, have to be tucked away for very long before you can open them. Nor will they last all that long: 10 years or so is usually the limit. Traditional massive, black Priorats will last much longer, even up to 25 years.

The *vins doux naturels* of Roussillon, since they are effectively mummified by the oxidation process they undergo in barrels and demi-johns, will happily last 20 or 30 years, and don't seem to change very much over a few more decades.

Australian 'ports', where there is also some Shiraz blended in with the Grenache, can age for decades, but they are drinkable sooner than their Portuguese counterparts, usually being a little sweeter and less fierce in their youth, and they achieve a tawny maturity much sooner.

The taste of Garnacha

Good Garnacha/Grenache has wild, unexpected flavours: roasted nuts, leather, blackcurrants, honey, gingerbread, black cherries, pepper, coffee, spices, even tar and black olives. As yields are progressively increased these fade into a gentle, soft leathery earthiness; the shock is that they fade rather early. That leatheriness is a telltale sign that the fruit won't be there for ever. Young, vigorous Grenache tastes of herbs and strawberries and raspberries, as well, with often a certain dustiness to the fruit that speaks of those arid vineyards in Spain and southern France. Too fanciful a connection? Probably. Perhaps not.

And Priorat: I mentioned earlier that it seems to attain a tarry, figgy character with age. The blackberry fruit fades, and you're left with something not unlike a Spanish version of Italian *amarone*. It must be something in the terroir as well as the grape, because you find it in Priorat of both old and new styles.

Rancio Garnacha/Grenache is leathery, certainly, but also nutty like tawny port and sour like cheese. Young rosado/rosé wines, by contrast, are all strawberries and cream.

Grenache was very much the forgotten grape of South Australia, dismissed as a junk grape only useful for making flagons of port. However, D'Arenberg realized that some of the McLaren Vale's oldest and most precious wines were Grenache and have been buying them up to preserve them and their fruit to make warm-hearted powerhouses like the Custodian. Finca Dofí is one of the great dark reds made by Alvaro Palacios on steep, infertile crumbly schist soils that yield almost no crop, yet the tiny harvest gives some of Spain's most perfumed wines.

MATCHING GARNACHA/ GRENACHE AND FOOD

This grape comes in so many styles that it's quite difficult to think of any meat dish it clashes with. Soft, low tannin versions can be good with spicy Indian food, where tannin doesn't work. Barbecued food also matches these well, as do stuffed peppers and aubergines. Bigger, more tannic versions will go with roast beef and lamb, pheasant and duck and with the most flavoursome casserole in your repertoire.

Chilled young Garnacha/Grenache can be a good summer red, while pink versions are great with vegetarian dishes and strongly flavoured fish such as grilled sardines.

CONSUMER INFORMATION

Synonyms & local names

Also known in Spain as Aragón, Aragonés, Garnacha Tinta, Garnacho Tinto, Garnatxa, Lladoner and Tinto Aragonés; as Alicante in France, as well as Grenache Noir; as Granaccia or Granacha in Italy and Cannonau in Sardinia.

Best producers

FRANCE/Southern Rhône Achiary, D & D Alary, Amouriers, Beaurenard, Henri Bonneau (Réserve des Celestins), Bosquet des Papes, la Bouissière, Bressy-Masson, Brusset, Cabasse, Cabotte, les Cailloux (Centenaire), Cassan, Cayron, Chapoton, D Charavin, la Charbonnière, L Charvin, Chaume-Arnaud, Clos du Caillou, Clos des Cazaux, Clos du Joncuas, Clos du Mont Olivet, Clos des Papes, Jean-Luc Colombo, Combes, Coriançon, Couroulu, Cros de la Mûre, Delas, Espiers, Espigouette, Font de Michelle, Font de Papier, Font-Sane, Fortia, la Fourmone, Galet des Papes, la Gardine, la Garrigue, les Goubert, Gour de Chaulé, Gramenon, Grand Moulas, Grand Tinel, Grand Veneur, Guigal, Jaboulet, la Janasse, Jérôme, Longue-Toque, Marcoux, Monadière, Mont-Redon, Montirius, Montmirail, Montpertuis, Montvac, la Mordorée, Moulin de la Gardette, Mourchon, Nalys, la Nerthe, l'Oratoire St-Martin, Palleroudias, Pallières, Pégaü, Pélaquié, Père Caboche, Roger Perrin, Piaugier, Rabasse-Charavin, Raspail-Ay, Rayas, Redortier, la Réméjeanne, Richaud, la Roquette, Roger Sabon, St-Cosme, St-Gayan, Ste-Anne, Sang des Cailloux, Santa Duc, Solitude, la Soumade, Tardieu-Laurent, la Tourade, Tourelles, Tours, Trapadis, Trignon, J-P Usseglio, P Usseglio, Raymond Usseglio, Verquière, la Vieille Julienne, Vieux Donjon, Vieux Télégraphe, de Villeneuve; **vins doux naturels** Casa Blanca, Cazes, Celliers des Templiers, Chênes, Clos des Paulilles, l'Étoile, Jau, Mas Amiel, du Mas Blanc, les Vignerons de Maury, la Rectorie, Sarda-Malet, la Soumade, la Tour Vieille, Vial-Magnères.
SPAIN/Cataluña/Priorat Clos Erasmus, Clos Mogador, Costers del Siurana, Ferrer Bobet, Fuentes, Mas Alta, Mas Doix, Mas Martinet, Alvaro Palacios, Rotllan Torra, Scala Dei, Terroir al Limít, Torres, Vall Llach; **other producers** Borsao, Celler de Capçanes, Martinez Bujanda.
ITALY/Sardinia/CannonaudiSardegna Argiolas, Sella & Mosca.
USA/California Alban, Bonny Doon, Hope Family, Jade Mountain, Sine Qua Non; **Washington State** McCrea Cellars.
AUSTRALIA Tim Adams, Charles Cimicky, Clarendon Hills, Coriole, D'Arenberg, Hamilton, Hardys, Henschke, Kilikanoon, Peter Lehmann, Charles Melton, Mitchelton, Penfolds, Rockford, Rosemount, Seppelt, Tatachilla, Turkey Flat, Veritas, Yalumba.

RECOMMENDED WINES TO TRY

Châteauneuf-du Pape wines

See Best Producers, left.

Ten other southern French reds

Dom. de Cayron Gigondas
Dom. le Clos des Cazaux Vacqueyras Cuvée des Templiers
Dom. les Goubert Gigondas
Dom. Gramenon Côtes du Rhône Ceps Centenaires
Dom. Lafon-Roc-Épine Lirac
Dom. de l'Oratoire St-Martin Côtes du Rhône-Villages Cuvée Prestige
Dom. de la Rectorie Collioure la Coume Pascal
Dom. St-Gayan Gigondas
Dom. Santa Duc Gigondas
Dom. la Tour Vieille Collioure

Ten other European reds

Argiolas Turriga (Italy)
Bodegas Borsao Campo de Borja (Spain)
Celler de Capçanes Tarragona Costers del Gravet (Spain)
Clos Mogador Priorat (Spain)
Costers del Siurana Priorat Clos de l'Obac (Spain)
Martínez Bujanda Rioja Garnacha Reserva (Spain)
Mas Martinet Priorat Clos Martinet (Spain)
Alvaro Palacios Priorat l'Ermita (Spain)
Rotllan Torra Priorat Amadis (Spain)
Sella & Mosca Anghelu Ruju (Italy)

Ten New World Grenache reds

Tim Adams The Fergus (Australia)
Charles Cimicky Grenache (Australia)
D'Arenberg The Custodian (Australia)
Langmeil The Fifth Wave (Australia)
McCrea Tierra del Sol (Washington State)
Charles Melton Grenache (Australia)
Rockford Dry Country Grenache (Australia)
Sine Qua Non Red Handed (California)
Turkey Flat Grenache (Australia)
Yalumba Old Bush Vine (Australia)

The colour looks wonderfully dark – but get Grenache into the winery and that colour can disappear with disconcerting ease. It's not the simplest of grapes to handle, and needs to be treated seriously if it is to produce serious wine.

Maturity charts

Grenache is increasingly beloved of wine producers who want to make rich red wines for early drinking but will keep too.

2010 Priorat (new style)

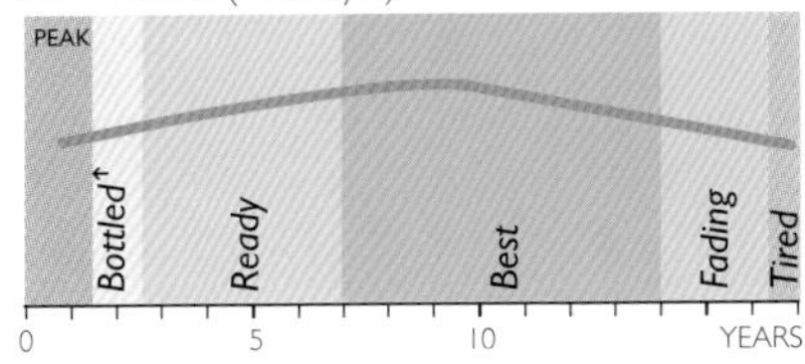

Incredible when first released, the wine has huge blackberry fruit which goes tarry two to three years later, and it develops an almost amarone style.

2010 Châteauneuf-du-Pape

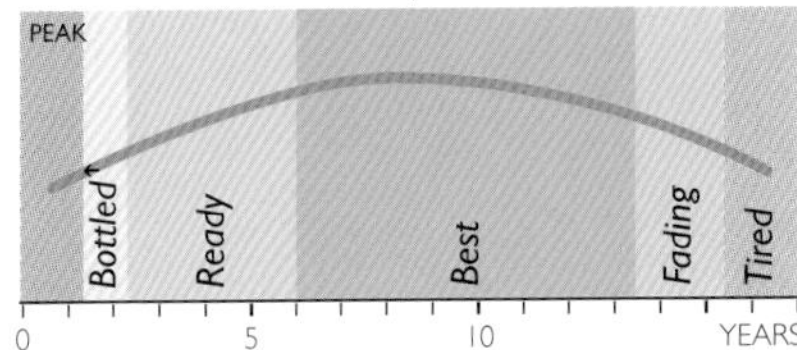

2010 was a very good vintage in Châteauneuf, with wines showing rich, heady depth but surprisingly fresh acidity. Delicious young, they will age impressively.

Vin doux naturel Banyuls Grand Cru

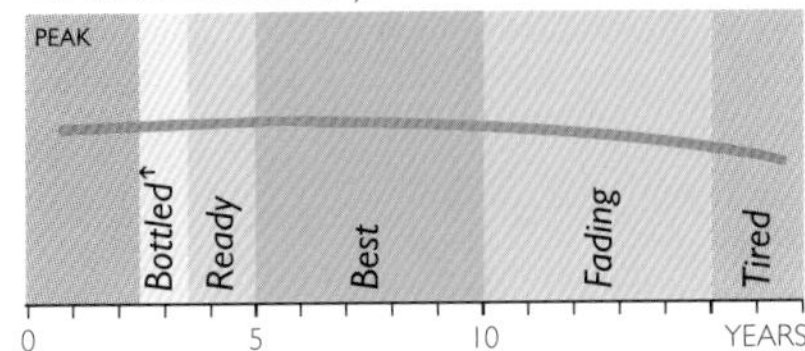

Vin doux naturel wines are bottled at various ages – 10, 20 and even 30 years old. They require no further aging.

Termeno
E R

GEWÜRZTRAMINER

It must be tough being a Gewürztraminer sometimes. All your life, all you've ever wanted to do is to please people. And not just the odd person. You want to please everybody, and you're willing to use every weapon in your armoury to win them over. So you doll yourself up to the nines. You preen in front of the mirror and spread out your vast panoply of make-up, especially mascara and rouge. And you ladle them all on to emphasize every single feature of your face. Might some features perhaps have been better left unadorned, possibly even muted a bit? Do you think your cheeks should be quite so pink, your lips quite so pouting and bright? Well, well, maybe not, but once you get started you sort of can't leave anything alone. Can you?

And scent. Aah, scent. What should it be? A dab of something mineral and restrained behind the ears? But who would smell that? Something high-toned and floral with citrus notes giving way to autumnal orchard ripeness? Mmm, yes, yes, but you want more. You want the sultry tones of passion and seduction giving way to the earthy notes of exhaustion and sleepy satisfaction. You want to guarantee that nostrils will quiver from the first moment you sweep through the door, you want clouds of Yves Saint Laurent's Opium, Calvin Klein's Obsession and Giorgio of Beverley Hills to billow out before you, announcing the arrival of the one grape that no one can resist. Yourself!

Ah, if only. Poor old Gewürz. Not all wine lovers can take it. For some, Gewürztraminer is a parody of perfume and powder that sashays around them. For them anything as scented, as seductive, as voluptuous as Gewürztraminer offers far too thrilling an experience to possibly be any good.

The Gothic-style, carved wooden spice cabinet, stork, and buttery, knot-shaped sweet pretzels suggest Alsace where the grape achieves its highest fame. The pomander, cloves, nutmeg and spice mortar are references to the word Gewürz, which means spice in German – but Gewürztraminer is a complex story. The wines are far more than just being fat and spicy, famous only for their heady perfume.

It is strange, isn't it? You get a grape that combines a most irresistible scent of lychees and tea rose petals with the lushness of tropical fruit, the bite of black pepper, and the intimate dressing room aroma of Nivea Crème, and instead of getting excited by the sheer sensuousness of the experience, they call it overblown, blowzy, boorish, clumsy and so on.

It's true that Gewürztraminer does have its problems for the winemaker. Do you go all out for perfume and lushness like Alsace? Or will that put people off? If you make the grape exercise a little restraint, will it then come over as half-hearted? And is there any point in being half-hearted? So far the world has yet to find a more convincing model of Gewürztraminer than the Alsace one and a sort of semi-scented middle road is proving surprisingly elusive. Some simply try to be too polite, as though they admit their scent is a little common. Well, grow something else, then, is my general view. But Gewürz can be delicate and scented. Italy's Alto Adige, Chile, California, British Columbia, Tasmania and New Zealand show it's possible, but not easy. Lurking on the edge of all Gewürztraminer is an oily fat beast with a strange, disappointingly hard edge to it, and fruit resembling leftover marmalade on the edge of the breakfast plate. And critics leap on this aberration as being the real Gewürz. But it isn't. And hang on. Bad Chardonnay is tasteless, lifeless stuff. Unripe Cabernet is stalky and tart. Poorly made Riesling is sulphurous and stale. Yet we don't judge these varieties on their bad examples but on their good ones.

We should do the same with Gewürztraminer: pack away our puritanism, rustle up some self-indulgence and enjoy it for what it is.

Gewürztraminer: from Grape to Glass

Geography and History page 114; Viticulture and Vinification page 116; Gewürztraminer around the World page 118; Enjoying Gewürztraminer page 120

GEOGRAPHY AND HISTORY

If you want your wine all polite and well-mannered, you shouldn't really be drinking Gewürztraminer at all. The magic of Gewürztraminer only really shines out when its sumptuous, exotic perfumes make your head spin and your thoughts go giddy with desire.

But for those of you who find such self-indulgence a bit embarrassing, there are some milder forms of Gewürztraminer available from wine regions all across the globe, as the map shows. But despite being spread widely, it is also spread thinly. In no region is Gewürz dominant, not even in Alsace, where it has a mere 18.6 per cent of vineyard plantings; yet this small strip of hillside in northeast France produces nearly all the great examples of the grape – luscious, weighty and laden with scent.

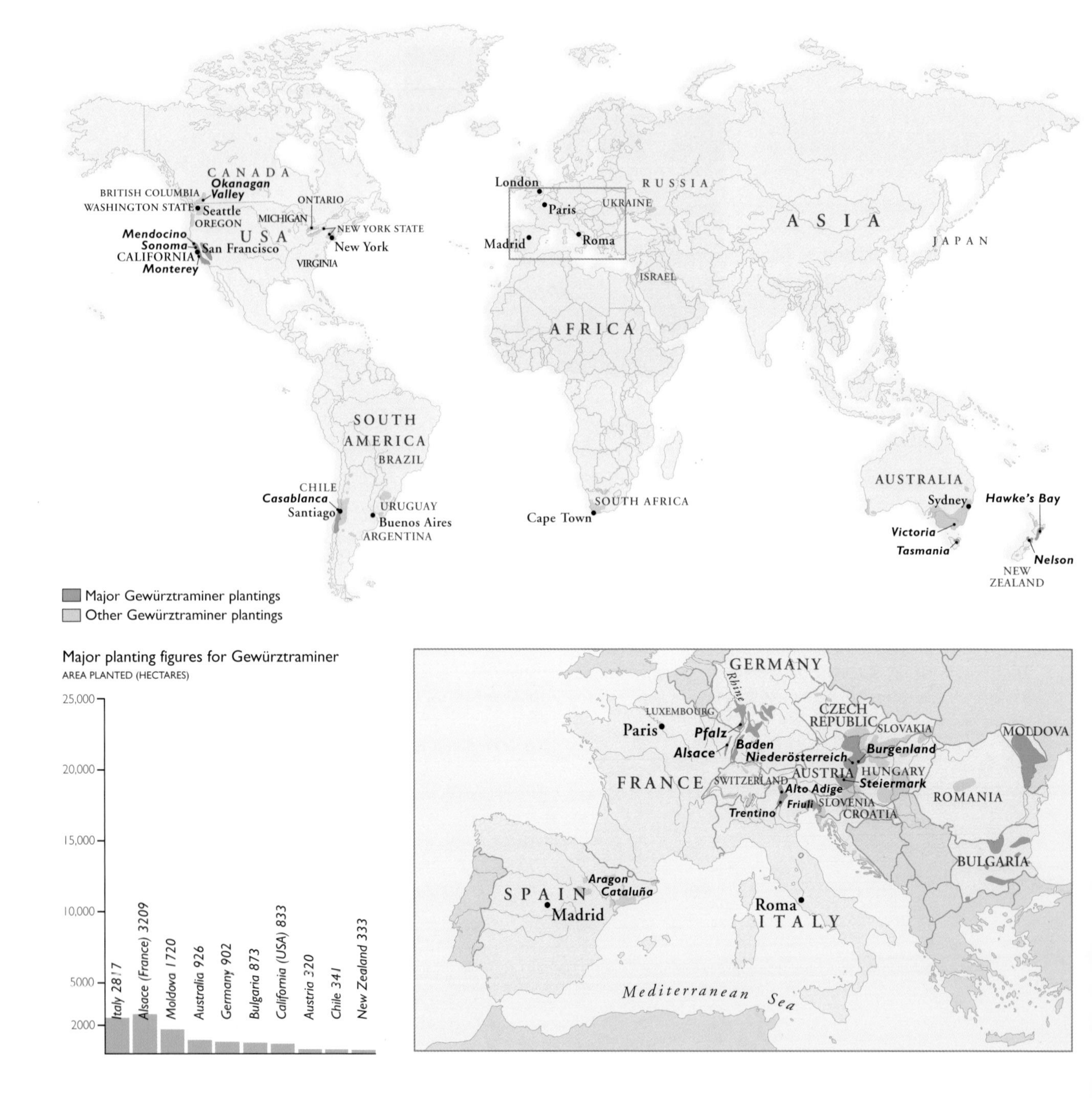

That very weight, that uncompromising, flavour-driven personality, is both Gewürztraminer's calling card and its big disadvantage. Tone it down and you get a wine hardly worthy of the name of Gewürztraminer. Play it up and you get a wine that outside its devoted fan club is difficult to sell. It's easier to sell high-priced Chardonnay than Gewürztraminer of equal quality, even at a much lower price.

So world plantings are not rising. It pops up all over eastern Europe, but results are mixed, to put it mildly. In the Alto Adige in northeastern Italy, its wines are lighter and more acidic, and much less exuberant and flavour-laden than in Alsace, Germany's can be a little heavy, New Zealand and the USA make some very good versions, and the first few efforts from Chile are promising.

But nobody makes good Gewürztraminer by accident. Those who choose to grow it do so because they love its richness and are prepared to accept low yields and relatively low returns. It's unlikely ever to cover the map: it can add fabulous scent to a dull white, but it is not an all-purpose wine. But with such an in-your-face personality, why would it ever want to be?

HISTORICAL BACKGROUND

DNA analysis has completely changed our understanding of Gewürztraminer. It is now known to be genetically identical to Savagnin. The flavours of Gewürztraminer, which comes in Blanc and Rose versions, are very different, but this is consistent with Savagnin being one of the oldest grape varieties we have: mutation goes with the territory. Since Savagnin seems to predate its Gewürztraminer form we would, if we were being correct, list this grape under Savagnin. But we're not – we're keeping it under Gewürztraminer because that's the grape that everyone knows best. Just so long as everyone remember that actually, it's Savagnin.

How did Savagnin begin, and when was its Gewürztraminer form first noticed? Savagnin seems to have arisen in the area between northeast France and southwest Germany, most likely as the result of an alliance between Pinot and some other vine. Because it was a chance crossing, and because references to grape varieties in early records are imprecise, it's impossible to be certain of very much, but Savagnin Rose seems to be a mutation of Savagnin Blanc, and Gewürztraminer seems to be a mutation of Savagnin Rose. Savagnin Blanc is also known as Traminer, the name commonly used for the much less aromatic form of the grape; Savagnin Rose, which has pink berries, is likewise not very aromatic. Gewürztraminer, which also has pink berries, is the super-aromatic form, and first appears in the records in Germany in 1827. From there it would have moved easily to Alsace, its current headquarters.

The Grand Cru Zinnkoepflé high above the village of Soultzmatt in the Haut-Rhin, Alsace. Gewurztraminer occupies less than one-fifth of the total Alsace vineyard area, and deep, rich, marly soils with some chalk, like the soil of Zinnkoepflé, suit it well. It is particularly happy if planted towards the lower part of a slope, where the soil is richest and deepest, but it also needs good drainage, so it won't thrive on the valley floor.

Ripe Gewurztraminer in the Grand Cru Kitterlé at Guebwiller in Alsace. It flowers early and ripens in mid-season, before Riesling; late-picked wines have concentrated sweetness and sensational perfume.

The fortified village of Bergheim, with its brightly painted, medieval, timber-framed houses, is famous for its magnificent Gewürztraminer, especially from its two Grands Crus, Altenberg and Kanzlerberg.

VITICULTURE AND VINIFICATION

Gewürztraminer is not the easiest of grapes to grow or handle. Its perfume is its *raison d'être* – all roses, face cream and lychees – but it must have some structure to back it up. Without structure Gewürztraminer speedily tumbles into cloying blowziness. And low acidity is its besetting fault.

It needs a long ripening season, and loves sunny and dry weather, but too much heat exacerbates the low acidity problem. All that perfume does not suit the flavour of new oak, but it can make glorious sweet wines, even though the action of *Botrytis cinerea* tends to destroy its varietal perfume.

Climate

If Alsace is the template for Gewurztraminer, then it needs mild springs, sunny summers and low rainfall. The last is the work of the Vosges mountains, which protect the vineyards from westerly rain-bearing winds: Colmar rates as one of the driest spots in all France. The sunshine continues well into the autumn, and there can be snow on the highest peaks of the Vosges while *vendange tardive* (late-harvest) grapes are picked lower down. Gewurztraminer buds early, which makes it susceptible to late frosts, and ripens in mid-season; too much heat will bring it on too quickly, resulting in a loss of acidity and aroma. But that's the beauty of Alsace: for all its sunshine, it still has a relatively cool climate – and loads of sunshine and a cool climate equals loads of sugar in the grapes but appetizing acidity and aroma.

Much of New Zealand also has a cool climate with loads of sun – although rainfall is more erratic here and there can be unwanted autumn deluges. During its first rebirth as a wine nation in the 1960s, visiting consultants from Germany recommended Germanic grapes as the route to success in New Zealand and that is why a fair amount of Gewürztraminer was planted. Indeed, New Zealand's first world-class wines may have been the early Matawhero Gewürztraminers from the warm, sunny though rainy, Gisborne region. Not surprisingly with the current advance of Sauvignon Blanc, Chardonnay, Pinot Noir and Merlot there's still not a huge amount of Gewürz in New Zealand – the planted area was a mere 332ha (820 acres) in 2013 – but Gisborne can still excel.

Germany's more northerly vineyards can be too cool; Gewürztraminer is a more natural choice for the warmer lands of the Pfalz and Baden. On the other hand, the high mountain vineyards of Italy's Alto Adige – again loads of sun but cool – are ideal; however, few growers limit the yields enough to make more than mildly aromatic examples. Most New World countries have a little Gewürztraminer, but the conditions are almost always too hot to avoid flabbiness. In Australia, Tasmania and cool spots in Victoria have produced delicious examples and cool spots in Chile have just started to shine.

Soil

Gewürztraminer is fairly unfussy about soil, though it may be that the richness of Alsace versions comes from the soil above all else. There is certainly a spiciness that pervades all Alsace's white varieties. Alsace producer Olivier Zind-Humbrecht MW, who has made a speciality of *terroir*, states that it is the high percentage of calcareous or limestone soil in Alsace that gives the region's wines their spicy character. Gewurztraminer also seems to like a proportion of clay, and is best suited to rich, deep but well-drained soils, with plenty of minerals. For every Alsace grower who stresses the importance of limestone, there is another who stresses the clay: Albert Mann, for example, believes that the reason that

Gray Monk vineyard and winery above Lake Okanagan in British Columbia, Canada. Gray Monk has just over half an acre (0.2ha) devoted to Gewürztraminer. The harshness of the Canadian winter is mitigated by the proximity of the lake and the angle of the hillside, which helps cold air drain from the vineyards.

acidity in Alsace Gewurztraminer can be low is because so many of the vines are planted on clayey soils.

Some of the best Grand Cru vineyards for Gewurztraminer are Altenberg and Kanzlerberg (in Bergheim), Eichberg (Eguisheim), Hengst (Wintzenheim), Kessler and Kitterlé (Guebwiller), Kirchberg (Barr), Mambourg (Sigolsheim), Schlossberg (Kientzheim), and Zinnkoepflé (Soultzmatt and Westhalten). Zind-Humbrecht adds that the soil type can affect the skin colour of Gewurztraminer as well as the taste: in his experience limestone soils give more orange skins with visible darker orange lines; acidic or gravelly soils give more purple colours.

Cultivation

Serious Gewürztraminer producers agree that there is a direct correlation between low yield and intense perfume. Many would put the crisis point for loss of quality at no higher than 40hl/ha. Alsace Grand Cru yields, now reduced to 55hl/ha from 60hl/ha, are still too high for good quality, and top growers take much less. Also, 55hl/ha is only the base yield (*rendement de base*) for Alsace Grand Cru: there is now a 'buffer' yield set at 66hl/ha, which has been introduced as a sop to the producers because an even bigger dodge has been removed. This was the system by which the growers could produce higher yields of one grape variety to compensate for low yields in another, providing the total did not exceed 60hl/ha.

Freshly ploughed soil ready for replanting in the Grand Cru Altenberg de Bergheim in Alsace. The mix of chalk and clay here is ideal for Gewurztraminer.

Many of the rootstocks in Alsace encourage vigour, while in Germany high-yielding clones usually compromise quality. Excess vigour increases the risk of stem rot, to which the vine is very susceptible.

In New Zealand high-quality Gewürztraminer cannot be cropped over 2–3 tons per acre. One grower points out that if you are a commercial grower, you are probably going to be able to sell your Gewürztraminer grapes for NZ$1000–1200 per ton. It is obviously far more profitable to get paid $2000 per ton for similar yields of the Mendoza clone of Chardonnay – or even $1400 per ton for Sauvignon Blanc that you have cropped at 6 tons plus per acre. The tendency is therefore to try and crop Gewürztraminer at higher yields, to get a higher return. But then quality suffers, and you get into a vicious circle of overcropped, faintly perfumed wine in a market that does not know much about Gewürztraminer anyway. This, basically, is why plantings aren't growing in New Zealand.

Clones

There is not a great deal of clonal variation reported in Gewürztraminer, though obviously there's a vast amount within Savagnin as a whole. The colour of the grapes is not necessarily a result of clonal variation, although Alsace producer Olivier Zind-Humbrecht points out that modern, high-yielding clones tend to have darker, more purple colours, and softer, thinner skins. But he attributes the yellow or grey colour of some Gewürztraminer to different growing conditions and insufficient time to mature the colour of the skins. Potential alcohol levels can be over 13 or 14 per cent, but if a hot climate has encouraged too rapid a development of sugar, the grapes will not be physiologically ripe at that level and the seductive hothouse scent simply won't have developed in the grapes.

However, Gewürztraminer ripens unevenly. It is not unusual to find berries with 15 per cent potential alcohol on the same cluster as green ones. This makes vinification difficult, even if the average ripeness looks satisfactory.

VINIFICATION

Olivier Zind-Humbrecht of Alsace describes his winemaking thus:

'All the quality of Gewürztraminer is in the skin, so it pays to press the grapes slowly and gently, in order to extract the maximum of flavour. But lack of flavour is not the main worry, so one very rarely finds skin contact in Alsace. Gewürztraminer tends to ferment quickly to high alcohol levels, so it is important to keep a clear must, ideally from whole cluster pressing, to avoid the extraction of tannin from the stems, and too much sediment. I am not a big fan of cold fermentations, and I believe that any wine must reach a certain temperature in order to express complexity and not just banana or strawberry character. The richer and more complex a wine is, the more it can stand higher temperatures of up to 23° or 24°C (73° or 75°F). Poor grapes must ferment cooler in order to preserve the little flavour they have. But temperature control is very important, even if it is used for only a day or two. Gewürztraminer can ferment very quickly and reach temperatures of over 25°C (75°F) that could damage the wine.

'Instead of lees stirring, I prefer to have fermentations that last three months or more, so that the lees are naturally in suspension in the wine, and stirring is not necessary. I prefer to have yeast alive in suspension in the wine, rather than dead yeasts that have to be stirred regularly.

'The malolactic fermentation is usually prevented in Gewürztraminer, but I let the wine do what it wants to do. If the grapes are very ripe and concentrated, and if the malo happens only late and slowly, perhaps after Christmas, it often makes better wines. I have often had trophy-winning wines that had been through the malo'.

Many New World fermentations are at far cooler temperatures than those in Alsace. But not all: in New Zealand, Dr Neil McCallum, founder of Dry River liked to ferment his Gewürztraminer at 16–17°C (61°–63°F), and let it rise to 20°C (68°F) as the fermentation progressed. Says McCallum, 'the Cool School [of fermentation temperatures] believe they retain more aromatics at low temperatures, and this therefore is in the wine's interest. On the other hand most Europeans feel that fermentations at higher temperatures yield a wine with more body, while not sacrificing undue varietal flavours. Studies of about five years ago seem to indicate that the major amount of volatile losses are in the first two to three days of the fermentation and it therefore seems possible to embrace both schools to some extent. Natural ferments in Alsace would mostly start at low temperatures'.

GEWÜRZTRAMINER AROUND THE WORLD

Gewürztraminer's fame is firmly based on its performance in Alsace, the thin sliver of vineyard land in northeast France bordered by Germany and the river Rhine. Here, Gewürztraminer can – when the most talented growers and producers put their minds to it – produce one of the most astonishing scented wine styles in the world. Such intensity is rarely repeated elsewhere, but the vine does turn up all over the world with varying degrees of success.

Alsace

The usual planting density here is 4400 to 4800 vines per hectare; the legal minimum is 4500. (The yields per vine in Alsace are high when compared with the 10,000 vines per hectare of the Côte d'Or.) Machine-harvesting is no longer allowed for Grand Cru wines. But over-production is a problem in Alsace, even for the relatively low-yielding Gewurztraminer, and so measures have been introduced to try and curb this. Instead of leaving the over-production on the vine, all vineyards must now be fully picked – the idea being that the thought of paying for all that unnecessary picking might encourage such growers to control their yields better. As always, though, bureaucracy is less inventive than quantity-minded growers, who have sometimes simply uprooted one row in three of their (non-Grand Cru) vineyards in order to allow space for mechanical harvesters and have higher yields per vine.

Acidification was used for the first time in the roasting vintage of 2003, but in 2011 the Grands Crus decided that neither chaptalization nor acidification should ever be used for their wines. They want to establish a proper difference of quality between themselves and standard Alsace – and not before time. There is also talk of a Premier Cru to sit between the two.

As well as intensely aromatic dry wines, Gewürztraminer makes some of Alsace's finest sweet wines, but it is not in fact very susceptible to *Botrytis cinerea*. It is easy to ripen Gewürztraminer to *vendange tardive* richness – the grape can get to 16 per cent potential alcohol without a scrap of noble rot – but it has rather thick skins, which tend to protect it against botrytis.

Germany

Official figures of 902ha (2230 acres) (2013) do not distinguish between Savagnin Rose and Gewürztraminer – both appear as Roter Traminer. But most are probably accounted for by Gewürztraminer, particularly in the southerly regions of Pfalz, Rheinhessen and Baden. Flavours tend to be fruitier and more flowery than those of Alsace, seldom reaching such heady spiciness. In Baden, which is effectively a continuation of Alsace across the Rhine, there is more Traminer planted in the village of Durbach than in any other village in the whole of Germany; here it is known locally as Clevner.

Austria

The grape is likely to be called Traminer here, though its flavour can be as redolent of roses and spice as that of Alsace. Sunny Steiermark produces both lean, bone dry and barely aromatic versions and rather riper ones. Traminer makes powerful sweet wines in Burgenland and both dry and sweet ones in Vienna and Niederösterreich. For most of its growers, however, it is a minority grape.

Italy

Here it goes by the name of Traminer Aromatico, though the theory that it originated in the Alto Adige village of Tramin now seems a lot less likely. Savagnin was not planted in the Alto Adige and Austria until the 19th century. In spite of its Aromatico suffix, few wines have traditionally been very scented here, though they can be attractively elegant to compensate. The vine comes in red, white and pink versions. Local wisdom is that 'roter' or red Traminer is less aromatic but plumper.

Rest of Europe

Traminer or Gewürztraminer adopts various names in eastern Europe, including Tramini in Hungary, Traminec in Slovenia and Traminac in Croatia. These last two countries made fat, scented wines in the old 'Yugoslavian' days and still do. It is widely grown by Lake Balaton in Hungary, where soils are rich and volcanic; and it can give full, fiery, lychee- and mango-flavoured dry whites of considerable character. It is also grown in Romania, Russia, Ukraine, Moldova, the Czech Republic and Slovakia. It can make some very attractive wines – light and elegant in Slovakia, surprisingly perfumed in the Czech Republic and substantial and scented in Romania and Moldova. But the serious socio-economic problems of most of these newly emergent democracies mean that the grape is regularly overcropped, leading to dull, dilute flavours, and is usually poorly

Domaine Weinbach

Weinbach makes exemplary Gewurztraminer from several different sites that are always among the best in Alsace.

Domaine Marcel Deiss

The Deiss style of Gewurztraminer is less flamboyant than Domaine Weinbach's, but there is the same emphasis on the importance of terroir.

Andreas Laible

Probably the best producer in south Germany, Andreas Laible makes Gewürztraminer in a weighty Baden style but with more elegance than is customary.

The vineyards around the village of Tramin or Termeno in Italy's Alto Adige, where the grape is thought to have originated. Termeno Aromatico is still one of the grape's synonyms, but it has been edged out of most vineyards in its birthplace by other, higher-yielding varieties. What is left makes light but fragrant wine.

vinified, leaving unattractive grubby odours where there should be seductive scent. And in countries where expertise and finance are in very short supply, all efforts are likely to be concentrated on wines that are easy-to-sell (Chardonnay, Sauvignon, Cabernet and Merlot) rather than unfashionable old Gewürz. There are small amounts planted in Switzerland, Luxembourg and Spain.

New Zealand

New Zealand's cool but sunny climate seems to offer a perfect fit to Gewürztraminer. The drawback, however, is that the wine is difficult to sell, and so plantings fell from 182ha (450 acres) in 1991 to 85ha (210 acres) in 1998. However, figures are creeping up again, with 332ha of vines in 2013, some of them highly prized, small berry selections making sultry, scented wines.

Style-wise, it is the direct opposite of New Zealand's star grape, Sauvignon Blanc; but whereas consumers seem happy to accept the high acidity of Sauvignon as intrinsic to the grape (albeit with a few grams of residual sugar to balance), they seem unwilling to embrace Gewürztraminer's low acidity with the same enthusiasm. I suspect that people relate its soft, perfumed style to the cheap, off-dry Müller-Thurgau that was the mainstay of New Zealand's early wine revival and which is now regarded with disdain. A pity. Acidification is quite common, and some producers put the wines through the malolactic fermentation, either wholly or partially (to produce wines of a different style). Most are dry or off-dry, and can be wonderfully scented. There are some late harvest wines, sometimes fermented in new oak.

USA

Plantings here are falling and, as in New Zealand, the fundamental problem seems to be that while it is a relatively expensive grape to grow, with the usual problems of uneven ripening and low yields, you simply can't get as much money for it as for the more popular, easier to grow varieties like Chardonnay. In the mid-1980s there were nearly 4000 acres/1620ha of Gewürztraminer, but by 2012 the figure had fallen to around 3000 acres/1214ha, most of which are in California. Here it suits the cooler areas of Monterey, Sonoma, Mendocino and Russian River Valley well, and styles are usually dry or off-dry, though Anderson Valley can produce some delightful dessert versions. It is found also in Oregon and Washington State, though many wines can be a little too dilute and a little too sweet.

Rest of the world

In Australia it may be called Traminer or Gewürztraminer. In warmer, highly productive regions, it is often blended with Riesling to improve acidity. As Gewürztraminer, from cooler areas of Tasmania and Victoria and from South Australia's Eden Valley, it can be thrillingly scented. Varietal wines range from dry through off-dry to late harvest.

Canada has the potential to make good Gewürztraminer, and Chile is making some very convincing examples, particularly from cool coastal and southerly vineyards. There are a few sweet versions from South Africa.

Te Whare Ra
Te Whare Ra in Marlborough, New Zealand makes some of the southern hemisphere's most lush, scented Gewürztraminer using old-clone vines..

Cave Spring
The sunny but cool conditions at Cave Spring overlooking Lake Ontario are perfect for ripening and preserving the fragrance of Gewürztraminer.

Cono Sur
Cool Casablanca first showed promise with Gewürz in Chile, but coastal areas such as San Antonio and Bío-Bío in the south are proving to be even better.

ENJOYING GEWÜRZTRAMINER

Gewürztraminer seems to be able to offer exceptions to the rule that white wines need high acidity in order to age well. Admittedly, in those Alsace vintages where the sun shone too warmly and the wines verge on flabbiness, long aging is not recommended because the wine can quickly develop a bitter edge and a taste of breakfast marmalade. It does need relatively high acidity to make it past its third or fourth birthday.

In years when the balance is good, and providing that yields are kept down, a Grand Cru Gewürztraminer can happily age for 10 years. At the end of that period it has lost its exuberant perfume and become winier and more subtle – less like Gewürztraminer, perhaps more like Pinot Gris. But it gains enough in honey and complexity to make the loss of roses and lychees worthwhile.

Few New World examples are made to be aged, though the best will improve for a few years. Late-harvest wines can age, though are so delicious young that it is tempting to drink them then and there. The most concentrated Vendange Tardive (late-harvest) and Sélection de Grains Nobles (super-ripe, sweet, usually botrytized) wines from Alsace will live for 20 years or more – but again, they need respectable, if not high, levels of acidity to begin with.

The taste of Gewürztraminer

Ripe, concentrated Gewürztraminer tastes like no other grape, though underripe or overcropped versions can resemble second-rate Muscat, and less aromatic versions, if underripe, can be no better than third-rate Riesling. But we are only interested in good Gewürztraminer here: low-cropped, cleanly made and with enough structure and acidity to give form and balance to all that headiness of perfume.

So what does this fabled perfume consist of? Lovely freshly plucked tea roses, the petals rubbed lasciviously between thumb and forefinger before you gorge yourself on their heady scent. And lychees too, fresh or tinned, it doesn't matter, their flesh almost slithery as your tongue and teeth worry it free of the stone and your palate is amazed by its scented fruit. Face cream, too – particularly Nivea Crème – and a whole range of other scents – cinnamon or lilac, orange blossom and citrus peel, tea and bergamot and honeysuckle. And, intriguingly, really good Alsace examples also have a fierceness like really fresh ground black peppercorns. In simple wines these perfumes are muslin-light; in weightier ones they can become oily. If acidity is lacking they become flabby and slightly greasy, and one is reminded of butter melting in the sun and marmalade left uneaten at the side of the breakfast plate; unbalanced Gewürztraminer can be every bit as unappealing and unrefreshing as that.

Late-harvest Gewürztraminers have a combination of sweetness and heavy perfume that can make them tricky to match with food, though bottle age helps here, by toning down the perfume to more manageable levels. Botrytis, where it is part of the picture, adds complexity and richness at the same time as it reduces the roses and lychees element.

There doesn't seem to me to be much point in making Gewürztraminer unless you set out to highlight its heady, sultry scent. Schoffit makes wines with sublime texture and perfume from the warm, sandy soils of the Harth vineyard near Colmar. Dry River makes a more powerful, exotic style, still perfumed but with denser fruit, from Martinborough in New Zealand's North Island, an area better known for savoury Pinot Noir reds than scented whites.

MATCHING GEWÜRZTRAMINER AND FOOD

Traditional Alsace food and wine pairings include rich, smooth duck or chicken liver pâté (with either dry or sweet examples), onion tart, smoked fish, roast goose, and pungent washed rind cheeses such as Munster. Or indeed, blue cheeses.

Young Gewürztraminer is an absolute delight to drink by itself, but it is also excellent with Indian and Chinese dishes, or fusion food redolent of ginger or other spices. Southeast Asian food, with its use of lemon grass, coriander and coconut, is often beautifully matched with Gewürztraminer. When the spicing is more subtle, mature Gewürz is an excellent partner for chicken.

CONSUMER INFORMATION

Synonyms & local names

Also known as Traminer; as Rotclevner, Traminer Musqué, Traminer Parfumé and Traminer Aromatique in Alsace; as Roter Traminer, Clevner or Klavner in Germany and Austria; as Traminer or Traminer Aromatico, Traminer Rosé or Traminer Rosso in Italy; in Switzerland as Heida, Heiden or Païen; as Traminac (Croatia); Tramini (Hungary); Traminec (Slovenia); Drumin or Prync in the Czech and Slovak Republics; Rusa in Romania; Mala Dinka in Bulgaria; and Traminer in Russia, Moldova and Ukraine.

Best producers

ALSACE Adam, Albrecht, Allimant-Laugner, Barmès-Buecher, Bechtold, Bernhard-Riebel, Beyer, Binner, P Blanck, Boesch, Bott-Geyl, Boxler, Burn, Deiss, Dirler-Cadé, Dopff & Irion, Dopff au Moulin, Pierre Frick, Hertz, Hugel, Hunawihr co-op, Josmeyer, Kientzler, Kreydenweiss, Kuentz-Bas, Seppi Landmann, J-L Mader, Albert Mann, F Meyer, Meyer-Fonné, Mittnacht-Klack, René Muré, Ostertag, Pfaffenheim co-op, Rieflé, Rolly Gassmann, Eric Rominger, Schaetzel, A Scherer, Schleret, Schlumberger, Schoffit, Gérard Schueller, Jean Sipp, Louis Sipp, Bruno Sorg, Pierre Sparr, Spielmann, Tempé, Trimbach, Turckheim co-op, Weinbach, Zind-Humbrecht.
GERMANY Fitz-Ritter, A Laible, U Lützkendorf, Heinrich Männle, Rebholz, Wolff-Metternich, Klaus Zimmerling.
AUSTRIA Fritz Salomon.
ITALY/Alto Adige Abbazia di Novacella, Caldaro co-op, Colterenzio co-op, Cornaiano co-op, Franz Haas, Hofstätter, Prima & Nuova/Erste & Neue, San Michele Appiano co-op, Termeno co-op, Elena Walch; **Trentino** Cesconi, Pojer & Sandri.
SPAIN Enate, Raimat, Torres, Viñas del Vero.
USA/California Adler Fels, Bouchaine, Clendenen Family Vineyards, Thomas Fogarty, Gundlach-Bundschu, Handley, Lazy Creek, Navarro; **Oregon** Amity, Bridgeview, Eola Hills, Foris, Henry Estate, Tyee; **Washington State** Covey Run; **New York** Bedell.
CANADA Gray Monk, Konzelmann, Mission Hill, Sumac Ridge.
AUSTRALIA Delatite, Henschke, Knappstein, Moorilla Estate, Pirie, Seppelt, Skillogalee, Audrey Wilkinson.
NEW ZEALAND Brancott, Cloudy Bay, Dry River, Forrest, Framingham, Hunters, Lawson's Dry Hills, Lincoln, Millton, Rippon, Saint Clair, Seifried, Spy Valley, Stonecroft, Te Whare Ra.
CHILE Concha y Toro, Viña Casablanca.
SOUTH AFRICA Delheim, Neethlingshof, Simonsig Estate, Zevenwacht.

RECOMMENDED WINES TO TRY

Ten Alsace dry or off-dry wines

Léon Beyer Cuvée des Comtes d'Eguisheim
Paul Blanck Furstentum Vieilles Vignes
Ernest Burn Cuvée de la Chapelle
Marc Kreydenweiss Kritt
Meyer-Fonné Wineck-Schlossberg
Mittnacht-Klack Schoenenbourg
Schaetzel Kaefferkopf Cuvée Catherine
Domaines Schlumberger Cuvée Christine
Trimbach Seigneurs de Ribeaupierre
Turckheim co-op Brand

Ten Alsace Sélection de Grains Nobles

Deiss Altenberg
Dirler-Cadé Spiegel
Hugel
Kuentz-Bas Pfersigberg Cuvée Jeremy
Albert Mann Furstentum
René Muré Vorbourg Clos St-Landelin
Schoffit Clos St-Théobald Rangen
Bruno Sorg
Weinbach Cuvée d'Or Quintessence
Zind-Humbrecht Rangen

Five Italian wines

Cesconi Trentino Traminer Aromatico
Hofstätter Alto Adige Gewürztraminer Kolbenhof
Caldaro co-op Alto Adige Gewürztraminer Campaner
San Michele Appiano co-op Alto Adige Gewürztraminer Sanct Valentin
Elena Walch Alto Adige Gewürztraminer Kastelaz

Ten New World wines

Concha y Toro Winemakers Lot Gewürztraminer (Chile)
Delatite Dead Man's Hill Gewürztraminer (Australia)
Eola Hills Gewürztraminer Vin d'Epice (Oregon)
Foris Rogue Valley Gewürztraminer (Oregon)
Henschke Joseph Hill Gewürztraminer (Australia)
Lawson's Dry Hills Gewürztraminer (New Zealand)
Mission Hill Gewürztraminer Icewine (Canada)
Pirie Gewürztraminer (Australia)
Stonecroft Gewürztraminer (New Zealand)
Te Whare Ra Gewürztraminer (New Zealand)

Gewürztraminer berries may be more orange, more purple or more red than these, depending on the growing conditions. The wine generally has a deep golden colour, though Traminers in the northern Italian mould may be paler.

Maturity charts

Most simple Gewürztraminer follows the New Zealand pattern: drink the wines early. Only top Alsace examples age successfully in bottle.

2011 Alsace Gewurztraminer Grand Cru (dry)

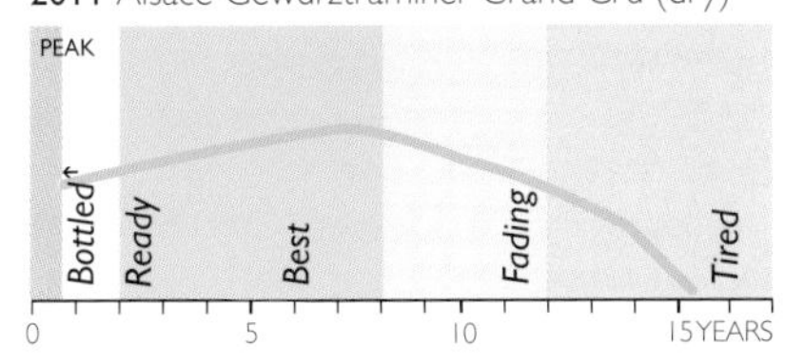

2011 was a good, early vintage in Alsace, with fragrant and well-balanced wines in most varieties. Top Gewurztraminers were attractively scented.

2010 Alsace Gewurztraminer Vendange Tardive

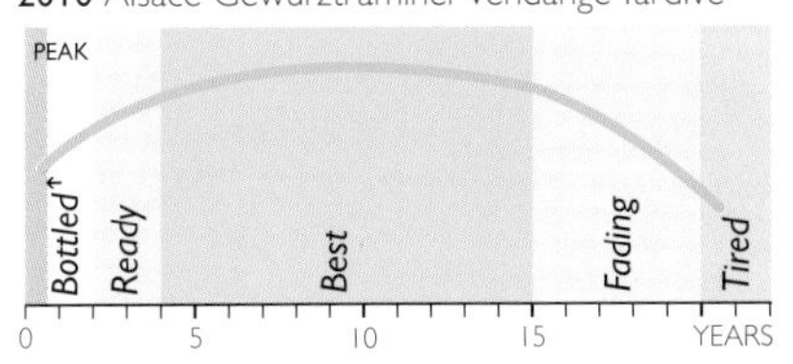

2010 was an exceptional year for rich Gewürztraminer. The top Vendange Tardive wines have tremendous richness and scent, but balancing acidity.

2014 New Zealand North Island Gewürztraminer

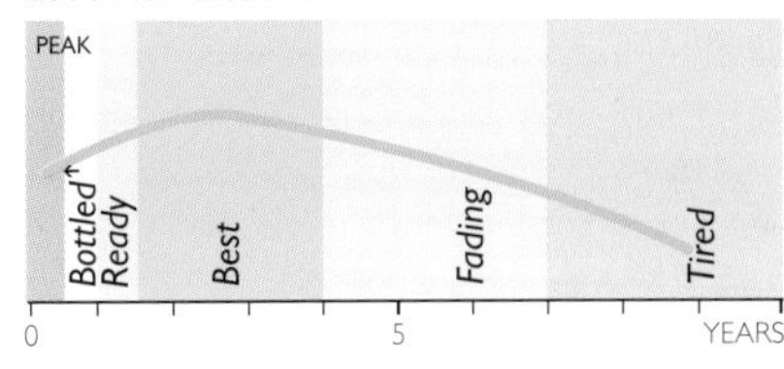

Most New Zealand Gewürztraminers are best drunk without further bottle age. The 2014 harvest was sunny and of good, scented quality.

GIRÒ

Mid- to late-ripening Sardinian black grape, DOC in Cagliari. It can be dry, sweet or fortified, or even dried for a local passito. Tannins are firm, and the fruit is attractively cherry-scented, but acidity tends to be lowish. The Girò Blanc grown on Mallorca, and atracting some attention there for its pineapple and mango fruit, good quality and power, is not related.

GLERA

This grape used to be called Prosecco, but its name was changed by the Prosecco producers' Consorzio so that they can legally protect the Prosecco name from being used elsewhere in the world. Prosecco's most familiar incarnation is as a dry or off-dry sparkling wine with good acidity and a lightly creamy flavour. It is a favourite in nearby Venice and the vicinity, and is the proper wine to use in a Bellini, the peach juice and sparkling wine cocktail served in Harry's Bar in Venice (and, inevitably, elsewhere). It is not high in alcohol or body, and is fresh and rather neutral rather than particularly aromatic – the ideal base for sparkling wine, in other words. There are occasional sweet versions, quite a lot of frizzante, or semi-sparkling, and a little still wine.

Virtually all Prosecco is grown in the Valdobbiadene and Conegliano zones, north of Venice, or in Colli Euganei near Padua. It ripens late, and it was this late ripening that originally gave rise to the spumante tradition: the fermentation tended to stop in the autumn, leaving some carbon dioxide and perhaps some residual sugar in the wine, which would begin to ferment again in the spring – if it hadn't all been drunk by then. Most of the wine today is made by the Charmat method. Best producers: (Italy) Adami, Bernardi, Bisol, Bortolin, Carpenè Malvolti, Col Vetoraz, Le Colture, Nino Franco, Ruggeri & C, Tanorè.

GODELLO

This rich, aromatic, high-quality Spanish grape is the same as Portugal's Gouveio, alias Verdelho do Dão, but is different to the true Verdelho grown on Madeira (see page 286). As Godello it is the best white grape in Valdeorras, though still manages to account for only a tiny per cent of the vineyard: Palomino and Valenciana, or Doña Blanca, far outdo it in terms of area planted. It also appears in other DOs of the far northwest, including Ribeira Sacra, Ribeiro, Monterrei and Bierzo.

Rafael Palacio
Godello is quite possibly the most scented of the newly fashionable white varieties in Spain's far north-west. Almost extinct in the 1970s, it is now flourishing in Valdeorras.

It must surely be in the running for the title of Spain's most interesting white grape; and Spain is now producing such excellent white wines that that is not a negligible title. It shares the softly aromatic apricot character of Albariño but has a silkier texture and a lusher fruit, rather like good Viognier. It seems to be native to the gorges of the Sil Valley, but nearly became extinct in the 1970s, and only determined efforts by the Consejo Regulador of Valdeorras saved it.

Godello is delicious as a varietal wine, but also blends well with other grapes, including Albariño, Treixadura and Doña Blanca. In Portugal, where it is called Gouveio, it's found mostly in the Douro and Dão regions, and in the former is mostly blended into quite rich, flavoursome table wine. Best producers: (Spain) Casal, Godeval, Guitian, Jesús Nazareno co-op, Santa Maria dos Remedios co-op, Señorío, La Tapada, Valderroa.

GOLDBURGER

A 20th-century cross used for making sweet wines in Austria. It makes quite powerful, straightforward wines.

GOLDMUSKATELLER

A strain of Muscat Blanc à Petits Grains (see pages 156–165) grown in Italy's Alto Adige is known as Goldmuskateller or Moscato Giallo. Best producers: (Italy) Lageder, Thurnhof, Tiefenbrunner.

GOUAIS BLANC

Ancient and almost extinct variety which is a parent of at least 81 western European vines. Chardonnay is Gouais Blanc x Pinot, as are Gamay and Aligoté; Riesling is the offspring of Gouais Blanc; so is Furmint, so is Blaufränkisch. Its aliases include Heunisch in Germany; it was, in its day, grown extremely widely. It wasn't, however, terribly good at making wine, merely at cross-breeding, with any other vine that came near, and by 1598 orders were being made to uproot it. By the 20th century, it was almost extinct in France. If you want to find one you might, if you look hard, spot it in northern Piedmont, where it is sometimes known as Liseiret. Georg Breuer has planted a bit of Heunisch in the Rheingau, and there's some planted in Switzerland under the name of Gwäss. In Australia, Chambers of Rutherglen has some. However, no one reports queues of enthusiastic ampelographers desperate to try the wine.

GOUVEIO

Gouveio is one of the many white varieties used to make dry white wines in Portugal's Douro Valley and it is also one parent of more than 80 varieties spread across Western and Central Europe. There seem to be two Gouveios, and one is the same as Spain's Godello. Best producers: (Portugal) Churchill, Martinez, Niepoort, Sogrape.

GRACIANO

A spicy, aromatic, intensely flavoured grape that is challenging Tempranillo as the most interesting red vine in Rioja, but which is unfortunately far more capricious to grow than the region's other vines. However, that doesn't stop the good guys from trying.

The success of a pioneering handful of varietal examples has encouraged plantings there, and just 15 per cent in a blend will add perfume and acidity. It is not unlike Petit Verdot in structure and flavour profile, with high acidity. It is less tannic than Tempranillo and tends to oxidize easily: it is not designed for long aging. It is late budding, susceptible to downy mildew and low yielding – the last two being characteristics not likely to endear the vine to the average grower. It prefers cool, clay-chalk soils and mild, damp climates; it dislikes drought and strong summer heat, and it is hardly found outside Rioja.

As Morrastel it is found rarely in the south of France, but in Spain the name of Morrastel is a synonym for the very different Monastrell, or Mourvèdre (see pages 152–153). What is called Morrastel in North Africa can be either grape. Portugal's Tinta Miúda, which is found in the Ribatejo and in Estremadura, is the same grape as Graciano, as is southern Spain's Tintilla de Rota. There is also a bit of Graciano in California, which is gaining attention.

The most encouraging signs are coming from Australia, where it shows clear affinity for the conditions and is producing award-

Greco Bianco growing near the town of Bianco in Calabria, Italy for (yes) Greco di Bianco – on, as it happens, very 'bianco' soil. Greco Bianco is mainly used for sweet wine and isn't the same as simple Greco, which is used to make some of Campania's best dry whites, such as Greco di Tufo.

winning reds. Best producers: (Australia) Brown Brothers, Rosemount; (Spain) Artadi, Campillo, Contino, CVNE, Viña Ijalba, La Rioja Alta, Dominio de Valdepusa.

GRANDE VIDURE

The old Bordeaux name for Carmenère (see pages 66–67), a grape almost extinct there but flourishing in Chile. Some wineries in Chile opted to call their Carmenère by this name but since 2001 its use has been banned under EU legislation.

GRASA

Some authorities maintain that this Romanian vine is the same as Furmint (see page 99), but I don't quite see why. Furmint is famously acid, Grasa is famously fat and soft. But the leaves are the same, and Grasa does, like Furmint, hold on to a green tint as it ages. Ah. They have some Grasa in Tokaj, where it's known as Kövérszölö and makes sweet wine, but is quite different to the Furmint. Does that settle it? Back to Romania. It is concentrated in the Cotnari region, where it makes rich, long-lasting, sweet, nobly rotten wines, often given a bit of extra zest by an admixture of Tămaîoasă Românească. Best producer: (Romania) Cotnari Cellars.

GRASEVINA

Welschriesling (see page 298) is called by this name in Croatia, where it is the most widely planted variety and can produce excellent, broad, honeyed wines.

GRAUBURGUNDER

German synonym for Ruländer (see page 220), or Pinot Gris (see pages 186–187). Best producers: (Germany) Bercher, Schlossgut Diel, Drautz-Able, Dr Heger, Müller-Catoir, Salwey.

GRECANICO DORATO

Sicilian name for Garganega, the mainstay of good Soave in Veneto. No one's quite sure how it got down to Sicily, but plantings are on the increase, and the wine has very good flavour, with some grassy pungency. Best producer: (Italy) Planeta.

GRECHETTO DI ORVIETO

An interesting central Italian grape that adds structure, richness and a pleasing nuttiness and leafiness to many Umbrian whites. Good Orvieto owes a lot to Grechetto; so do Antinori's Torgiano and Cervaro whites. It is generally blended: with Chardonnay in Cervaro, with Trebbiano, Malvasia and Verdello in others. It is low yielding but sufficiently disease-resistant to compensate. It makes good Vin Santo. Grechetto di Todi is a separate variety and a synonym for Pignoletto. Best producers: (Italy) Antonelli, Avignonesi, Barberani-Vallesanta, Arnaldo Caprai, Falesco, Palazzone, Castello della Sala.

GRECO

The name Greco has been given to a multitude of Italian grapes, and DNA profiling is only now sorting them out. Any link with Greece, ancient or modern, seems to be fallacious: they seem to have no genetic relationship with any known Greek vine, which they would do if they had originated there. The Greek implication of the name is more likely to mean that they made Greek-style wines, i.e. sweet or sweetish.

Plain Greco is the same vine as Greco di Tufo, Asprinio and various other Grecos; Greco Bianco, from Calabria, makes sweet wines and is a different vine; Greco Bianco di Gerace is the same as Malvasia di Lipari and Greco Bianco di Novara is the same as Erbaluce.

Greco Nero, likewise, covers umpeteen varieties, mostly in Calabria.

So: Greco. The flavour is slightly peachy and leafy, fresh and brisk, full-bodied and sometimes herbal. Examples vary in their degree of aroma, which could be winemaking or vineyard: higher altitudes tend to favour aroma. But it's a good grape, and one we're seeing more of now that Italians are getting keener on aromatic wines.

GRENACHE BLANC

A white mutation of Grenache Noir (see Garnacha). Intrinsically a rather dull grape which, like its black counterpart, Grenache Noir, oxidizes easily and is rather low in acidity. However, it contributes a certain

Bonny Doon

Le Cigare Blanc is a Grenache–Roussanne blend from the wonderfully named Beeswax Vineyard in Monterey County, California. Made as a homage to white Châteauneuf-du-Pape, the wine manages to be citrussy, rich and chewy.

softness and fleshiness to a blend and is widely planted across the south of France. It features heavily in much white Châteauneuf-du-Pape and Côtes du Rhône and to a limited extent (10 per cent, to be precise) in some Côtes du Rhône-Villages. It also appears in some Rivesaltes Vin Doux Naturel from Roussillon. It is also now being used as a varietal, and modern winemaking wizardry can work wonders in turning it into very attractive and occasionally thrilling wine. Low temperature fermentation can produce a fresh, dill-scented wine to drink very young. Alternatively, low yields, maceration on the skins, oak barrel aging and/or fermentation, lees aging and perhaps a dash of something more aromatic like Muscat or Viognier, can give it more memorable quantities of flavour and a broad, rich, satisfying texture.

As Garnacha Blanca (see page 101) it is found in north and northeast Spain, where, like Garnacha Tinta, it originated. South Africa and California have a little. Best producers: (France) Beaucastel, Caraguilhes, Clavel, Font de Michelle, la Gardine, Mont-Redon, la Nerthe, Rayas.

GRENACHE NOIR

See Garnacha pages 102–111.

GRIGNOLINO

A pale but not hugely interesting red grape that originated in Piedmont in north-west Italy and has never strayed; and the wine itself is made and mainly drunk locally.

The acidity and tannin are high, and the flavour varies between fresh leafiness and herbal rasp, and tired vegetable stock, according, apparently, to the luck of the draw. There is great clonal variation, which may account for some of the differences in style. The vine also tends to reflect its soil type quite strongly. At its best it is quite attractive though curious – not exactly a wine you'd want every day. It is best drunk young. There are three DOCs: d'Asti, del Monferrato Casalese and Piemonte. For some reason the Heitz vineyard in California's Napa Valley grows a bit – don't ask me why. Maybe it's to remind them how good their Cabernet is. Best producers: (Italy) G Accornero & Figli, Braida, Bricco Mondalino, Dante Rivetti, Cantine Sant' Agata, Vinchio e Veglio Serra co-op.

GRILLO

One of the better grapes for Marsala, Sicily's fortified wine, Grillo reaches high levels of potential alcohol – always a plus point for

Tasca d'Almerita
Grillo was for a long time simply a source of Marsala's base wine, but the grape variety shows real character as a lemony table wine to be drunk young.

makers of fortified wine, because it saves on (expensive) brandy. Plantings have begun to rise again as more Grillo is used for dry whites. It's a potentially interesting grape, though, with lemony fruit, robust weight and structure. Best producers: (Italy) De Bartoli, Florio, Pellegrino.

GRK

Very good Croatian white grape that seems to be a parent or offspring of Zinfandel, alias Primitivo, alias Tribidrag. It's one of a number of closely related vines from the same area. Its wine are rich and structured with good acidity. Best producers: Bartul Batistíc, Stipan Cebalo, Milano.

GROLLEAU

The most interesting thing about Grolleau or Groslot is that, according to Galet, its name is derived from the old French word *grolle* or *grole*, meaning 'crow'; the comparison is with the deep black of Grolleau's berries. It is curious but appropriate that while the more aristocratic Merlot grape is said to be named after the melodious blackbird, or *merle*; poor old plebeian Grolleau gets the crow. Grolleau, perhaps depressed by all this, is on the decline in its Loire homeland. It is allowed only into rosé wines, not red, and plays a part in Anjou, Saumur and Touraine Rosé. Its wines are light, low in alcohol, a bit earthy and high in acidity. Grolleau Gris is a pink-berried version. Best producers: (France) Bouvet-Ladubay, Pierre-Bise, Richou.

GROS MANSENG

An intensely flavoured grape of southwest France, now beginning to be revived as winemakers learn how to get the best from it. It is found in Béarn, Jurançon and Gascony, where it is considered a *cépage améliorateur* for its high acidity and floral, spicy, apricot-quince fruit. It is probably the offspring of Petit Manseng (see page 180).

It is quite a good yielder, and gives up to 70 or 80 hl/ha seemingly without any adverse effect on quality. What problems it does have are in the winery. It has deep, golden-coloured grapes, and the skins have high levels of tannins, colour and polyphenols. If the grapes are treated at all roughly in the winery the results will be coarse: it must be pressed extremely delicately, and carefully settled so that only clean juice goes to the fermentation vats.

It can produce interesting, though different results at a wide spectrum of ripeness. If picked at 11.5–12 per cent potential alcohol, the wine will be fresh, flowery and sharp. The grapes are also easier to work with at this level. But to get the full potential out of the grape it is necessary to pick at 12.5–13.5 per cent potential alcohol. Then the wines are much more powerful, but still with high acidity, and still dry.

It can also make very good sweet wines, and recent experiments with new oak have shown wines full of personality not unlike oaked Sémillon, much valued locally for drinking with foie gras. These wines may be from late-harvest or noble-rotted grapes. It strongly reflects its terroir, giving fresh wines on chalky clay, richer, fatter ones on sandy clay, and light, delicate ones on sand. Best producers: (France) Aydie, Brana, Cauhapé, Clos Lapeyre, Clos Uroulat, Souch.

GRÜNER VELTLINER

A versatile Austrian grape that, at its best, is now in the forefront of quality and, given its long Germanic name, has become surprisingly fashionable in restaurants and bars, with its name often shortened to Grüner, or simply GV. At its simplest, it forms Austria's everyday jug wine, is fresh and peppery, and best drunk within months of the vintage. The best growers in the Wachau and Kamptal regions of Niederösterreich produce much more structured, ripe wines, however, which bring out the lentil, celery and white pepper aroma of the grape, with sometimes a touch of citrus and often honeyed weight, sometimes with a little too much alcohol for balance. It is never very aromatic, but can age well in bottle, becoming more honeyed after a few years and yet staying peppery. It is site-sensitive and favours loess soil but is planted pretty well anywhere. Yields are high, up to 100hl/ha for everyday wine.

It was the only grape permitted for the first DAC (Districtus Austriae Controllatus) wine, from Weinviertel. This premium designation only applies to a particular style: with no discernible wood on the palate, and with a light yellow or greenish-yellow colour, and delicate, peppery fruit.

Grüner Veltliner is also successful in the Czech Republic and Hungary. Its names here are Veltlin Zelene or Veltlinkske Zelené, and Zöldveltelini, respectively. Austria also has three other Veltliner vines: Grauer Veltliner is a colour mutation of Grüner Veltliner, Brauner Roter and Frühroter Veltliner are unrelated. All three are dark-skinned and while the first two are planted only in very small quantities, Frühroter Veltliner (red-skinned rather than black) is slightly more widely planted, mostly in the Weinviertel region. It has lower acidity than Grüner Veltliner. Its trendiness and easy drinkability have encouraged other countries to have a go, led by New Zealand, but with interesting examples from Australia, California and Oregon. Best producers: (Austria) Bründlmayer, Fred Loimer, Franz Hirtzberger, Josef Jamek, Emmerich Knoll, Lenz Moser, Familie Nigl, Nikolaihof, Willi Opitz, Franz Xaver Pichler, Prager, Dr Unger, Domäne Wachau.

GUTEDEL

Gutedel, or Chasselas, as it is better known, is far less famous than its offspring: by virtue of an unscripted encounter with Riesling it is a parent of Müller-Thurgau. 'Unscripted' because until recently the parents of Müller-Thurgau were believed to be Riesling and Silvaner, which is presumably what Dr Müller of Thurgau himself thought. DNA fingerprinting has demonstrated the truth: clearly there is no limit to what vines will get up to, left to themselves. Gutedel is less important in Germany than Chasselas is in Switzerland (see page 80). It is found almost entirely in the Markgräferland region of Baden, where it makes typically neutral, slightly earthy wine. There is also a dark-berried version, Roter Gutedel. Best producers: (Germany) Blankenhorn, Dörflinger, Pfaffenweiler co-op, Schneider.

HANEPOOT

Muscat of Alexandria is known by this name in South Africa. See pages 156–165.

HARRIAGUE

Tannat (see page 255) may be called by this name in Uruguay, especially when referring to Tannat descended from original 19th-century plantings rather than new clones from France.

It doesn't look sunny – and it isn't. These are the vineyards around Bilbao on the rainy, windy north coast of Spain, looking out onto the Bay of Biscay. I've been visiting the area since I was a student and I reckon I've seen about one day of sunshine – but that's why the local Hondarrabi grape makes such sharp wine – which just happens to go brilliantly with the local seafood.

HÁRSLEVELÜ

Tokaj's second main grape variety is found in other parts of Hungary as well, but so far few examples have the delicate pollen scent and elderflower fruit which the grape can boast when made as a dry wine. In Tokaj its sweet wines are typically smoky and spicy, robust and intense, and fat sometimes to the point of being slightly oily. Hárslevelü is also found in Slovakia and in South Africa. Best producers: (Hungary) Château Megyer, Château Pajzos, Disznókö, Oremus, Royal Tokaji Wine Co.

HEIDA

Savagnin/Traminer (see page 242) takes this name in Switzerland.

HIMBERTSCHA

A Swiss rarity, rescued from extinction in the 1970s by Josef-Marie Chanton, who is still almost the only grower. The wine is crisp, lemon-zesty and attractive.

HONDARRABI BELTZA

The dark-skinned grape of Spain's Basque country, where it features in the Chacolí de Guetaria and Chacolí de Vizcaya DOs (or Getariako Txakolina and Bizkaiko Txacolina if Basque is your preferred tongue). Hondarrabi Beltza's wines are fresh, acidic, with some leafy black fruit, and best drunk young. The vine is believed to have originated in the Pamplona Valley. Best producer: (Spain) Txomin Etxaniz.

HONDARRABI ZURI

No one quite seems to know precisely what is and isn't Hondarrabi – this is hardly surprising given the scattered and rudimentary nature of most vineyards in Spain's Basque country. There are two Pyrenean varieties – Crouchen and Courbu – and an American hybrid called Noah. These all also travel under the Hondarrabi title in the Basque – though how anyone could think the powerfully scented Noah can have anything in common with the neutral Crouchen is beyond me. Makes you wonder whether Hondarrabi Zuri actually exists at all. Even so, it/they dominate(s) the Basque vineyards of Chacolí de Guetaria /Getariako Txakolina and Chacolí de Vizcaya/Bizkaiako Txacolina. The former is slightly fuller and richer, but being light, lemony and acidic both wines are ideal for the local seafood. Drink young.

While the wines are prized locally they are seldom seen outside the region. The vines are trained on a modified version of pergolas;

yields average less than the legal maximum of 93.6hl/ha, and generally work out at around 60hl/ha. The wines generally contain an admixture of Folle Blanche and perhaps a little Chardonnay, Sauvignon Blanc or Riesling. Best producers: (Spain) Eizaguirre, Txomin Etxaniz, Virgen de l'Orea.

HUMAGNE BLANCHE

One of the less aromatic Swiss varieties. The wine does, however, have some attractive plumpness and freshness and can age interestingly. It has been cultivated in the Valais region since at least the 12th century.

HUMAGNE ROUGE

See Cornalin page 92.

HUXELREBE

Perfumed German crossing, propagated in 1927, which produces grapes with high sugar levels and plenty of flavour, but little elegance. It is found in declining though still substantial quantities in the Pfalz and Rheinhessen. It can produce enormous crops, and if it does the flavour will be accordingly diluted. In England it can produce delightful grape juice and elderflower-scented, relatively dry wines but is declining. Best producers: (Germany) Kurt Darting, Geil, Koehler-Ruprecht, Johann Ruck, Schales; (England) Biddenden.

INZOLIA

A good-quality, low-yielding grape found in Sicily, where it is part of the blend, along with Catarratto and perhaps Trebbiano, in many of the island's white wines. Its wine is fresh and rather racy at its best. It is part of the Marsala blend (at least, the genuine Marsala blend as understood by leading producer De Bartoli), along with Grillo, and the grapes may be partially dried and made into sweet wine. It is also planted in Tuscany, and is sometimes known as Ansonica or Anzonica. Best producers: (Italy) Colosi, De Bartoli, Florio, Pellegrino, Duca di Salaparuta, Tasca d'Almerita.

IRSAY OLIVER

There's nothing very complex about this eastern European crossing of Pozsonyi with Pearl of Csaba – and if its parents sound unfamiliar it is because it was bred as a table grape. It has a pleasant, rather Muscatty perfume, and lowish acidity. It is found in Slovakia and Hungary, where it is called Irsai Olivér. If you like grapy dry whites, these are never expensive. Best producers: (Hungary) Balatonboglar Winery; (Slovakia) Nitra Winery.

JACQUÈRE

Jacquère is found in the French region of Savoie, where it produces light, mountain-fresh white wines for drinking young. Best producers: (France) Blard et Fils, Pierre Boniface, Charles Gonnet, Philippe Monin, Jean-Claude Perret.

JAEN

Early-ripening, easy-to-grow variety that is the second most planted grape, after Baga, in Portugal's Dão region and is, in fact, Mencia from Spain's northwest. Its colour is good and it can have a delicious raspberry/loganberry fruit, but its acidity and tannin are low, so it is usually blended with other varieties, often the more powerful Touriga Nacional, or Alfrocheiro Preto. Best producers: (Portugal) Bacalhõa, Quinta das Maias, Quinta dos Roques, Sogrape.

JOÃO DE SANTARÉM

Another name for Castelão (see page 65) or Periquita (see page 180). Best producers: (Portugal) Horta de Nazaré, Quinta de Lagoalva de Cima, Herederos de Dom Luís de Margaride.

JOHANNISBERG

Swiss name for Silvaner (see pages 256–257) from the Valais region, where it is sometimes planted in the best sites. The wines can be rather plump and very tasty. Best producers: (Switzerland) Caves Imesch, du Mont d'Or.

JOHANNISBERG RIESLING

The name, or one of the names, given to Riesling in California (see pages 204–215). Best producers: (USA) Callaghan Vineyards, Chateau St Jean, Gainey.

KADARKA

This Hungarian vine is on the decline partly because it is late ripening and prone to grey rot, and partly because the export market has turned so decisively against what was its main wine, Bull's Blood from Eger. It can make rather good weighty, lush red but it needs careful cultivation for this, and its natural vigour needs to be kept in check. If yields are too high, which they almost always are, and if it is picked before it is fully ripe, which it almost always is, it makes dull, dilute wine. It is still grown in Hungary, especially on the Great Plain and in Szekszárd, but plantings are smaller than they were.

It is found in small quantities in Austria's Neusiedlersee, in Romania (as Cadarca) and in Vojvodina. It plays a bigger part in Bulgaria, where it is called Gamza and is found in the northwest. There is also a white grape called Izsáki, or White Kadarka, which is grown on Hungary's Great Plain too, but it is unrelated. Best producers: (Hungary) Hungarovin; (Bulgaria) Domaine Boyar.

KALECIK KARASI

Turkish grape widely grown in Central Anatolia. Phylloxera nearly destroyed it, but it was propagated again in the 1970s and 1980s to make soft, fresh reds with agreeable cherry and red-fruit flavours, best drunk young. Best producers: Pamukkale, Kavaklidere.

KÉKFRANKOS

The Hungarian name for Austria's Blaufränkisch (see page 48). It produces reasonable quality in Hungary. Best producers: (Hungary) Hungarovin, Akos Kamocsay.

KÉKNYELU

Potentially aromatic and high-quality Hungarian grape, now hardly grown. At its best in Badacsony on the shores of Lake Balaton where the wines were traditionally fiery and golden, high in acid but waxy and scented.

KÉKOPORTO

Old Hungarian name for Blauer Portugieser (see page 48), no longer used.

KERNER

A reasonably good-quality vine bred in Germany in 1969. It is one of the better modern crossings (Trollinger x Riesling), which perhaps is not saying a great deal, and German growers have loved it from the start for its combination of high sugar levels and high yields. From the consumer's point of view, the flavour is not bad, being less vulgarly perfumed than most crossings and rather nearer to Riesling in its balance and character. But it does not hold a candle to good Riesling. It is concentrated in the Pfalz and Rheinhessen and at its best is worth bottling as a varietal. There are also plantings in Württemberg, Franken, Mosel-Saar-Ruwer and Saale-Unstrut. It has spread around the globe a bit. Italy, Switzerland, Canada and Japan have some, and England has a bit, too. Best producers: (Germany) Jürgen Ellwanger, W G Flein, Geil, Karl Haidle, Jan Ulrich.

KIRÁLYLEÁNYKA

Hungarian white grape making attractively aromatic, grapy, light wines.

KISI

Georgian white variety often blended with Rkatsiteli (see page 216). It's a difficult grape to grow but rewards with intense aromas of baked apples. If fermented in Georgia's traditional *qvevri*, it will be richer, broader but less fresh. If it survives the local erratic cellar hygiene, it can be very good.

KLEVNER

A name applied in Alsace to several grape varieties, among them the Pinot family and Chardonnay; the same usage is common in German-speaking regions. It can be spelt Clevner. It is not the same as Alsace's Klevener de Heiligenstein, which is Savagnin Rose (see Gewürztraminer pages 112–121). Best producers: (France) Marc Kreydenweiss; (Germany) Andreas Laible.

KOSHU

Japanese *Vitis vinifera* grape. It's not clear whether it was introduced or bred there. It is much promoted now in export markets and has acquired a following, though possibly mostly for its novelty value: in style it is quite fresh, light, neutral and unmemorable.

KOTSIFALI

Cretan variety giving soft, spicy, broad wines and best blended with something with a bit more backbone. Best producers: (Greece) Archanes co-op, Peza co-op, Miliarakis Bros.

LACRIMA DI MORRO D'ALBA

Wonderful wild strawberry and rose-scented red grape saved from extinction in the 1980s, and DOC in Morro d'Alba in central Italy. The wines are dry, with high acidity, though less high than they were a few years ago. A keen local market means that prices are high. The locals drink it as an apéritif, or with salami.

LAGREIN

Interesting, mouthfilling Italian variety found in Trentino-Alto Adige. It has a flavour of sour plums with a touch of grass and bitter cherries, some black chocolate richness and a deep, dark colour, but it is not very tannic, and so it's criticized for not being ageworthy.

Well, first, how long do you want it to age? Its low tannin gives it that rare combination of good chewy dark fruit and easy drinkability. And second, I've seen deep, satisfying chocolate and plum 10-year-old examples that had lost nothing and gained quite a bit by the sojourn in bottle. Some producers give the wine more substance by aging it in *barrique*, but it doesn't really need it. Without some added oak the style of Lagrein is very much in the mould of other northeastern Italian reds, such as Refosco dal Peduncolo Rosso, Marzemino and Teroldego (it is actually related to these last two). If you like them, you'll have a soft spot for Lagrein, too.

Lagrein Dunkel or Scuro is the name of the red wine; Kretzer or Rosato is pink, though usually pretty dark pink. Lagrein di Gries or Grieser comes from just outside Bolzano in Alto Adige. So far Australia and California have taken more interest than New Zealand, where it might also have potential. Best producers: (Italy) Viticoltori Alto Adige co-op, Barone de Cles, Colterenzio co-op, Muri Gries, Gries co-op, Hofstätter, Lageder, Josephus Mayr, Niedermayr, Hans Rottensteiner, Santa Maddalena co-op, Thurnhof, Tiefenbrunner, Peter Zemmer.

LAIRÉN

Not the same as Airén, though often confused with it. Lairén is to be found in Andalucia, southern Spain; Airén is the white grape of La Mancha. Lairén is also known as Malvar and makes an interesting contribution to Vinos de Madrid. Its wines are quite high in alcohol, reasonably high in acidity and can be good in late-harvest versions.

LAMBRUSCO

The Italian grape and the wine have the same name, and in addition the name covers a large number of vines, some of which may have their names specified on the label for DOC wines. But as well as differences between different Lambrusco vines, there are also differences between different Lambrusco wines. The main division is between 'proper' Lambrusco which is quite a challenging, bracing mouthful – fairly dry, slightly sweet-sour, frothy, strawberry-fruited and with a bitter twist on the finish, low in tannin and ideal for drinking with the fatty, pork-based diet of the Modena region in Emilia-Romagna – and industrial Lambrusco, sold in screwcapped bottles, usually sweetened, and nothing like the refreshing character of real Lambrusco.

Proper Lambrusco comes in cork-stoppered bottles, and is DOC. It is also more expensive and more difficult to find outside the region, and especially outside Italy: sweetened Lambrusco was, from the late 1970s, exported to the USA in huge quantities. Pink, white and low-alcohol versions were also invented. There is no white Lambrusco grape: Lambrusco Bianco is made by vinifying the grapes without the skins.

On to the vines. They seem to be native to the region and indeed may have evolved naturally from wine vines in Emilia Romagna

Lambrusco vines are trained high, on pergolas, for Lambrusco wine. In times past they used to be trained up tree trunks: even today, poplars with vines climbing up them are a familiar sight in Italy.

and Piedmont. In any case, they've been there long enough to develop a lot of clonal variation. The best subvarieties (and there are many) are: Lambrusco Marani, which gives good colour; Grasparossa, which is found around the village of Castelvetro, south of Modena; Salamino (its clusters are supposed to resemble the shape of salami sausages), which is found at Santa Croce, east of Sorbara; and the most highly prized, Sorbara, from the village of that name – but there are lots more.

There are four DOCs which, needless to say, do not coincide precisely with the subvarieties. DOC Lambrusco di Sorbara must be made from Sorbara and Salamino grapes; DOC Lambrusco Grasparossa di Castelvetro must be 85 per cent Grasparossa; DOC Lambrusco Salamino di Santa Croce must be 90 per cent Salamino; and DOC Lambrusco Reggiano is made from a blend of Salamino with the lesser Marani, plus the even lesser Maestri and Montericco. (Maestri gives fleshy, bubblegum wine, and is taking over from the higher-acid Marani.) DOC Lambrusco Reggiano is often made *amabile*, or lightly sweet, this sweetness coming from up to 15 per cent of a concentrated grape must coming from a dark variety called Ancellotta. There is a lot of Ancellotta in the region; more, in fact, than of any individual strain of Lambrusco. Lambrusco Reggiano, the most produced wine, is seldom the most interesting of the DOCs. Best producers: (Italy) Casimiro Barbieri, Francesco Bellei, Casali, Cavicchioli, Moro Rinaldo Rinaldini, Vittorio Graziano.

LASKI RIZLING

The name under which Welschriesling (see page 298) is frequently encountered in the countries of the former Yugoslavia. The branded version exported for many years by the state-run winemaking operation gave the grape a bad name; undeservedly so, since the style of that wine owed more to poor equipment, bad handling and the desire to sell the wine at the lowest possible price than it did to the grape's potential. Best producers: (Slovenia) Slovenjvino, Stanko Curin.

LEÁNYKA

Workhorse white grape, normally found in the northeast of Hungary (the Eger region, particularly), producing mild, soft wines of no great excitement. It used to be linked with Romania's rather better Fetească Albă (see page 98). Some growers are taking it more seriously than of yore. Best producer: (Hungary) Hungarovin.

Kiona

Kiona makes one of the few Washington State Lembergers. The grape is well suited to the local dry, sunny conditions and produces a bright blackberry and pepper-flavoured wine on Red Mountain, which is generally a citadel of Cabernet power.

LEMBERGER

Blaufränkisch (see page 48) is known by this name in Germany, New York and in Washington State. Limberger is an alternative German name but less common. In Germany it is found in Württemberg, where it produces pale-coloured wines that are light but have decent acidity. It is often blended with Trollinger (see page 284). Best producers: (Germany) Graf Adelmann, Drautz-Able, Fürst zu Hohenlohe-Öhringen; (USA) Covey Run, Hogue Cellars, Kiona.

LEN DE L'EL

The name of this grape, native to southwest France, means 'far from sight' – it is a corruption of 'Loin de l'oeil'. A curious name for a vine, you might think: the clusters have long stalks, and so are a long way (well, relatively) from the eye, or bud, from which they sprang.

It's true that plantings have declined, but there is a minimum (15 per cent) of Len de l'El required in the blend for white Gaillac (although Sauvignon Blanc is allowed as an alternative), so it is to be hoped that this characterful grape will not disappear completely from view. It is low in acidity and the grapes rot easily, but the wine is powerful, weighty and good. Best producers: (France) de Causses-Marines, de Labarthe, Labastide-de-Lévis co-op.

LIMBERGER

A less common German name for Lemberger (see above) and the Blaufränkisch of Austria (see page 48).

LIMNIO

One of Greece's most important red varieties, this is still found on its original home of Lemnos, an island in the northern Aegean Sea, though for some reason is not used for appellation wine there. It is also found in the northeast of Greece, particularly Macedonia and Thrace, where it seems to perform better, and it contributes colour, weight, acidity and a flavour of bay leaves to a blend. Best producer: (Greece) Domaine Carras.

LIMNIONA

With luck, we'll be seeing more of this Greek black grape, because it has lovely spicy aromas and flavours and a good silky texture. It also ages well. There's not much of it planted at the moment, but more and more producers are looking at it. Best producers: Christos Zafirakis, George Tsibidis.

LISTÁN

Another name for Palomino, one of the world's most boring grapes – but one that paradoxically can produce extraordinarily complex flavours when turned into sherry. It is found also in France's Languedoc and in the Armagnac vineyards of Gascony, but it is being replaced here by better varieties. As Listán Bianco it is important in the Canary Islands. See Palomino Fino page 178.

LISTÁN NEGRO

Unrelated to Listán, this is found on the Canary Islands, where it is blended with such others as Negramoll (the Tinta Negra Mole of Madeira), Prieto, Tintilla, Malvasia Rosada and others. In Tacoronte-Acentejo on Tenerife it may make up most of the blend, or be made as a varietal.

It is the main red vine on Lanzarote, where the vines are planted in hollows in the black volcanic soil, and sheltered from the strong humid winds by low stone walls. Its wines are mostly soft and appealing, without tremendous weight, and may be made more substantial by oak aging, or softer and lighter by carbonic maceration. There is also a small amount of sweet red made on Tenerife from Listán Negro. Best producers: (Spain) Insulares Tenerife, La Isleta, Monje, La Palmera.

LLADONER PELUD

Another name for Spain's Garnacha Peluda, or hairy Garnacha. A southern French name is Grenache Poilu or Velu. The only difference is that the underside of the leaves have more down on them than normal Garnacha Tinta/Grenache Noir (see pages 102–111). It is specified in many ACs in Languedoc-Roussillon but doesn't make much of an impression. The grapes are less prone to rot, but the flavour of the wine is very similar to that of Garnacha/Grenache though with rather better acidity.

LOURENO

Aromatic grape used for Vinho Verde in the north of Portugal and, as Loureira, for the very similar wines of Rías Baixas, over the border in northwest Spain.

Its aroma is reminiscent of bay leaves, its taste of apricots, with high acidity and low alcohol. It seems especially suited to the cooler parts of the region around Braga and near the coast; the flavour can sometimes remind one of young Riesling. Yields are high and the wine is either used as a varietal or blended with Trajadura and Paderña in Vinho Verde, or, in Rías Baixas, made as a varietal or blended with Albariño and/or Treixadura. Best producers: (Portugal) Quinta da Aveleda, Quinta da Franqueira, Sogrape, Solar das Bouças, Quinta do Ameal, Quinta do Tamariz.

MACABEO

A non-aromatic variety found across northern Spain and, as Maccabéo or Maccabeu, in France's Languedoc-Roussillon. It brings neutrality, a resistance to oxidation and sometimes good palate weight to many white blends, but rarely flavour or character.

As Viura (see page 298) it is the backbone of white Rioja, along with lesser quantities of Malvasia and Garnacha Blanca. Here, it produces a powerful custardy, orange blossom flavour when fermented in oak, and is also beginning to show form with crisp, unoaked wines to drink young.

Macabeo is used for Cava both in Cataluña and elsewhere in northern Spain: Catalan Cava contains an admixture of Parellada and Xarel-lo, though the small amount made in other regions tends to be pure Macabeo.

In Languedoc-Roussillon Macabeo is used for *vins doux naturels*, and is blended into many white wines in Minervois and Corbières. Its affinity for hot, dry climates – it is prone to rot in wet ones – means that it has also been grown in North Africa. Best producers: (Spain) Agramont, Cosecheros Alaveses, Castellblanch, Cavas Hill, Codorníu, Franco Españolas, Freixenet, Marqués de Monistrol, Masía Barríl, Pirineos; (France) Casenove, Vignerons Catalans, Les Vignerons du Val d'Orbieu.

MADELEINE ANGEVINE

A bit of confusion here. There's a dual-purpose table and wine grape vine known as Madeleine Angevine, which has pretty much died out in France but does pop up in cool, wet places because it does ripen very early. I've even had Danish and Swedish examples. And there's Madeleine x Angevine 7672 – you'd think this was just another Madeleine Angevine, but no, it seems to have come from Germany, where they don't always get round to naming their new vine crosses. Because it is very early-ripening, it has had some success in England, where – correct me if I am wrong – they usually just call it Madeleine Angevine. Well, wouldn't you? Best producers: (England) Stanlake Park, Three Choirs.

MAGLIOCCO

There are two Magliocco grapes, both found in Calabria, and unrelated. Magliocco Dolce is the main one, Magliocco Canino the other. Both feature tannin pretty heavily.

MAIOLINA/MAJOLINA

Italian black grape local to the Brescia region of Lombardy. It has deep red berries ('pigeon's blood,' says Cantina Majolini, which makes a varietal – the name of the cellar is coincidental) and makes wine of relatively light body and lots of cherry perfume. As a vine it has big clusters of thin-skinned grapes and dislikes both excessive sun and excessive rain. At the time of writing, fewer than 1000 plants exist, but it's worth rather more attention.

MALAGOUSIA/MALAGOUSSIA

Very delicious Greek variety making a comeback. The grapes are large, thin-skinned and susceptible to rot, and it's an early ripener. The wines are gloriously aromatic without being flashy, though poor examples can be heavy: rose petals, ripe citrus, basil are all keynotes. Some producers practise skin contact and a few ferment in barrel. Acidity is on the low side, and sometimes it's blended with Assyrtiko to remedy that. On its own, it will age successfully for several years in bottle.

MALBEC

See pages 130–131.

MALBECK

An occasional Argentinian spelling of Malbec (see pages 130–131).

MALMSEY

More a style of wine than a grape, Malmsey is a corruption of Malvasia (see pages 132–133), and was in the past used generically for the sweet wines of Greece and the eastern Mediterranean. It is now used specifically for the sweetest style of Madeira.

However, true Malmsey or Malvasia vines have almost disappeared from the island. Only 13–14 per cent of the vines on Madeira are noble varieties – that is, Sercial, Verdelho, Bual or Malmsey – with the rest being Negramoll (previously known as Tinta Negra Mole) or hybrids. True Malmsey Madeira is thus a fairly rare wine, though any bottle that calls itself Malmsey on the label should be made from Malvasia grapes. When young the wine has an attractive floral orange blossom smell; after the characteristic Madeira aging process it becomes pungent, intense and caramelly, rich and smoky, but never loses its acid tang.

The usual Malvasia grown on Madeira is Malvasia de S. Jorge, a modern variety only introduced to Madeira in 1970; it rapidly took over from the traditional Malvasia Candida which had been the island's Malvasia since the 15th century when the Portuguese first arrived. It is subject to irregular fruit set, but achieves good size and acid levels in early ripening, being picked at between 11.5 and 15 per cent potential alcohol. Best producers: (Portugal) Barros e Sousa, Blandy, Cossart Gordon, Henriques & Henriques, Leacock, Rutherford & Miles.

Domaine Gerovassilou

Malagousia is one of several almost extinct Greek varieties that are now making some of Greece's most individual wines. The Gerovassiliou version is wonderfully scented and lightly oaked.

Blandy's

Malmsey wine now has to come from 85% Malvasia grapes, rather than from the inferior Negramoll. Even so, less than 40ha (100 acres) of Malvasia remain on Madeira.

MALBEC

Well, this grape is a native of southwest France. But try telling that to a modern wine drinker. It's far more likely that today's enthusiast will say 'Argentina'. And they'd be right. But 10 years ago when asked to respond to the question – 'What is Malbec'?, you'd have been met with a blank stare and a distinct lack of interest. Thanks to Argentina, Malbec is now a seriously thrilling star in the red wine firmament.

In its birthplace of Bordeaux and its traditional French home base of Cahors, it was going nowhere – fast. It's a soft, juicy grape that gives lovely dark, damsony, perfumed, purple wine in a dry warm climate – but in Bordeaux, it was not replanted after the severe frost of 1956. So its ability to soften the Bordeaux blend has been supplanted by the Merlot – which is less susceptible to *coulure* and rot and gives a more regular crop.

It is a blending component in many southwest reds, but leads the field only in Cahors, where it must comprise at least 70 per cent of the blend. In a warm dry year (not that common) it can give deep dark wines tasting of damson skins and tobacco leaf. It is grown to a limited extent in the Loire, where it is called Cot, but it rarely ripens fully there.

It was first propagated in Argentina in 1852 (the oldest vines still growing are from 1861), but the cuttings came from Bordeaux, not from Cahors, and the Malbec in Argentina now seems to be rather different from that in Cahors. There is far greater clonal diversity in Argentina than in France, with everything from vines with small berries and lots of concentration to the opposite. The ideal spot for it in Mendoza is 1000m (3280ft) altitude and above – which covers a lot of land including the Uco Valley, because the ground rises to the Andes very gradually. Vistalba, Luján de Cajo, Agrelo are leading areas near Mendoza City; La Consulta, Tupungato and Gualtallary are important in Uco. The newest, highest-altitude wines, planted at 1440m (4700 ft) and above, have elegance, finer tannins, better acidity and fabulous, dark-scented fruit.

A vine-pull scheme in Argentina, in the 10 years to 1993, uprooted a lot of Malbec, most of it more than 50 years old.

Chile already has some Malbec, and it's doing surprisingly well in cool spots. California uses it as a blender and Washington State in its own right. There are some historic plantings in South Australia, and a few plots in New Zealand, where it contributes richness and scent.

THE TASTE OF MALBEC

At its best, carefully grown and skilfully vinified in Argentina, Malbec has a dark purple colour, a thrilling damson and violet aroma (this violet aroma comes from a molecule called beta-ionnona, since you asked), a lush fat dark fruit flavour and a positively soothing ripe tannic structure. It can take new oak aging, but it's a pity to smother its natural delicious ripeness with wood. In Cahors the flavour is more likely to be raisins, damson skins and tobacco. In both Chile and Australia you occasionally get the violet perfume and you usually get the soft, ripe, lush texture, but Argentina, by taming the tannins and maximizing the perfume and the lush, chubby texture, is leading the way.

Nicolás Catena can take a lot of the credit for propelling Malbec onto the world stage. He returned to his family winery in Argentina from the USA keen to make wine in the style of California's Napa Valley Cabernets, but, realizing that by far the best supplies of mature vines in Argentina were Malbec, had no trouble in focusing his attention on them. Mendoza is, in fact, ideally suited to Malbec, with excellent vineyard sites rising from about 1000m (3280ft) in height up to 1500m (4900ft). At higher altitudes, the big difference between day and night temperatures makes the grapes take several more weeks to ripen compared to the hotter, lower altitude Mendoza sites, giving wines of great concentration, balance and scent.

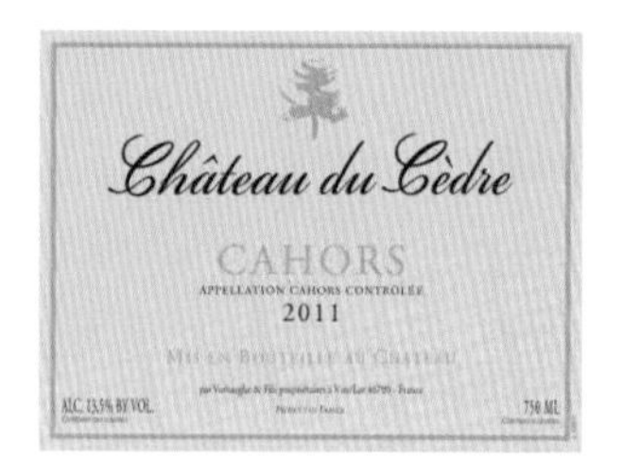

Château du Cèdre

Malbec, alias Auxerrois, is the mainstay of Cahors in southwest France: this cuvée, called Le Cèdre, is 100% Auxerrois from 30-year-old vines. Aged in new oak barrels for 20 months, it will last for at least 10 years.

Juan Benegas

The Benegas family are one of the pioneers of Argentine wine. This Malbec marries grapes from 100-year-old Maipu vines with high-altitude Gualtallary fruit in a fine example of modern Argentine Malbec.

Left: Malbec vineyards in the region of Luján de Cuyo in the province of Mendoza, Argentina with the Andes in the background. The cold nights found at high altitudes here extend the growing season, allowing crucial extra ripeness while still conserving acidity. Even so, it's difficult to get soft tannins at less than 14% alcohol. And while Malbec rots in wet weather, there is precious little rain this side of the Andes, so rot is not a problem. ***Above:*** *Malbec may be blended with Cabernet and Merlot in small amounts in New Zealand and Australia, for extra aroma and fatness of texture.*

CONSUMER INFORMATION

Synonyms & local names

Known in the Loire Valley and the South-West of France as Cot or Côt – in Cahors it is called Auxerrois but this is not the same as the white Auxerrois of Alsace. On Bordeaux's Right Bank, around Libourne, it is called Pressac. In Argentina, where it is called Malbec, the spelling Malbeck is also found.

Best producers

FRANCE Armandière, la Caminade, du Cèdre, Clos la Coutale, Clos de Gamot, Clos d'un Jour, Clos Triguedina, Cosse-Maisonneuve, les Croisille, de Gaudou, Lamartine, Mas des Etoiles, Mas del Périé, du Prince, la Reyne, les Rigalets.
ARGENTINA Achaval-Ferrer, Alamos Ridge, Alta Vista, Altos Las Hormigas, Altos de Medrano, Altos de Temporada, Andeluna, Anubis, Balbi, Juan Benegas, Luigi Bosca, Bressia, Humberto Canale, Catena Zapata, Colomé, Domaine Vistalba, Etchart, Fabre, Finca El Retiro, O Fournier, La Agricola, La Rural, Medalla, Nieto Senetiner, Noemía, Norton, Salentein, Terrazas de Los Andes, Michel Torino, Trapiche, Weinert.
CHILE Altamana, Viña Casablanca, De Martino, Concha y Toro, Lomo Larga, Montes, MontGras, Morandé, Odfjell, Valdivieso, Viu Manent.
AUSTRALIA Jim Barry, Bleasdale, Cullen, Delatite, Ferngrove, Leasingham, Wendouree, Zonte's Footstep.
NEW ZEALAND Esk Valley, Mills Reef.
SOUTH AFRICA Druk my Niet, Fairview.

RECOMMENDED WINES TO TRY

Fifteen Argentinian wines

Alta Vista Serenade
Altos Las Hormigos Terroir
Andeluna 1300 Malbec
Anubis Mendoza Malbec
Benegas Estate Malbec
Finca El Retiro Malbec Mendel Finca Remota
Catena Zapata Nicasia Vineyard Malbec
Bodegas Fabre Grand Vin
La Agricola Santa Julia Malbec Reserva Nieto
Noemía J Alberto
Norton Reserva Malbec
O Fournier Alfa Crux Malbec
Senetiner Cadus Malbec
Terrazas de Los Andes Las Compuertas
Zuccardi Aluvional

Ten Cahors wines

Ch. du Cèdre Cahors le Cèdre
Clos la Coutale Cahors
Clos de Gamot Cahors Cuvée Vignes Centénaires
Clos d'un Jour Un Jour
Clos Triguedina Cahors Prince Probus
Dom. Cosse-Maisonneuve Le Sid
Ch. les Croisille Cahors Tradition
Ch. de Gaudou Cahors
Ch. Lamartine Cahors Cuvée Particulière
Ch. la Reyne Cahors

Nine other wines containing Malbec

Jim Barry McCrae Wood Cabernet Sauvignon Malbec (Australia)
Viña Casablanca San Fernando Miraflores Estate Malbec (Chile)
Cullen Mangan East Block (Australia)
Esk Valley Hawkes Bay Reserve The Terraces (New Zealand)
Montes Reserve Malbec (Chile)
MontGras Reserva Malbec (Chile)
Morandé Chilean Limited Edition Malbec (Chile)
Valdivieso Reserve Single Vineyard Malbec (Chile)
Wendouree Cabernet/Malbec (Australia)

MALVASIA

I think the most gorgeous Malvasia I ever had was pink and frothy, heady with the scent of rose petals and hothouse grapes, and it was from northern Italy. We were all entranced by it, but nobody could really say why it was like that – or even if it had ever been like that before. But that's the thing about Malvasia – it's not so much a single grape as a whole family. France uses the name of Malvoisie as a synonym for a whole range of grapes, none of which are Malvasia Bianca. And the various Malvasias are mostly unrelated; so think of the name as being a style rather than a family. Most Malvasia is white, but there are a sufficient number of black versions to produce a range of pink and red wines.

Italy and Iberia are its main homes. As a dry white it is blended with Trebbiano all over central Italy, generally to Trebbiano's benefit because while Trebbiano is often marked by a singular lack of flavour or texture, Malvasia is generally quite fat and soft and should add a bit of peachy plumpness when young and nutty depth when mature. But not too mature: one or two years old is mature in these sorts of wines. In Lazio's Frascati, where Malvasia certainly improves the blend, the local version is Malvasia di Candia, though the more aromatic Malvasia Puntinata or Malvasia del Lazio is regarded more highly and makes wines with less of a tendency to flabbiness. Emilia's Malvasia is also called Malvasia di Candia. There is also a Malvasia di Toscana, which is usually only fit for uprooting. In Friuli there is Malvasia Istriana, which is well structured and flavoursome, producing bright, fresh, scented whites here and in Slovenia and Croatia. A few producers barrel-age it and that can produce wines of surprising intensity.

Passito or fortified sweet white Malvasia – the latter have more aroma – is a speciality of southern Italy, though many wines are almost extinct. The few survivors are really good. Malvasia delle Lipari is the best known; others are Malvasias di Bosa, di Grottaferrata, di Planargia and di Cagliari. Black Malvasia Nera is found in Puglia and Tuscany, and adds a very interesting dark, rich quality to Tuscan Sangiovese reds as well as a chocolaty, sometimes grapy warmth to Puglian blends. There is also some in Piedmont.

Malvasia is found in Rioja and Navarra in Spain and old-fashioned, barrel-aged white Riojas benefit from the Malvasia fatness fleshing out the skinnier Viura. You find it in different forms in Portugal, especially the Douro (Malvasia Fina and the inferior Malvasia Rei, which could be the same as Palomino), Beiras, Estremadura and Madeira, where it produces the rich, smoky, acidic Malmsey (see page 129). There is a scented version in California, and a little in Greece, and knowing Malvasia, it'll keep turning up in a good few more places yet, under some pseudonym or other.

THE TASTE OF MALVASIA

With so many subvarieties it is difficult to generalize about Malvasia's flavour, but where lesser versions can be flabby and low in acidity, tending to oxidize easily, better Malvasias are characterful and aromatic, though not as exotically so as Muscat, with flavours of peaches, apricots and white currants. Istrian Malvazija from Croatia is fresh and scented with wild flowers and peach blossom. Sweet ones are apricotty and rich, red ones dark and chocolatey. In Madeira, where it's usually called Malmsey, the taste is intense, smoky and treacly but cut with a very insistent acidity.

The word 'Malvasia' is rarely found on Madeira labels. The traditional name for Madeira wine made using Malvasia grapes is Malmsey. Indeed, for a time the name Malmsey simply referred to the darkest and richest style of Madeira wine and other grapes – in particular Negramoll – could be used. But despite Malvasia grapes being rare on the island, a serious independent producer like Justino will be able to ensure his wines are made of the grape stated on the label. About time too.

La Stoppa
One of Malvasia's many variants. Ths one from La Stoppa in the Colli Piacentini in Emilia-Romagna is sweet, fizzy and low in alcohol.

Henriques & Henriques
Henriques & Henriques is one of the only Madeira wine companies to have significant vineyard holdings. The wines are always finely focused.

Left: *Barrels of Malmsey at the Madeira Wine Company. Because Madeira gains its character from the aging process more than from the grape or from the winemaking, Malmsey Madeiras do not have the peachy, apricot aromas of most Malvasias. Malmsey is the sweetest style of Madeira but even so never loses its acid bite and smoky tang. In Madeira several different strains, or even different grapes, go by the name of Malvasia. Plantings of all Malvasias put together on the island are very small.*
Above: *Malvasia is one of the hardest of all vines to pin down, simply because it isn't just a single vine, and not all vines called Malvasia are necessarily related. They're not even necessarily the same colour.*

CONSUMER INFORMATION

Synonyms & local names

Widely planted in Italy and Iberia, there are many different subvarieties of Malvasia Bianca, usually called Malvasia this or that. Not to be confused with the name Malvoisie which is rarely used as a synonym for Malvasia. Malmsey is a style of wine using Malvasia grapes on the island of Madeira.

Best producers

ITALY/Friuli-Venezia Giulia Borgo del Tiglio, Ca'Ronesca, Paolo Caccese, Cormons co-op, Sergio & Mauro Drius, Edi Kante, Lorenzon/I Feudi di Romans, Eddi Luisa, Marega, Alessandro Princic, Dario Raccaro, Schiopetto, Villanova; **Emilia-Romagna** Forte Rigoni, Luretta, Gaetano Lusenti, La Stoppa, La Tosa, Vigneto delle Terre Rosse; **Tuscany** (Vin Santo) Avignonesi, Badia a Coltibuono, Fattoria di Basciano, Bibbiano, Bindella, Cacchiano, Capezzana, Fattoria del Cerro, Corzano e Paterno, Fontodi, Isole e Olena, Romeo, San Felice, San Gervasio, San Giusto a Rentennano, Selvapiana, Villa Sant'Anna; **Lazio** Castel del Paolis; **Sicily** Colosi, Carlo Hauner.
SPAIN Carballo, El Grifo, Abel Mendoza.
PORTUGAL/Madeira Barbeito, Barros e Souza, H M Borges, Henriques & Henriques, Vinhos Justino Henriques, Madeira Wine Company (Blandy, Cossart Gordon, Leacock, Rutherford & Miles), Pereira d'Oliveira.
CROATIA Coronica, Kabola.
USA/California Bonny Doon, Robert Mondavi, Sterling Vineyards.

RECOMMENDED WINES TO TRY

Ten dry Italian wines

Borgo del Tiglio Collio Malvasia
Castel del Paolis Frascati Superiore Vigna Adriana
Sergio & Mauro Drius Friuli Isonzo Malvasia
Forte Rigoni Colli di Parma Malvasia
Edi Kante Carso Malvasia
Luretta Colli Piacentini Malvasia Boccadirosa
Dario Raccaro Collio Malvasia
Schiopetto Collio Malvasia
La Tosa Colli Piacentini Malvasia Sorriso di Cielo
Vigneto delle Terre Rosse Colli Bolognesi Malvasia

Five sweet Malvasia Italian wines

Colosi Malvasia delle Lipari Passito di Salina
Carlo Hauner Malvasia delle Lipari Passita
Isole e Olena Vin Santo
La Stoppa Colli Piacentini Malvasia Passita Vigna del Volta
San Giusto a Rentennano Vin Santo

Three Spanish Malvasia wines

Carballo La Palma Malvasía Dulce (sweet) (Spain)
El Grifo Lanzarote Malvasía Dulce (sweet) (Spain)
Abel Mendoza Rioja (Blanco) Fermentado en Barica (Spain)

Five Madeira wines

Barbeito Malmsey
Blandy 15-year-old Malmsey
Cossart Gordon 1920 Malmsey
Henriques & Henriques 15-year-old Malmsey
Justino 10-year-old Malvasia

MALVASIA NERA

A black form or forms of Malvasia (see pages 132–133). In Italy Malvasia Nera is found in the province of Bolzano in Alto Adige, and in Puglia, Sardinia, Basilicata, Calabria, Tuscany and Piedmont. It may be blended with Negroamaro in Puglia, and Sangiovese in Tuscany. It often has a black plum richness and hint of floral scent. In Piedmont it has two DOCs, Malvasia di Castelnuovo Don Bosco and Malvasia di Casorzo. Both can be sweet, sparkling and gently aromatic. Best producers: (Italy) Francesco Candido, Leone de Castris, Roda del Golfo, Cosimo Taurino.

MALVOISIE

This ought to be a French synonym for Malvasia, but it isn't. Instead, it is a name applied locally to several different varieties in France, among them Pinot Gris (Malvoisie is also the name by which Pinot Gris is known in Switzerland's Valais), Maccabéo, Bourboulenc, Clairette, Torbato and Vermentino (also known by the name of Malvoisie in Spain and Portugal). Best producers: (France) Vignerons Catalans, Jacques Guindon.

MAMMOLO

Mammole is the Italian word for 'violets', of which Mammolo is said to smell. It is planted in such small quantities now that few Tuscan reds owe much to it. It is permitted in Chianti but has largely been bypassed in the rush to improve Sangiovese. There is also a little planted in the Vino Noble di Montepulciano region and it appears in Corsica as Sciacarello. Best producers: (Italy) Antinori, Boscarelli, Contucci, Dei, Poliziano, Castello di Volpaia.

MANDILARIA

Widely planted Greek vine that makes dark, powerful, tannic wine, and is found on various islands, including Crete. Best producers: (Greece) CAIR, Archanes co-op, Paros co-op.

MANTONICO

A Calabrian grape good for both dry and sweet passito wines. Montinico is a synonym for Pagadebit (see page 178). Best producer: (Italy) Librandi.

MARATHEFTIKO

Cypriot variety traditionally grown to beef up the less interesting Mavro, but now being taken a bit more seriously. The wines have lots of colour and body, and are nicely aromatic. Best producers: Aes Ambelis, Ezousa, Vasa.

MARÉCHAL FOCH

Early ripening and with good resistance to winter cold, this French hybrid is popular in Canada and New York State, where it may be made as a varietal. It makes attractively soft, sometimes jammy, sometimes smoky red wines, and may be aged in oak for more substance. Carbonic maceration is used to make lighter wines. It is named after the First World War French general. Best producers: (Canada) Stoney Ridge, Quails' Gate; (USA) Wollersheim.

MARIA GOMES

The name given to Fernão Pires (see page 98) in the Bairrada region of northern Portugal. Best producers: (Portugal) Caves Aliança, Luís Pato, Messias, Quinta de Pedralvites.

MARSANNE

See pages 136–137.

MARSELAN

A 1961 French crossing of Cabernet Sauvignon and Garnacha favoured by a few growers in southern France and Spain, and also the Americas and China. They have good reason: it makes structured, balanced wines of good colour and aroma.

MARZEMINO

Don Giovanni operatically enjoys Marzemino just before being swallowed up in Hell, and while Marzemino may not be everyone's idea of the perfect last wish, it is very attractive in less desperate circumstances.

It is found in Trentino and to a lesser extent in Veneto and Lombardy, and has typically northern Italian grassy acidity and cherryish fruit, combined with good colour and ripeness.

Armando Simoncelli
Marzemino is one of the most refreshing and mouthwatering of northern Italy's red grapes – making perfect mountain picnic wine.

In Trentino it is usually made as a varietal wine for drinking young; in Lombardy it may also be blended with Sangiovese, Barbera, Groppello and Merlot, though probably not all at once. Sweet *passito* versions also exist, perhaps blended with other grapes. DNA fingerprinting has shown that it is closely related to Teroldego, and fairly closely to Lagrein. Best producers: (Italy) Battistotti, La Cadalora, Cavit, Concilio Vini, De Tarczal, Isera co-op, Letrari, Mario Pasolini, Eugenio Rosi, Simoncelli, Spagnolli, Vallarom, Vallis Agri.

MATARO

An Australian name for Mourvèdre (see pages 152–153), now little used.

MATURANA

An old Riojano vine which practically died out in the early part of the 20th century, this has now been rediscovered and authorized for the Rioja blend. There is also a red version, Maturana Tinta, not related and probably the same as Trousseau and Bastardo. Both colours are regarded in Rioja as of high quality.

MAUZAC

A grape of southwest France with a rustic, green apple skins flavour that is best in sparkling versions, Mauzac is found principally in the Gaillac and Limoux regions. It probably takes its name from either the town of Mauzac in Haute-Garonne, or from that of Meauzac in Tarn-et-Garonne. It is usually blended with other grapes – Len de l'El in Gaillac, Chardonnay and Chenin Blanc in Limoux. In Gaillac it appears in many guises, from dry to sweet, still to sparkling. In Limoux it has been partly supplanted by Chardonnay.

In Limoux it is made into sparkling Blanquette de Limoux and Crémant de Limoux, though usually by the Champagne method rather than the old *méthode ancestrale* (this basically relied on the fermentation dying down in the autumn – Mauzac ripens late and so the fermentation was not finished before the winter – and then starting again in the spring, by which time the only partially fermented wine would have been bottled). Now the grapes are more likely to be picked earlier for acidity and the Champagne method is used to make a good sparkling wine. Mauzac can produce berries that vary in colour from green through russet and pink to black; Mauzac Noir from the Tarn, however, is a different vine altogether. Best producers: (France) l'Aigle, Gineste, Maison Guinot, Labarthe, Robert Plageoles.

MAVRO

The main red grape of Cyprus is of pretty low quality, and there may be limits to what the most inventive modern winemaking can do with it. Since the name merely means 'black' it may, however, cover several different grapes. It is used in the local sweet Commandaria.

MAVRODAPHNE

A Greek vine which makes sturdy, rich red wines, in particular the porty Mavrodaphne of Pátras. It is nearly always made sweet, and treating it this way seems to bring out all its aroma; dry versions, are, however, now being made, with more concentration and extraction. Best producer: (Greece) Achaia-Clauss.

MAVROTRAGANO

Greek vine found on Santorini; the name is probably used for more than one variety. It was on the point of dying out when it was rediscovered, and while still grown in tiny amounts, its wines are impressive. It's best made into a dry wine, but the tannins need to ripen fully to soften their grip; the wines are dark and spicy, very aromatic, and with a vibrant personality. Best producer: (Greece) Hatzidakis.

Hatzidakis

The Mavrotragano variety covers only 2% of Santorini's vineyards, but its dark, spicy, powerful wine is impressive, by itself or in blends.

MAVRUD

Mainly found in Bulgaria, late-ripening, small-berried Mavrud makes weighty, solid red wines of character but no particular elegance and it adds a broad-shouldered rustic power when blended with Cabernet or Merlot. It is high in tannins and colour and takes well to oak aging. Best producers: (Bulgaria) Assenovgrad, Maxxima, Santa Sarah.

MAZUELO

The Rioja name for Carignan/Cariñena (see page 64). It brings acidity, colour and tannin to the blend, but is less prized than Tempranillo. Best producers: (Spain) Amézola de la Mora, Berberana, Martinez Bujanda, CVNE, Muga, La Rioja Alta, Marqués de Riscal.

MELNIK

Bulgarian vine capable of quite high quality wines. It is found mainly around the town of the same name, near the Greek border, and it needs plenty of heat to ripen. Its wines have very good colour and balance and sometimes assertive tannins, and a certain rich warmth, particularly with oak aging. They can be smoky, coffeeish and toffeeish and very appealing, and can improve in bottle. There is a softer, earlier ripening clone that can be easier to drink.

MELON DE BOURGOGNE

This is the one and only grape behind Muscadet, and so closely is it identified with the wine that the grape's other name is Muscadet. But as one might gather from this name, it originally came from Burgundy and travelled west to reach the western end of the Loire Valley around the city of Nantes. It had been much planted in Burgundy, until its destruction was ordered in the early 18th century whereupon, according to the ampelographer Galet, the growers, in order to save such a useful vine, engineered confusion with Chardonnay and managed to get some Chardonnay torn up in its place. There is still some, albeit only a little, planted on the Côte d'Or.

The Burgundy growers favoured Melon so much because of its resistance to cold – although it buds early, so is apt to be hit by spring frosts – and its generous and regular yields. Its secondary buds are fertile, so even after frost it may be able to produce some sort of a crop. It is, however, subject to some of the growers' worst nightmares, in the form of downy and powdery mildew and grey rot, and may have to be picked before complete maturity to avoid the worst effects of these.

In the Pays Nantais, where it was introduced after the freezing winter of 1709 which pretty much destroyed whatever grapes were there, it has a poor reputation for high acidity and low flavour, and the best that is usually said of it is that it goes well with the local *fruits de mer*. But in fact winemaking standards have been rising – skin contact, lees stirring and barrel fermentation for the best wines (not usually new barrels, which can unbalance the flavours) all help to give greater weight and richness. There are even some *vins de garde* being made, and these can improve in bottle for a decade, becoming honeyed with flavours of quince and ripe greengage. Such wines are, of course, not the norm. Muscadet is still at heart a simple wine for drinking young, but even the *négoçiants*, who still control 80 per cent of the market, and supermarkets have started to stock better examples. In an era where we are turning against oaky styles and too much aggressive Sauvignon, there's something to be said for the mellow, reflective charm of decent Muscadet. Growers' wines are the best bet. *Sur lie* wines, bottled off their lees, have the greatest depth, and the schist soils of the Sèvre-et-Maine region give the best quality.

The Coteaux de la Loire appellation has good *terroir* but is small, and the sandy soil of the Côtes de Grand Lieu appellation produces light wines for very early drinking.

In California and Australia there has been considerable confusion between Melon and Pinot Blanc, with even cuttings from the University of California at Davis being wrongly identified as Pinot Blanc. Plantings are more or less sorted out now, however. Best producers: (France) Chéreau-Carré, Luc Choblet, l'Ecu, Pierre Luneau, Louis Métaireau, Ragotière.

MENCÍA

This grape of northwestern Spain traditionally gives light, fresh, acidic reds with a raspberry and blackcurrant leaf flavour not unlike a slightly raw Cabernet Franc, and good tannin. It has become increasingly fashionable and is now making many darker examples, often heavily influenced by oak. Which is a pity, since Mencía is delicious drunk young, without oak. It is a rather reductive variety, however, and an enthusiastic adoption of screwcaps has created too many rather sulphidic examples. More careful winemaking will cure this.

It is found in Valdeorra, Ribeiro and, above all, in Bierzo, where it is the main red grape and where it may be blended with Garnacha Tinta. It is delicious young, and doesn't need the oak it is sometimes smothered in. It is the same vine as Portugal's Jaen where it can also make delightful juicy reds (see page 126). Best producers: (Spain) Jesús Nazareno co-op, Moure, Priorato de Pantón, Vire dos Remedios

MENU PINEAU

A grape of the Loire Valley, also known as Arbois, now in decline. The wine is relatively low in acidity and supple.

MARSANNE

I used to think that Marsanne and Roussanne were the Siamese twins of white Rhône grapes; one was never mentioned without the other, but one was also clearly more favoured. Roussanne was always described as far more fragrant and refined, and the clumsier Marsanne was chided for marching in and taking over most of Roussanne's best sites. Well, it's true Marsanne has largely supplanted Roussanne in the northern Rhône – it's a far more reliable cropper which inevitably influences growers' decisions – but it's a good grape. And if the only Marsanne you had ever tasted was Tahbilk's in Australia or the mighty, throat-filling beautiful beast of a mature Hermitage you'd say it was a superb grape. Anyway, it dominates the white blends in Hermitage, Crozes-Hermitage, St-Joseph and St-Péray but is not permitted in Châteauneuf-du-Pape, whereas Roussanne is.

If Marsanne is to have character it must be grown in the right place, and then vinified with care. In too hot a climate the wine produced becomes flabby; in too cool a spot it has an undeveloped simple, bland flavour. And it's another of those grapes you can't overcrop – it becomes neutral and characterless. As for new oak, more than just a splash squashes the grape's personality.

Most Marsanne is made to be drunk young, within just a few years of the harvest. Grapes for this style are often picked before full ripeness to retain acidity, but you risk having a wine without any real taste. That's why in the Languedoc Marsanne is often blended with the more aromatic Viognier. If you do let the grapes ripen fully on the vine, as they often do in Hermitage and sometimes in Australia, you get high-alcohol wines that can age a good while, and though they often go through a dumb phase after a few years when they taste flat and unforthcoming, after about 10 years they emerge darker in colour and more complex in flavour, and with an oily, weighty, honeyed character accompanied by nuts and quince fruit. Hermitage and the better wines of the Rhône are the best candidates for this sort of aging; there are also a few suitable examples in Australia, where it is planted in small quantities in Victoria for high-quality wine.

There is some Marsanne in Switzerland, where it produces good wine called Ermitage in the Valais region, and there's some in France's Savoie, where it is known as Grosse Roussette or Avilleran. California has a bit of Marsanne, but so far the wine tends to be dull and gluey in flavour.

THE TASTE OF MARSANNE

When young it has a minerally edge, often with a citrus, peachy flavour. With age this matures into a rich palate of honeysuckle and jasmine, acacia honey and perhaps a touch of apricot or quince; it is aromatic, quite oily, nutty, and surprisingly heavyweight. Accordingly, it can partner rich food very well when it's mature. As a young wine it's equally good with or without food.

The Tahbilk winery in central Victoria is one of the most beautiful in Australia, and has Australian National Trust classification. Tahbilk's association with Marsanne can be traced as far back as the 1860s although none of these plantings have survived. Vines dating from 1927 are still in use and reputed to be the oldest Marsanne vines in the world. Chateau Tahbilk Marsanne was served to the young Queen Elizabeth II when she visited the winery in 1953. The wine has a distinct honeysuckle aroma and flavour and is very good young as well as improving with up to 12 years' aging in bottle.

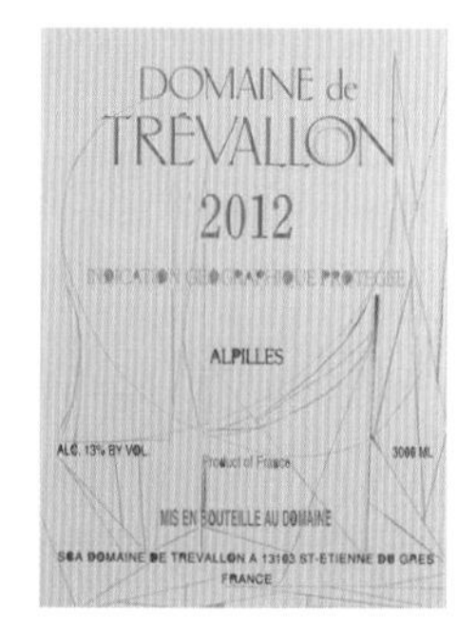

Domaine de Trévallon

This intense wine from Provence is a blend of 45% each Marsanne and Roussanne and 10% Chardonnay; since it fits into no AC rules, it is classified as IGP.

Guigal

A powerful but mildly scented Marsanne wine from Hermitage in the Rhône Valley, impressive when young but also capable of aging for decades.

Left: Picking Marsanne grapes in Chapoutier's Chante-Alouette vineyard at the top of the Hermitage hill in the northern Rhône. Chapoutier uses Marsanne to make several single-vineyard Hermitage wines as well as a Hermitage Vin de Paille. For the vin de paille the grapes are picked extra ripe and then left to dry indoors for two months before being fermented into an intensely sweet dessert wine. Above: Marsanne can be a bit of a love-it-or-hate-it grape; but much depends on how well it is handled. Bland, gluey Marsanne is no one's idea of a good time; but get the balance right and you'll be rewarded with one of the most unusual whites of all, with weight and complexity and the ability to transform from a bright honeysuckle, fresh youth to a deep, waxy, quincy maturity.

CONSUMER INFORMATION

Synonyms & local names

Known in Savoie as Grosse Roussette and in Switzerland as Ermitage or Ermitage Blanc.

Best producers

FRANCE/Rhône Beaucastel, Belle, Chapoutier, Chave, Coursodon, Cuilleron, Delas, Entrefaux, Florentin, Gaillard, Gripa, Grippat, Guigal, Jaboulet, Lionnet, l'Oratoire St-Martin, Perret, Pradelle, Remizières, Marcel Richaud, Sorrel, Cave de Tain, Trollat, Villard; **Provence** Clos Ste-Magdelaine, la Ferme Blanche, Pibarnon, Trévallon; **Languedoc-Roussillon** Alquier, Estanilles, Fabas Augustin, Jau, Lascaux, Pech-Latt.
SWITZERLAND Gilliard, Caves Imesch.
ITALY Bertelli, Casòn Hirschprunn.
SPAIN Celler Mas Gil.
USA/California Cline, Qupé, Tablas Creek; **Washington State** McCrea Cellars; **Virginia** Horton.
AUSTRALIA All Saints, Cranswick, D'Arenberg, Marribrook, Mitchelton, Tahbilk, Turkey Flat.
CHILE Errázuriz.

RECOMMENDED WINES TO TRY

Eleven northern Rhône white wines

Chapoutier Ermitage Cuvée de l'Orée
Chave Hermitage Vin de Paille
Pierre Coursodon St-Joseph Blanc le Paradis St-Pierre
Yves Cuilleron St-Joseph Blanc le Lombard
Delas Hermitage Blanc Marquise de la Tourette
Pierre Gaillard St-Joseph Blanc
Grippat Hermitage Blanc
Guigal Hermitage Blanc
Jaboulet Crozes-Hermitage Blanc la Mule Blanche
Sorrel Hermitage Blanc les Rocoules
Cave de Tain Hermitage Au Coeur des Siècles

Ten other French wines containing Marsanne

Domaine Alquier Roussanne/Marsanne
Ch. de Beaucastel Côtes du Rhône Blanc Coudoulet de Beaucastel
Clos Ste-Magdelaine Cassis
Ch. des Estanilles Coteaux du Languedoc Blanc
Ch. Fabas Augustin Minervois (Blanc)
Ch. de Jau Côtes du Roussillon Blanc de Blancs
Ch. de Lascaux Coteaux du Languedoc Pierres d'Argent
Domaine de l'Oratoire St-Martin Côtes du Rhône-Villages Blanc Haut-Coustias
Ch. de Pibarnon Bandol (Blanc)
Domaine de Trévallon Vin de Pays des Bouches-du-Rhône (Blanc)

Eight other Marsanne wines

Bertelli St-Marsan Bianco (Italy)
Casòn Hirschsprunn Contest (Italy)
Cline Cellars Marsanne-Roussanne (California)
Robert Gilliard Ermitage Réserve Choucas (Switzerland)
Caves Imesch Ermitage du Valais (Switzerland)
Celler Mas Gil Clos d'Agon Blanco (Spain)
Tahbilk Marsanne (Australia)
Turkey Flat Butcher's Block White (Australia)

MERLOT

So what reason would you give for Merlot becoming the world's most popular red wine style? Would it be because the great Bordeaux reds of St-Emilion and Pomerol, the lush and heady and insanely expensive wines like Pétrus, Le Pin and Pavie are all based on Merlot? Sounds like a good reason. Or would it be because you can make large amounts of red wine out of Merlot which has no hard edges and very little evident flavour so that millions of people who don't really like red wine at all, can somehow swallow this innocuous brew? Well, I hate to say it, but the second reason is why you'll find it planted anywhere on the planet where a red grape can ripen.

It all began in 1991, when the prime time USA television show '60 Minutes' covered a phenomenon they titled 'The French Paradox', based on the fact that the French eat large amounts of fat and yet have far less heart disease than nations who eat less. So what's the secret? Supposedly, red wine. Knock back a couple of glasses of red every day and your ticker will glide faultlessly towards a healthy, vigorous old age.

Ever since Prohibition ended in 1933, American wine producers have been trying to turn the USA into a nation of wine drinkers – with limited success. But there's nothing the Americans like better than being told you can live forever and if the man on the telly tells you you must knock back a couple of glasses of red a day – suddenly millions of Americans who have never drunk wine before are queuing up for their daily dose. American consumption of red wine quadrupled within the year. But these were novice wine drinkers. They didn't necessarily like the flavour of red wine very much – and what they needed was something that was soft, easy and mild, and yet discernibly, undeniably red. One wine fitted the bill perfectly – California Merlot. It was even easy to pronounce. Suddenly Merlot's perceived weaknesses were its unique selling point. California couldn't plant the stuff fast enough, and sold out every vintage, making oceans of, frankly, forgettable but glugglable mellow red. Yet Merlot's background is a lot more serious than this.

Merlot's name is said to be derived from merle, French for blackbird, which apparently loves its sweet, early-ripening fruit. Being planted like fury around the world, Merlot has done particularly well in the Napa Valley where the California poppy, as orange as a blackbird's bill, grows wild.

Merlot has traditionally been Bordeaux's 'other' red grape, overshadowed by Cabernet Sauvignon, the dark aristocratic grape that dominates Medoc and Pessac-Léognan. Merlot was planted here too, but simply to soften the Cabernet. Only in St-Émilion and Pomerol was Merlot dominant. When American critics began to swoon over these wines in the 1980s, realizing that their ability to mix the fabled reputation of Bordeaux with a wine you could glug back at only a couple of years old would play brilliantly back home in the States, Merlot finally stepped out from Cabernet's shadow.

Merlot's sudden dash to stardom wasn't lost on the less prestigious Bordeaux regions which always had trouble ripening their Cabernet Sauvignon. As the worldwide rush for red swept through Bordeaux, resulting in a current planting ratio of 88 per cent red to 12 per cent white in a region that was traditionally 50-50 – Merlot was the favoured variety because it didn't mind damp, clay soils, it didn't need a massive amount of sun, and it usually gave you a nice, plump bank manager-pleasing crop of something reasonably soft, if a little bit earthy. Much more attractive than thin stalky Cabernet. Bordeaux is now 64 per cent Merlot.

If Bordeaux went mad for Merlot, so did nearly everywhere else. France's Languedoc and northern Italy have tons of it, mostly making wines that risk being a byword for nothing very much. Spain has a lot, though Merlot is hardly suited to the warm conditions. More encouragingly, it's really taken off in Eastern Europe – Bulgaria, Romania, Moldova, Croatia all make large amounts of eminently enjoyable Merlot: easy-going red. It's become a star grape in Chile and New Zealand, Washington State and British Columbia – and California: Napa and Sonoma make some pretty serious stuff. Where else? Oh, pretty much everywhere, really: New York, Brazil, Malta, Bolivia – I'm not joking. Australia has a lot, but has never really got it to work. And the elephant in the room? China. Already the fourth biggest plantings in the world. What will it taste like? We'll soon know.

Merlot: from Grape to Glass

GEOGRAPHY AND HISTORY

At first glance the map showing where Merlot is found in the world might look remarkably similar to that of Cabernet Sauvignon. And why not? After all, in Bordeaux they collaborate to excellent effect, and many a Cabernet Sauvignon grower in South Africa, California, Australia and Eastern Europe has found that the traditional Bordeaux blend seems to work pretty well.

And yet there are differences. Merlot is saved from a role of eternal bridesmaid to Cabernet Sauvignon by two factors: its liking for cooler climates than Cabernet Sauvignon, and its ability to produce soft, rich-textured, low-acid, low-tannin red. The latter appeals to many red wine drinkers who don't care for the rasp of tannin and acidity, but love the appetizing black

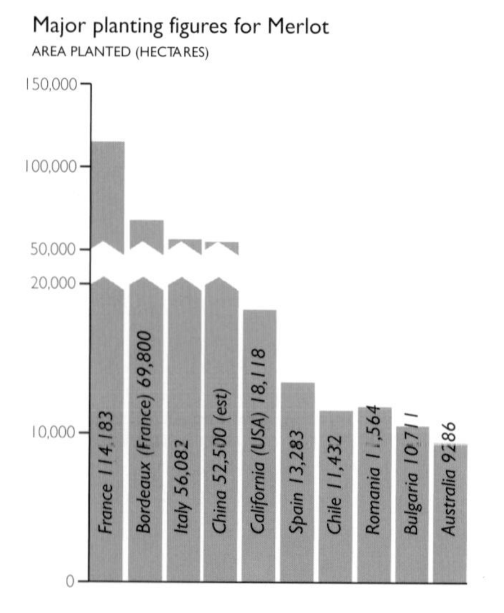

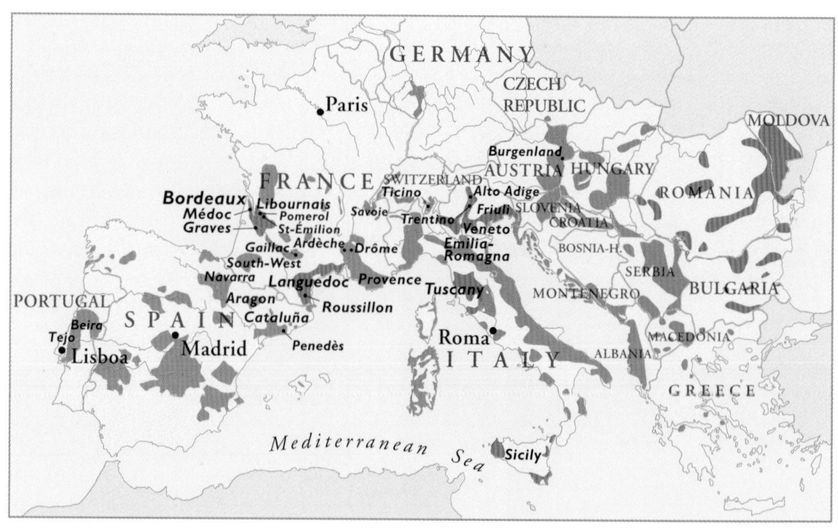

cherry and chocolate and fruitcake flavours with which Merlot is packed. The former appeals to winemakers in climates that are warmer than typical Pinot Noir conditions (i.e. barely warm enough to ripen any red varieties), but just too cool for Cabernet Sauvignon. In these sites, where Cabernet Sauvignon will turn out too green and raw for fun, Merlot can ripen to soft, silky sweetness. So while the maps of the two varieties might look much the same, they disguise differences of emphasis. Above all, Merlot inhabits the cool fringes and the damper soils where growers look at Cabernet Sauvignon every year and wonder whether to persevere with it. Here Merlot excels.

Merlot likes warmer conditions too, but doesn't prosper with too much heat because it actually ripens too fast and never develops its lovely plummy personality. This would explain why a lot of budget-priced Californian and Australian Merlot lacks guts. And it needs a firm hand in the vineyard, since it will overcrop like mad if you let it. If you've ever stared in disappointment at a glass of pale, flavourless north Italian Merlot, you're looking at the results of a massively overcropped vineyard. It is far from being as forgiving of unskilful viticulture as Cabernet Sauvignon. Merlot may have wide appeal, but it doesn't work as an all-purpose grape for lackadaisical producers.

HISTORICAL BACKGROUND

Merlot seems to be native to Bordeaux, but it gets scant attention from commentators on the region until 1784, when a local official called Faurveau named it as one of the best vines in the Libournais region on the Right Bank of the Dordogne; by the mid-19th century it was being planted in the Médoc on the Left Bank of the Gironde for the sake of its ability to blend well with Cabernet Sauvignon and Malbec.

Where did it spring from? It's part of a local family of grapes which seem to have been interbreeding for centuries. Merlot's father is Cabernet Franc (which probably originated in the Basque country) and its mother is a terminally obscure little number called Magdeleine Noire des Charentes discovered in an abandoned vineyard near St-Malo in Brittany. I say 'abandoned' – St-Malo hasn't been a wine area for hundreds of years; it's famous for cider! Honestly, you marvel at the persistence – and the luck – of those scientists who spend their lives trying to unravel the rampant promiscuity of ancient grape varieties. Magdeleine Noire had various names, usually referring to its early ripening, which was around the Feast of Ste Magdelaine on 22 July. So at least we can see where Merlot's early ripening habits come from. Cabernet Franc and Carmenère are the two vines with which Merlot is most easily confused, and in Chile the confusion was only sorted out in 1997.

Merlot seems to have been noted in Italy at much the same time as it was moving into the Médoc. It is first recorded in the Veneto (as Bordò) in 1855, and in the Swiss canton of Ticino between 1905 and 1910. However, it was brought to Switzerland from Bordeaux, not from Italy.

The limestone côte or hillside on the edge of the town of St-Émilion is one of the best spots for Merlot in the whole of Bordeaux. Merlot is often thought of as a soft, mild wine, but from these steeply sloping vineyards the wines can rival those of the Haut-Médoc on the other side of the Gironde for complexity and longevity, and yet they do retain a delightful suppleness from youth through to maturity.

The glitzy new château building at Château Pavie shows the flamboyant self-confidence of Bordeaux's modern Right Bank. Pavie's Merlot-dominated, rip-roaring wines create controversy with every vintage.

The Clos de l'Oratoire barrel chai in St-Émilion. Merlot comprises about 80% of the blend here, the balance being Cabernet Franc, and the result is an impressively juicy, fleshy, robust wine.

VITICULTURE AND VINIFICATION

It is unusual for a grape to become so fashionable without there being any real consensus about how it should be grown and made. If there is a benchmark style, it is presumably that of Pomerol and St-Émilion, but in much of Bordeaux Merlot has long been as much of an insurance policy as a grape appreciated for its own distinct virtues. Its failings as a varietal wine include poor flavour profile: it can be a doughnut wine, with a hole in the middle of the palate.

Clonal variation is important, and there is undoubtedly more work to be done here. It is possible that only 60 or 70 per cent of Merlot's potential has so far been realized, as anyone who has suffered a lean, mean Merlot vin de pays will testify.

Climate

The climate in the Libournais – the Right Bank of the Dordogne in Bordeaux, where Merlot is concentrated – is actually warmer than that of the Médoc across the Gironde on the Left Bank. The Libournais has less rain and warmer days, but more chance of frost and colder nights. One might think that such conditions would favour Cabernet Sauvignon, with its greater resistance to cold, and greater need of warmth for ripening, and no doubt the French wine authorities reasoned the same way when they insisted on Cabernet Sauvignon being planted on the lower slopes here in the 1970s. But warmer weather can't compensate for colder soil – at least, not in Bordeaux – and the soils in the Libournais are mostly cold, damp clay. Merlot's early budding, early ripening habit means that it can cope with cold soils and still ripen, though early budding increases the risk of frost damage. It is also sadly subject to rot because of its thin skin: the vintages of 1963, 1965 and 1968 were even more rot-ridden in St-Émilion and Pomerol than elsewhere in Bordeaux, and if autumn rains threaten, growers fearfully eye their Merlot crop since the grapes could begin to split and rot within a couple of days. Its thin skin also makes it dangerously susceptible to rot in humid climates.

But at least it buds and ripens up to two weeks earlier than Cabernet Sauvignon, which makes it a better bet in most parts of New Zealand; the same is true of Austria, Germany, northern Italy, Switzerland, New York State and Canada.

Though its reputation is for soft, plummy flavours, it can all too easily exhibit a herbaceous green streak. A touch of herbaceous greenness was long thought of as part of the St Émilion style, until the present vogue for lush, ultra-ripe wines took over; Washington State, New Zealand, Switzerland and the Veneto in northern Italy have all had problems with greenness, which can often be overcome by better canopy management. Even Australian examples are far from being perfectly ripe all the time. Much Californian Merlot avoids any risk of underripeness by being planted in the hot Central Valley – but the downside is that the wines frequently taste baked and lacking in freshness.

Soil

According to Galet (1998), Merlot likes cool soils that retain their humidity in summer; in dry soils, says Galet, the grapes don't fill out properly. Jean-Claude Berrouet, the previous technical director of Château Pétrus in Pomerol (his son Olivier has now taken over), believes that, on the contrary, water stress is more important than soil temperature for Merlot, and that it performs far better on well-drained soil than at the foot of a slope. Well, they're both experts. Take your pick. Château Pétrus itself is located on a tiny blister of clay that bulges through the topsoil of ferruginous sand and gravel, and that imparts more than usual structure to the wine; it is clay and iron that give the wines of Pomerol the backbone that most Merlots lack – helped, usually, by a little Cabernet Franc for perfume and acidity. The percentage of clay in some Pomerol vineyards is as high as 60 per cent.

In the sand, clay–limestone (*argilo-calcaire*) and gravel-over-limestone soils of St-Émilion, too, the traditional blend is with Cabernet Franc. Styles vary with the *terroir*: sandier soils give looser-knit wines; clay gives more substance, limestone more elegance and perfume.

Clones – and confusion

In much of the New World, clone and climate currently make more difference to style than soil type. Stephan von Neipperg of Château Canon-la-Gaffelière in St-Émilion believes there are only two or three really good

I find it difficult to believe this impressive, elegant building is now Le Pin. When I first visited, you literally had to kick the chickens out of the cobwebby garage where the wine was made. But in 2011 this beauty was built. The vineyard is bigger than on my first visit – all of 2.5ha (6 acres) of surprisingly pebbly clay soil where the Merlot ripens particularly fast. And the pine trees? They're the same.

clones of Merlot, though many more than that are available. Some give structure; others have larger berries, larger yields and give soft, early-drinking wine. The right choice of rootstock can also help to control the vigour of this naturally vigorous vine and help keep its cropping under control.

Some plantings in California, Australia and Austria are strongly suspected of being Cabernet Franc rather than Merlot: in the Napa Valley, Duckhorn's best block turned out to be Cabernet Franc; and Christian Moueix of Bordeaux confirms that as long ago as the early 1980s he'd thought that some of the Merlot in California was Cabernet Franc. One grower even reckons he can't tell the difference between the two varieties until their third year. In Chile there's been confusion with Carmenère since the 19th century.

Chris Camarda makes some of Washington State's most successful Bordeaux blends under his Andrew Will label. The relentless sunshine in the Washington vineyards would make you think it would be too hot for the cool-loving Merlot – but check out that heavy coat Chris is wearing as he considers his Merlot harvest. It may be sunny, but it isn't hot at vintage time in the near-desert conditions of eastern Washington.

Yields

Overcropped Merlot produces a thin, stringy, weedy wine, light in colour and vegetal in flavour. Conversely, some *vins de garage* in St-Émilion may have yields of as low as 26hl/ha; normal Bordeaux yields reach 60hl/ha, and are generally higher than those of Cabernet Sauvignon unless reduced by the vine's tendency to poor fruit set. In the Midi yields may rise to 80hl/ha; over 100hl/ha quality falls rapidly. In New Zealand yields are between 8–10 tonnes/ha; in California's Napa Valley, between 4–4.5 tons/acre; in Argentina, where yields are generally high, 12 tonnes/ha is considered low, and 18 tonnes/ha is not unusual.

Michel Rolland, worldwide consultant winemaker, Merlot specialist, and advocate of ultra-ripe, lush wines, believes that yields of the grape throughout the world are too high. 'It doesn't like too much sun, and it is difficult to make very expressive, complex Merlot if the climate is too hot. But if you reduce the yield you can improve quality'.

When to pick

This is one of the main focuses of debate between winemakers around the world. A characteristic of Merlot is that once ripe, it must be picked quickly. It will rush from ripeness to overripeness in a matter of days, sometimes even a matter or hours; in this it is less accommodating than Cabernet Sauvignon.

There are two schools of thought about when to pick. Michel Rolland favours late picking, and welcomes fruit with an element of overripeness. 'The picking date is vital. Get that wrong and everything else will be wrong; you will lose 80 per cent of the potential. Many properties in Bordeaux pick between a few days and a week too early. But the crop level is the main factor. If the crop is too big, you can wait until Christmas, and it won't ripen. The ideal crop is 40–50hl/ha. At that level you can successfully delay picking for eight to 10 days more. And if you remove some of the leaves, there is less risk of rot, so you can wait.'

The other side of the Merlot coin, that which advocates traditional elegance, finesse and potential for longevity above richness and lushness, is represented by Christian Moueix of JP Moueix who points out that overripeness means loss of acidity. 'We want light, elegant wines for drinking, not body-building wines for winning tastings.'

WINEMAKING: THE GREAT DEBATE

At the higher quality levels, Merlot is a wine with an identity crisis. It needs elegance, complexity, fine tannins and silky texture – but can winemakers best achieve that ideal by treating the grape like Pinot Noir, or by treating it like Cabernet Sauvignon?

The issues revolve around extraction and oak. Should Merlot be heavily extracted and tannic, and aged in a high proportion of new oak? Or should it be lighter, more delicate? Fashion is changing. Until a few years ago, it dictated heavy extraction and tannin, soupy textures and high alcohol, but now lighter reds are finding favour again. Napa still focuses on weight, and St-Emilion is divided, but the rest of the world is returning to elegance, recognizing that not all Merlot is naturally as structured as Pétrus, or as concentrated as Valandraud; and that if it's not, it cannot be made so by long aging in new oak.

Consultant winemaker Michel Rolland recommends extracting the phenols (colour and flavour compounds) at the beginning of the fermentation process with a shorter, faster maceration with frequent pumping over. Researchers in Bordeaux are trying to build up a better picture of Merlot by analyzing the structure of its tannins and phenolic compounds. By compiling a dossier of the aromas and flavours most frequently found in the wine, they hope to arrive at a clearer understanding of its potential.

I have to say, some of the most beautiful St-Émilions I've enjoyed have been made in the lighter style, because Merlot gains a gorgeous soft butter and toffee smoothness with age if you don't try to force more weight and colour out of the grape than it naturally possesses. However, if you're after something altogether more powerful and lush, Michel Rolland clearly has the answer in Bordeaux and around the world.

MERLOT AROUND THE WORLD

How many great varietal Merlot wines are there in the world? A lot depends on what you mean by 'great'. If you mean dark, muscle-bound bruisers built for the long haul, or slow-developing intellectual beauties that take a generation to reveal their treasures, then the answer is – there aren't many. But if greatness can be measured in a hedonistic way, if succulence, sensuous fruit and a perfume that fills your head with sweet ripeness could be called great, there'd be quite a few. High alcohol, dark colour, low tannins and acidity and a cauldron-full of nice, ripe fruit is what Merlot is best at. Pleasure, in other words. But are they great? Look. I'm too thirsty to argue. Hand me the bottle. I'll drink, you fret.

Bordeaux

Merlot has quietly been staging a takeover in Bordeaux, spurred on by the worldwide demand for softer, earlier-drinking wine. It now constitutes some 56 per cent of plantings overall – but that figure disguises its minority role in the better parts of the Médoc and Graves, where it makes up an average of 25 per cent of the blend, and its dominance in St-Émilion and Pomerol. In the outlying areas of Bordeaux – the areas collectively known as the Côtes – it is very often the majority grape.

In St-Émilion it accounts for more than 60 per cent of plantings; in Pomerol about 80 per cent. In both these appellations it is commonly blended with some Cabernet Franc; there is a little Cabernet Sauvignon in St-Émilion, and some, though less, in Pomerol.

The standard Bordeaux blend of Merlot, Cabernet Sauvignon and Cabernet Franc, with or without the addition of Malbec or Petit Verdot, has a lot to recommend it. Merlot adds softness and fleshiness, Cabernet Franc adds perfume and Cabernet Sauvignon adds the all-important structure. Merlot on its own is generally less long-lived, although St-Émilion and Pomerol can be tight and closed when young, and like Médoc and Graves reds, can need some years to open. It is the *crasse de fer* – the iron-rich clay subsoil – that imparts structure to Pomerol's Merlot; the very structure that can be so wanting elsewhere. Château Pétrus is made from 95 to 100 per cent Merlot in most years; Le Pin and Gazin are 90 per cent Merlot. But at some Pomerol châteaux, for example Lafleur with its particularly gravelly, mineral-rich soil, the amount of Merlot in the vineyards goes down to 50 per cent with old Cabernet Franc taking up the rest.

The heart and the muddy boots of Michel Rolland are most at home in the clay-clogged vineyards of Bordeaux's Right Bank where he grew up, but much of his year is spent travelling since he is the most successful and famous wine consultant in the world, with rich, mouthfilling Merlot reds very much a calling card.

In St-Émilion the *vin de garage* phenomenon – wines made on so small a scale that one could envisage making them in one's garage – was fuelled by Merlot. Yields of 30hl/ha or less are the rule here, from tiny vineyards (Gracia is 3ha/7.4 acres; Clos Nardian is 1.52ha/3.75 acres; La Mondotte is a positive giant at 4.3ha/10.6 acres). Picking may be by individual berry; viticulture may be organic or biodynamic; oak will almost certainly be 100 per cent new, and may even be 150 per cent new, this remarkable figure being achieved by putting the wine into 100 per cent new oak to begin with, and then racking it into more new oak.

These wines have enormous and seductive richness and a few have joined the mainstream; otherwise the fashion has peaked amid question marks over longevity.

Vins de garage could be made anywhere: they're the equivalent of a bigger estate creaming off the very best vat, or couple of vats; something that bigger estates are usually loath to do because of its detrimental effect on the rest of the blend. However, the principle has been accepted across the globe by winemakers and growers (not necessarily of Merlot) who want to make a fast impact.

Other French Merlots

Merlot has now overtaken Carignan and Grenache as France's most planted grape variety. It is a recommended variety for *appellation*

Château Palmer
Palmer often contains 40% Merlot, high for a top Haut-Médoc wine. The result is classic Margaux perfume married with seductive softness of texture.

Château Pétrus
If anyone doubts that Merlot can age, show them a mature Pétrus: nowadays it's 100% Merlot. But it does come from a unique, Merlot-friendly terroir.

Le Macchiole
You'd expect Bolgheri on the Tuscan coast to be too warm for the cool-loving Merlot, but some of Italy's most successful Merlots are from here.

Merlot vineyard at Tenuta dell' Ornellaia owned by Marchese Lodovico Antinori at Bolgheri. so near the sea that locals in this part of Tuscany used to reckon that the wines from round here tasted of salt. That was before the super-Tuscan brigade got cracking. This Merlot, called Masseto, is one of the world's best.

contrôlée wines and vins de pays in Provence, the Languedoc, the Southwest, the Ardèche, the Charente, the Corrèze, the Drôme, Isère, the Loire, Savoie and Vienne, though in the South it is a fairly recent arrival, having only been recommended there since 1966. In the Southwest it is a traditional part of many *appellation contrôlée* blends. In the Languedoc most Merlot goes into vins de pays, both varietals and blends. Most are soft, fruity wines for early drinking.

Italy

Merlot fever has hit Italy in a big way: in the mid-1990s it was Italy's fifth most planted grape, but current plantings have pushed it to fourth, even ahead of Barbera. In the northeast it has traditionally been regarded as a high-volume producer, and clones were chosen primarily for their yield. It wasn't given the best sites, and the wine was easy-drinking stuff, not intended to be taken seriously. This is true of the Veneto, Trentino and Alto Adige, though Merlot from Alto Adige is benefiting from better clones and more attention to site. The best wines come from Friuli, Tuscany and Umbria, and they are very good indeed. Some is varietal; a lot is blended with Sangiovese, and is valued for adding richness without dominating the blend in the way that Cabernet Sauvignon does. In Tuscany it is increasingly being planted in the warmer Maremma, and alcohol levels can easily reach 14 per cent, though acidity doesn't seem to suffer; perhaps it is added. In more northerly parts of Tuscany, Merlot's low acidity is often seen as a useful softener for the more acidic local grapes.

Rest of Europe

Almost all of Europe grows Merlot. In Switzerland's southern canton of Ticino, Merlot accounts for 85 per cent of production. It needs warm sites here, and vineyards no higher than 450m (1480ft), to produce wines of any substance; the best are well structured, and have some oak aging. Some Merlot here is made 'white' – pale pink, in other words.

Elsewhere it is concentrated in the cooler areas: it has made little impact in Iberia, though it has experimental status in Rioja. (There are moves to get practically everything on to the list of approved grapes in Rioja.) There is some in Slovenia, Croatia, Serbia & Montenegro, some in Austria and Germany, and more in Romania, Bulgaria, Hungary and Moldova; much is blended with Cabernet Sauvignon.

California

Experience is telling here. More and more Merlot is being planted, and once the current plantings come on stream, California will be the world's fourth largest producer of Merlot, after Bordeaux, Italy and the Languedoc. Post-phylloxera plantings are on more suitable rootstocks than before, and much has been learnt about trellising: Merlot grows particularly long shoots, which need to be tied, often to an extra moveable wire. Site selection is also better. Cooler spots that wouldn't suit Cabernet Sauvignon are being chosen – Carneros, for example, gives good bright flavours to Merlot, though Cabernet doesn't ripen well there. Parts of Napa, like Oakville and Stags Leap, give richer wines, and hillside sites like Howell Mountain, where Cabernet is structured and long-lived, impart similar qualities to Merlot.

These, and other parts of California like Mendocino, Alexander Valley, Dry Creek Valley, Sonoma Valley and the warmer areas of Monterey, are where the serious Merlots are coming from. Some are huge wines, so massive that they make Pomerols look feeble. Extreme ripeness – the US market recoils at

Castello di Ama
Castello di Ama already makes one of the most sensuous and approachable of Chianti Classico wines. This Merlot is in the same delicious vein.

Leonetti Cellar
This Merlot brought attention to Walla Walla, an area in Washington State that is now making some of the Pacific Northwest's most impressive reds.

Merryvale Vineyards
Merryvale is based near St Helena in central Napa, but the grapes for this Merlot come from the cooler vineyards down toward Carneros and San Pablo Bay.

any hint of herbaceousness – is obtained by leaving the grapes on the vines for as long as possible, and alcohol levels are high. But those who want light Merlot (or perhaps Merlot lite) are also being catered for. A lot is being planted in the hot Central Valley for high-volume, low-intensity wine. And by low, I mean low.

North America

Cabernet Sauvignon is the leading red grape in Washington State with almost 10,293 acres/4165ha planted, with Merlot following behind in second place with 8235 acres/3332ha of vines in 2011. Merlot gives richer flavours here than Cabernet Sauvignon, and while the wines are riper and plummier than they were, they still tend to be drier than Californian examples, with more freshness and often more personality. Site selection, control of irrigation and later picking have all helped. Cold eastern Washington winters remain a problem for the variety. Classic Bordeaux blends are also successful, and at the moment seem the best way of using Merlot here. Oregon seems generally too cool, and fruit set is poor. New York State is more promising: Long Island's long growing season means that Merlot ripens reliably here, and Merlot is now the most planted red variety in Virginia. Merlot is important in Canada, and is British Columbia's most planted variety, as Canadians strive to make grand reds.

South America

The term 'Chilean Merlot' used to be somewhere between a euphemism and a nickname, but Chile's Merlot vineyards are now well sorted out. The reason was that much Chilean Merlot was a field blend of Merlot and Carmenère: the original cuttings were taken to Chile from Bordeaux before phylloxera and propagated as Merlot ever since. Nobody had thought to question the vines' identity until it was realized in the early 1990s that a lot of Chilean Sauvignon Blanc was in fact the less aromatic Sauvignonasse, or Sauvignon Vert. French ampelographers were then invited to look over the vineyards, and came up with the unsettling discovery that a great deal of the 'Merlot' was not Merlot at all.

The two grapes are now separated in new plantings, but it will be years – decades, even – before the existing mixed plantings are superseded. New varietal Merlots on the market are now genuine, but some of the best still boast the extra flavours of Carmenère.

According to the latest official figures (2011), Chile has 11,432ha (28,250 acres) of Merlot and 10,040ha (24,810 acres) of Carmenère, respectively the third and fifth most planted varieties in the country. The problem is not that Carmenère is an inferior grape – indeed, it can have greater quality in Chile than Merlot – but that they ripen at different times. Merlot ripens some three weeks before Carmenère, so unless you have each vine marked in the vineyard and can pick the two varieties separately, you will either end up with ripe Merlot and underripe Carmenère, or overripe Merlot and ripe Carmenère, but whereas over-ripe Merlot spoils a blend by being too jammy, under-ripe Carmenère adds a fascinating and refreshing green peppery, leafy, savoury character if mixed with ripe Merlot, which is a uniquely Chilean combination.

Site selection is fashionable now in Chile, and Merlot often seems to be less interesting than Pinot Noir and Syrah to those producers planting in the cooler regions of the far north and south (the north counts as cool when it's coastal and is cooled by sea breezes; it's actually cactus country). The Central Valley regions are still Merlot's stronghold

Merlot in Chile: a tale of two vines

'We were the origin of the discovery that a lot of Chilean Merlot is actually Carmenère,' say Paul Pontallier and Bruno Prats. These leading Bordeaux winemakers (Pontallier makes the wine at Château Margaux, and Prats used to own Cos d'Estournel) joined forces in the late 1980s to make wine near Santiago in the Maipo Valley – the estate, now called Viña Aquitania, is located at 700–800m (2300–2625ft).

They bought two batches of Merlot vines to plant, and soon realized that if one was Merlot, the other had to be something else. An ampelographer friend from France eventually identified the mystery vine as Carmenère. 'The latter blends well into Chilean reds,' says Pontallier, because 'Merlot doesn't behave in Chile as it does in Bordeaux: it is harsher than in Bordeaux, whereas Cabernet Sauvignon is softer here. So the classic Bordeaux blend may not be the best thing for Chile.'

The differences between the varieties seem to be glaringly obvious if you look at this post-harvest vineyard scene at Casa Lapostolle's Clos Apalta in the Colchagua Valley. Look at the different colours of the vine leaves – the green ones are the late-ripening Cabernet Sauvignon, the red ones are Carmenere whose name refers to the redness of the leaves and is derived from an old French word for 'red' – carmine. The yellow leaves are Merlot, so how did they ever mix up Carmenère and Merlot?

Hawkes Bay is one of New Zealand's best regions for Merlot, the country's most successful 'Bordeaux' red variety. It ripens earlier and more completely than the other varieties, and provides crucial richness and softness to blends in conditions that are still a little on the cool side for Cabernet Sauvignon.

and plantings outside the regions of Maipo, Cachapoal, Colchagua, Maule and Curicó are small or non-existent.

Merlot in Argentina is less planted than Cabernet Sauvignon or Malbec, and can be trickier to grow. High yields may well be part of the problem. The best, most concentrated wines come from the Mendoza hills at around 1100m (3610ft) and above. Merlot is important in Uruguay for softening the toughness of Tannat, and there's a fair bit in Brazil.

Australia

Recent years have seen a substantial increase in plantings and Merlot is now the third most popular red variety, but is still some way behind Shiraz and Cabernet. At the moment a lot is still blended with Cabernet Sauvignon, but the number of varietal wines is increasing as growers understand it better. Paul Lapsley, of Coldstream Hills in the Yarra Valley, believes that the clones in Australia are particularly poor. But new clones, released over the last few years, should help to improve flavours and the bane of the Merlot grower's life, poor fruit set.

New Zealand

New Zealand growers like Merlot because, unlike Cabernet Sauvignon, it's a reliable ripener in cool climates. Merlot is now the country's fifth most planted grape, and second most popular red after Pinot Noir. It is at its best in the North Island, with the South Island being a little chilly even for Merlot. Over half the plantings are in Hawkes Bay, particularly in Gimblett Gravels and the Ngtarawa Triangle, where it produces dense, dark wines that dominate the local Bordeaux blends. Most of the rest is in Auckland and Marlborough: it may have a good future in cool-climate Marlborough, where Cabernet Sauvignon is a poor ripener. Blends are usually with the other Bordeaux varieties; some producers prefer a Cabernet Franc and Malbec blend.

South Africa

By 2013 Merlot accounted for just under 7 per cent of total plantings; Cabernet Sauvignon, by contrast, was top with 11 per cent. Since there was no Merlot planted in the country at all until the 1980s, that is a rapid increase – and expect plantings to continue to rise.

Two basic styles of Merlot are emerging in South Africa: the soft, juicy, easy-drinking version, and something altogether more impressive, powerful and potentially long-lived. One-third of the vines are planted in Stellenbosch, with a fair number also in Paarl, followed by Worcester and Malmesbury and cool-climate Constantia. New virus-free clones have helped its advance in recent years.

Asia

China has taken to Merlot in a big way as part of the classic red Bordeaux blend and may well have the largest Merlot plantings in the world after France by the time you read this – things are moving at lightning speed in Chinese

Bedell Cellars
Bedell make some of New York's best reds on the sandy loam soils of Long Island's North Fork. The elegant label shows their love of modern art. .

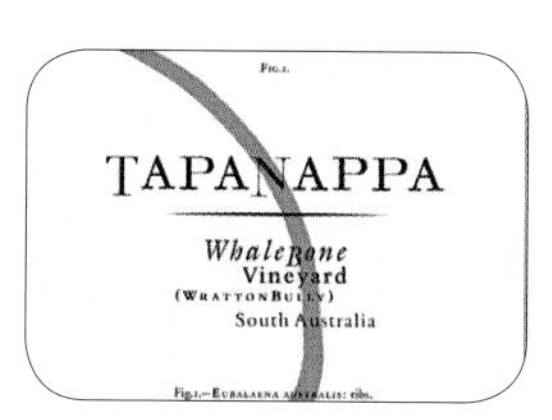

Tapanappa
Australia's largest limestone area is South Australia's Limestone Coast, with Wrattonbully at its heart. Tapanappa's Whalebone Vineyard takes full advantage to produce soft but beautifully balanced red wines.

Craggy Range
Craggy Range make use of Merlot's suitability for the warm Gimblett Gravels in New Zealand's North Island to produce the rich, sensuous but powerful Sophia Merlot.

Vergelegen
Vergelegen's Helderberg vineyards are cooled by strong southerly winds whipping in from False Bay, creating one of South Africa's freshest, juiciest Merlots.

ENJOYING MERLOT

A great deal of the time, the question of how well Merlot ages is irrelevant. Most is made to be drunk young, even immediately: modern Merlot is the epitome of the red wine that gets all the bottle age it needs on the way back from the wine shop.

That is true both of simple, juicy Merlots and of big, concentrated ones. Merlot's low tannins and low acidity make it a less than obvious wine for the cellar; the majority of Merlots not only don't need bottle age, but often positively don't want it. Age these wines and your patience will be rewarded by fading fruit and disintegrating structure.

Yet Merlot can age perfectly well as part of the Bordeaux blend – and not just because it has Cabernet Sauvignon to prop it up. St-Émilions and Pomerols, usually with little or no Cabernet Sauvignon in the blend, will easily last a decade, and the very best wines will see in their 30th birthday. Such wines need time to come round in the first place: from two or three years to longer, depending on the property and the vintage. Top Pomerols and St-Émilions with decades ahead of them are, however, exceptions even in Bordeaux: normally if a Bordeaux proprietor simply wants to make earlier-maturing wine, he plants more Merlot and doesn't prune too hard.

The structure of Merlot from Bordeaux is also found in the best examples from California, Italy and elsewhere. There is no question about these balanced wines having the grip and backbone to last at least 10 years and perhaps more; the question is how many get the chance.

The taste of Merlot

This is where one has to raise the biggest question mark of all. Merlot tastes of all sorts of things: it can be anything you want, from light and juicy through Pinot Noir-silky to Cabernet-oaky and extracted. What should it taste like? That's the very question that winemakers across the world are asking themselves. The lack of benchmarks means that there is no easy answer.

At its best it is succulent and silky, with velvety tannins: 'smooth' is the word that has cropped up most often in US market research. Fruit flavours can be of strawberry, raspberry, black cherry, blackcurrants, plums, damsons, figs or prunes. Fruitcake flavours are often found in Bordeaux. There can be spice, too: cinnamon and cloves and a touch of sandalwood, and truffles, tobacco, liquorice and toasted nuts. Warm climates can give stewed fruit flavours; cool ones, or overcropped vines, can give minty, lean, herbaceous notes. It can be gamy, chocolatey or coffeeish; yet for every Merlot that exhibits such appetizing complexity there are 20 examples that aim for no more than sweet juicy fruit and that seductively velvety texture.

To many a Californian winemaker, Merlot is about texture more than flavour. Texture is what they believe consumers notice most; and if Merlot mania is a fair guide, perhaps they're right. But what they are often really saying is 'lack of texture': blandness, total smoothness, with a flavour as mild as can be, sometimes helped along with a little sugar. This is lowest common denominator 'wine without tears', and the phenomenon has been massively boosted by the claims of red wine being good for your heart. These claims meant that millions of people who didn't like red wine, and never drank it, would drink it for their health – so long as it didn't taste like red wine. Many of the budget-priced commercial American Merlots are merely trying to satisfy that market. Good luck to them. May it keep their customers healthy. I'll stay healthy by drinking equal amounts of something slightly more interesting.

Merlot was the wine that made Chile famous, but Merlot wine wasn't necessarily made from Merlot grapes in those days – the Merlot vineyards were co-planted with Carmenère, a rare Bordeaux grape at one time thought to be extinct. Indeed, Carmenère used to be called Merlot, while Merlot was called Merlot-Merlot! Nowadays, Carmenère sails under its own colours, as does Merlot, and 20 Barrels is a high quality special selection of true Merlot. Angelus has made one of the top St-Émilions for 20 years, and was promoted to the highest classification of 1er Grand Cru Classé (A) in 2012. Merlot is St-Émilion's leading grape variety, and Angelus is usually 90% Merlot.

MATCHING MERLOT AND FOOD

Merlot makes soft, rounded, fruity wines that are easy to enjoy without food, but it is also a good choice with many kinds of food. Herby terrines and pâtés, pheasant, pigeon, duck, goose and spicier game dishes all blend well with Merlot. It can also partner subtly spiced curries and tandoori dishes. Meaty casseroles made with wine are excellent with top Pomerol châteaux; and the soft fruitiness of the wines is perfect for savoury foods with a hint of sweetness such as ham, savoury pancakes and gratins.

CONSUMER INFORMATION

Synonyms & local names

Also known as Merlot Noir, and in Hungary as Médoc Noir. Not to be confused with the unrelated Merlot Blanc, also found in Bordeaux but now almost extinct.

Best producers

FRANCE/Bordeaux/St-Émilion Angélus, Ausone, Beau-Séjour Becot, Canon, Canon-la-Gaffelière, Clos Fourtet, Grand-Mayne, Magdelaine, Monbousquet, la Mondotte, Pavie. Pavie-Macquin, le Tertre-Roteboeuf, Troplong-Mondot, Valandraud; **Bordeaux/Pomerol** le Bon Pasteur, Certan-de-May, Clinet, la Conseillante, l'Église-Clinet, l'Evangile, la Fleur-Pétrus, Gazin, Hosanna, Latour-à-Pomerol, Petit-Village, Pétrus, le Pin, Trotanoy.

ITALY/Veneto Maculan; **Friuli** Borgo del Tiglio, Livio Felluga, Graf de la Tour, Russiz Superiore; **Tuscany** Castello di Ama, Avignonesi, Castelgiocondo, Ghizzano, Le Macchiole, Ornellaia, Petrolo, San Giusto a Rentennano, Tua Rita; **Lazio** Falesco; **Sicily** Planeta.

SPAIN Can Rafols dels Caus, Pago del Ama, Torres (Atrium).

SWITZERLAND Brivio, Huber, Werner Stucky, Tamborini, Vis, Zanini, Christian Zündel.

ISRAEL Barkan, Dalton, Margalit.

USA/California Arrowood, Beringer, Buccella, Chateau St Jean, Frog's Leap, Matanzas Creek, Merryvale, Newton, Pahlmeyer, Rubissow, Shafer, Truchard; **Washington State** Andrew Will, Canoe Ridge, L'Ecole No 41, Leonetti, Long Shadows, Northstar, Quilceda Creek, Woodward Canyon; **New York** Bedell, Lenz, Macari, Paumanok, Shinn Estate, Wölffer.

CHILE Carmen, Casa Lapostolle, Viña Casablanca, Concha y Toro, Cono Sur, Errázuriz, Gillmore, Viña Leyda.

AUSTRALIA Brand's, Coldstream Hills, Elderton, Fermoy Estate, Irvine, Parker Coonawarra Estate, Tapanappa, Tatachilla, Yalumba.

NEW ZEALAND Alluviale, Craggy Range, Esk Valley, Fromm, C J Pask, Sacred Hill, Trinity Hill, Villa Maria.

SOUTH AFRICA Bein, Eagles' Nest, Laibach, Shannon, Steenberg, Thelema, Vergelegen.

RECOMMENDED WINES TO TRY

Ten classic Bordeaux red wines (as well as the Best producers, left)

Ch. Canon-de-Brem Canon-Fronsac
Ch. Canon-la-Gaffelière St-Émilion
Clos Fourtet St-Émilion
Ch. la Conseillante Pomerol
Ch. Gazin Pomerol
Ch. Grand-Mayne St-Émilion
Ch. Fontenil Fronsac
Ch. Latour-à-Pomerol Pomerol
Ch. Pavie-Macquin St-Émilion
Ch. Roc de Cambes Côtes de Bourg

Seven good-value Bordeaux reds

Ch. d'Aiguilhe Côtes de Castillon
Ch. Annereaux Lalande-de-Pomerol
Ch. Carsin Premières Côtes de Bordeaux Cuvée Noir
Ch. Pitray Castillon-Côtes de Bordeaux
Ch. la Prade Côtes de Francs
Ch. Segonzac Premières Côtes de Blaye
Ch. Siaurac Lalande-de-Pomerol

Six top Italian Merlots

Castello di Ama Vigna l'Apparita
Falesco Montiano
Livio Felluga Colli Orientali del Friuli Rosazzo Merlot Riserva Sossó
Le Macchiole Messorio
Tenuta dell'Ornellaia Masseto
Tua Rita Redigaffi

Seventeen New World Merlots

Andrew Will Ciel du Cheval Merlot (Washington State)
Bedell Merlot (New York)
Buccella Merlot (California)
Casa Lapostolle Cuvée Alexandre Merlot (Chile)
Cono Sur 20 Barrels (Chile)
Craggy Range Sophia (New Zealand)
Fermoy Estate Merlot (Australia)
Leonetti Merlot (Washington State)
Long Shadows Pedestal (Washington State)
Sacred Hill Brokenstone Merlot (New Zealand)
Shafer Merlot (California)
Steenberg Merlot (South Africa)
Tapanappa Whalebone Vineyard (Australia)
Thelema Merlot Reserve (South Africa)
Vergelegen Merlot (South Africa)
Wölffer Christian Cuvée (New York)
Woodward Canyon Merlot (Washington State)

Merlot magic: put it on the shop shelf and watch it disappear. But Merlot's commercial success goes hand in hand with an identity crisis. Merlot winemakers around the world can never agree about what the wine should taste like.

Maturity charts

Merlot, even premium Merlot from Napa or Sonoma, does not fit into a single pattern. Not all will age well in bottle, or be meant to.

2010 St-Émilion Premier Grand Cru Classé and top Pomerol

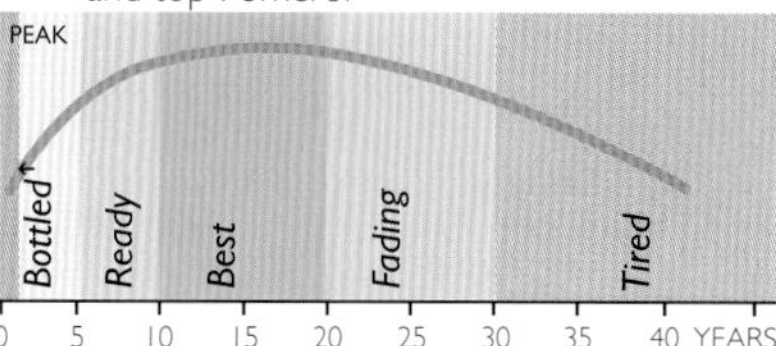

A classic Merlot year: the wines are lush and drinkable early, but their excellent structure ensures a long peak of pleasure and a gentle decline.

2012 Top Napa/Sonoma Merlot (Reserve)

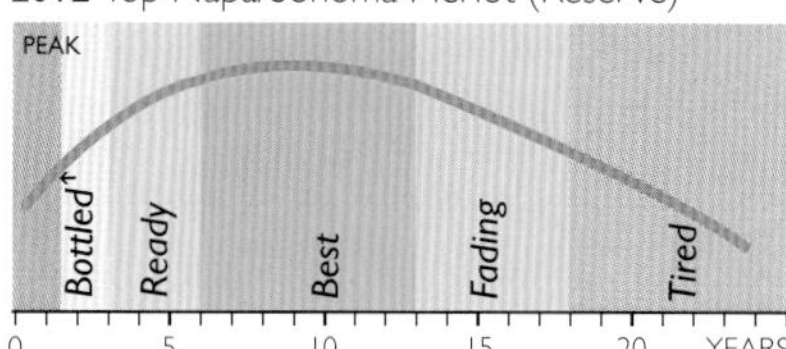

Reds ripened a little more slowly than usual and the grapes have excellent colour and concentration. The wines are rich, dark and supple with good acidity.

2013 Chile Merlot (Reserve)

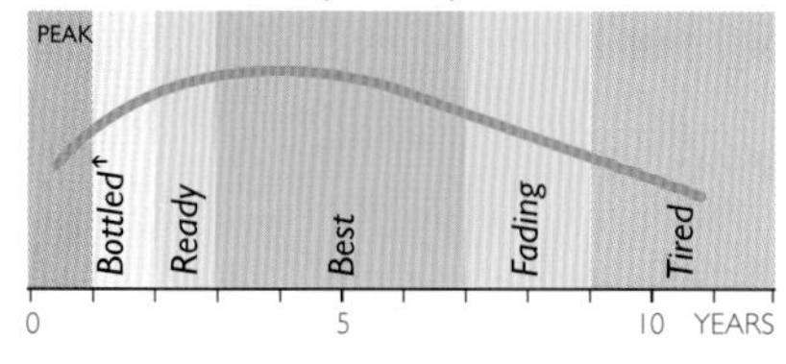

An unusually late vintage with bright flavours and fair concentration. The wines have excellent balance and varietal flavours for remarkable aging.

MERWAH

Lebanese white grape best known for its role in the white wine of Château Musar. Owner Serge Hochar picks it in October, after the black grapes, at 11–15% potential alcohol. The untrained bushvines grow there on chalky gravel, give yields of 10–20hl/ha, and are ungrafted. The wine, which is a blend of Merwah and Obaideh (alias Chardonnay, perhaps) is complex, citrus and rich, with notes of hay and honeysuckle.

MISKET

Bulgaria has three different Misket grapes: two are modern crosses, with Misket Cherven (Red Misket) by far the most important. The berries are pink rather than red, and it is not related to Muscat, though its white wines do have a simple mild, grapey perfume. Often blended with Dimiat. Best producer: (Bulgaria) Domaine Boyar.

MISSION

If you were an early Spanish settler in the Americas, Mission was your everyday wine. It may seem odd that the Franciscan missionaries who brought it from Spain, didn't take the trouble to bring something better. However, the southwest of Spain wasn't exactly awash with decent black varieties – it was Palomino and Muscat country – and since Mission also had the traditional name of Palomina Negra, it may have been what passed for a local black grape in Andalucia. It turns out that Palomina Negra is actually Listán Prieto, a traditional variety from the centre of Spain. It isn't there anymore, but it is grown on the Canary Islands and used to be called Moscatel Negro. And anyway vine exports from Spain to her colonies were banned, so the missionaries took whatever they could find.

Mission has the advantage of producing good crops in drought conditions, and because of the wine's high resveratrol content it is said not to turn easily to vinegar. Both of these attributes could account for its survival in large quantities to the present day. In Chile, País is mostly grown in the south of the country, in Maule and Bío-Bío, where it may also be called Negra Peruana. The vines are bush trained and head pruned, and can yield 3000 kilos per hectare of grapes on the unirrigated hills, and 10 times that amount on the irrigated plains. There are some attempts to make serious País (Movi are important here), but while it can have some character, and soft plums and sweet chocolate fruit, it also keeps its red currant acidity and rustic bite.

In Argentina the lighter-skinned Criolla Chica is less commonly planted than Criolla Grande, and is of marginally higher quality.

Mission can still be found in California, but it is in decline. It would be nice to think that some will always be made, however: it is part of California's history. And it was the first vinifera vine to reach the Americas. Best producers: (USA) Sobon, Story.

MOLETTE

A neutral grape grown in the Savoie region of France, albeit in declining quantities, for sparkling wine. Altesse is often added to the blend to give more character. Best producer: (France) Varichon et Clerc.

MOLINARA

A minority grape in the Valpolicella blend, Molinara is easy to drink and light in body but pale in colour and contributes little except acidity. It's being gradually phased out. Best producer: (Italy) Quintarelli.

MONASTRELL

The Spanish name for Mourvèdre – although, since the variety originated in Spain, it would be more accurate to say that Mourvèdre is the French name for Monastrell. Be that as it may, the grape is best known by its French name.

Quality in Spain has risen, having been pretty dismal for years. The use of new and better virus-free clones and some serious investigation into how best to handle Monastrell – when to pick it, how long to macerate it, how hot or cool to ferment it and when to bottle it – have paid off. The grape may be made as a varietal (perhaps with carbonic maceration for early drinking) or it may be blended with Tempranillo, Cabernet Sauvignon and Merlot. The wines of Yecla, where Monastrell is also the majority grape, are not yet up to those of Jumilla, but improvements continue to be made. Garnacha and Merlot are commonly blended in here. Best producers: (Spain) Agapito Rico, Castaño, Salvador Poveda, Primitivo Quilés, Señorio del Condestable.

MONDEUSE NOIRE

Mondeuse was perfectly content with being a fairly obscure but, in my experience, really tart and spicy variety grown in the seductive savage wilds of Savoie's mountain valleys. Sure, history implied it might have something to do with the Rhône's Syrah grape, and indeed one of its names was Grosse Syrah (the big-berried Syrah), but people thought it more likely to be the same as Refosco dal Pedunculo Rosso, which made pretty clunky stuff round the Adriatic coastal curve of Italy's Friuli and Slovenia. Certainly, the best Mondeuse reds do have an Italianate dark-scented, plum and blackberry fruit, spice, and a bitter cherry twist to the end. Well, it may taste Italian, but the grape is pretty solidly French, nothing to do with Refosco but a good deal to do with the mighty Syrah, perhaps its grandparent – which in grape-breeding terms is about as direct a family connection as a grandad is for you and me. And suddenly I do find myself thinking, Hmm, that spice, that blackberry, that ... well, it is quite like Syrah, after all. The tremendous worldwide popularity of Syrah has actually pushed Mondeuse just a little way into the limelight. Which is a good thing because it's a really good grape we should hear more of, not just from its Savoie hidden sites. So far, there's a tiny bit in California and Australia, but it would be nice to see that increase. Best producer: (France) Rocailles.

MONEMVASSIA

Right, get ready. Monemvasia is the name of a Greek island in the Pelopponese. It's also, we're always told, the name which was corrupted into Malvasia, that famous sweet wine of the Middle Ages. Malvasia, you will be glad to hear, was a blend, and might have had nothing to do with the Malvasia grape. Nor is the Malvasia grape anything to do with the Monemvassia grape, although the latter probably originated in Monemvasia, and somewhere along the line gained an extra 's'.

On to happier matters: it's one of the indigenous Greek vines currently enjoying a revival. It tends to be big and fat, and needs cool spots to keep its acidity. It can be made dry or sweet.

MONICA

Monica is found in Sardinia, where it makes easy-to-drink everyday red, often with low tannin and low acidity. It is sometimes said to be the same as Mission (see left) and País and to have been taken to the Americas from Spain. Well, it certainly doesn't taste like Mission to me, because it produces nice tart, cranberryish red wine to drink young. It is said to originate in Spain, although it is no longer found there, but Spain was at one time the overlord of Sardinia, so that's presumably how it got there. Best producers: (Italy) Argiolas, Dolianova co-op, Santadi co-op, Trexenta co-op.

Valentini
Valentini make a big, powerful but sumptuous Montepulciano d'Abruzzo that benefits from several years' aging.

MONTEPULCIANO

Montepulciano the grape should not be confused with Vino Nobile di Montepulciano, which is a wine made from Sangiovese from an area around the town of Montepulciano in Tuscany. Montepulciano the grape is planted in central Italy (including Umbria and southern Tuscany) and the South. Its best known form is as Montepulciano d'Abruzzo, a round, plummy and weighty red with ripe tannins, good acidity and a low price tag. Its most useful contribution is probably as a dominant partner in Rosso Conero in the Marche.

It is also found blended into many reds in much of the southern half of Italy. It is late ripening and perhaps would not be successful north of Tuscany and Umbria. It ought to do well in California and Australia. There's a little now in New Zealand. Best producers: (Italy) Boccadigabbia, Cornacchia, Garofoli, Illuminati, Masciarelli, Moroder, Nicodemi, Saladini Pilastri, Le Terrazze, Umani Ronchi, Valentini.

MORELLINO

A synonym for Sangiovese in the Scansano region just inland of the Maremma in southern Tuscany. Morellino di Scansano is traditionally an appealingly fruity, cherryish, fairly light red with soft tannins; not as elegant or refined as the wines of the best Chianti estates, but with plenty of charm. However, some producers are now making more gutsy examples with sweet sour fruit and a herbal rasp.

Scansano has the advantage of an almost perfect mesoclimate for Sangiovese: it ripens easily here, and gives good concentration and structure providing yields are kept well under the legal maximum of 84hl/ha. True, the summers can be very hot and dry, but irrigation is a possibility for the future, which should solve any ripening problems induced by drought. Up to 15 per cent of other varieties are permitted under the DOC rules: Alicante and Spagna are the vines most commonly grown in the area, apart from Sangiovese; Merlot and Cabernet are also used for more international flavours. It's now being pioneered in Australia. Best producers: (Italy) Erik Banti, Carletti, Cecchi, Il Macereto, Mazzei, Montellassi, Moris Farms, Poggio Argentaria, Le Pupille. See also Sangiovese pages 222–231.

MORILLON

Chardonnay goes by this name in Austria's Steiermark region; Morillon was also a synonym for Chardonnay in much of France. Best producers: (Austria) Frühwirth, Neumeister, Polz, Erwin Sabathi, Manfred Tement.

MORIO-MUSKAT

This over-scented German crossing is now mercifully on the decline. It used to be a standby of the Liebfraumilch industry, and was much grown in the Rheinhessen and Pfalz, where most Liebfraumilch originates. But demand for Liebfraumilch has slumped and much Morio-Muskat has been uprooted. It has a coarse aroma and flavour, as grapy as Muscat but without any of the latter's elegance or lightness, and in fact does not seem to be related.

It is at its most aggressive, mixing musky perfume with grapefruit acidity, when picked slightly unripe, which it usually is because it needs good sites to ripen properly, and good sites are quite rightly reserved for the best grapes. It is subject to downy mildew, oidium and rot. Very small amounts are planted in South Africa, Japan and Canada. Don't ask me why.

MORISTEL

Light-coloured red grape found in Somontano, Spain. It is nothing to do with Morrastel, alias Graciano (see below), or Monastrell (see facing page). (Moristell, however, is a synonym for Graciano.) Its wines are aromatic, quite light, best unoaked and young, and often blended with Tempranillo and others. Best producers: (Spain) Borruel, Monclús, Pirineos.

MORRASTEL

See Graciano (see page 122). Morrastel Bouchet is a separate grape, a 19th-century crossing of Graciano and Petit Bouschet bred in southern France.

MOSCATEL

Iberian name for Muscat (see pages 156–165). Wines labelled Moscatel are basic, sweet and golden and usually made from Muscat of Alexandria, or Moscatel de Alejandría; another name for this grape is Moscatel Romano. Muscat Blanc à Petits Grains is Moscatel de Grano Menudo. Moscatel de Malaga is believed to be a variety local to southern Spain. Best producers: (Spain) Gandía, López Hermanos, Ochoa.

MOSCATELLO

Muscat Blanc à Petits Grains sometimes takes this name, sometimes Moscato, in Italy. See Muscat (pages 156–165) for more details.

MOSCATO DI ALEXANDRIA

Muscat of Alexandria in Italian. See Muscat (pages 156–165) for more details.

MOSCATO BIANCO

Italian name for Muscat Blanc à Petits Grains (see pages 156–165). It is found pretty well all over Italy, and most regions have a tradition of lightly sweet, low-alcohol aromatic Moscato. The South and the islands produce sweeter, richer wines, and Trentino-Alto Adige makes good crisp, dry versions.

MOSCATO GIALLO

A high-quality, golden-berried strain of Muscat found in northeastern Italy. See Muscat pages 156–165 for more details.

MOSCATO ROSA

A high-quality pink-berried strain of Muscat found in north-eastern Italy. See Rosenmuskateller page 217 for more details.

MOSCHOFILERO

An aromatic Greek grape of rather good quality, used either as a varietal or as part of a blend. Its aroma is somewhere between that of Gewürztraminer and that of Muscat, but with better acidity than either. Its grapes are pink, and it is used either for white wine or for pink. It is genetically unstable, and has many versions, like Asprofilero, Xanthofilero and the black Mavrofilero, which may look, behave and taste different to each other. Best producer: (Greece) Cambas.

MOURISCO

One of the lesser Portuguese grapes used in the Port blend and best suited to warm spots. The wine is light in colour and quite high in acidity; it is not one of the five favoured varieties planted in the newer vineyards of the Douro Valley. It crops up elsewhere in Portugal and Spain under a variety of names.

MOURVÈDRE

The French call it Mourvèdre, the Australians and Californians may call it Mataro, but the Spanish call it Monastrell, the grape from the monastery. Why? Well, the grape originated in Murviedro, a wine port near Valencia now called Sagunto, but called Morvedre in Catalan. The name suggests it was originally grown by monks. It was probably shipped to France from Murviedro, but also from the nearby coastal town of Mataró. Hopefully, this explains the different names. No one except the Spanish uses the Monastrell name and, frankly, Mourvedre is much more honoured away from its birthplace. In Spain it's the key to many reds from south of Madrid, giving high alcohol and plentiful tannins but so far rarely bringing much distinction in its wake. The growers here are being persuaded to time their picking better and recently there has definitely been an improvement in fruit flavour, particularly in such DOs as Alicante, Jumilla, Valencia and Yecla, injecting much needed life into what had been a forgotten part of vinous Spain.

Mourvèdre is, however, best known by its French name. It is very much a grape of the South – it won't ripen north of Châteauneuf-du-Pape, and it doesn't even ripen properly there in cool years. But in Bandol on the Mediterranean coast, which is substantially warmer, it produces big, hefty wines which are nevertheless not without finesse. It is a finicky grape to get right: it needs the warmest, south-facing sites, but cool, shallow, clay soil to tame its vigour. Growers must get a minimum of 13 per cent alcohol because it simply has no flavour at a lower strength; they have to pick in a small window when the grape finally has body and fruit but before it becomes too pruney.

In the winery and in bottle it also has its idiosyncrasies. It is very reductive, and distinctly farmyardy more often than not. This means that it's a surprisingly good partner for the lush, fat Grenache, which is prone to oxidation and which benefits from the rather meaty aromas of Mourvèdre. What fruit it has is blackberryish and there's generally quite a rasp of herbs. It's easy to think a young Mourvèdre is faulty, but after an even more farmyardy middle age, it usually emerges into a rich, leathery maturity after five years or more.

It's only quite recently that people have started to take Mataro/Mourvèdre seriously in South Africa, California and Washington, and Chile is testing its first plantings. In Australia, where a lot of the clones come from Spain, old bush vines are increasingly revered, and give dark, herby, rich but rustic, and tannic reds that are part of the classic blend: GSM, or Grenache-Shiraz-Mourvèdre.

THE TASTE OF MOURVÈDRE

Young Mourvèdre, picked at low yields, has a fairly wild mix of rasping hillside herbs, more than a hint of the farmyard, and, if you're lucky, flavours of blackberries and bilberries. It is solid in style, with high alcohol and tannin. Most is blended, often with other southern French grapes like Grenache and Syrah. Blended or not, Mourvèdre will always add a farmyardy, herby roughness for a few years before developing flavours of leather, gingerbread and game.

Californian examples of this grape tend to be less broodingly tannic than examples from Bandol in southern France, but J. Lohr's version still has a fair amount of beefy personality, although the warm growing conditions of Paso Robles, south of San Francisco, also create pretty dark, rich fruit to balance it out.

Domaine Tempier

One of the two top Bandol estates (the other is Pibarnon), Tempier uses a high proportion of Mourvèdre in all its reds; this one is 100% Mourvèdre and can age 20 years.

HEWITSON
Baby Bush
2012
BAROSSA VALLEY
Mourvèdre

Hewitson

Barossa Valley is famous for old vine Shiraz, but Mourvèdre was also planted in the 19th century, and a few vines survive. Sometimes called Mataro, it used to be made into Port, but is now sought after for strong, dry reds.

Left: Mataro/Mourvèdre growing in the Barossa Valley in South Australia. The name Mourvèdre is supplanting that of Mataro in Australia as demand for old vine fruit soars. Aficionados reckon that if you call it Mataro, you don't take it seriously. And wines from old bush vines like these are very serious indeed: they're not as immediately likeable as Shiraz or Grenache but can add a fascinating extra something to red blends. **Above:** *Mourvèdre is not a variety that is terribly easy to grow successfully: it is said to like its face in the hot sun and its feet in the water, which is a tall order even for many of the Bandol vineyards perched on the cliffs high above the Mediterranean. It's an even taller order for most of Australia.*

CONSUMER INFORMATION

Synonyms & local names

Called Monastrell, Morrastel or Morastell in Spain. Mataro is a synonym used in the New World (Esparte is also used in Australia). One of its more picturesque French names is Estrangle-Chien or 'Dog strangler'.

Best producers

FRANCE/Southern Rhône Beaucastel, l'Oratoire de St-Martin, Richaud, Dom. Ste-Anne; **Provence** la Bastide Blanche, la Bégude, la Courtade, d'Esclans, Frégate, Galantin, Jean-Pierre Gaussen, Gavoty, Gros' Noré, l'Hermitage, Mas Redorne, Moulin des Costes, la Noblesse, Pibarnon, Ch. de Pradeaux, Ray-Jane, Roche Redonne, Romassan, la Rouvière, Ste-Anne, Salettes, Simone, la Suffrène, Tempier, Terrebrune, la Tour du Bon, Vannières; **Languedoc-Roussillon** Jean-Michel Alquier, Canet-Valette, la Grange des Pères, l'Hortus, Mas Blanc.

SPAIN Agapito Rico, Bodega Balcona, Castaño, Julia Roch Melgares, Torres; **Alicante Fondillón** Bodegas Alfonso, Bocopa, Bodegas Brotons, Salvador Poveda, Primitivo Quiles.

USA/California Bonny Doon, Cline Cellars, Edmunds St John, Jade Mountain, J Lohr, Qupé, Ridge, Sine Qua Non, Tablas Creek, Sean Thackrey, Zaca Mesa, Villicana; **Virginia** Horton; **Washington State** McCrea Cellars.

AUSTRALIA Boireann, Grant Burge, Cascabel, D'Arenberg, Henschke, Hewitson, Langmeil, Charles Melton, Mitchelton, Penfolds, Pikes, Rosemount, Seppelt, Spinifex, Torbreck, Veritas, Wendouree, Yalumba.

SOUTH AFRICA Beaumont, Fairview.

RECOMMENDED WINES TO TRY

Ten southern French wines

Domaine de la Bastide Blanche Bandol Cuvée Fontanieu
Ch. de Beaucastel Châteauneuf-du-Pape
Domaine de l'Hermitage Bandol
Ch. de Pibarnon Bandol
Ch. de Pradeaux Bandol
Domaine Ray Jane Bandol
Ch. de la Rouvière Bandol
Domaine Tempier Bandol Migoua
Domaine de Terrebrune Bandol
Ch. Vannières Bandol

Five New World Mourvèdre wines

D'Arenberg The Twenty Eight Road Mourvèdre (Australia)
Hewitson Barossa Valley Old Garden Mourvèdre (Australia)
Spinifex Indigène (Australia)
McCrea Cellars Mourvèdre, Ciel du Cheval Vineyard, Red Mountain AVA (Washington)
Ridge Vineyards Evangelo Mataro (California)

Ten New World Mourvedre blends

Bonny Doon Le Cigare Volant (California)
Cline Cellars Oakley Cuvée (California)
Edmunds St John Les Côtes Sauvage (California)
Jade Mountain La Provençale (California)
Charles Melton Nine Popes (Australia)
Qupé Los Olivos Cuvée (California)
Rosemount GSM (Australia)
Sine Qua Non Red Handed (California)
Villicana Winemaker's Cuvée (California)
Yalumba Antipodean (Australia)

MTSVANE KAKHURI

Go to Georgia and just about every white wine you see that's not Rkatsiteli will be Mtsvane. Georgia boasts, in theory, hundreds of indigenous grapes: it's just that it seems to forget to grow them. This one finds favour because – well, I'm not quite sure. It yields quite well, gets high sugar levels and tastes quite nice: some citrus and floral notes, a bit of tropical fruit, decent acidity. But it's difficult to see it setting the world on fire.

MÜLLER-THURGAU

It is a curious fact that all the world's greatest grape varieties have happened by accident. Man may be a dab hand at the selective breeding of pigs or cattle, but when it comes to grape vines nature has always (so far, anyway) done better on her own.

Müller-Thurgau is a case in point. It was bred in 1882 by Dr Hermann Müller-Thurgau, working at the Geisenheim Institute in Germany. His aim was, as usual in Germany, to produce a crossing that would combine all the advantages of Riesling – high quality, elegance, complexity – with the earlier ripening of Silvaner. Müller-Thurgau proved so popular that by the early 1970s it was the most planted grape in Germany – though it is hard to believe that anyone thought the vine had one iota of the quality of Riesling. It is now in decline, but has a long way to go yet, and still accounts for 12.6 per cent of total plantings (2013).

Ironically, it has now been shown not even to be the Riesling x Silvaner cross that it was long believed to be. DNA fingerprinting has revealed its parents to be Riesling and Madeleine Royale (first tests suggested that they were Riesling and Chasselas de Courtillier). Madeleine Royale is no longer grown, and was itself a 19th-century crossing of Pinot and Schiava. It seems to have inherited little from its Riesling parent, however, being merely dull, aromatic in a rather unspecific way, and with little structure. But it is easy to grow and yields extremely generously, easily giving 200hl/ha, and that, in an era of Liebfraumilch and associated wines – Piesporter, Niersteiner and Bernkasteler – was a great attraction for the growers of most German regions. It even made inroads to the tune of 20 per cent in the Mosel.

It is true that if yields are restricted and the vineyards are well managed, then Müller-Thurgau is capable of producing respectable wine, attractively scented and with a pleasant acidity. But the problem is that to produce a top Müller-Thurgau you need to plant it in a top Riesling site, which is a terrible waste of a top Riesling site. In a third-rate site you'll get a third-rate Müller-Thurgau – but then Riesling from such a site wouldn't be worth drinking at all. It is also a grape that seems to be at its best in categories not higher than Spätlese since it seems to lose what definition it has as it gets riper. For most cheap German wines it is blended with something more assertively aromatic, like Morio-Muskat.

Its disadvantages (apart from the above) in Germany are that it is less resistant to winter cold than Riesling, and susceptible to rot, downy mildew and a painful sounding fungal affliction called Rotbrenner.

A great deal of Müller-Thurgau was planted, on German advice, in New Zealand in the 1950s and 1960s, when that country was just beginning to think seriously about modernizing its wine industry. Plantings have declined to less than 2.6ha (6 acres) in 2013 – but in fact New Zealand examples can have reasonable balance, are less overcropped than German ones and can be rather good. The same can be said of Müller-Thurgau from northern Italy, where the high-altitude vineyards of the Alto Adige region keep good acidity in the wine.

It is found across Europe from Austria eastwards, but seldom produces anything of note. England and Luxembourg (where it is called Rivaner), on the other hand, have cool climates, and produce some quite nice snappy wines, but it is falling out of favour here, too, due to unreliable cropping. Oregon and Washington State have produced some decent ones, but are unlikely to continue doing so for much longer. Best producers: (England) Breaky Bottom, Chapel Down, Three Choirs; (Germany) Karl-Heinz Johner, Juliusspital, Dr Loosen, Markgraf von Baden (Salem/Durbach); (Italy) Nilo Bolognani, Caldaro co-op, Casarta Monfort, Cortaccia co-op, Graziano Fontana, Pojer e Sandri, Pravis, Enrico Spagnoli, Tiefenbrunner, A & R Zeni.

Tiefenbrunner

When Müller-Thurgau struggles to ripen, its wine can be deliciously sharp and mouthwatering. Tiefenbrunner Feldmarschall from vines 1000m (3280ft) up in the Dolomite mountains is a prime example.

MÜLLEREBE

A German name for Pinot Meunier (see page 183). The name – 'Miller's grape' – means the same as the French 'Pinot Meunier'; both names derive from the floury appearance of this downy-leaved vine. Best producers: (Germany) Drautz-Able, Fürst zu Hohenlohe-Öhringen, von Neipperg.

MUSCADELLE

Muscadelle is unrelated to Muscat, although its flowery, grapy aroma might suggest some connection. It is found in the sweet wines of Sauternes as a minor part of the blend: just 2–3 per cent Muscadelle added to the Sémillon and Sauvignon Blanc can give this sweet wine an attractive lift and some extra finesse. Further east in Monbazillac it is the second most planted grape after Sémillon, but is losing ground to the latter.

In dry white Bordeaux from Entre-Deux-Mers, Premières Côtes de Bordeaux and other areas it can form up to 40 per cent of the blend but seldom does, since it is an irregular yielder and rots easily. In the gravelly south of Entre-Deux-Mers it can be picked earlier and is often quite good, but these regions are mad for red wine these days and are busy uprooting their white vines of all types.

Where Muscadelle reaches its apogee, however, is at Rutherglen and Glenrowan in northeastern Victoria, Australia, where it used to be known as Tokay and made the dark, sticky fortifieds known as Liqueur Tokay that are the counterpart to the Liqueur Muscats produced there from the Muscat grape (see pages 156–165). Now, in a gesture of appeasement to Hungary and the EU, the name of Tokay has been changed to the faintly silly one of Topaque, though the wines are the same. With age, Muscat and Muscadelle become more like each other, though Muscadelle always has a more malty, butterscotch flavour which contrasts with the tea rose scent and raisininess of Muscat, and Muscadelle is the less fleshy and syrupy of the two wines. Best producers: (Australia) All Saints, Baileys of Glenrowan, Campbells, Chambers, Peter Lehmann, Morris, Stanton & Killeen.

MUSCADET

This grape is just as well known by this name as it is by the name of Melon de Bourgogne (see page 135). Although it originally arrived in the Pays Nantais having travelled from Burgundy, it is now associated almost entirely with the Muscadet wine produced in the flat

Chéreau
Muscadet is often thought of as being a simple, neutral, bone-dry glugger to drink as young as possible, But Chéreau makes wines from old vines – in this case centenarians – which can age successfully for years.

lands at the mouth of the Loire around the city of Nantes, which goes so well with the local seafood. Best producers: (France) B Chéreau, Chéreau-Carré, Luc Choblet, Gadais, l'Ecu, Pierre Luneau, Louis Métaireau, Ragotière.

MUSCARDIN

This grape comprises a small part of the blend for red Châteauneuf-du-Pape in the southern Rhône Valley. It appears to have no synonyms, which is extremely rare for a vine, and little is known of its origins. The wine is light in colour, but with an appealingly floral aroma and faint acidity.

The colour of these old casks of Liqueur Muscats and Tokays (now renamed Topaque to avoid upsetting the Hungarians) in the corrugated iron winery of Morris Wines, Rutherglen, Victoria, Australia, echoes the colour of the wine. It can stay in cask for decades, becoming ever blacker, more concentrated and stickily sweet.

MUSCAT

See pages 156–165.

MUSCAT OF ALEXANDRIA

Muscat of Alexandria has less finesse than Muscat Blanc à Petits Grains, but is widely grown throughout the world. See Muscat pages 156–165.

MUSCAT BLANC À PETITS GRAINS

The strain of Muscat that almost invariably produces the best, most elegant, most finely aromatic wines. See Muscat pages 156–165.

MUSCAT CANELLI

Californian synonym for Muscat Blanc à Petits Grains derived from the name Moscato di Canelli or Moscato Bianco. See Muscat pages 156–165.

MUSCAT DE FRONTIGNAN

French synonym for Muscat Blanc à Petits Grains when used for Muscat de Frontignan, the Languedoc's main fortified wine. See Muscat pages 156–165.

MUSCAT GORDO BLANCO

Australian name for Muscat of Alexandria. See Muscat pages 156–165.

MUSCAT OF HAMBURG

A dual-purpose wine and table grape, but more often grown for the latter. As a wine grape it is found mostly in eastern Europe. See Muscat pages 156–165.

MUSCAT OTTONEL

An addition to the Muscat family bred in 1852 in the Loire Valley, and with less intensity than either Muscat Blanc à Petits Grains or Muscat of Alexandria. It is the main Muscat variety grown in Alsace unfortunately. See Muscat pages 156–165.

MUSKADEL

South African name for Muscat. It generally refers to Muscat Blanc à Petits Grains. See Muscat pages 156–165.

MUSKAT-SILVANER

Sauvignon Blanc (see pages 232–241) is sometimes called by this name in German-speaking countries. Best producers: (Austria) Alois & Ulrike Gross, Lackner-Tinnacher, Erich & Walter Polz, Otto Riegelnegg, Sattlerhof.

MUSKATELLER

German name for Muscat (see pages 156–165). It generally refers to Muscat Blanc à Petits Grains: Gelber Muskateller, the gold-skinned version found in Austria's Steiermark region and in Slovenia, and Weisser Muskateller are both forms of this strain of Muscat. Rotmuskateller is the Austrian Roter Veltliner and Grün Muskateller is Grüner Veltliner – neither of which taste like Muscat, but then Austria sometimes calls Sauvignon Muskat-Silvaner.

MUSKOTÁLY

Muscat takes this name in Hungary: most of the local Muscat planted there is Muscat Ottonel, but there is some Muscat Blanc à Petits Grains, which goes by the name of Sárga Muskotály, or 'Yellow Muscat'. See Muscat pages 156–165.

NARINCE

Turkish variety, also a table grape, with good citrus or floral flavours. Can age well, and takes happily to some oak aging.

E.R

MUSCAT

We had this old greenhouse at home when I was a kid. Out past the potting shed and the wood pile, just before you got to the potato plot. For most of the year I didn't bother with it much; it was grimy, unkempt, with a great sprawl of branches and shoots pressing against the windows and cutting out most of the light. I'd venture in once or twice in the summer, when the place was like a forest and a canopy of broad green leaves added to the gloom. But come the autumn, just before I had to go back to school, that's when I'd creep inside and breathe in an air so muggy and hot and thick with scent I could hardly open my throat wide enough to draw it in. The perfume was so heady, so exotic you'd think it should have been coloured purple and gold and been as impenetrable as clouds. But it wasn't. There was just enough light for me to look above my head and see vast clusters of purple black grapes, so ripe and fat that they threatened to tumble and crush me if I so much as touched them.

These were our Black Muscats or Muscats of Hamburg. Just once a year, I'd gorge myself on their succulent wildly aromatic flesh before stumbling back to the house with no plausible excuse as to why I was too full to eat tea. And the next day the man from the Savoy Hotel would come and measure the grapes. All the complete bunches of all the biggest grapes, he'd buy. It was usually worth making myself pretty scarce when my mother discovered how many he *wasn't* going to buy. No money for the cinema *that* week.

But the perfume in that hothouse and the flavour of those succulent grapes has never left me. The scent of Muscat haunts me in an entirely pleasurable way, and I can't smell a top-notch Muscat wine without being wreathed in the happy memories of childhood. And it makes me wonder. If Muscat was the original wine grape – and there's a fair amount of opinion suggesting that it was – what on earth made producers move on to other lesser, leaner varieties when they already had this irresistible headspinning grape? As it was, Muscat led the wine grape's conquest of the Mediterranean – spreading from the East through Greece and Italy, then on to southern France and down the Spanish coast to Málaga, within sight of the shores of Africa. And for hundreds of years it was highly prized for its fabulous perfume and for the fact that, of all the wine varieties, it was just about the only one that made wine that tasted of grapes.

The wine-dark sea around the island of Samos, a home of the Muscat grape, lies in the distance. Was this the wine drunk by Dionysus, the Greek god of wine? The ancient Greeks clearly loved wine and all its paraphernalia. Here is a mixing jar, a wine jug and a beautiful kylix or two-handled wine cup. And, goodness me, some hooligan has been scrawling his name on the side of the cup. Disgraceful.

And yet today it's treated with casual disregard by the majority of wine drinkers. This marvellously adaptable grape with the unmistakeable scent of the vine in full bloom is even sneered at. What does a grape have to do to earn a little respect? Producing utterly delicious wines for anyone with half a palate and a heart that beats obviously isn't enough. Having a gorgeous hothouse scent that makes you swoon with delight doesn't seem to do the trick. Being so versatile that it can make outstanding wine sweet, medium or dry, sparkling or fortified – is that enough? No, it doesn't seem to be.

There are some signs of hope. Moscato has suddenly become very popular in the United States. Fizzy, fruity, scented, quite sweet; the perfect party pop, surely. Brazil's known this for ages. Some of the best Muscat fizzy is made in Brazil, Party Central of whatever hemisphere you find yourself in. Maybe this explains why Italy – famous for sparkling Muscat-like Asti Spumante – is enjoying a planting boom while most of the world's plantings of Muscat are in genteel decline. In tough times, they still like to party in Italy and Asti does the trick better than most.

But elsewhere, I'm afraid, Muscat is commonly the victim of a current white wine obsession driven by the distinctly unexotic charms of Chardonnay, the tangy attack of Sauvignon and the all-pervasive influence of vanilla-scented oak. But when these fashions have all faded away, Muscat, the original wine grape, will still be there to remind us why this whole wine thing got started in the first place.

Muscat: from Grape to Glass
Geography and History page 158; Viticulture and Vinification page 160; Muscat around the World page 162; Enjoying Muscat 164

GEOGRAPHY AND HISTORY

If there is a more complicated family of grapes than the Muscats, it has yet to be discovered. There are over 200 vines in the world called Muscat Something-or-Other, and some are related, some not. Muscat Blanc à Petits Grains, Muscat of Alexandria and Muscat Hamburg – the three best-known Muscats – are related, and their capacity for mutation can make the Pinot family look positively regimented. Their berries can be golden, pink or black, and Muscat Blanc à Petits Grains is notorious for its ability to produce one colour one year and another the next. If they were human the Muscat family would be regarded as hopelessly feckless. Which just goes to show that orderliness is not everything.

Muscats of different sorts are planted all around the Mediterranean and across Central and Eastern Europe, as well as in Australia, New Zealand, parts of North and South America, and South Africa. They make light dry wines, light,

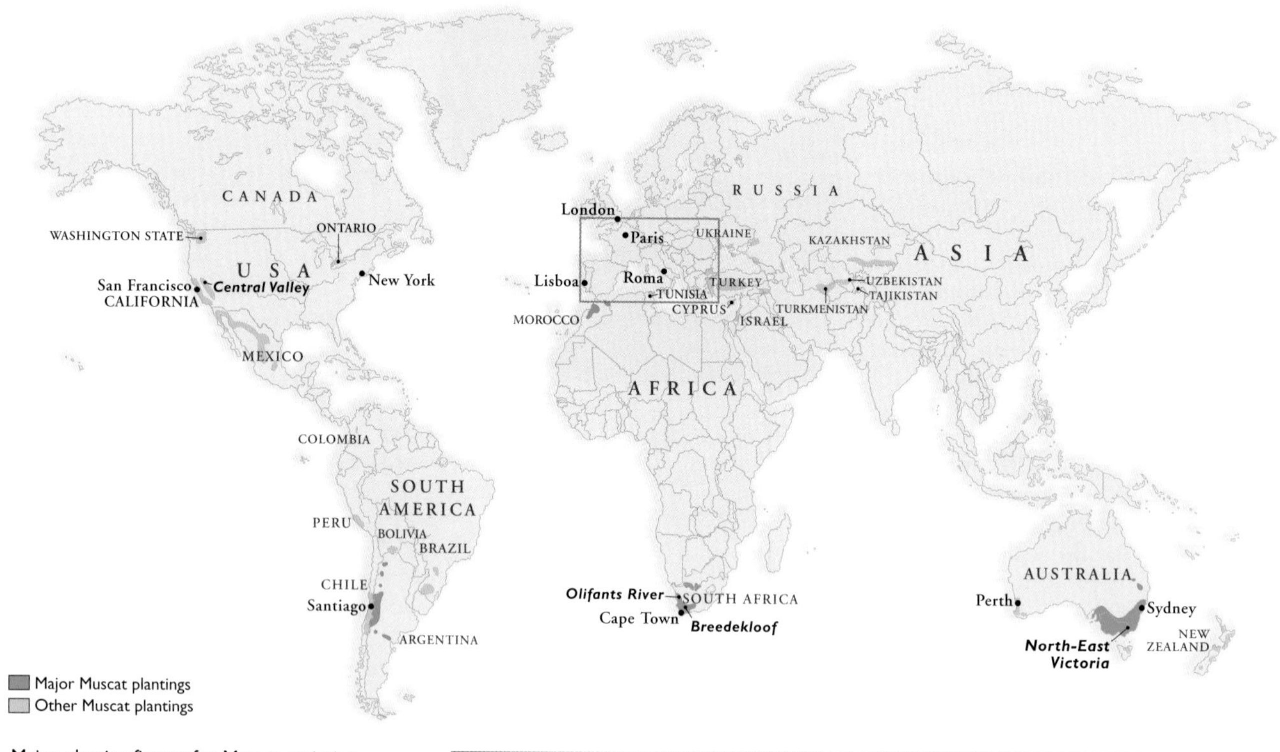

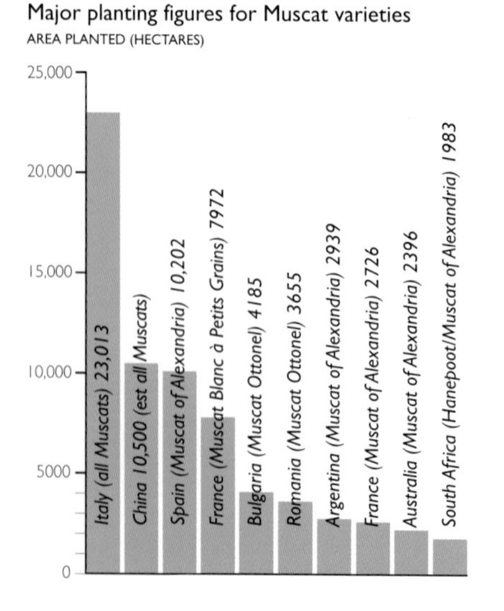

sweet sparkling ones, late harvest and sweet fortified wines. In theory Muscat could make nobly rotten sweet wines, though in practice it generally doesn't because the *Botrytis cinerea* fungus would destroy its extraordinary aroma.

But don't think that this blanketing of the vinous globe with Muscat means that it produces some of the world's most popular wines. It doesn't. Some of the world's most delicious wines – yes. Wines that express the pure perfumed essence of the grape in a way no other variety can approach – yes. But these wildly aromatic styles – dry, medium and sweet – are not in fashion right now although Moscato has enjoyed a sudden vogue in the USA. Its superb, heady, grapy fragrance makes it one of the very few wine grapes that is good eating. Indeed Muscat of Alexandria and Muscat Ottonel are widely grown for table grapes or raisins, and these plantings can distort the picture if one is only looking at wine grapes. Much of the Muscat of Alexandria in Chile and Peru, in addition, is grown for distillation into pisco brandy.

HISTORICAL BACKGROUND

Muscat Blanc à Petits Grains may well be the oldest grape variety we have. The ampelographer Galet (1998) identifies it with the *Anathelicon moschaton* vine of the Greeks, and the *Apianae* vine of the Romans because of the fondness of insects, particularly bees (*apis* in Latin) and wasps, for the grapes. Fairly thin evidence, you might think; but there is more romance connected to Muscat than to most, and that's saying something.

This Muscat, the oldest member of the family, was probably brought to the province of Narbonne by the Romans, though the Greeks seem to have known it earlier and might have brought it to Marseille. Frontignan was exporting its Muscat wines by the time of Charlemagne. As Muskateller it is recorded in Germany in the 12th century, and it had certainly reached Alsace by the 16th century.

Muscat of Alexandria was also spread around the Mediterranean by the Romans, probably being a Muscat cross from the Eastern Mediterranean, as its name suggests. Its Australian synonym is Lexia.

Muscat Ottonel is a more recent arrival, having been bred in the Loire in 1852. Its parents were Chasselas and the undistinguished Muscat d' Eisenstadt.

The most recent addition to the family is a cross between Muscat Ottonel and Pinot Gris, bred (for greater resistance to *coulure*) in Alsace and so known, for the time being at least, as Muscat Cendré. The grape is still at an experimental stage. One just hopes fervently it was bred with beauty and fragrance in mind, not just high yield and disease resistance.

Other members of the Muscat family include Italy's red Aleatico and Moscato Rosa del Trentino, and Argentina's Torrontés Riojano.

Muscat vines on the slopes of the Dentelles de Montmirail at Lafare, Beaumes-de-Venise in the southern Rhône. The appellation's name is a combination of the old Provençal word for grotto – beaumes – and a corruption of venaissin, an Old French word denoting church property. In the Middle Ages the Avignon Popes claimed all this area as their property.

A sign for the local Muscat at Lunel in France's Languedoc. According to the local growers, Napoleon drank Muscat de Lunel in exile on St Helena. Well, it's a nice story, but it didn't seem to do him much good.

Harvesting Muscat at Beaumes-de-Venise. The most aromatic wines come from the highest altitudes; the wines from Beaumes-de-Venise are in any case lighter than other fortified Muscats.

VITICULTURE AND VINIFICATION

One might think, from the spread of Muscat across the globe, that it is easy to please in terms of soil and climate. In some ways this is true, since the same type of Muscat will make different styles of wine, from light and dry to rich and fortified, according to where it is grown. But it is far from being an easy grape to grow. It tends to crop erratically, and acidity is low. In the winery new oak is anathema, and blending with other grapes unusual, though a little Muscat may well be blended in with other varieties – red or white – when a bit of fruit and fragrance is needed.

Muscat is the opposite of Chardonnay: flamboyantly itself where Chardonnay is all things to all winemakers; yet temperamental and sickly where Chardonnay is robust. And it's far from being a name on every wine drinker's tongue.

Climate

All Muscats like a fair amount of warmth. Mediterranean climates, especially those of the south of France, Spain, southern Italy and Greece, are its natural habitats, and its most famous New World incarnations, Australia's fortified Muscats, come from a pretty hot, though not scorching, region.

The varying susceptibilities of the different Muscats limit their suitability to different regions: Muscat Blanc à Petits Grains buds early and ripens late, and so needs a long growing season. It is a martyr to leaf roll, oidium, downy mildew and grey rot, and dislikes too much humidity; Muscat of Alexandria needs warm weather during flowering to avoid *coulure* and *millerandage*, and is also susceptible to mildew, though it endures drought well, so is happy in hot, dry regions. Muscat Ottonel is the earliest ripening of the three, and so is often planted in relatively cool Alsace (cool relative to the south of France, that is), though it nevertheless suffers from *coulure* here. In Alsace all Muscat needs sites with the best exposure to the sun.

Soil

It is not hard to guess that this multi-faceted grape is planted on a huge variety of soils throughout the world. In Alsace it favours loamy, sandy or calcareous soils, and Muscat Ottonel likes soils that are deep and rather damp; in Mireval and Frontignan, in Languedoc, the soils are red clay and limestone. In Piedmont it grows on chalk and limestone; in Australia's Rutherglen and Glenrowan soils vary from alluvial loam over gravel to deep, friable granite-derived soils. Soil type probably has little influence on the fortified Muscats of Rutherglen, however; much of the character of this wine comes from the aging process, and the availability of well-aged stocks of rich, sticky wine to blend in with younger stuff is more important.

Cultivation and yields

The maximum legal limit for French *vins doux naturels* is 28–30hl/ha, though on deep fertile soils even the medium-vigour Muscat Blanc à Petits Grains can produce double this. However, its sugar level then falls and its aroma disappears, and aroma and sugar are essential to these wines. In Rutherglen, in northeast Victoria, some growers take around 2.5 tonnes per hectare, others irrigate and get much higher yields. It's not difficult to tell which producer does which when you taste a sumptuous mouthful of grapy gorgeousness next to a simple, sweet quaffer, both from the same region. Muscat of Alexandria is more vigorous, more robust and higher yielding, but less perfumed and beguiling; Muscat Ottonel is low vigour and reasonably scented.

At the winery

Muscat may be made into a dry wine, as in Alsace; a light, sparkling one, as in Asti and Brazil; or a sweet, fortified one, as in the French *vin doux naturel* or the Liqueur Muscat of Rutherglen. It may also be made into a late-harvest sweet wine, or made sweet from *passito* (dried and shrivelled) berries. You very occasionally see a noble-rotted (Sélection de Grains Nobles) sweet Muscat from Alsace.

Neither the sparkling nor semi-sparkling Moscato wines of Piedmont are made by the traditional method – i.e. the Champagne method. Instead the must, fermented to up to 2–3 per cent alcohol, is kept in pressurized steel tanks called autoclaves at about 0°C (32°F) until it is needed; the fermentation is then continued until over half the sugar is

Asti and Moscato d'Asti have never pretended that they're made by anything other than industrial methods. And why should they? They wouldn't be nearly so refreshing if these large pressurized steel tanks, known as autoclavi, didn't keep the partially fermented juice fresh until it's ready to finish, bottle and ship.

Brown Muscat grapes starting to raisin at Rutherglen in Northeast Victoria, Australia. Nobody really knows why Muscat was first planted in this flat, unpromising region, but probably the gold rush was the reason: goldminers and prospectors tend to be thirsty, and where they go wine tends to follow.

WHICH MUSCAT?

Of the three main Muscat grapes used for winemaking, Muscat Blanc à Petits Grains is the star. It produces the most refined wines, with the greatest elegance, and with aromas that are both the most intense and the most delicate and complex. Its name, however, is deceptive: its berries are certainly small, but to describe them as white flies in the face of all the evidence. Golden yellow is their most common colour, although they can be all shades of pink and can even have dark skins, though seldom dark enough to make red wine.

Australians, typically, do not beat about the bush: the Muscat Blanc à Petits Grains that grows in Rutherglen is known simply as Brown Muscat. The astonishingly rose-scented deep pink or red Rosenmuskateller of Italy's Alto Adige and Trentino is also a variety of Muscat Blanc à Petits Grains. Galet lists a Muscat de Rivesaltes which has downy leaves; this is a mutation of Muscat Blanc à Petits Grains.

Muscat of Alexandria is less refined. Its wines are sweet and intense, but they lack the complexity of Muscat Blanc à Petits Grains, and they tend to be clumsier, lacking the balance and focused perfumed beauty of the best Muscat. Muscat of Alexandria's more assertive and less subtle aroma is accounted for by a higher proportion of geranium-scented geraniol, and a lower proportion of fragrant nerol. It is much used for grape concentrate, raisins and brandy, as well as for table grapes.

Muscat Ottonel has less powerful aromas than the previous two Muscats. It is also paler in colour.

Muscat of Hamburg is always black, and is best as a table grape. Its quality is lower than the previous three, though it is planted to some extent in Eastern Europe for light red wine, and is found elsewhere as well.

The names of all these Muscats seem to have mutated at an even faster rate than the vines. We list the main synonyms of each variety on page 165; but it is worth noting that while anything called Moscato, Moscatel, Moscadello, Muskateller or Muscadel is Muscat, any vine called Muscadelle, Muscadet or Muscardin is not. Muscat Bailey A is a hybrid grown in the Far East. Morio Muskat is a flashily scented German variety unrelated to Muscat. Muskat-Sylvaner is an Austrian synonym for Sauvignon Blanc.

fermented. The grapes have been picked at a potential alcohol level of around 11 per cent, and the rest of the sugar remains in the wine as sweetness. Asti is fully sparkling, Moscato d'Asti semi-sparkling, and the minimum alcohol content of Asti is 7 per cent. Lightness is everything with these wines: light sweetness, elegant fruit. They should be full-flavoured, yet always delicate. As befits a Party Animal place like Brazil, they make excellent Asti-type Moscatos there to lubricate carnival. And, of course, youth in the wine, as in the Carnival, is everything.

The southern French fortified wines or *vins doux naturels* (VDN) are made by *mutage*, which involves adding spirit halfway through fermentation to knock out the yeasts – the same technique that is used to make port. For VDN the fermentation is allowed to continue until the alcohol has reached about 6 per cent, and the addition of alcohol brings the total to 15 per cent or slightly over. At around 95 per cent, the alcohol used for VDN is stronger than that used in the Douro Valley for port, which is usually 77 per cent.

Some skin contact may be used, particularly in Rivesaltes, where the vineyards are dominated by Muscat of Alexandria. This increases the perfume, but must be used sparingly if heaviness, even bitterness, is not to result.

Rutherglen's methods of producing its fortified Muscats vary. Some producers pick their grapes at potential alcohol levels of 16 per cent, others prefer to pick at 20 per cent. In some years 30 per cent is possible. Some like to pick raisined grapes, some don't.

A typical method of winemaking is to ferment the must for a mere 24 hours before fortifying so that the juice changes enzymatically, but the alcoholic fermenation is barely started. Some producers operate a gradual solera blending system, or a version thereof, generally employing hogsheads or large casks for aging, though as the wine ages and loses volume to evaporation, barrels are likely to get smaller and smaller; others blend selected wines before bottling. Some add alcohol to the unfermented must, so that no fermentation at all takes place. Whatever the method, the final figures are around 17–18 per cent alcohol and 9–14° Baumé residual sugar. Aging, usually in pretty torrid conditions under the tin roofs of the wineries, may be for years, decades or even longer: some wineries boast a barrel or two of dark, treacly wine from the 19th century, used in tiny quantities to give depth and viscosity to younger wine.

MUSCAT AROUND THE WORLD

There are vast swathes of Muscat of Alexandria growing in the warm vineyards of the world. But remember, a considerable percentage is not destined for winemaking but for eating. There's less Muscat Blanc à Petits Grains planted, but most of it is destined for winemaking and it makes some of the most exotically scented wines in the world.

France

Both Muscats are used in southern France's cluster of *vins doux naturels*. Several of the ACs specify Muscat Blanc à Petits Grains: it should be the only variety in the delicious, scented Muscat de Beaumes-de-Venise (whose residual sugar level is 110g/litre instead of 125g/litre in the other ACs). Mireval, St-Jean-de-Minervois, Lunel and Frontignan should also be Muscat Blanc à Petits Grains but aren't always. Muscat de Rivesaltes is dominated by Muscat of Alexandria grapes, although Muscat Blanc à Petits Grains is on the increase here thanks to new clones which yield more reliably. Rivesaltes also produces VDNs called simply Rivesaltes, from blends of grapes: both Muscats may be mixed with Grenache, Maccabéo and Malvoisie du Roussillon (alias Torbato), and these wines turn out every imaginable colour, and in styles from young and fresh to *rancio*. Naturally, they do not have the pure Muscat aroma of Muscat de Rivesaltes.

There is also some dry, unfortified Muscat in the South, usually with more body and substance but less ethereal perfume and crunchy fruit than the dry versions from Alsace. Muscat Blanc à Petits Grains also dominates the blend in light, sweet and sparkling Clairette de Die Tradition, often to the virtual exclusion of Clairette and to the great benefit of the wine.

In Alsace, Muscat takes up only about 3 per cent of the vineyard, and while Muscat Ottonel is the preferred variety, it never produces such delightful wine as Muscat Blanc à Petits Grains, which also makes good fresh wines on Corsica, some fortified.

Italy

Moscato (mainly the superior Muscat Blanc à Petits Grains) is grown all over the peninsula. In Piedmont, where it is grown for light, sweet, sparkling Asti and Moscato d'Asti, Muscat Blanc à Petits Grains is planted up to 550m (180ft), with the higher, steeper vineyards giving more acidity and the lower vineyards more body. But even at relatively low potential alcohol levels (about 11 degrees for Moscato d'Asti) there is no shortage of aroma. Moscato Rosa, or Rosenmuskateller, is a variation of the same Muscat, used for intensely rose-scented (and even rose-coloured, if the roses in question are red) still wines in Trentino and Alto Adige; Moscato Giallo or Goldmuskateller is another variant, and makes orange-blossom-scented dry or sweet wines in Trentino and Alto Adige. Sweet red Aleatico is a close relation of Muscat Blanc à Petits Grains.

On the island of Pantelleria, between Sicily and Tunisia, Muscat of Alexandria, often under the name of Zibibbo, is grown for amber-coloured Moscato di Pantelleria, made from *passito* grapes. Sardinia, Basilicata and Puglia also have traditions of sweet Moscato, though they're hardly flourishing.

Spain and Portugal

Simple, sweet, grapy Moscatel is grown all over Spain, and usually the grape in question is Muscat of Alexandria (Moscatel de Alejandría), though there is some Muscat Blanc à Petits Grains (Moscatel de Grano Menudo) in the North. Moscatel de Málaga is a speciality of the South, and is probably a synonym for Muscat of Alexandria. Moscatel de Setúbal is Portugal's best known, made from Muscat of Alexandria, but crisp, dry Muscats are made in the same area.

Austria

Gelber Muskateller or Muscat Lunel (both synonyms for Muscat Blanc à Petits Grains) is the grape behind Austria's best Muscats, though Muscat Ottonel takes up more of the vineyard. Muscat Ottonel is, however, currently being uprooted to make way for red vines. Muscat Ottonel makes good late-harvest wines in the Neusiedlersee, while Muscat Blanc à Petits Grains appears as dry wine in the Wachau and Südsteiermark. There is also some in Burgenland.

Greece

Muscats Blanc à Petits Grains and Alexandria are grown here, though only the former is used for the high-quality sweet wines of Samos, Pátrai and Kefallinía; these may be unfortified, fortified in the manner of *vin doux naturel*, or made from grapes dried in the sun. Muscat of Alexandria rules in Cyprus. Greece has obvious potential, given its current talent for making crisp, dry, characterful whites, for aromatically dry Muscat.

Rest of Europe

Muscats are grown in Germany, Bulgaria, Romania, Croatia, Slovenia, Moldova, Uzbekistan, Kazakhstan, Tajikistan and Turkmenistan, and in Crimea (now back in Russia), where the Massandra winery continues its tradition of making both white

Domaine de Durban
One of the principal producers of Muscat de Beaumes-de-Venise, Domaine de Durban's wine has a style that is delicate but beautifully scented.

Dirler-Cadé
Muscat is one of the four noble varieties that can be planted in Alsace's Grand Cru vineyards. The sandy soil here gives this Muscat lightness and perfume.

La Spinetta
Bricco Quaglia is the Rivetti family's top Moscato from its La Spinetta estate near Asti. 'Bricco' is a Piedmontese dialect word meaning 'hilltop'.

and pink Muscat. The former is aged for six years before release, the latter for 20. Hungary has some Muscat in Tokaj, which is picked late (and, yes, botrytized) for sweet Tokaji, but which also has potential as a dry wine.

Australia

Of the two great fortified wines made in Rutherglen and Glenrowan, in northeastern Victoria, one is made from Muscat grapes. This is, not surprisingly, known as Rutherglen Muscat, or Liqueur Muscat; the other wine, made from Muscadelle, known locally as Tokay, is now called Topaque. Yes, lovely name, isn't it? Not.

Both wines darken with age until they are thick, brown and viscous, but both start out much paler. Young Muscat tastes grapy/raisiny and rich, often with a tea-rose scent; with increasing maturation in barrel it becomes more concentrated and complex, and takes on flavours of figs and blackberries, coffee and chocolate, and the acidity at length becomes more apparent, though Muscat always tastes sweeter and fleshier than Topaque. A four-tier quality system puts Rutherglen Muscat at the basic level. Above it, in ascending order, are Classic, then Grand, then Rare. There are no specific criteria for any of these categories; peer pressure, in this small community of growers, is supposed to ensure high quality at every level.

Not all Australian Muscat is used for such styles. Both Muscat of Alexandria (Muscat Gordo Blanco, or Lexia) and Muscat Blanc à Petits Grains (White Frontignac or Brown Muscat) are used to make light, fruity bulk wine sold in wine casks. Plantings are declining slowly.

South Africa

The star Muscat is Vin de Constance. It's made from Muscat de Frontignan, aka Petits Grains, picked really shrivelled and sunburnt, and it has elegance, delicacy, raciness, concentration – everything you want, really. It's a reinvention by Klein Constantia of the style that made South Africa famous right at the start – and since there are apparently no records of Muscat vines having been imported since the early days of the colony, those that make Vin de Constance are quite possibly descendents of the originals.

Otherwise Hanepoot, alias Alexandria, is concentrated in Worcester and Olifants River, but is found in most regions and is used for everything: dry, sweet and fortified wine, table grapes, raisins and grape juice. Quality is generally dull, excepting a few top fortifieds. Muscadel (Muscat Blanc à Petits Grains) is the grape behind Jerepigo, made by adding grape spirit to unfermented grape juice.

Terraced vineyards on the Greek island of Samos. Muscat is grown at altitudes of up to 800m (2620ft), and grapes from the highest, coolest spots may be picked as much as two months after those grown lower down.

USA

Plantings of Muscat of Alexandria outnumber those of Muscat Blanc (Muscat Canelli, or Muscat Blanc à Petits Grains), but are falling fast. It grows mostly in California's Central Valley, where it was planted for raisins and also makes bulk wine for blending. Muscat Canelli is planted both in the Central Valley and in the coastal regions, and is generally made light, medium-sweet and slightly sparkling. There are also some fortified versions.

California also has small amounts of orange-flower-scented Orange Muscat, which appears to be related to Muscat Blanc à Petits Grains, and some rose-scented Black Muscat, which is presumably Muscat Hamburg. Both can make exciting sweet fortified wines in the right hands.

Rest of the world

Brazil makes good use of Muscat – usually Petits Grains – for high-quality sparkling wines. Elsewhere in South America large quantities of Moscatel of different sorts are grown for pisco brandy, table grapes and bulk wine. Tunisia manages to make some dry rosé from Muscat of Alexandria.

Marco De Bartoli
Moscato grapes are allowed to dry naturally in the autumn sunshine for Marco De Bartoli's passito wine. Zibibbo is the local name for Muscat of Alexandria.

José Maria da Fonseca
This wine is a blend of Moscatel with Bual and Malvasia; other Moscatels de Setúbal from the same house may be different blends, or pure Muscat.

Klein Constantia
Constantia wine was first made in South Africa in the late 17th century. It is an unfortified but intensely sweet wine with a lime and orange zest acidity.

ENJOYING MUSCAT

Enjoying Muscat is so easy, I feel a bit silly even considering writing down any lists of dos and don'ts. But it is that very fresh, fragrant, blossomy beauty which is at both the heart of the pleasure and the core of the decline. It rarely lasts. It fills the glass with the heady scents of summer ripeness, the sweet perfumes of a Mediterranean bower laden with fruit – and then, just as summer always crumbles into autumn, the scent becomes dulled, the youthful beauty becomes coarsened and fat. Summer is gone. And so is the succulent, sybaritic fleeting beauty of Muscat. It would be hard to bear were there not another harvest every year: another chance to revel in young Muscat yet again.

Well, this is the case with most Muscat. Most dry examples have a beautiful but fleeting perfection – though some Alsace ones can age to a fascinating scented decadence. Even the sweet *vins doux naturels* of southern France gain nothing from aging, though they decline gracefully.

Lack of acidity is the problem. Perfume almost always needs balancing acidity in wine to develop past its first bright-eyed flush of youth, and Muscat simply doesn't have the requisite acidity. If the aroma tires, Muscat wines quickly fade to an oily sullenness. Yet there are a few exceptions – fortified wines made from small crops of grapes, their flavours intensified by overripeness and shrivelling on the vine, their dark sensuous fruit arrested and preserved by the fortification process, and their intoxicating scent married with the mystical tastes of age and decay by long barrel-aging and blending. The greatest of these are the Muscats of Glenrowan and Rutherglen in Australia, though South Africa can occasionally come up with the goods too.

The taste of Muscat

Light young Muscat tastes of rose petals spiked with orange blossom, sometimes elderflower, and perhaps a touch of orange zest; it is this slight citrus tang that can help to balance the wine's low acidity. Sometimes the roses dominate; sometimes there is a peachy richness; sometimes the orange comes to the fore. With so much variation even within Muscat Blanc à Petits Grains vines, it is hardly surprising that the wines can taste different.

Oh! And grapes. There is always a crisp, crunchy grapiness about good examples of Muscat Blanc à Petits Grains, and it is this crispness, as well as greater finesse and subtlety, that makes Muscat Blanc à Petits Grains wines such sublime drinks. Muscat of Alexandria is simpler, broader, less perfumed, less beautiful; Muscat Ottonel less powerfully scented, sometimes verging on the vegetal.

Dark, aged Muscats that have been sitting in old oak barrels for years acquire an oxidative bouquet of figs, blackberry and coffee, prunes, treacle, nuts and chocolate, immensely complex and pungent. These wines can be so thick and viscous that they coat the glass with a thick film of translucent brown; the smell will remain in the glass long after the wine has been drunk.

And, of course, most Muscat is sweet. Sweetness is certainly what such a perfume leads you to expect – and it is generally what you get – a sweetness so fruity that such wines are the best all-purpose wines to go with dessert. Even so, with Alsace in the lead, the number of wines of crystalline fragrant dryness is increasing.

Chambers Liqueur Muscat is dark, rich and very sweet indeed, although it is by no means the most heavyweight Muscat to come from Rutherglen in Northeast Victoria, and it retains its glorious scent. By contrast, Goldert is an east-facing Grand Cru vineyard in Alsace and Zind Humbrecht make a golden and full-bodied and wonderfully perfumed wine from its grapes.

MATCHING MUSCAT AND FOOD

These fragrant, grapy wines range from delicate to downright syrupy. The drier ones are more difficult to pair with food, but are a good choice with delicately spiced Thai or Indian food. The sweeties come into their own with most desserts (try chilled Liqueur Muscat with really good vanilla ice-cream). Chilled Moscato d'Asti, delicious by itself, goes well with rich Christmas pudding or mince pies.

CONSUMER INFORMATION

Synonyms & local names

Muscat Blanc à Petits Grains, the most important variety, is also called Muscat de Frontignan, Muscat Lunel, Muscat Blanc, Muscat d'Alsace, Muscat Canelli, Muskateller in Germany, Brown Muscat or Frontignac in Australia and Muskadel or Muscadel in South Africa; the Italians call it Moscato or Moscato Bianco, Moscato Canelli or Moscato d'Asti. Moscato Giallo (Goldmuskateller) and Moscato Rosa (Rosenmuskateller) are mutations with darker berries from Alto Adige and Moscadello is a Tuscan variant. Spanish synonyms are Moscatel de Grano Menudo or Moscatel de Frontignan. Tamîioasa is the Romanian name. The less thrilling Muscat of Alexandria or Muscat Romain is called Moscatel in Spain (local names include Moscatel de Alejandría, Moscatel de España, Moscatel Gordo Blanco and Moscatel de Málaga), Moscatel de Setúbal in Portugal and Moscato di Alexandria in Italy (Zibibbo in Sicily). It is called Muscat Gordo Blanco in Australia and Hanepoot in South Africa. Muscat Ottonel is Muskat-Ottonel in Germany and Muskotály in Hungary. None of these should be confused with Muscadelle or Muscadet.

Best producers

FRANCE/Alsace J-M Bernhard, Bott-Geyl, Burn, Dirler-Cadé, Kientzler, Kuentz-Bas, René Muré, Ostertag, Rolly Gassmann, Schoffit, Sorg, Trimbach, Weinbach, Zind-Humbrecht; **Rhône/Midi** Achard-Vincent, Bernardins, Cazes, Clairette de Die co-op, Durban, Jaboulet, Jau, la Peyrade, Pigeade, Poulet, Raspail, Sarda-Malet, Vidal-Fleury.
GERMANY Bercher, Dr Heger, Huber, Müller-Catoir, Rebholz.
AUSTRIA Gross, Alois Kracher, Lackner-Tinnacher, Opitz, F X Pichler, Polz, Sattlerhof, Tement, H Tschida.
ITALY Bava, Bera, Braida, Caldaro co-op, Castel de Paolis, Cascina Fonda, La Caudrina, Col d'Orcia, G Contratto, La Crotta di Vegneron, De Bartoli, Forteto della Luja, Gancia, Franz Haas, Lageder, Icardi, Ivaldi, Conti Martini, Murana, Perrone, Saracco, Schloss Salegg, La Spinetta, Tiefenbrunner, Thurnhof, Vignalta, Voyat, G Voerzio.
SPAIN Bocopa co-op, Camilo Castilla, Chivite, Gandía, Garcia, Juan Gil, Gutiérrez de la Vega, Lustau, Enrique Mendoza, Ochoa, Miguel Oliver, Ordóñez, Primitivo Quiles, Telmo Rodriguez, Torres, Valdespino.
PORTUGAL/Moscatel de Setúbal Bacalhôa, José Maria da Fonseca.
GREECE Samos co-op.
USA/California Robert Mondavi, Quady, St Supery, Philip Togni; **Washington State** Chateau Ste Michelle, Covey Run, Kiona.
CHILE De Martino.
AUSTRALIA All Saints, Baileys of Glenrowan, Brown Brothers, Buller, Campbells, Chambers, Crittenden, Kosovich, McWilliams, Morris, Pfeiffer, Seppeltsfield, Stanton & Killeen, Talijancich, Yalumba.
SOUTH AFRICA Delheim, Klein Constantia (Vin de Constance), Nederburg, Nuy.
BRAZIL Courmayeur, Monte Paschoal, Salton.

RECOMMENDED WINES TO TRY

Five Alsace Muscat wines

Albert Boxler Muscat Brand
Ernest Burn Muscat Goldert Clos St-Imer
Dirler-Cadé Muscat Spiegel
Schoffit Alsace Muscat Vendange Tardive
Zind-Humbrecht Muscat Goldert

Five other European Muscat wines

Bercher Burkheimer Feuerberg Muskateller Spätlese Trocken (Germany)
Alois & Ulrike Gross Steiermark Muskateller Kittenberg (Austria)
José Maria da Fonseca Moscatel de Setúbal 20-year-old (Portugal)
Alois Kracher Muskat Ottonel TBA (Austria)
Salvatore Murana Moscato Passito di Pantelleria Martingana (Italy)

Five French Vins Doux Naturels

Bernardins Muscat de Beaumes-de-Venise
Durban Muscat de Beaumes-de-Venise
Jaboulet Muscat de Beaumes-de-Venise
Sarda-Malet Muscat de Rivesaltes
St-Jean de Minervois co-op Muscat de St-Jean-de-Minervois

Five Australian fortified Muscats

All Saints Rutherglen Museum Release Muscat
Campbells Rutherglen Merchant Prince Muscat
Chambers Rutherglen Special Liqueur Muscat
Morris Rutherglen Old Premium Muscat
Seppelt Rutherglen Show Muscat DP63

Five Italian sparkling wines

Bera Moscato d'Asti Canelli
Caudrina Moscato d'Asti La Caudrina
G Contratto Asti De Miranda
Saracco Moscato d'Asti
La Spinetta Moscato d'Asti Bricco Quaglia

Muscat is the most protean of vines, and exists in many local variants around the world. The 'standard' colour of the berries may be white – though golden would be a better description – but many Muscats have pink, red or even brown berries.

Maturity charts

This is not a grape to age in bottle. Fresh, aromatic young Muscats will only lose their aroma and gain nothing in its place.

2014 Alsace Muscat (dry)

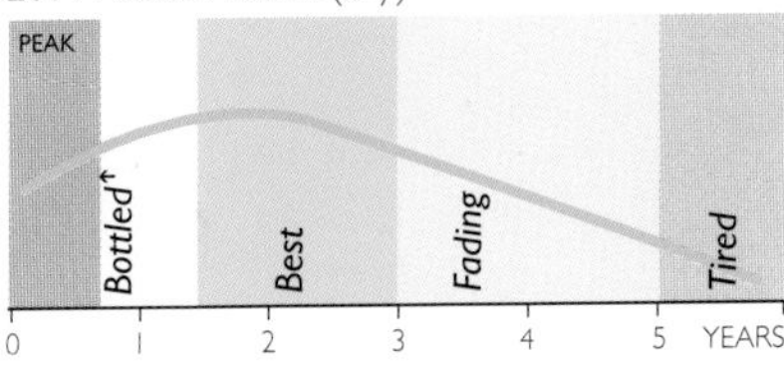

Alsace Muscat should be drunk young while the wine is fresh and floral.

2014 Southern France *vin doux naturel* Muscat

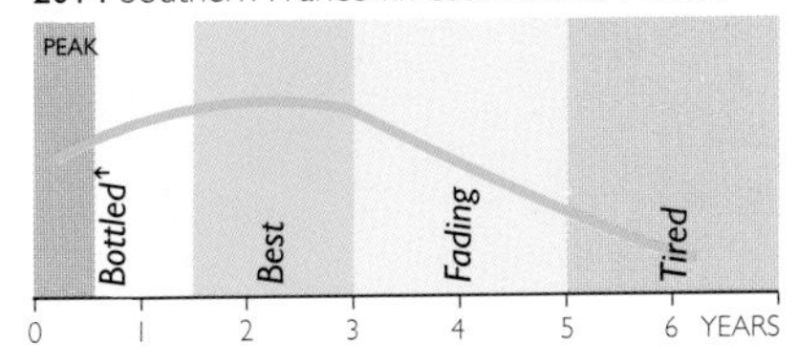

Another Muscat wine for which bottle age is not an advantage. Again, these wines lose aroma and freshness if kept longer than a year or two.

North-East Victoria Muscat (Rutherglen)

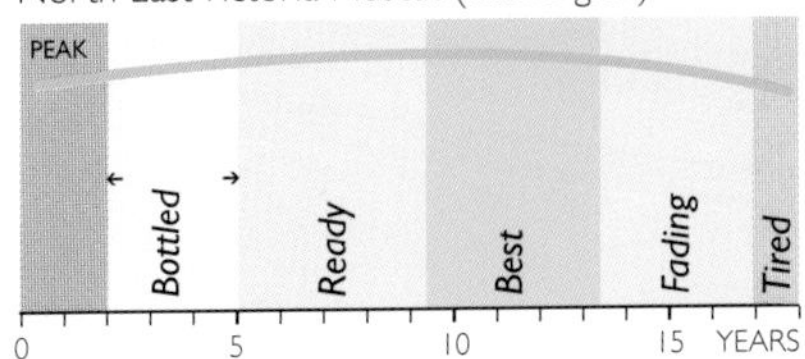

Bottled after years in barrel, these wines hardly change in bottle. Dark, sticky and super-raisiny, they will stay much as they are for up to a couple of decades.

NEBBIOLO

I don't know. I tried hard enough. But for years I just couldn't get to grips with Nebbiolo. Oh, I used to leap at the chance to taste it – ideally in its Barolo or Barbaresco manifestations. I've hummed and hawed, nodded my head as I listened to my elders and betters, all the while trying to stifle a grimace on my face as my gums withered at the brutal assault of tannin and acidity from whatever wine they'd brought. Phew! Expectoration was never such sweet pleasure. And slowly the life would creep back into my battered palate. What were those people going on about, with all their reverential verbiage about tar and roses and blackberry and sloes. And how it was time for lunch.

Well, in my callow, know-all state of mind, I rarely stayed for the food. I should have done. For two reasons. First, many of the old-timers would never dream of bringing out their true treasures – their 20- and 30-year-old Riservas that had lain undisturbed at the back of the cellars since they were bottled – without platefuls of Piedmontese fare to accompany them. Barolo and Barbaresco, more than almost any other great wines, demand the convivial company of hearty food – *bollito misto*, pheasant, pigeon or hare, great chunks of beef smouldered in lakes of rich red wine sauce – and did I mention white truffles?

Dusty purple grapes and bright red leaves seen against a background of snow and fog – this is the Piedmontese town of Alba, famous for red wine and white truffles, in autumn. Against a backdrop of snow-capped Alps, the town's medieval towers emerge from the late autumn fogs, or nebbie, *that have given the Nebbiolo grape its name. In late autumn truffle hunters and their specially trained dogs set off into the oak forests bordering the vineyards in search of the revered delicacy.*

And, of course – age. That unfashionable, curmudgeonly state of mind that is known as age. Did I say 20 or 30 years? Certainly – and more when the vintage was perfect and the producer devoted to his craft. Then, at last the crusty, argumentative, abrasive nature of Nebbiolo slips quietly from the wine to reveal something old but magnificent, prepared to give of its all if the food is good and the company warm. Suddenly you sit bolt upright with the glass halfway to your lips – and there it is – the haunting, beautiful aroma of sloes, and blackberry, and roses, and tar.

So the old-timers were right. And thankfully, I did occasionally get lucky over time. But things have changed, too. There's a new generation at work in Piedmont, who are too impatient to sit for 30 years waiting for the wines to bloom, but also too knowledgeable and world-savvy to think that it's necessary. They used to say you had to age your Bordeaux for 20 years before you could drink it, before those beetle-browed Cabernet tannins mellowed in the autumn of their lives. They don't now. They said the same in Hermitage, in Spain's Rioja and Ribero del Duero, in the Port lodges of Portugal. They don't now. In these areas, and many others where the traditional wine styles were rough or raw and only time could tame them, modern vineyard management, up-to-date winemaking techniques designed to maximize their flavour, scent and succulence not their sullen surliness, have transformed the wine styles. And, of course, let's not forget climate change with its ongoing, and for now, effect on the ripeness levels in vineyards. Piedmont was late to get in on the act and the tannins and bitterness that lie embedded in the Nebbiolo's skins and pips have proved a real challenge. But the fruit in the Nebbiolo, the scent and, increasingly in the 21st century, the optimistic beauty of youth is starting to show in the wines as well.

But don't expect floods of great Nebbiolo. It's still a tricky creature, whose favours are by no means universally granted. Just look at the New World. Why aren't there far more plantings in California, in Australia, in South Africa or Chile? Why have plantings in Argentina dropped by two-thirds this century? Because even the toughest and most talented of the young guns rarely want to take on the challenge. But there are now some stars in Australia – most of the best fruit being grown in Victoria's King Valley by guys with very Italian names – some in Argentina, and in Virginia, of all places. It's just a couple of handfuls of courageous, creative wines. But do search them out. If their remarkable flavours and perfumes – parodoxical, argumentative, but arresting in their totally original beauty – are a sign of Nebbiolo's future brilliance, the entire wine world should cheer.

Nebbiolo: from Grape to Glass

Geography and History page 168; Viticulture and Vinification page 170; Nebbiolo around the World page 172; Enjoying Nebbiolo page 174

GEOGRAPHY AND HISTORY

British wine merchants, it seems, 'discovered' the Nebbiolo wines of Piedmont in northwest Italy early in the 18th century, when Britain and France were at war and French wine was, to some extent, out of bounds to British drinkers. But the local duties were prohibitive, transport all but impossible, and the project of exporting Nebbiolo to Britain was abandoned. But if the opposite had happened, would Nebbiolo now be established in a much wider area than its Piedmontese stronghold? Surely the great wines of Burgundy, Bordeaux, sherry and port became famous through being disseminated along the world's trade routes. France sold wine to all of northern Europe; Piedmont to almost no one. France became the undisputed world leader in wines while Piedmont became a connoisseur's curiosity.

Nebbiolo is grown in North and South America and in Australia as well. But don't let the map imply there are massive plantations. There aren't. And very few of the smattering of

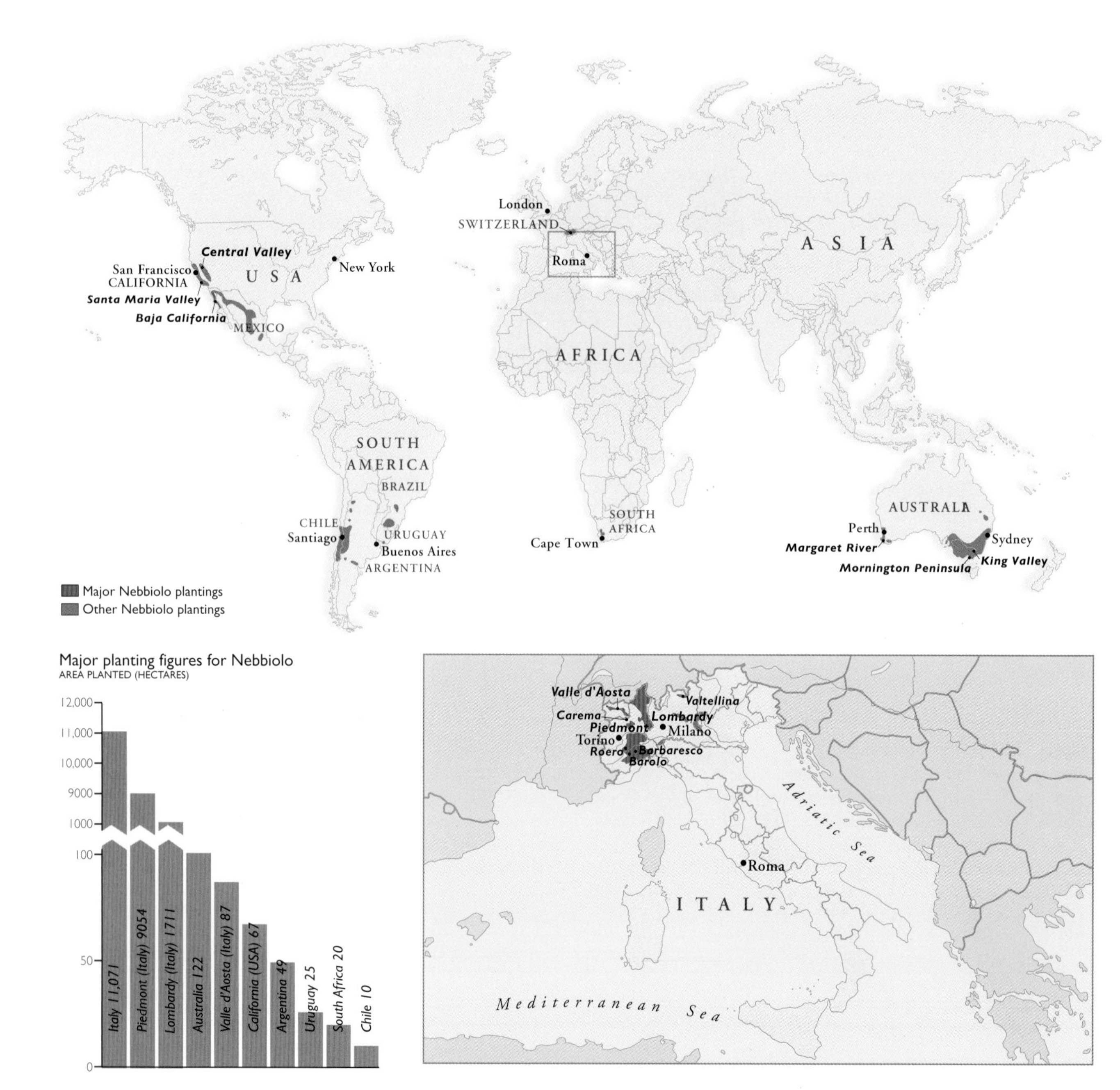

examples have so far shown much class. But the beginnings are there, and the best wines in Australia and Virginia are showing tremendous acidity, fascinating scents and flavours, a greater approachability when young, but an inspiring ability to age. None of these New World titans pretends Nebbiolo is easy, but they are loving the challenge.

This is not to say that great Nebbiolo will never be made outside Piedmont. But it will take time to learn which are the best sites and which the best winemaking techniques to apply to which clones, since these matters are still heatedly argued over in Piedmont itself. It is a vine that bears more than a little resemblance to Burgundy's Pinot Noir in its refusal to be treated casually in vineyard or winery; it took New World winemakers a generation to crack Pinot's code. Nebbiolo will prove even more difficult, but the gauntlet has been thrown down.

HISTORICAL BACKGROUND

Genetically unstable vine varieties that mutate readily are often old varieties; and Nebbiolo can demonstrate considerable clonal variation. It has been with us for a long time.

What it cannot demonstrate, of course, is precisely when it came into being. The name is usually said to derive from the word for fog – *nebbia* – either because the grape ripens late, sometimes well into November, by which time it's foggy, or equally likely because of the white bloom on the ripe grapes. It seems to have originated in the Novara hills, between Turin and Milan in northern Piedmont, and the first written record of it by name was in 1268, when 'Nibiol' is mentioned as growing at Rivoli, which is now a suburb of Turin. In 1303 Canale d'Alba names it as 'nebiolo', and in Petrus de Crescentiis' 1304 *Liber ruralium commodorum* there is a mention of 'Nubiola'. In the 15th century anyone in the commune of La Morra guilty of cutting down a Nebbiolo vine was liable to be fined, have a hand amputated or be hanged.

In the 19th century it was far more widely planted in Piedmont than it is today – and not just because it became possible to get rid of it without being hanged or having your hand chopped off. (Appellation laws in Europe have become more enlightened over the years, believe it or not.) When phylloxera wiped out the vineyards, the growers took the opportunity to replant with Barbera, which yields more generously and more reliably. Nebbiolo now covers only about 6 per cent of the vineyards in Piedmont, yet is far and away the most prestigious vine of the region, being the sole variety allowed in the majestic DOCG wines of Barolo and Barbaresco, from the Langhe hills around the town of Alba. It almost always hogs the warmest sites – and it needs them: its colleagues Barbera and Dolcetto can be picked and fermented before Nebbiolo is ready to harvest.

The commune of Castiglione Falletto with, sloping down from the village, part of the Rocche subzone. Top Barolo these days comes from single vineyards, or crus, within the subzones, though a great deal of local politicking remains to be done before everyone agrees on an official classification.

Mauro Mascarello seems to epitomize tradition as leans on his giant botti, but in fact he was a great modernizer in the 1980s, who has now seamlessly married good tradition with good modernity.

Davide Rosso makes some of Barolo's most balanced, attractive – and generally affordable – wines despite being based in Serralunga d'Alba, famous for powerful, long-lasting, sturdy reds.

VITICULTURE AND VINIFICATION

Nebbiolo is a grape at odds with its reputation. For many Barolo is the epitome of the brutal, tannic wine that needs years in bottle before it is tamed into relative politeness. Modern Barolo, and modern Nebbiolo, is a very different creature. It is softer and more forward but still keeps its tannic rasp. It can exhibit a divine rose petal scent, black chocolate and prunes richness, tarry intensity – all lovely, but all in danger of becoming just one more braggart on the new-oaked circuit of the international superstars. I have to say that almost without exception I prefer the new styles. And yet, shining among a sludge of bitter, fruitless old-style Barolos and Barbarescos, I have tasted a few ancient jewels far more thrilling, more individual, more indelible in taste, than anything the modernists have yet produced.

Climate

Nebbiolo is extremely fussy about its climate, and getting the climate wrong is probably the biggest single reason why so few New World Nebbiolos have been successful. In Piedmont it needs the best exposed, south- and southwest-facing sites, at altitudes of between 150 and 300m (500 and 990ft). Higher than that is too cool. Nebbiolo suffers horribly from *colatura* (which sounds like a plague of sopranos, but is merely poor fruit set at flowering) and so must be sheltered from cold and wind, especially since it is the earliest of all Piedmontese varieties to bud. Wet weather at flowering doesn't help, either.

Nor does wet weather in September and October help. These are relatively rainy months in the Langhe and Roero hills of southern Piedmont, and it is notable that the best Nebbiolo vintages are those in which September is fairly dry. Weather conditions in spring and summer affect the quantity of a vintage; weather after *véraison* affects quality.

The Piedmont climate is markedly continental, with hot summers and cold winters, and its shortish growing season means that Nebbiolo (early budding, late ripening) is at the limit of its cultivation here. Experience in Australia, however, suggests that planting in much warmer climates doesn't work. Research there points out that Piedmont has a curiously contradictory climate: continental, but with relatively little variation from year to year. Australian growers trying to match Piedmont's growing conditions won't have it easy.

Soil

Nebbiolo seems to be able to cope with a wide variety of soils in northwest Italy, though it produces its greatest quality only on the clay and limestone of the Albese, the area around Alba, on the south bank of the Tanaro river. The soil type, as well as the climate, is reflected vividly in the wine: soils in Roero, on the other side of the Tanaro from the Langhe, are sandy, with a neutral or slightly acid pH, and produce earlier-maturing wine. The hills between Novara and Vercelli, where lighter Gattinara and Ghemme come from, are porphyry and acidic; Valtellina's soils are schist, and those of the lower Aosta Valley are granitic. Good drainage is important, and the presence of ores such as potassium and magnesium can affect the style of the wine.

Cultivation and yields

Around 40 different clones of Nebbiolo have been identified, but the three that are most widely grown are Lampia, Michet and Rosé, though the last is genetically different and is therefore presumably a different variety.

Angelo Gaja was one of the first in Piedmont to turn to international winemaking practices, determined to make the world take notice of his Barbaresco. He was also one of the first to charge sky-high prices – and achieve them. And he's a fanatic for detail: the extra-long corks he specifies proved such a problem for restaurant wine waiters that he had to become the Italian agent for Screwpull. You don't normally catch him looking as relaxed as this, but he and his daughter Gaia now make a formidable pair.

Lampia has large three-lobed bunches with good colour and yields well, sometimes too well. Michet has smaller, more compact bunches and gives lower yields; it is a genetic mutation of Lampia, the mutation having been caused by a virus. Rosé is paler in colour, as its name suggests, and is unpopular for this reason. Nebbiolo is often problematically light in colour even without this clone. New clones based on these three have been developed, and almost all growers prefer to grow a mix of clones rather than confine themselves to one or two. Different regions have their specialities, too: in Aosta and the far north of Piedmont the local clone is called Picoutener, or 'tender stem', and in Valtellina, in Lombardy, it is the Chiavennasca.

Unless weakened by viruses Nebbiolo is vigorous, but since the basal buds are not very fertile it is usual to leave a long cane with ten to 12 buds on a guyot training system. But the better growers of the Langhe maintain that about 40–60 quintals (4–6 tonnes) per hectare is the maximum for top quality, even though the DOCG regulations allow 80 quintals. So they might cut the cane back to seven to ten buds. A smaller crop also has the advantage, in this marginal climate, of ripening earlier.

At the winery

In terms of vineyards and clones, the Nebbiolo of Barolo is probably further ahead than the Sangiovese of Tuscany – but then it had less far to come. The quality of the grapes is generally very good. All the controversy hinges on the winemaking. In Barolo (and, indeed, in Barbaresco) you are a traditionalist or a modernist. Time was when the divisions between the two were as sharp, and as sharply felt, as between Montague and Capulet. One winemaking family even built a wall physically to divide the traditionalist sibling from the modernist.

Now modernism has affected (infected, some might say) most of the traditionalist producers to some extent. Most Barolo is, accordingly, more forward and fruity than it was, but that is a question of degree. Let's look at Barolo winemaking as it was before anything changed.

Nebbiolo is late ripening, and so by the time it is picked the weather, and the cellars, are getting cold. Starting the fermentation was not easy in the past: it could take eight or 15 days before there was any sign of action in the must. Delayed fermentations like this, in old wood or cement where hygiene was not absolute, could lead to bacterial infection. That was the cause of the off-flavours that necessitated opening the bottles 24 hours in advance. Once the fermentation began the temperature of the must could rise extremely high, to 35° or 38°C (95° or 101°F); this tended to reduce aroma and fruit flavours. But the weather was getting colder all the time: the fermentation would slow, and drag out for two or three months. The amount of tannin extracted was enormous, and had to be softened by leaving the wines in big old casks. or *botti*, for five years or more, which in turn could cause problems of volatile acidity.

Today's traditionalism is based on long macerations of 20 to 30 days and long aging. Technical improvements such as starting the fermentation swiftly, and hygiene, can be taken for granted. Modernism, meanwhile, centres on the use of *barriques*, on shorter fermentation and maceration times and on earlier bottling. The idea is that with temperature control you can keep the temperature down to 28° or 30°C (82° or 86°F) to preserve fruit and aroma. Rotovinifiers are fashionable in Barolo: 1996 was the first vintage in which they were generally used. They give better extraction, so that a short maceration of seven to 10 days may be all that is necessary. Heating the cellars towards the end of the alcoholic fermentation will encourage the malolactic one, which in turn means less aggressive acidity.

The point of small oak *barriques*, for fermentation, aging or both stages, is not to add tannin but to soften the wine for two years or so with their gentle oxidative effect. But they do, of course, give a certain sameness to the wines. Barolo is cleaner than it was, easier to understand, fruitier, more supple and less unpredictable to buy and drink. But new oak can cover the wonderful rose scents of Nebbiolo with its vanilla richness. And tamer wines, by definition, are less exciting wines. Modernism is both a gain and a loss.

If Nebbiolo derives from nebbia, *Italian for 'fog', this picture of La Morra, hidden in an autumn fog, says it all. Nebbiolo is such a late ripener, fog frequently coats the vineyards before and during harvest.*

BLENDING NEBBIOLO

In Piedmont most Nebbiolo is made as a varietal wine. Barolo is 100 per cent Nebbiolo; so is Barbaresco. Across the Tanaro river from Barolo and Barbaresco comes Roero which is supposed to contain between 2 and 5 per cent Arneis, but often doesn't these days. In Gattinara, Ghemme, and the less familiar wines of northern Piedmont, blending is also officially required, the other grapes being Vespolina, Croatina and Bonarda, but not all producers opt for obeying the law on this point. In Lombardy's Valtellina, Nebbiolo (called Chiavennasca here) may be blended with Rossola, Pignola, Prugnolo, Pinot Nero and Merlot for basic Valtellina, and with other red grapes up to 5 per cent for Valtellina Superiore. That's the dull bit.

The interesting bit, for those curious about how things are really done, as opposed to how they're supposed to be done, is whether or not other grapes are blended into Barolo and Barbaresco. Officially, of course, this is completely forbidden, and if you want to add a dash of Barbera or whatever, you must demote the wine to Vino da Tavola. But some tasters have noted smells of Cabernet Sauvignon or Syrah in Barolo; others have pointed to unusually deep colours in wines from this rather poorly coloured grape. Some go so far as to point out that Barbera was always the traditional blending grape in Barolo, because of its greater colour and lower acidity. Purely hypothetical, of course. No one ever admitted to doing it, though an amused smile could be seen playing on the lips of some successful producers.

In 1998 there was a proposal to make it legal to add 10–15 per cent of other grapes to Barbaresco – the grapes could have been Cabernet Sauvignon, Merlot or anything else. They would have pepped up the wine in a bad vintage, and made it rounder and richer; more saleable, in other words. The growers of Barolo were invited to join in the party, but they declined, and then the idea got a bad press in the Italian media. It was then discarded in the interests of (as the official line went), maintaining the individuality of Barbaresco in the face of increasing international standardization. Well, good for them for making a stand. Especially if, as some commentators believe, they're doing it on the quiet anyway.

NEBBIOLO AROUND THE WORLD

Attempts at making Nebbiolo worldwide seem so far to be having only limited success, and while it's true that even in its native Piedmont quality can vary hugely, it is Barolo that is the ultimate target of every ambitious producer in California or Australia. At the moment I'd be thrilled if most of them could make anything as good as simple Langhe Nebbiolo.

Italy

Nebbiolo is extremely sensitive to differences in soil and mesoclimate, yet pinpointing the differences between the wines from neighbouring Barolo and Barbaresco is not easy. Barbaresco is said to be the lighter, softer wine, though differences between communes in Barolo can be as great as differences between Barolo and Barbaresco. The regulations are a little less stringent in Barbaresco: slightly shorter minimum aging (21 months, including nine months in oak for *normale*, compared to Barolo's three years total, and one year in wood; Barbaresco Riserva must have 45 months' total aging, compared to Barolo Riserva's four years and nine months) and a slightly lower minimum alcohol level (12.5 per cent to Barolo's 13 per cent).

Within Barolo – which is a tiny zone, though three times the size of Barbaresco – Serralunga is the latest-ripening commune. Picking begins here some 15 days after it starts in La Morra, even though the latter is higher; the closer proximity of the river in La Morra makes for a slightly warmer mesoclimate. The wines of Serralunga are among the weightiest in the Barolo region; those of La Morra are at the other extreme: silky, and appealing when young. The other communes in the Barolo zone are Castiglione Falletto, Grinzane Cavour, Monforte, Novello, Verduno, and of course Barolo itself. Castiglione Falletto makes concentrated, powerful wines of great finesse; Monforte wines are firm and long-lived; Barolo wines combine structure with supple fruit. Well, that's the theory anyway.

The main difference in soil is between the sand and limestone in the east, in the communes of Castiglione Falletto, Monforte and Serralunga, which give burly, aggressive wine, and the chalky marl in the west, in La Morra and Barolo, where the wines are less robust and more perfumed. But clay is found everywhere, and it is the clay of Barolo (and Barbaresco, too) which gives the wines their remarkable tannin. Elsewhere Nebbiolo is seldom quite as tannic.

Luca Paschina grew up in the Nebbiolo vineyards of Piedmont, but now leads the way at Virginia's Barboursville Vineyards in showing the variety's potential in North America.

The best (and certainly the most expensive) wines are from single vineyards or crus. There is no established cru system, though this has been mooted for a while, and vineyard names are used on labels. So far only individual subzones in the communes are legally recognized.

In theory Nebbiolo is planted on the best sites here. But by 2000 optimism was running so high in the region that other, less suitable sites were also being planted – the lower slopes, which are usually given over to Barbera, or vineyards at too high an altitude, or with poor exposure. The same thing has happened before, in times of boom; such vineyards were abandoned later, and will no doubt be abandoned again.

Roero has traditionally made lighter, less tannic, more supple wines than those from Barolo and Barbaresco south of the Tanaro river. Occasional examples are bigger and more complex, however. Nebbiolo d'Alba, from the same area, is simpler and earlier-maturing; Langhe Nebbiolo is a catch-all DOC that covers anything that doesn't fit into the other DOCs.

In the Novara and Vercelli hills in northern Piedmont, Nebbiolo is known as Spanna, and in Carema it goes by the name of Picoutener. Gattinara and Ghemme, from the former, tend to be somewhat earthy; Carema can be more perfumed, but also less ripe. In Valtellina Nebbiolo becomes Chiavennasca, and all too often becomes a touch stringy. Greater ripeness would help, and perhaps lower yields. Frankly, I think a bit of blending with softer,

Giacomo Conterno
Giacomo Conterno is an ultra-traditional Barolo producer and produces Nebbiolo wines that will last for decades.

Elio Altare
A modernist par excellence, Altare seeks balanced wines that are supple in youth yet will age. This cru spends time in both barrique and botte.

Bruno Giacosa
Giacosa is a semi-traditionalist: his macerations are long, but shorter than they used to be, and he ages his wines in French, not Slavonian, botti.

Those fogs again. This is the Bien Nacido vineyard in California's Santa Maria Valley. It's one of California's coolest sites, significantly aided by the morning fogs that roll in off the Pacific and blanket the vines. Frequently they don't clear till the afternoon, when the brighter sunshine attempts to make up for all those lost hours. Nebbiolo thrives in the fogs of Piedmont in northern Italy. Attempts to find suitable sites elsewhere in the world have met with little success, but Jim Clendenen of Au Bon Climat makes a fine Nebbiolo from Bien Nacido fruit, and it just might have something to do with those Pacific fogs.

earlier-ripening varieties would also help, but there you go. There is also an Amarone-type red called Sfursat or Sforzato which, when it's good, is very good. But the wines from here are seldom exported further than Switzerland.

Some is grown in Lombardy for Franciacorta Rosso, for which it is blended with Cabernet Franc, Barbera and Merlot.

Australia

Nebbiolo should have great potential here, but finding the right sites for this very site-specific vine has been tricky. Clones have also been a problem, usually giving too much vigour and not enough colour. According to research in Australia by Alex McKay, Garry Crittenden, Peter Dry and Jim Hardie, Margaret River (Western Australia) and Mornington Peninsula (Victoria) can match the Langhe in terms of sunshine hours, relative humidity and rainfall. If heat summation is to be matched, however, then Clare Valley, Adelaide Hills, Mudgee and Bendigo might be appropriate. Despite the problems, S.C. Pannell produces some of the greatest non-Italian Nebbiolo I have tasted.

USA

Given the Italian influence in establishing America's vineyards, there was a reasonable amount of Nebbiolo in the old days. But it never had much success. I've done a tasting of just about every varietal Nebbiolo in California, mostly produced by a variety of old codgers I'd never heard of, and I might as well have stepped back 50 years, the wines were so lumpen and stewed. A bit like traditional Barolo, perhaps, but not at all like modern Barolo. It's strange: you'd expect California to be able to apply first-rate winemaking principles to the grape, and you'd have thought they could identify the right location and viticultural techniques to get concentration, colour and ripe tannins in the finished wine. But no – in this Cabernet- and Merlot-obsessed world, little thought has been given to the notoriously moody Nebbiolo. There are only about 200 acres/70ha of Nebbiolo in California and most of the original plantings are scattered forlornly about the bulk-producing Central Valley, but with some in the Sierra Nevada foothills, Paso Robles and Santa Barbara. The best wine so far is from Jim Clendenen at Au Bon Climat in the Santa Maria Valley; he has planted lower-yielding clones and sanguinely talks about moving from making good wine to great wine – next year. And next year. The Cal-Ital cooking movement, a fusion of Italian and West Coast styles, should have given Nebbiolo a shove, but, I could have told them: Nebbiolo doesn't like being pushed around.

Nebbiolo has been equally problematical in Washington State, where the climate ought at least to be warm enough for it. Probably its greatest success has been in the warm yet humid conditions of Virginia, where Barboursville and Breaux have produced a few superb examples.

Rest of the world

Experiments with Nebbiolo in Chile have produced wines with poor colour and high acidity, perhaps partly because the clones were less than ideal. Argentina has a little planted, but yields are high. Mexico has made some rather attractive examples in Baja California, up towards the US border. Austria's Mittelburgenland has a little.

Matteo Correggia
The Roero hills are across the river Tanaro from Barolo and Barbaresco. Their wines make up for less depth and power with elegance and enticing scent.

Nino Negri
This 100% Nebbiolo, aged in barrique, is one of the stars – in fact, it's probably the greatest star – of the Valtellina region in northern Lombardy.

S C Pannell
Steve Pannell made the first Nebbiolo from outside Piedmont that persuaded me the grape was capable of great things worldwide.

ENJOYING NEBBIOLO

The traditional view is that not only does Nebbiolo age well, but that it positively must be aged for decades if it is to be even drinkable. Well, there's no doubt Nebbiolo can age well. If it had the ripeness in the first place and wasn't wrecked by incompetent winemaking – this still happens, but far less than before – Nebbiolo should last a decade at the least. But tough, dry Nebbiolo will usually remain just that – tough and dry, but older. However, you don't have to age modern Nebbiolo. The grape actually has a rather beautiful perfume and beguiling black cherry, sloes and damson fruit when carefully made. Add a splash of new oak to this and you've got a very attractive though sturdy drink that is challenging but enjoyable at only three or four years old.

The lighter sort of Nebbiolo – from Carema or Langhe – can certainly be drunk within a few years of the vintage and will last from five to eight years or so; perhaps longer in the case of the most concentrated Caremas. Gattinara is slower maturing and longer lived: it often needs six to eight years before it reaches its best, and will live in bottle for another 10 or more. Roero is nearly always much lighter, and should be drunk within a few years, but you'll sometimes find weightier examples from adventurous producers.

Barbaresco is supposed to be lighter than Barolo, but there are as many differences between wines and communes in Barolo as there are between Barolo and Barbaresco themselves. Lighter wines from both regions may be at their best before their 10th birthday; the top, most concentrated cru wines should probably not be opened until they are about eight years old, and will easily live for 20 or 30 years.

The wines made in Australia and California so far have aged relatively rapidly compared to the wines of Piedmont, but Virginia's Nebbiolo ages rather well.

The taste of Nebbiolo

The classic description of the scent of Nebbiolo is tar and roses, immortalized in Michael Garner's and Paul Merritt's 1990 Barolo study of the same name. It sounds an improbable combination, but then Nebbiolo is an improbable grape, combining as it does high levels of tannin with high acidity and (in its best examples) some of the most complex, exotic scents to be found on any red wine.

To tar and roses one might add cherries, damsons and mulberries, leather, herbs both fresh and dried, spice, liquorice and dried fruit. The tannins mature to a powdery softness; the acidity should be ripe, and there should be sufficient weight of fruit and alcohol to balance the tannin and acidity while the wine ages. There is often a certain austerity about Barolo and Barbaresco, though this is less common in wines from modernist producers where the fruit is fleshier and rounded out by aging or fermentation in new oak *barriques*. Even in the most traditionalist examples, mature Barolo should be supple.

It is a pity if Barolo or Barbaresco smells strongly of new oak. Any wine from anywhere in the world can do this: a whiff of new oak is one of the easiest of all aromas to achieve in the winery. But only Nebbiolo has this extraordinary, haunting aroma. To mask it with add-on scents is to lose the point of the wine completely.

Bussia is one of Barolo's most renowned vineyards, divided into three distinct parts, with Aldo Conterno the most famous producer. The grapes for this wine come from sheltered southwest-facing vines angled towards the afternoon sun and looking out towards the village of Barolo itself. Steven Pannell proves it is just possible to make great Nebbiolo away from Italy by producing scented, stony, appetizing wines from grapes grown in the northern part of Adelaide Hills.

MATCHING NEBBIOLO AND FOOD

Fruity, fragrant, early-drinking styles of Nebbiolo wine are best with local salami, pâtés, bresaola (wind-dried beef sliced wafer-thin) and lighter meat dishes. The best Barolos and Barbarescos need big, substantial food to stand up to them; *bollito misto* (a hotchpotch of boiled meats served with a green garlicky sauce), jugged hare, spiced beef casseroles and brasato al Barolo (a large piece of beef marinated then braised slowly in Barolo) are just the job in Piedmont, or anywhere else for that matter.

CONSUMER INFORMATION

Synonyms & local names

Also known in Piedmont as Spanna (especially in the north in the provinces of Novara and Vercelli) and Picoutener (especially in Carema and the Aosta Valley); and as Chiavennasca in Lombardy.

Best producers

ITALY/Piedmont/Barolo/Barbaresco/ Langhe Nebbiolo/Nebbiolo d'Alba/ Roero Alario, Fratelli Alessandria, G Alessandria, Altare, Azelia, Produttori del Barbaresco, Boglietti, S Bolmida, Bongiovanni, Brezza, Bricco Maiolica, Brovia, Burlotto, G Canonica, Cappellano, Cascina Chicco, Cascina Fontana, Cascina Luisin, Cascina delle Rose, Ceretto, Cigliuti, Clerico, Aldo Conterno, Giacomo Conterno, Conterno-Fantino, Corino, Correggia, Stefano Farina, Fenocchio, Fontanabianca, Gaja, Bruno Giacosa, Elio Grasso, Marchesi di Gresy, Lano, Poderi Marcarini, Manuel Marinacci, Bartolo Mascarello, Giuseppe Mascarello, Massolino, Monfalletto-Cordero di Montezemolo, Fiorenzo Nada, Castello di Neive, Andrea Oberto, Oddero, Vigneti Luigi Oddero, Paitin, Parusso, Pelissero, Pio Cesare, Pira, E Pira di Chiara Boschis, F Principiano, Punset, Renato Ratti, Revello, F Rinaldi, G Rinaldi, Rizzi, Roagna, Albino Rocca, Bruno Rocca, Roccalini, Rocche dei Manzoni, G Rosso, Sandrone, M Sebaste, Sottimano, La Spinetta, Trediberri, Vajra, Castello di Verduno, Vietti, G Voerzio, R Voerzio; **Lombardy/ Valtellina** ArPePe, La Castellina, Enologica Valtellinese, Fay, Nino Negri, Nera, Rainoldi, Conte Sertoli Salis, Triacca, .
USA/California Clendenen Family Vineyards, Renwood, Viansa; **Washington State** Cavatappi, Morrison Lane.
MEXICO L A Cetto.
AUSTRALIA Coriole, Crittenden, Dromana, First Drop, Happs, Henschke, Maglieri, S C Pannell, Pizzini, Primo Estate, Scaffidi, Trentham Estate.
SOUTH AFRICA Morgenster, Steenberg.

RECOMMENDED WINES TO TRY

Twenty classic Barolo wines

Altare Brunate
Azelia San Rocco
Brovia Villero
Domenico Clerico Ciabot Mentin Ginestra
Aldo Conterno Gran Bussia
Giacomo Conterno Riserva Monfortino
Luigi Einaudi Nei Cannubi
Fenocchio Pianpolvere Soprano
Bruno Giacoso Villero di Castiglione Falletto
Elio Grasso Ginestra Vigna Casa Maté
Bartolo Mascarello Barolo
Giuseppe Mascarello Monprivato
Monfalletto-Cordero di Montezemolo Enrico VI
Francesco Rinaldi Cannubbio
Roagna La Pira
Giovanni Rosso Cerretta
Luciano Sandrone Cannubis Boschis
G D Vajra Albe
Gianni Voerzio La Serra
Roberto Voerzio Cerequio

Ten classic Barbarescos (or equivalent)

Produttori del Barbaresco Barbaresco
Cascina Luisin Barbaresco
Gaja Barbaresco
Bruno Giacosa Barbaresco Santo Stefano
Marchesi di Gresy Barbaresco Martinenga Gaiun
Castello di Neive Barbaresco
Punset Barbaresco
Albino Rocca Barbaresco Vigneto Brich Ronchi
Bruno Rocca Barbaresco
Roccalini Barbaresco

Fifteen other Italian wines

Silvano Bolmida Langhe Nebbiolo
Brezza Nebbiolo d'Alba
Brovia Nebbiolo d'Alba
Burlotto Nebbiolo d'Alba
Carema Cantina dei Produttori Nebbiolo di Carema
Michele Chiarlo Langhe Barilot
Aldo Conterno Langhe Il Favot
Conterno-Fantino Langhe Monprà
Matteo Correggia Nebbiolo d'Alba la Valle dei Preti
Luigi Ferrando Carema Black Label
Mauro Franchino Gattinara
Gaja Langhe Sperss
Nino Negri Valtellina Sfursat 5 Stelle
Luigi Oddero Langhe Nebbiolo
G D Vajra Langhe Nebbiolo

Five non-Italian Nebbiolo wines

Barboursville Reserve Nebbiolo (Virginia)
Breaux Nebbiolo (Virginia)
Clendenen Family Vineyards Nebbiolo (California)
S C Pannell Nebbiolo (Australia)
Pizzini Nebbiolo (Australia)

Delinquent or just misunderstood? Nebbiolo is proving to be one of the most frustrating grapes to grow outside its Piedmont heartland. Growers are examining the minutiae of soil and climate, but the grape still obstinately refuses to play ball.

Maturity charts

Even Barolo is drinkable much earlier than it used to be. Tannins are softer these days, and the fruit is more to the fore.

2007 Barolo (top Cru)

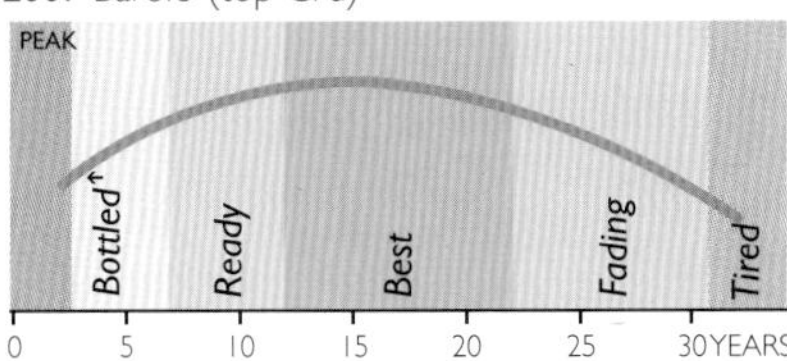

2007 was a hot, dry year for Barolo, but the vines in the best vineyards thrived and the wines are deep and concentrated.

2009 Barbaresco

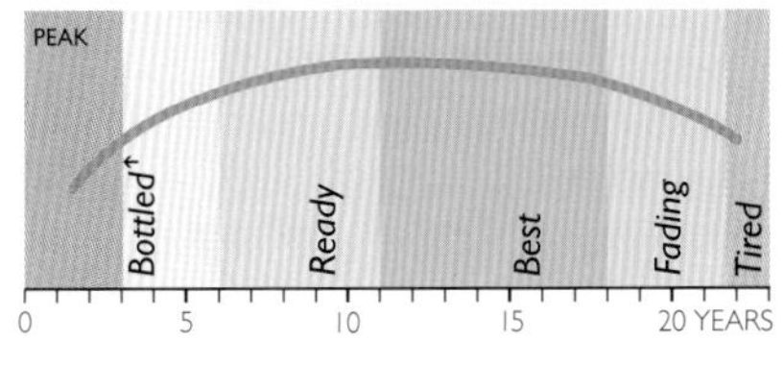

Barbaresco had a good vintage in 2009 – a warm and dry summer producing rich but balanced reds to drink or age.

2011 Nebbiolo d'Alba

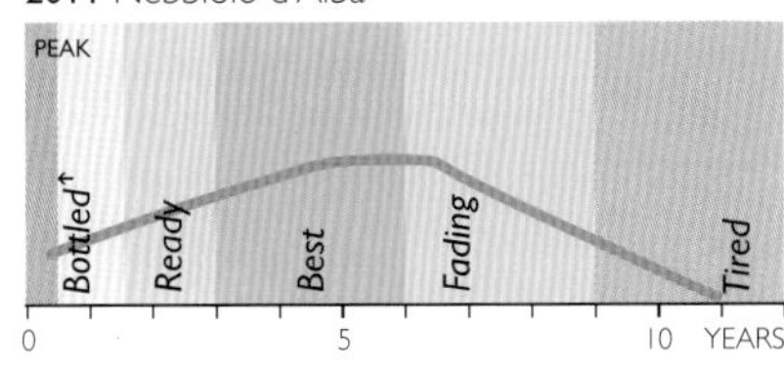

The hot dry 2011 harvest produced only small quantities of ripe lush wines for early drinking.

NEGOSKA

Soft, richly fruity grape found in northern Greece and blended in small quantities with the much tougher, [illegible] (see page 299) to make Goumenissa. Best producer: (Greece) Boutari.

NEGRA MOLE

The most important grape of the Atlantic islands off the coast of Africa, either as Negramoll in the Islas Canarias or as Tinta Negra Mole on Madeira (Tinta Negra since 2000). It even reaches the Azores, where it is called Saborinho. Mole means 'soft' and though this may refer to the berries, the vines are generally light, vaguely aromatic – and soft. It is identical to Mollar, a variety known in Andalucia since the 18th century.

NEGRAMOLL

This is the new name for Tinta Negra Mole, a high-yielding vine grown on the Portuguese island of Madeira which had all but replaced the four noble varieties of Sercial, Verdelho, Bual and Malvasia or Malmsey. Efforts to revive the production of these four continue, and they have increased to around 12 or 14 per cent of the vineyard. Negramoll (or Tinta Negra) is the basis of most Madeira wine and covers about 68 per cent of the total plantings. The discrepancy is explained by the presence of hybrids.

Only since 1993 has it been required by law that any bottle bearing the name of one of the noble varieties should contain at least 85 per cent of that grape; nowadays, if the label does not mention a grape name, it will have been made from Negramoll.

Wine from the four noble varieties is superior in quality to that from Negramoll but if Negramoll were cropped at the 4 tonnes per hectare that is usual for the noble varieties, instead of the 10 tonnes or more (sometimes much more) that it is currently encouraged to give, its quality might be much improved. It is picked early, at 9.5–10 per cent potential alcohol, because viticultural practices on Madeira tend to encourage rot. It is, however, pretty disease-resistant, and this was the reason it became so widely planted on Madeira after phylloxera. The Madeira shippers are trying hard to get the law changed so that they may put the name of Negramoll on Madeira labels since several producers such as Barbeito are turning it into serious, well-aged Madeira. It is permitted rather than recommended for table wines.

Although initially thought to be an unlikely crossing of Pinot Noir and Grenache – where would that have occurred? – it now seems to be an ancient variety from Cadiz in Andalucia, a port from which many ships would depart and [illegible] to the Americas. It's also found on the Canary Islands where it has long been called Negramoll and, as Saborinho, in the Azores – again, trading routes would explain that.

NÉGRETTE

A grape of southwest France, found in the Côtes du Frontonnais appellation, where it must constitute 50 to 70 per cent of the blend. Négrette contributes a wonderfully silky texture and raspberry perfume to the best wines. It is also found in other wines from north of Toulouse. Common blending partners include Fer and Syrah, but inevitably there is a fair bit of Cabernet around as well, which tends to hijack the style of wines to which it is added. Admittedly, Négrette's tendency to a slightly high level of volatile acidity means it sometimes needs a bit of stabilizing, but too often Cabernet ejects the silky Négrette baby with the bath water. Négrette is usually best drunk young and without oak aging. The hot, dry climate of the region suits the grape well – it gets botrytis bunch rot rather easily in wet conditions. Best producers: (France) Bellevue-la-Forêt, la Colombière, Ferran, Dom. Montauriol, la Palme, le Roc.

NEGROAMARO

A rather odd-tasting grape found in the south of Italy, particular in Puglia. It could hardly be more Italian in its combination of perfume and bitterness – the name means 'black bitter' and the wine is indeed very dark in colour. It is sturdy in structure and can be both slightly farmyardy in flavour and have a distinctly medicinal edge. The combination takes a bit of getting used to but is certainly interesting.

It is far more easy to appreciate, however, if blended with a little of the far more scented, succulent Malvasia Nera (see page 134). Some of Puglia's best reds, notably Salice Salentino, feature this double act. Best producers: (Italy) Giuseppe Calò, Michele Calò, Candido, Leone de Castris, Pervini, Taurino, Vallone.

NEGRU DE DRAGASANI

A new Romanian crossing which seems promising. One of its parents is Saperavi, one of Eastern Europe's best grapes (see page 221).

Château Bellevue-la-Forêt
Negrette is usually blended with Syrah or Cabernet Sauvignon, so it is a a delight to find this 100% 'Pure Negrette' taking the chance to express the variety's delightful silky texture and lush but acidic raspberry fruit.

NERELLO

A Sicilian grape that comes in two versions, Nerello Mascalese and the less commonly planted and less good Nerello Cappuccio. Both contribute high alcohol to a blend, though they lack the concentration of Nero d'Avola, with which they are usually blended. Even so, the growers around Mount Etna are getting very excited by the wines' transparent character, which allows the sometimes ferocious lava-esque minerality and chewy herb flavours to shine through. Best producers: (Italy) Benanti, Palari, Duca di Salaparuta.

NERO D'AVOLA

Increasingly fashionable Sicilian grape which is leading the revival in Sicilian quality reds due to its ability to provide deep colour, robust but gentle texture and an irresistible combination of floral scent and blackberry and black cherry fruit. It can benefit from a little oak aging, but doesn't need it, and for such an immediately appetizing grape variety, matures surprisingly well. One or two growers in Australia and California are trying it out, and they should succeed. Best producers: (Italy) Abbazia Santa Anastasia, Calatrasi, COS, Cusumano, Donnafugata, Morgante, Planeta, Duca di Salaparuta, Spadafora, Tasca d'Almerita.

NERO DI TROIA

This is the new name for Uva di Troia, a grape found in central and northern Puglia. It can be complex and can age well if properly handled. It is part of the blend in several DOCs, and brings deep colour and substantial alcohol to the blend, but its potential for quality is outweighed, as far as most growers are concerned, by its low yields. Best producers: (Italy) Rivera, Santa Lucia, Torrevento.

Looking across the Feiler-Artinger vineyards to the little town of Rust and the shallow Lake Neusiedl, you can almost feel the sticky late season warmth as the humid mist hovers over the vineyards and the lake. Noble rot attacks the vines naturally every year, making this the world's most consistent supplier of intensely sweet wines. The Neuburger grape makes a fairly broad-boned dry white, but is very susceptible to noble rot and is best suited to syrup-sweet wines.

NEUBURGER

Austria's Neuburger grape gives rich, spicy wines, sweet, semi-sweet or dry, which are not unlike a more intense version of Pinot Blanc, although in fact it is a cross between Austria's Roter Veltliner and Silvaner. The broad, flat neutrality of Silvaner might explain Neuburger's general lack of the thrill factor. It is happily susceptible to noble rot and its best wines are sweet, but dry or off-dry versions also age quite well in bottle, becoming nicely honeyed and nutty, but always lacking a little zing. Best producers: (Austria) Beck, Feiler-Artinger, Haider.

NIAGARA

Vigorous, high-yielding hybrid grown especially in New York State and Michigan, where its ability to withstand winter cold comes in useful. It was bred in 1872 from a crossing of Concord (see page 92) and Cassady, and has a strong dose of the so-called 'foxy' aromas that mark many *Vitis labrusca* grapes. It is popular in Brazil.

NIELLUCCIO

This robust, tannic Corsican grape is identical to the Sangiovese of Italy. It is generally blended with the less substantial though more complex Sciacarello: the two of them constitute the best of Corsica's red grapes. Other varieties on the island are often those, like Cinsaut and Carignan, which were planted by immigrants returning from North Africa in the 1960s. However, these are on the decline, and Nielluccio and other indigenous grapes are on the increase. So, inevitably, are international varieties.

Nielluccio is more widely planted than Sciacarello, and has its heartland in Patrimonio in the North; the chalky clay here suits it particularly well. It is also grown in the South, around Porto Vecchio. Its acidity and tannins are good, though the tannins can be somewhat harsh and need careful handling. It also makes excellent rosés. Best producers: (France) Alzipratu, Antoine Aréna, Orenga de Gaffory, Leccia.

NOSIOLA

An unusually neutral grape found in Trentino-Alto Adige – unusual because most of the white vines in this part of northeastern Italy have more character. Nosiola is crisp, light and ephemerally aromatic, although some producers are now finding a nutty depth in it. It is probably at its best in terms of personality when made into Vino Santo, the Trentino sweet wine made from dried grapes. It is grown only in small and diminishing quantities. Best producers: (Italy) Concilio Vini, Pojer & Sandri, Pravis.

NOUVELLE

A South African crossing of Crouchen and Ugni Blanc, which when picked slightly underripe gives a pungent leafy, green peppery flavour that can add welcome freshness and tang to fat or soft examples of Sauvignon Blanc. Some critics in South Africa decry it, but I think it's rather a useful performer.

NURAGUS

Prolific Sardinian vine which makes light, neutral, perfectly attractive and perfectly unmemorable wine. Best producers: (Italy) Argiolas, Dolianova, Santadi, Trexenta.

OJO DE LIEBRE

The Catalan name for Tempranillo (see pages 270–279). The name means 'hare's eye'.

ÖKÜSGÖZÖ

Turkish grape which, when blended with Bogazkere, is responsible for most of the country's good reds. On its own it is bright, fresh and aromatic

OLASZ RIZLING

The Hungarian name for Welschriesling (see page 298). For all its poor reputation in some countries, the grape can make very attractive wine if well handled. Hungary's problem was that it exported mainly the dirtier and cheaper examples, thus comprehensively destroying the grape's reputation in its export markets.

ONDENC

Almost extinct grape found occasionally in southwest France. Nowadays it is unpopular for its low yields and susceptibility to rot but is hanging on in Gaillac, primarily because noble rot can produce excellent sweet wine there. It used to be grown in Australia, most recently in Victoria, where it made a very dry, cold-hearted base for sparkling wine, and Barossa Valley. Best producers: (France: Causses-Marines, Robert Plageoles.

OPORTO

Slovakian name for Blaufränkisch (see page 48), where it is also known as Frankovka Modrá and is the most planted black variety.

OPTIMA

Mercifully declining German crossing, produced in 1930 from a Silvaner x Riesling vine crossed with Müller-Thurgau. Its virtue for the grower is that it ripens very early, and reaches high sugar levels; it is hard to think of any advantages it has for anyone else. Its wines are blowsy, dull and crudely sweet, but its late-budding habit endeared it in the past to growers in the Mosel. The Rheinhessen was its main field of operations; I have had one or two fairly rich dessert examples, but it is declining there as well. Best producer: (Canada) Quails' Gate.

ORANGE MUSCAT

There is a little grown in Australia and California, of which the most famous example is made by Andrew Quady as a fortified wine called Essencia. He picks early for aroma and lightness, and stops the fermentation by adding grape spirit. The wine, which is elegant with a tangily sweet finish, has an intense orange-flower and orange peel aroma and flavour. Best producers: (USA) Quady; (Australia) Brown Bros.

ORMEASCO

Ligurian name for Dolcetto (see pages 94–95).

ORTEGA

A modern German crossing (Müller-Thurgau and Siegerrebe) which has little more than high sugar levels to recommend it. It has plenty of flavour but no complexity, which rather misses the point of the Spätlese, Auslese and higher Prädikat levels that it reaches with such ease. Acidity can be low, and the vine is prone to various diseases. In Germany it is mostly found in the Rheinhessen.

To be honest, the best Ortegas I've had have been from England, where the vine only just ripens in the cool conditions and gives wine with good grapefruit acidity and breezy elderflower scent with good potential for aging. Best producers: (England) Biddenden, Denbies, Stanlake Park, Throwley.

ORTRUGO

A grape found in the Colli Piacentini zone in Emilia, central Italy. The wine is deep in colour, with high alcohol and decent acidity, and may be blended with Malvasia. It also comes as frizzante. Best producers: (Italy) Gaetano Lusenti, Pernice.

OSELETA

An old variety reckoned newly interesting in Valpolicella in northeast Italy, with small berries and little juice. It gives acidity, colour and body to the blend, though it can have rather too much of all those, plus a good helping of rusticity, to be much fun on its own. Best producer (Italy): Masi.

PAGADEBIT

Found on Italy's Adriatic coast, where its name refers to its ability to pay a grower's debts. But Pagadebit is a name given to several varieties, one of which is Bombino Bianco (see page 49), and another of which is Mostosa.

PAIEN

Gewürztraminer (see pages 112–121) may be called by this name in the Valais area of Switzerland.

PAÍS

See Mission page 150.

PALOMINO FINO

One of the dullest grapes in the world, which just happens, when grown in a small area of white, chalky *albariza* soil in the south-west tip of Spain, and aged in a solera system, to produce wines of matchless complexity and pungency. The wine is, of course, sherry – a name that cannot, under EU law, be applied to the fortified wine of any other country or region. Palomino Fino is effectively the only grape grown in the sherry region and has replaced the lesser Palomino de Jerez or Palomino Basto.

The Pedro Ximénez grapes that may be used for sweetening some sherry nowadays largely comes from nearby Montilla-Moriles. A few bodegas opt for the rarer but fruitier Moscatel as a sweetening wine, but Moscatel's intense grapy perfume doesn't make for great sherry. Various ways of concentrating the sweetness of Palomino – leaving the grapes in the sun or concentrating the must – may also be used to produce sweetening wine, according to the desired quality and style of the final result.

The wine is always fermented to dryness, and fortified later. Considering the aridity of the climate and the lack of irrigation, yields are on the high side – between 75hl/ha and perhaps 150hl/ha. High yields are aided by the nature of the region's *albariza* soil, which has the capacity to retain 25 per cent of its volume in water. Not surprisingly, vineyards on less good soils have now been uprooted.

The vines are generally trained as low bushes, though training on wires is increasingly popular. Rot is less of a problem with wire-trained vines. In addition, wire training is necessary for mechanical harvesting, and while this has never so far been permitted in the region it is by no means impossible that it won't sometimes be allowed. It was a ban imposed for social reasons, to provide seasonal work for unskilled labour, but nowadays the number of pickers looking for work is smaller each year.

However, high yields do not really seem to be a problem for quality. Some 90 per cent of the style and quality of sherry comes from the aging, and the position of the bodega and how the solera is managed are far more important than the base wine.

Sherry is divided into two basic styles: those that grow *flor*, and those that do not. Those that do are Fino and Manzanilla, and by extension Amontillado, since proper Amontillado is Fino or Manzanilla on which the flor has, after some years, died. Flor is a yeast, or rather several yeasts, and each bodega, indeed each solera, may have its own particular balance of yeast strains in its *flor*.

Flor grows on the surface of the wine in a porridgey white layer perhaps 1cm (⅜in) thick. It feeds on the wine, keeping it free of contact with oxygen (thus preventing oxidation while keeping it fresh-tasting) but it also changes the chemical make-up of the wine so that it develops a characteristic pungency.

Oloroso sherry does not grow flor. Wines for Olorosos are selected at an early stage – they may well come from the third pressing of the grapes, and thus be higher in phenols and coarser in flavour. (The most delicate wines are used for Finos and Manzanillas, and

Ripe Palomino Fino grapes growing in the almost pure chalk soil of Jerez. Palomino's neutrality becomes a virtue here – but only because sherry gains nearly all its flavour from the solera system in the winery.

generally come from free-run juice.) Oloroso gains its dark colour and pungent flavour of nuts, prunes and coffee from long aging in solera, during which time the wine oxidizes and becomes more concentrated.

The solera system is a system of fractional blending in which wine is drawn off for bottling at one end, with the barrels in the last stage of the solera being replenished from those in the next stage back, and so on. New wine is added to the barrels in the first stage. Wine is drawn off for bottling perhaps four times a year, with perhaps a quarter of each barrel being taken at any time. All sherry is thus a blend of old and much younger wine; it is this that helps give it its balance of freshness and mature depth.

Palomino is also used for producing table wine in the three sherry towns of Jerez de la Frontera, Puerto de Santa María and Sanlucar de Barrameda. Many sherry bodegas now produce such a wine, and without exception they prove the utter dullness of the Palomino grape. The wine is low in acidity unless acidified in the winery, and neutral in flavour.

Palomino is also grown elsewhere in Spain, for table wines as well as for fortifieds and *rancio* styles: in Condado de Huelva, west of Jerez towards the Portuguese border, it is increasingly popular at the expense of Zalema, but in northern Spain, in Rueda and Galicia, it is fading in importance. It is a key variety in the Canary Islands, where its wine even manages to have a little character.

Where it is drunk as a table wine, the solution is sometimes to make it deliberately *rancio* by leaving it in glass demijohns in the sun to oxidize. It then acquires a taste reminiscent of sherry, even if not very good sherry. It is drunk this way in Rueda. And you wonder why Palomino is disappearing in Rueda.

It is found in decreasing quantities in western France, where it goes under the name of Listán or Listán de Jerez. It is the same grape as Portugal's Malvasia Rei.

Outside Europe it has almost vanished from South Africa, where it was known as Fransdruif. There are relatively small plantings in California's San Joaquin Valley, where the variety used to be known as Golden Chasselas, and in Australia, where it makes sherry-style fortified wines. There are still smaller plantings in New Zealand and Cyprus, where it is used for fortified wines, and Argentina. Best producers: (Spain) Argüeso, Barbadillo, Delgado Zuleta, Díez Mérito, Domecq, Garvey, Gonzalez-Byass, Hidalgo, Lustau, Osborne, Valdespino.

PAMID

Bulgarian grape that comes in various colours, from white through pink to red. The wine is light in colour and body, low in acidity, and isn't going to cause anyone to die of excitement. It's also a table grape.

PANSA BLANCA

This is the name given to Xarel-lo (see page 298) in Spain's Alella region, on the coast north of Barcelona, where it seems to acquire some extra bite and aroma. It ripens to perhaps 14 per cent potential alcohol without losing its lime cordial aroma and flavour, and seems to suit the region's sandy *sauló* soil. Best producers: (Spain) Marqués de Alella, Parxet, Roura.

PARDILLO

There's a lot of this neutral white grape in central Spain, and it's really very dull.

PARELLADA

One of the three grape varieties grown for Cava, the Spanish name for Champagne-method fizz, Parellada contributes lemony freshness to the blend, providing that it is not overcropped: it tends to be high yielding in fertile soils. It is outplanted in the Catalan region of Penedès by the other Cava varieties, Macabeo and Xarel-lo. It also makes still wines, light, fresh and gently floral, with good acidity and, for Spain, lowish alcohol (between 9 and 11 per cent). Drink it as young as possible – and I mean young – while it still has the benefit of freshness.

It can be blended with other varieties, notably Chardonnay and Sauvignon Blanc, and is also found in the Costers del Segre region inland of Barcelona. Best producers: (Spain) Castellblanch, Codorníu, Freixenet, Juvé y Camps, Marqués de Monistrol, Parxet, Raimat, Torres.

PASSERINA

An old Italian variety rediscovered some 20 years ago. It has quite large berries, highish yields and a long ripening period. It makes appealing wines with clear, focused fruit and is DOC in Offida, in the Marche.

PECORINO

And you thought Pecorino was a cheese. Well, so did I. This Pecorino is native to the Marches and Abruzzo in Italy, and is DOC in

Offida. It's low yielding, and for this reason fell out of favour some 50 years ago as growers concentrated instead on [illegible], but it will ripen at high altitudes, and though there are different clones it hasn't been studied in great detail. It's early ripening and has a good deep, rich, aromatic, nutty character.

PEDRO GIMÉNEZ

This, logically, should be Pedro Ximénez (PX) under a different name, but the large plantings of Pedro Giménez in Argentina and Chile are a different variety from Spain's PX and are probably related to the first importation of European vines in the 16th century. In Argentina it makes everyday wines which never get exported; in Chile it is grown for pisco, the local brandy.

PEDRO XIMÉNEZ (PX)

Along with Palomino Fino (see page 178), this used to be the other grape of the Jerez sherry region in southern Spain. It was used for sweetening purposes and sometimes for bottling on its own as a sweet, dark, dessert sherry. But the more disease-resistant Palomino has all but replaced it, and almost all the PX used in sherry these days is brought in from the nearby Montilla-Moriles area, where it is the main variety.

Traditionally it is picked and then left to dry and shrivel in the sun to concentrate its sweetness. It is also used for Málaga, and is found across Andalucía, Valencia, Extremadura and in Portugal's Alentejo. One legend attributes its introduction to a 17th-century Spanish soldier (called, presumably, Pedro Ximénez), who brought it from the Rhine on his return from the Netherlands. Why he thought that a vine that grew in the chilly Rhine would grow in Montilla – the hottest part of Spain – is not clear. Nor can one think of any current Rhine vine with much resemblance to PX, even allowing for the difference in climate. Still, it's a nice story.

Varietal PX sherry is one of the most immediately seductive of dessert wines. It has low acidity and a thick, silky, syrupy texture; its flavour is grapy and raisiny, with only the bare minimum of acidity to prevent it from cloying. In solera it becomes dark to the point of blackness, and can be very concentrated; examples of 60 years old or more, freshly bottled, nevertheless taste very similar to, though more intense than, younger examples. It is far less successful as a table wine grape, and gives flabby, dull-tasting wines with little acidity and no character.

It is grown in Australia, but in even smaller quantities than Palomino, with which it is lumped both in the statistics and, usually, in the blending vat. Although Chile distils most of its production into brandy or pisco, Falernia in the Elqui Valley makes a tasty but hardly memorable dry white wine version. Best producers: (Spain) Alvear, Pérez Barquero, Domecq, Garvey, Gonzalez-Byass, Lustau, Toro Albalá, Valdespino.

Gonzalez-Byass
Pedro Ximénez like this example has intense grapy flavours – but actually the wine develops very little over time, and while old versions are more concentrated the flavours are the same.

PERIQUITA

The 'correct' name (if such a thing exists) for this soft, raspberry-flavoured Portuguese variety is Castelão (see page 65), though it is also known as João de Santarém, and in the Alentejo used to be called Trincadeira. But Trincadeira is now known to be a separate variety. Best producers: (Portugal) Bacalhôa, J M da Fonseca, Pegos Claros.

PERLAN

One of the many names of Chasselas (see page 80). It is sometimes known by this name in the canton of Geneva in Switzerland, though Chasselas de Genève is more common.

PERRICONE

Sicilian grape grown for deep-coloured, alcoholic red wines.

PETIT COURBU

Not related to Courbu, with which it shares vineyard space. It is one of the many old and obscure vines found in Gascony, in southwest France, where it is an ingredient in Pacherenc du Vic-Bilh, Béarn, Jurançon and Irouléguy. It adds richness and lemony, honeyed fruit to the blend. Courbu Noir in Béarn is not related. Best producers: (France) Arretxea, Bru-Baché, Clos Thou, Clos Uroulat, Ilarria, Souch.

Domaine Cauhapé
Petit Manseng is the main grape used for sweet Jurançon wine. The berries dry out and shrivel on the vine and the sweet is marked by apricot and pineapple richness and particularly vivid acidity.

PETIT MANSENG

Petit Manseng is very similar to its cousin Gros Manseng (see page 124) and is found in the same places – namely between Gascony and the Pyrenees in southwest France. It has smaller berries, as its name might suggest, and is more extreme in its flavours, and more difficult to handle in the winery. It is very low yielding, sometimes giving less than 15hl/ha and like Gros Manseng, able to reach high levels of potential alcohol.

It is even more suitable for sweet wines than its Gros cousin, with the grapes regularly being left on the vine to become *passerillé*, or shrivelled in Jurançon; Gros Manseng, of which larger quantities are at present planted, may well be more suitable for dry wines than Petit Manseng, but the latter has similar intense floral, spicy fruit and high acidity, and these flavours, together with its considerable finesse, are making it increasingly popular with growers in other regions. It is appearing more and more in Languedoc, where its acidity is prized. Elsewhere, Virginia has had considerable success with it, and both Uruguay and Argentina have a little, as does Australia. One fascinating New Zealand example is made by Churton in Marlborough. Best producers: (France) Arrextea, Bru-Baché, Cauhapé, Clos Lapeyre, Clos Thou, Clos Uroulat, Montus, Souch.

PETIT MESLIER

Old Champagne variety now attracting some attention. It has piercing acidity and is difficult to handle in the winery, but Bollinger, Duval Leroy, Tarlant, René Geoffroy, Moutard and L Aubry Fils all have some. Irvine Wines in Australia's Eden Valley have some, too.

PETIT ROUGE

An obscure variety from Italy's Valle d'Aosta region. It is usually blended.

PETIT VERDOT

Petit Verdot may be planted only in small quantities in Bordeaux, but it is often highly valued for its colour, structure and lovely violet scent – and it is proving successful in Chile, Australia, California and even Virginia, both as a useful seasoning for Cabernet Sauvignon, and an exciting wine in its own right.

Petit Verdot fell from favour in Bordeaux because it is so late ripening: it ripens even later than Cabernet Sauvignon, which makes it an impossibility for the Right Bank regions of St-Émilion and Pomerol. It is, however, found in the Médoc and especially in Margaux, where the soils give lighter wines that need the extra tannin and colour provided by Petit Verdot. Château Margaux itself has 6 to 7 per cent Petit Verdot in the vineyard, and most of it goes into the *grand vin*; it has wonderful scent, but it supposedly lacks elegance, so winemaker Paul Pontallier says he would never want more than 10 per cent in the blend. In the 19th century, by contrast, 30 per cent of the vineyard at Château Margaux was planted with Petit Verdot. In Bordeaux, generally, it was reckoned to reach full ripeness in only one year in five, which made châteaux in less-favoured spots somewhat wary of planting it; in the 1980s this probably changed to one year in three, and more like one in two in the 2000s.

Pontallier describes his Petit Verdot as having a banana aroma when young, and developing violet aromas later. Violets are also a keynote elsewhere: in Tuscany, Spain (where it handles the La Mancha heat well and is also grown on Mallorca), Virginia, Long Island, Chile (Errázuriz is enthusiastic about its potential), New Zealand, South Africa, and Australia. It is of rapidly growing importance in South Africa's warmer regions. In Australia's warm, irrigated Riverland, Petit Verdot can give even better results than Cabernet Sauvignon, with better acid retention and fresher flavours. Even Australia, however, has areas too cool to ripen it.

There is an unrelated and less good variety known as Gros Verdot, which may still occasionally be found in Argentina. Best producers: (Australia) Kingston, Pirramimma; Trentham, Zilzie; (Italy) Castello dei Rampolla; (South Africa) Simonsig; (Spain) Dominio de Valdepusa; (USA) Araujo, Benziger, Cain Cellars, Jekel, Newton, Virginia.

Virginia has proven to be a happy hunting ground for Petit Verdot. Its warm yet humid conditions seem to suit this late-ripening, thick-skinned variety. These vines are at the Afton Mountain Vineyards in Monticello AVA with the seductive Blue Ridge Mountains gracing the horizon.

PETITE ARVINE

A high-quality grape of the Swiss Valais, also called Arvine, which gives wines of elegance and finesse, with tense acidity and unusual minerally, leafy flavour. It demands, and gets, the best sites, and if the grapes are left on the vine until November or even December it can get noble rot. Otherwise sweet or medium-sweet wines can be made from shrivelled berries.

There are also dry versions; all age well in bottle. The vine is also found over the Italian border in the Valle d'Aosta. Best producers: (Switzerland) Charles Bonvin, Caves Imesch, du Mont d'Or.

PETITE SIRAH

This grape is sometimes wrongly, and confusingly, spelled Petite Syrah – confusingly because some growers in the Rhône Valley refer to a small-berried version of their (true) Syrah as Petite Syrah.

Well, Petite Sirah is not real Syrah. It is, in fact, a cross between a little grown French vine called Peloursin and Syrah, but the name of Petite Sirah has become attached to several different grape varieties all of which have been traditionally planted together in California, which is where it, or they, are mostly found today.

Petite Sirah is the same grape as Durif (see page 96), though it is sometimes confused with Peloursin. Petite Sirah produces tannic wines which are even darker in colour than those of true Syrah; they have a savoury, almost meaty character and dense blackberry fruit. Its powerful style has long made it a useful blending wine, especially for Zinfandel: Paul Draper of Ridge Vineyards swears by an admixture of 10 to 15 per cent of Petite Sirah in his Zinfandels.

It can also be made as a varietal if the powerful tannins are handled well: it is intense and usually rustic, but good examples can be surprisingly supple and ageworthy, lasting up to 20 years in bottle. It is at its best in Sonoma and Mendocino, where unirrigated vineyards of often very old vines produce wines of considerable depth, backbone and brutal power. Not only is the area under vine on the increase, with Lodi and Paso Robles showing interest, but the wine is becoming a bit of a cult in California: its fan club is called – wait for it – PS I Love You.

Mexico is another source of good Petite Sirah; Argentina and Brazil also have some. In Australia, it is generally called Durif. Best producers: (Mexico) L A Cetto, (USA) David Bruce, Carmen, De Loach, Edmeades, Fetzer, Fife, Foppiano, Frick, La Jota, Parducci, Ravenswood, Ridge, Rockland, Stags' Leap Winery, Sean Thackray, Turley.

PETITE SYRAH

Some Rhône producers like to use this name for a smaller-berried strain of Syrah (see pages 258–269). Petite Sirah, a different variety (see page 181), is sometimes incorrectly spelled this way.

PICOLIT

An overpriced northeastern Italian variety which may, on occasion, be ravishing enough to justify the hype. But only on occasion, and prices are high. The wines are sweet, with apricot and peach flavours and firm, sometimes appley acidity. There may be a flowery note, too, and *passito* wines are full of dried apricot and candied fruit flavours. Some may be aged in oak *barriques*.

The most intense versions are the *passito* wines, made from grapes picked late and then left to shrivel before being pressed and fermented. There are also late-harvest versions, made from grapes which are picked even later in the autumn. Either way yields are tiny: the vine is prone to very poor fruit set, which is one reason why there is not much of the variety planted. It demands good sites, too, on hillsides with volcanic soil.

In the past Picolit has had periods of great popularity, long before its current fashionability: in the 18th century the wine was exported to the courts of Tuscany, Austria, Holland, Russia, Britain, Saxony and France. Best producers: (Italy) Cà Ronesca, Dario Coos, Dorigo, Livio Felluga, Davino Meroi, Primosic, Bernarda Rocca, Paolo Rodaro, Ronchi di Cialla, Ronchi di Manzano, Ronco del Gnemiz, Le Vigne di Zamò.

PICPOUL

Picpoul or Piquepoul is found in the Languedoc region, usually in its white form, though there is also a Picpoul Noir and a Picpoul Gris. All are old varieties. Picpoul Noir is aromatic and reaches high levels of potential alcohol, but is pale in colour and doesn't age. It is usually only found in blends.

Picpoul Blanc is noted for its high acidity (*piquepoul* means 'lipstinger') and lemony fruit, and any grape with high acidity can be an asset in the hot Languedoc climate. It used to be widely grown for vermouth. It is becoming quite popular and is particularly useful for blending, but can be delicious and lemony when in AC form as Picpoul de Pinet from around the village of Pinet near the Étang de Thau on the Mediterranean coast north of Agde. Quality of Picpoul de Pinet has greatly increased recently. Best producers: (France) la Grangette, Montagnac co-op, Pinet co-op, Gaujal, Genson, St-Martin de la Garrigue; (USA) Bonny Doon.

Félines Jourdan

Picpoul de Pinet is becoming a surprising star wine. Pale, fresh, lemony, scented with blossom, it comes from very warm conditions on the Mediterranean, which the very acid-retentive Picpoul seems to relish.

PIEDIROSSO

This grape, planted in Campania in southern Italy, is part of the blend for the wines Lacryma Christi del Vesuvio and Falerno Rosso, but with the increase in fashionability of Campanian varietal wines, you can now find attractive unblended versions, usually quite light and fresh for easy drinking, but with attractive red fruit and herbal scent. Best producers: (Italy) Feudi di San Gregorio, Galardi, Cantine Grotte del Sole, Luigi Maffini, Mastroberardino, Ocone, Giovanni Struzziero, Villa Matilde.

PIGATO

A grape found in the Riviera di Ponente zone in Italy's Liguria region which makes sturdy wine with plenty of fruit. It is the same grape as Piedmont's Favorita and the Vermentino/Rolle commonly found all around the Mediterranean coastline (see page 287).

It gains its name, which means 'spotted', from the appearance of the ripe grapes. Best producers: (Italy) Anfossi, Bruna, Maria Donata Bianchi, Cascina Feipu dei Massaretti, Parodi, Terre Rosse, La Vecchia Cantina, Claudio Vio.

PIGNOLO

Pignolo is found in Friuli where it seems to be having something of a revival, albeit a fairly small and quiet one. It deserves to be revived. It balances good acidity with rich blackberry and plum flavours and silky tannins, and it seems to suit *barrique* aging. It is not the same as Pignola, which is a black grape grown in the Valtellina region in Lombardy. Best producers: (Italy) Dorigo, Davide Moschioni, Le Vigne di Zamò.

PINEAU D'AUNIS

A grape of the Loire Valley, now in decline but still blended into the reds and rosés of Anjou and Touraine. It has good fruit and adds a certain zesty pepperiness to blends, but other reds, notably Cabernet Franc, are more fashionable and have more commercial potential.

It is an old vine, and has been known since the Middle Ages; it takes its name from the Aunis priory near Saumur, and King Henry III of England imported its light red wine to England in 1246. Anyone who wanted darker coloured wine blended it with something else: the local Teinturier du Cher would have given the quickest results. Best producer: (France) Hautes Vignes.

PINOT BEUROT

Burgundian synonym for Pinot Gris (see pages 186–187). A small amount is still grown on the Côte de Beaune and Hautes-Côtes de Beaune, where it is blended with Chardonnay and adds extra complexity and chubbiness to the wine. Odd patches survive in some pretty famous domaines, such as Simon Bize, Joseph Drouhin and Comte Senard. Rumour has it that there are still vines in the Grand Cru Corton-Charlemagne. Best producers: (France) l'Arlot, Simon Bize, Bruno Clair, Coche-Dury, Joseph Drouhin, Comte Senard.

PINOT BIANCO

Italian name for Pinot Blanc (see pages 184–185). It is grown mostly in the northeast, and gives particularly good crisp, light wines in the Alto Adige, the Veneto and Friuli, and makes useful sparkling wines in Lombardy.

Until the mid-1980s the names Pinot Bianco and Chardonnay were considered interchangeable, and the two varieties are still interplanted in some vineyards. The grape name that then goes on the label might, in these circumstances, just possibly owe more to market forces than to strict accuracy. In which case, buy whichever wine is cheaper.

PINOT BLANC

See pages 184–185.

PINOT GRIGIO

Italian name for Pinot Gris (see pages 186–187). Pinot Grigio is often more highly regarded than Pinot Bianco (see above) in Italy, though the quality of the wine is at much the same level. It grows in the northeast of the country, and is picked for its acidity rather than for the plump richness that distinguishes the variety in Alsace, but the best do have a light coppery tinge and some decent nutty weight. But in general Italian wines from this grape are accordingly some of the lightest and crispest around, with delicate spice. If vineyard yields are too high, as they frequently are, this delicacy turns to blandness. This doesn't stop Pinot Grigio being archetypal Italian restaurant wine, and some of it is really very pleasant, but it could explain why, with a little added sugar, it has been such a runaway success in the lower echelons of the export markets. New World countries generally use the title Pinot Gris. When you find a wine labelled Pinot Grigio from, for instance, Australia or New Zealand, it is usually in a lighter, fresh, not quite dry style.

PINOT GRIS

See pages 186–187.

PINOT LIÉBAULT

A local variation of Pinot Noir found in Burgundy. Its main characteristic seems to be its rather higher crops. See pages 188–199.

PINOT MEUNIER

Pinot Meunier, or Meunier, may be related to Pinot Noir but this is unproven. The leaf looks very different and far more indented. The leaves are also downy on the underside, giving them the floury appearance from which the vine takes its name: *Meunier* means 'miller', and many of the the vine's other synonyms – Farineux or Noirin Enfariné, or Müllerebe (see page 154) or Müller-Traube in Germany, right down to Dusty Miller in England or Miller's Burgundy in Australia – derive from the same characteristic. But it is possible to find canes bearing completely hairless leaves.

It is best known as a blending partner for Pinot Noir and Chardonnay in Champagne, and it is popular with growers there because it buds late and ripens early. Both of these are useful attributes in chilly Champagne, and it is found in the cooler, more frost-prone parts, especially in the Vallée de la Marne, where neither Chardonnay nor Pinot Noir would be a safe bet. It is useful in the blend because it matures much faster than the other two wines and provides softness, fatness and appealingly round fruit at an early age: ideal for wines intended to be sold and drunk young.

Generally it is not thought to age well, and is not regarded as having as much finesse or quality as the other two grapes. As a result most producers are somewhat shy of talking about the amount of Meunier they include in their blends. The exception is Krug, which uses a fair proportion (though still much less than either of the other grapes) in its very long-lived Champagne.

In the 19th century it was the great standby of vineyards all over the north, from the Paris basin as far east as Lorraine – regions where no vineyards exist now. It is still occasionally found in the Loire and makes a pleasant smoky pale pink Vin Gris near Orléans, but there is much more in Germany, where it grows in Württemberg, Baden, the Pfalz and Franken under the names of either Müllerebe, Müller-Traube or Schwarzriesling. In Württemberg it is used for the local pink speciality, Schillerwein; it also makes white sparkling and still red in Germany. The colour is fairly light – lighter than that of Pinot Noir – and the wine is often slightly higher in acidity and smoky in taste. In Germany there is a local variant, called Samtrot, found in Württemberg.

There is some in Austria and in German-speaking cantons of Switzerland. It has been cultivated on a small scale for many years in Australia. Some bright, aromatic reds which can sometimes age well are made and it is also used in Champagne-style fizz. New Zealand and Oregon have small plantings. Best producers: (Australia) Best's, Seppelt, Taltarni; (Canada) Tinhorn Creek; (England) Nyetimber; (France) Billecart-Salmon, Blin, Charles Heidsieck, Alfred Gratien, Krug, Laurent Perrier; (USA) S Anderson, Roederer Estate, Schramsberg.

PINOT NERO

Italian name for Pinot Noir (see pages 188–199). So far Italian examples lag behind Burgundy, California, New Zealand and the best of Australia. One reason may be that many Italian nurseries offer high-yielding clones more suited to sparkling wine production. The Alto Adige region in the northeast seems to have the most potential.

PINOT NOIR

See pages 188–199.

PINOT ST GEORGE

Obscure Californian grape, included here because it is a synonym for Négrette (see page 176) and thus not a Pinot at all.

Bride Valley, tucked into a south-facing fold in the Dorset hills in southern England, grows Pinot Meunier, Pinot Noir and Chardonnay on intensely chalky soil to make delightful, fragrant sparkling wine in the Champagne style.

PINOT BLANC

It's hard to think of a region where Pinot Blanc is regarded as a star grape, though Italy's Franciacorta comes pretty close. Pinot is widely enough grown, but seldom plays the leading role. It does well in northern Italy generally, making very nice bright, dry, creamy whites in the Alto Adige, and in Alsace, where, although the label on the bottle won't even mention its name, much of the best Crémant d'Alsace fizz is based on Pinot Blanc. But otherwise it is one of the wine world's genuine Cinderellas, and if there's going to be a ball to attend, Pinot Blanc's invitation hasn't arrived yet.

Yet it's got a fairly decent family tree. It's a mutation of Pinot Gris, which is itself a form of Pinot Noir, and its flavour closely resembles a mild Chardonnay, especially in northern Italy where the Chardonnay is often in any case light, and Pinot Blanc just tastes like an even lighter version. In Germany's Baden, Saale-Unstrut and Sachsen regions it is called Weissburgunder and often *barrique*-aged and in Austria's Burgenland it makes excellent sweet botrytized wines. In Eastern Europe it is widely grown though with no great distinction.

So we're left with Alsace. But even here it plays at best fifth fiddle to Riesling, Gewurztraminer, Muscat and Pinot Gris, and is often blended with Auxerrois – again, the two grapes look very similar. The ampelographer Galet calls the more productive Alsace version of the vine Gros Pinot Blanc, to distinguish it from other Pinot Blancs.

The New World has not taken to Pinot Blanc in any great way, and clearly prefers the greater glamour of Chardonnay. California has some, though much of what was thought to be Pinot Blanc there turned out to be Melon de Bourgogne, better known as Muscadet. Californian views about how Pinot Blanc should taste vary, so that some wines are as fat as Chardonnay while others are lighter and less assertive. Oregon and Canada have shown a bit more idea of what to do with it, and British Columbia has definitely shown considerable promise.

So, to be honest, it's difficult to know how good Pinot Blanc could be. In a way its very lack of assertiveness, its mild but bright drinkability in a world awash with Chardonnay, is actually one of its most important characteristics.

THE TASTE OF PINOT BLANC

Pinot Blanc in Alsace has a touch of spice to its round, creamy fruit, but not too much. It is not assertively spicy, or assertively aromatic. In Italy it becomes lighter and more minerally, sometimes with a reasonable pear and apple freshness. In Germany it takes submissively to new oak and makes a fair stab at a Chardonnay style with decent acidity and just enough body to cope with the wood.

Rudolf Fürst
German examples can have higher acidity than most Pinot Blanc wines, especially when they are Trocken, or dry. This one is from the Franken region and while the acidity is high it has good, creamy, nutty concentration.

Great Britain is one of the countries showing the clearest evidence yet of climate change transforming vineyard conditions. Not long ago, southern England was regarded as too cold to ripen the classic French varieties enough to make decent table wine, yet now Chardonnay, Pinot Gris and Pinot Blanc are all producing lovely examples, primarily in Kent and Sussex. Stopham Estate is in Sussex and this Pinot Blanc is fresh, light but scented with elderflower and delightfully tangy.

WillaKenzie Estate
Pinot Gris was the first white variety to make its mark in Oregon, but the cool, often damp conditions of the Willamette Valley are producing fresh, tasty Pinot Blanc too.

Left: Once it was seen that Pinot Noir did well in Oregon's Willamette Valley, local producers started to plant other members of the Pinot family including Pinot Blanc and Pinot Gris. This is the WillaKenzie Estate whose Pinot Blanc vines are planted on Willakenzie soil, known for its excellent drainage.
Above: Pinot Blanc's greatest virtue is its subtlety; but the trouble is convincing wine lovers that subtlety on its own is a quality worth shelling out money for. I see their point. 'Subtlety' can so easily become a euphemism for dilution and blandness. Even so, Pinot Blanc is brilliant with food: there are plenty of dishes that don't require fireworks in the wine.

CONSUMER INFORMATION

Synonyms & local names

Sometimes called Clevner or Klevner in Alsace. Known as Pinot Bianco in Italy and Weissburgunder or Weisser Burgunder in Germany and Austria. In Slovenia and the Balkans it may be called Beli (White) Pinot.

Best producers

FRANCE/Alsace Adam, Blanck, Bott-Geyl, Boxler, Deiss, Dopff & Irion, Hugel, Josmeyer, Kreydenweiss, Kuentz-Bas, Meyer-Fonné, René Muré, Ostertag, Rolly Gassmann, Schaetzel, André Scherer, Schlumberger, Schoffit, Turckheim co-op, Weinbach, Zind-Humbrecht; **Burgundy** Maurice Écard, Jadot, Louis Lequin, Daniel Rion.
GERMANY Bercher, Bergdolt, Schlossgut Diel, Fürst, Dr Heger, Karl-Heinz Johner, Franz Keller, U Lützkendorf, Rebholz, Schloss Reinhartshausen, Sasbach co-op, Dr Werheim.
AUSTRIA Feiler-Artinger, Walter Glatzer, Hiedler, Lackner-Tinnacher, Hans Pittnauer, Polz, Fritz Salomon.
ITALY Buonamico, Colterenzio co-op, Marco Felluga, Jermann, Masùt da Rive, Ignaz Niedriest, Querciabella, Alessandro Princic, Puiatti, San Michele Appiano co-op, Schiopetto, Terlano co-op, Vallarom, Vignalta.
ENGLAND Stopham Estate.
USA/California Arrowood, Au Bon Climat, Byron, Chalone, Laetitia, J Lohr, Saddleback, Robert Sinskey, Steele, Tyee, Wild Horse; **Oregon** Adelsheim, Amity, Archery Summit, Bethel Heights, Domaine Serene, Elk Cove, Erath, St Innocent, WillaKenzie, Ken Wright.
CANADA Blue Mountain, CedarCreek, Gehringer, Inniskillin Okanagan, Konzelmann, Mission Hill, Sumac Ridge, Wild Goose.

RECOMMENDED WINES TO TRY

Ten dry or off-dry Alsace wines

Blanck Alsace Pinot Blanc
Boxler Alsace Pinot Blanc
Josmeyer Alsace Pinot Blanc les Lutins
Kreydenweiss Alsace Pinot Blanc Kritt
Meyer-Fonné Alsace Pinot Blanc Vieilles Vignes
Réné Muré Alsace Pinot Blanc Tradition
Rolly Gassmann Alsace Pinot Blanc Auxerrois Moenchreben
Schlumberger Alsace Pinot Blanc
Schoffit Alsace Pinot Blanc Cuvée Caroline
Zind-Humbrecht Alsace Pinot d'Alsace

Five German/Austrian wines based on Weissburgunder

Bercher Burkheimer Feuerberg Weissburgunder Spätlese Trocken Selektion
Fürst Franken Weissburgunder Spätlese Trocken
Dr Heger Achkarrer Schlossberg Weissburgunder Spätlese Trocken
Lackner-Tinnacher Südsteiermark Steinbach Weissburgunder
Fritz Salomon Weissburgunder

Five Italian Pinot Bianco-based wines

Fattoria del Buonamico Vasario
Colterenzio co-op Alto Adige Pinot Bianco Weisshaus
Querciabella Batàr
San Michele Appiano co-op Alto Adige Pinot Bianco Schulthauser
Schiopetto Collio Pinot Bianco

Five US wines

Au Bon Climat Santa Maria Valley Reserve (California)
Chalone Vineyard Chalone (California)
Elk Cove Willamette Valley (Oregon)
Saddleback Cellars Napa Valley (California)
WillaKenzie Estate Willamette Valley (Oregon)

PINOT GRIS

Where have I had my most memorable Pinot Gris? Was it in Alsace? Was it in Germany? Or in Italy, under its alter ego of Pinot Grigio? Well, these are the three most likely suspects. But might it have been from Switzerland? Might it have been from New Zealand, or Canada or Romania? Ever since I tasted a remarkable St Helena Pinot Gris '87 from New Zealand's South Island, I had a feeling Pinot Gris would shine there. Pinot Gris from both Ontario and British Columbia looks as though it could be the white variety that manages best to ride the 'hot short summer, freezing long winter' that makes Canada so intriguing yet challenging. And Romania? If you'd experienced the sumptuous, honeyed, hedonistic flavours of nobly rotten, sweet Pinot Gris from Murfatlar on the Black Sea coast, you wouldn't be smirking behind your hand when I say Pinot Gris has the potential to be one of the world's great white grapes, in some of the world's most unlikely places.

But I still go back to Alsace to find what makes it tick. Here, it makes wine from gloriously sweet to dry but always mellow. And the reason no other region makes Pinot Gris like Alsace has a lot to do with climate. Alsace's dry autumns make long hang times possible; you can pick late for dry or off-dry wines, and even later for sweet ones.

In Alsace, and indeed in Baden, just over the Rhine, it needs ripeness – at least 12.5 per cent potential alcohol at picking but usually much more – in order to have character. It needs low yields, too, to have quality: no more than 60hl/ha for good wine, or 40hl/ha for great.

In Germany, Pinot Gris does well in Baden and Pfalz, called Ruländer or Grauer Burgunder – or Pinot Gris. It usually keeps the honeyed quality but rarely reaches Alsace levels. Italy grows a lot, though rarely impressively, with the exception of a few growers in Collio and Alto Adige in the northeast. Why? Well, everyone's making lots of money out of oceans of Pinot Grigio. That's the same variety, minus much of its ambition. For the best Pinot Grigios, head for Croatia, Slovenia, Hungary and Romania. In the New World, especially in Canada, Oregon and New Zealand, Pinot Gris, because of its pear and apple flesh fruit, and honeyed richness, is seen as an excellent alternative to oaked Chardonnay – and it certainly is. Australia mostly focuses on a crisp, fresh Pinot Grigio style.

THE TASTE OF PINOT GRIS

I adore Pinot Gris' ability to produce fabulously honeyed wines, with the richness of brazil nut flesh and the merest suggestion of something slightly unwashed. (Don't flinch – many of the most memorable wines have something slightly 'incorrect' about them.) Alsace Pinot Gris revels in spicy, musky, honeyed and exotic flavours yet northeast Italy's copper-tinged, relentlessly popular Pinot Grigio with lightly spicy, minerally flavours makes a sort of superior glugger to go with the *fritto misto*. The Canadians and New Zealanders are too new at the game to court the dangerous flavours of decay, and Oregon, in the Pacific Northwest, is happy to exploit the honeyed quality of the grape, but is rapidly showing that a bright, almost spritzy style of white may actually prove to be the state's best wine. In Oregon and New Zealand, fruit flavours prevail – pear and apple, mango and spring flowers – but the best examples always have a flickering honeysuckle scent while Switzerland, Hungary and Romania still hold on to the richer, honeyed style.

Oregon is best known for delicate red Pinot Noirs, but has rapidly made a name for itself with mellow, vaguely honeyed Pinot Gris, which many drinkers enthusiastically adopted as a reasonably full-bodied but unoaked alternative to Chardonnay.

Vie di Romans
This shows the serious side of Pinot Grigio – yellow-coloured pear and pastry-flavoured full dry wine from stony, glacial soils in sunny but cold Isonzo in Friuli-Venezia-Giulia, north-east Italy.

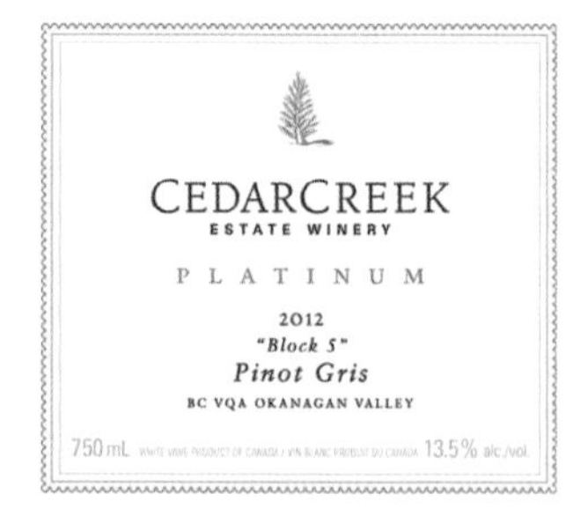

Cedar Creek
Pinot Gris is fast becoming one of British Columbia's top white wine varieties. Vanilla, honey and pear flavours blend in this barrel-fermented example.

Left: The glowing gold of autumn sunshine in the Brand vineyard above the village of Turckheim vividly suggests the luscious honeyed wine these late-harvest Pinot Gris grapes will create. ***Above:*** *Pinot Gris grapes, shown here in the early stages of* Botrytis cinerea, *are more susceptible to the fungus than any other noble Alsace variety. There are two strains of the vine here: Gros Grains, which has pretty well taken over, gives much higher yields than the smaller-berried Petits Grains, which has almost disappeared. But Petits Grains gives stellar quality and the new breed of ambitious young growers won't let it disappear completely.*

CONSUMER INFORMATION

Synonyms & local names

French synonyms include Pinot Beurot or Burot (Burgundy), Malvoisie (Loire and Savoie) and until recently in Alsace Tokay-Pinot Gris. Called Pinot Grigio in Italy and in Germany and Austria Grauburgunder (for dry) and Ruländer (for sweet wines). Switzerland's Valais calls it Malvoisie and Hungary Szürkebarát.

Best Producers

FRANCE/Alsace Adam, Albrecht, Barmès-Buecher, Beyer, Blanck, Bott-Geyl, Boxler, Burn, Deiss, Dopff & Irion, Hugel, Josmeyer, Kientzler, Koehly, Kreydenweiss, Kuentz-Bas, Albert Mann, Ernest Meyer, Meyer-Fonné, Mittnacht- Klack, René Muré, Ostertag, Pfaffenheim co-op, Rolly Gassmann, Schaetzel, Schlumberger, Schoffit, Bruno Sorg, Trimbach, Turckheim co-op, Weinbach, Zind-Humbrecht; **Burgundy** l'Arlot; **Loire** Henri Beurdin.
GERMANY/dry Bercher, Dr Heger, K-H Johner, Müller-Catoir, Zimmerling; **sweet** Salwey.
AUSTRIA Feiler-Artinger, Gross, Schandl.
ITALY Cesconi, Livio Felluga, Lageder, Le Monde, Ronco del Gelso, Russiz Superiore, Schiopetto, Vie di Romans, Villa Russiz, Elena Walch.
USA/California Flora Springs, Long Meadow Ranch, Seghesio; **Oregon** Adelsheim, Archery Summit, Chehalem, Evesham Wood, Eyrie, King Estate, Ponzi, Sokol Blosser, WillaKenzie; **Washington State** Chateau Ste Michelle.
CANADA Blue Mountain, Burrowing Owl, Gehringer, Mission Hill, Pelee Island.
NEW ZEALAND Astrolabe, Ata Rangi, Dry River, Gibbston Valley, Greywacke, Kumeu River, Lawson's Dry Hills, Neudorf, Palliser, St Helena, Seresin.
SOUTH AFRICA L'Ormarins, Nederburg, Van Loveren.
AUSTRALIA Tim Adams, Bay of Fires, Brokenwood, Crittenden, Heartland, Henschke, Mount Langhi Giran, Pike & Joyce, Pizzini, Primo Estate/Joseph, T' Gallant.

RECOMMENDED WINES TO TRY

Five dry or off-dry Alsace wines

Kientzler Vendange Tardive
Albert Mann Vieilles Vignes
Ostertag Muenchberg Vendange Tardive
Schoffit Cuvée Alexandre
Zind-Humbrecht Clos Windsbuhl VT

Five Alsace Sélections de Grains Nobles

Bott-Geyl Alsace Pinot Gris
Hugel Alsace Pinot Gris
Kuentz-Bas Alsace Pinot Gris Cuvée Jeremy
Weinbach Alsace Pinot Gris
Zind-Humbrecht Rangen Clos St-Urbain

Five German/Austrian Grauburgunders

Bercher Burkheimer Feuerberg Spätlese Trocken Selektion
Alois & Ulrike Gross Südsteiermark
Dr Heger Ihringer Winklerberg Spätlese Trocken 3 Sterne
Müller-Catoir Haardter Herrenletten Spätlese Trocken
Klaus Zimmerling Landwein Trocken

Five top Italian Pinot Grigio wines

Livio Felluga Colli Orientali del Friuli Pinot Grigio
Lageder Alto Adige Pinot Grigio Benefizium Porer
Ronco del Gelso Friuli Isonzo Pinot Grigio Sot Lis Rivis
Schiopetto/Poderi dei Blumeri Colli Orientali del Friuli Pinot Grigio
Vie di Romans Friuli Isonzo Pinot Grigio Desimis

Five New World wines

Burrowing Owl Pinot Gris (Canada)
Chehalem Pinot Gris (Oregon)
Evesham Wood Pinot Gris Estate (Oregon)
Gibbston Valley Pinot Gris (New Zealand)
T'Gallant Pinot Grigio (Australia)

E·R

PINOT NOIR

We should rejoice. There is now more beautiful Pinot Noir being made in the world than ever before. California, Oregon, New Zealand, Chile, Australia and Germany are all finally getting to grips with a grape that has been regarded as the ultimate challenge in the catalogue of red wine varieties, the heartbreak grape, the variety that seems to follow no rules which would allow you to produce a consistent crop of quality fruit that you could then turn into tasty balanced wine and build up a reputation for yourself. The world is cracking the Pinot Noir code.

Oh, and did I say that Burgundy is also producing more high-quality wine than ever before? Silly me. I forgot. Well, it is. Modern Burgundy has more peaks of beauty and grace in its wines than ever before, but also far, far more wines at the middle and lower rank that reliably give pleasure. And what's the relevance of that? Until very recently – i.e. about yesterday – the success of Pinot Noir worldwide was judged on its similarity to Burgundy, which does happen to to be the original Pinot Noir centre of excellence, but which, over the centuries, had produced precious few brilliant bottles to justify the reputation either of the region or of its grape – Pinot Noir.

Yet all round the world, Pinot Noir was supposed to ape Burgundy and in most cases this led to decades of disappointment and self-delusion as even the most talented producers glumly gazed at the non-Burgundian results of yet another harvest. Until enough growers and winemakers said, Look, we don't have Burgundy conditions. Indeed, who would want them? Damp, drab, icy winters, fickle springtimes, erratic summers which come all too often to a sudden, sodden end. Who would want that? And yet, and yet . . . So many winelovers could rhapsodize about a bottle of red Burgundy as being their most memorable lifetime red. But what is this Burgundian idea of Pinot Noir?

Is it pale, ethereal, with a scent as sweet and wistful as half-forgotten childhood summer memories hovering over the glass? Is it sensual or heady, swirling with the intoxicating excitement of the super-ripe cherries and strawberries and blackberries of the gardens of Paradise? Is it muscular, glowering, dark as blood, bittersweet as black cherries and liquorice and yet even within this brutish cave, invaded with an insidious, exotic scent? Is it any of these?

Seen through an arched Gothic window typical of those found at the Hospices de Beaune in Burgundy, Pinot Noir's homeland, is the château at Gevrey-Chambertin, one of the Côte de Nuits' best known wine villages. The moonlit scene is a tribute to the word Nuits. *On the windowsill are various references to Pinot Noir's many facets – the traditional silver Burgundian* tastevin, *or tasting cup, a Champagne cork and wire cage, a cone from the Oregon pine and a Knave of Spades to symbolize Pinot Noir's capriciousness both in the vineyard and the wine cellar.*

I can come up with a dozen more thoughts on what Burgundy tastes like. And then a dozen more again. I could describe these flavours to many great winemakers round the world who swoon about Burgundy and then say, But I love *your* Pinot Noir just as well – and it's not like any of the Burgundy sensations I've related – it's yours, it's unique – and I love it. And, at last, that's where we are with Pinot Noir.

One of the benefits of no one really knowing quite how to coax the best flavours from Pinot Noir is that you never know when you're suddenly going to be faced with a delicious glass of Pinot from a completely unexpected quarter. And another of the benefits is that whereas it's relatively simple to put a stylistic straitjacket around the two world favourites, Chardonnay and Cabernet, putting a stylistic harness on Pinot is as tricky as wrestling with an eel. While a host of talented but mainstream producers are perfectly happy to try to excel at Cabernet and Chardonnay in accepted styles, Pinot Noir attracts a much wilder bunch. A crowd who don't like being told what to do, a crowd who don't like a marketing manager to have more say in a wine than a winemaker. A self-indulgent crew of men and women who love flavour, who love perfume, who love the silky tactile experiences of a wine like Pinot, seductive, sultry, steamy, sinful if possible, but always solely there, solely made, to give pleasure.

And that's why I love Pinot Noir and enthusiastically stalk it round the world. And if that isn't a good enough reason to grow grapes and make wine – and then drink it – well, what the devil is?

GEOGRAPHY AND HISTORY

You'll find more uncertainty among growers and winemakers about what they think they should be doing with Pinot Noir than with almost any other grape. How to trellis it, prune it, crop it; when to pick it, early or late; what style of wine to make, light or dark; and using which winemaking method, new or old. Doesn't anyone have the definitive method of producing great wine from this grape?

You only have to look at the myriad styles of wine produced in its homeland, Burgundy, and the regular changes of fashion every decade or so, to realize it's not just the newcomers who are confused. And the grape is a most finicky traveller, its inconsistency abroad an amplification of that which it displays at home, to the fascination and despair of those who grow it, vinify it and indeed drink it.

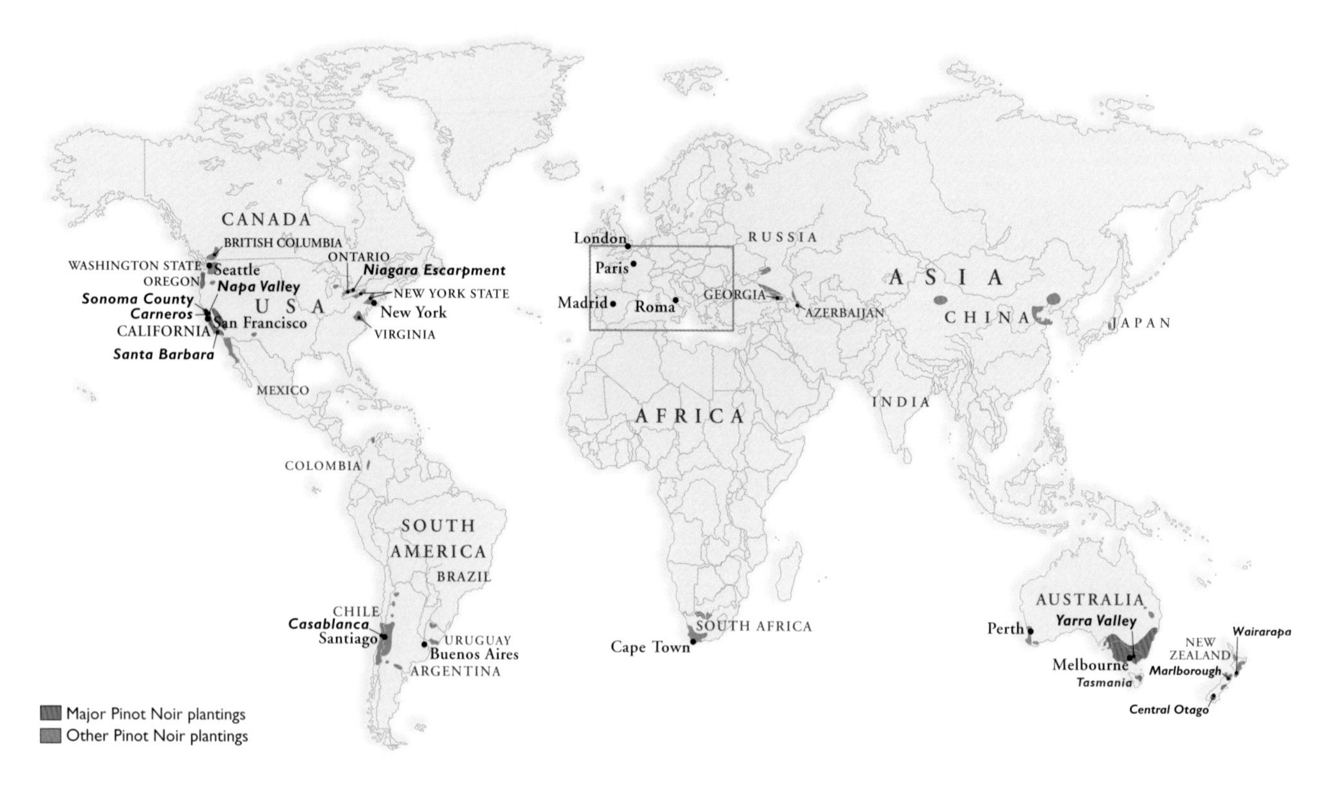

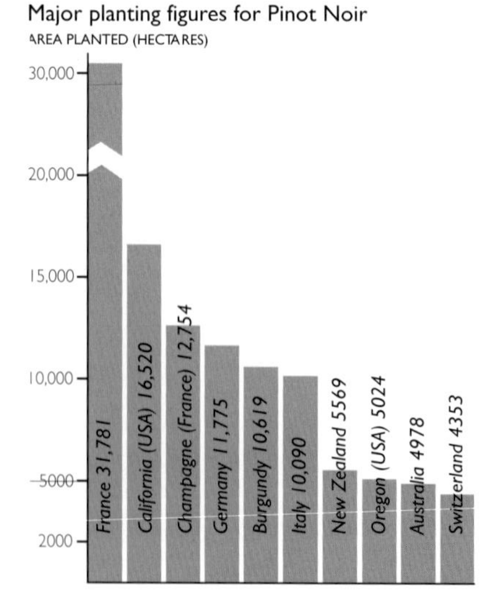

It claims to dislike hot climates, though as an Oregon grower says of Pinot Noir in the Margaret River, Australia, 'One has to rethink one's biases when one sees regions that have Shiraz and Pinot Noir growing next to each other, but producing world-class Pinot.' Too right. Certainly it is supposed to be a cool climate grape, yet some of its most exciting wines in Burgundy have come from seriously hot vintages. Everyone said that cool Oregon would be ideal for Pinot, but some of the USA's most exciting Pinots have come from warmer California. And so it goes on. Warm climate Pinot should be jammy and flat, but isn't always. Cool climate Pinot should be delicate and perfumed, but is equally often green and raw. This unpredictability ought to limit its spread around the globe. But everybody wants to have a go.

The truth is that everybody wants to make Pinot because, like true love, great Pinot is elusive but (perhaps) worth the effort. There is still more promising Pinot in the world than there is really great Pinot – yet the 2000s saw a vast increase in the numbers of very good Pinots from the Old and, especially, the New World. True love is commoner than it was.

HISTORICAL BACKGROUND

This is one of the oldest cultivated vines in existence. It seems to have already been in cultivation in Burgundy when the Romans arrived, some 2000 years ago; it probably originated in either northeast France or southeast Germany, and its parents or grandparents were probably wild vines. In an era when science is solving so many of wine's conundrums, it's a delight to report that scientists have compiled Pinot Noir's complete genetic code – nearly half a *billion* DNA letters – yet they still have no real idea where it came from. My wild romantic theories of where grape varieties come from can live to breathe another day.

The first written record of it dates from 1375, and just 20 years later Philippe the Bold, Duke of Burgundy, issued an edict banning the inferior Gamay from the Côte d'Or in favour of Pinot.

Its age is the reason for its extreme genetic instability. The Pinot family (Pinot Noir, Pinot Blanc and Pinot Gris are, to ampelographers, all the same grape, because they share the same DNA) is prone to constant mutation. On the one hand, this can be to the grower's advantage, since it means that vines will tend to adapt themselves to local conditions; on the other hand, it makes it harder to retain the desirable characteristics in particular clones.

Its descendants are legion. Work at the University of California at Davis, in collaboration with French specialists, has determined that Pinot Noir is the ancestor of 16 modern grape varieties, including Chardonnay, Gamay Noir, Aligoté, Melon de Bourgogne (alias Muscadet) and Auxerrois, and in total it's related to 156 varieties, and counting. Syrah, for example, is some sort of 10th cousin five times removed.

In the foreground (above) is the Romanée-Conti vineyard in Vosne-Romanée; behind the wall is Romanée St-Vivant. The desire to maintain the differences in taste given by such different but neighbouring terroirs has been a spur to the growth of biodynamism on the Côte d'Or.

They don't waste wine at Domaine Sylvain Cathiard. Son Sebastien pours back a tasting sample of one of his exquisitely balanced Vosne-Romanées to make sure there's enough for the next lucky visitor.

The queen of Chambolle-Musigny. Ghislaine Barthod's subtle and scented wines from a variety of Chambolle premier cru vineyards are some of Burgundy's most seductive reds.

VITICULTURE AND VINIFICATION

The aim of most Pinot Noir producers around the world is to make red wine which rivals that of Burgundy's Côte d'Or. That much is straightforward: there is, as yet, no other benchmark style than that of Burgundy, though both California and New Zealand are beginning to display distinctly non-Burgundian virtuosity.

It is less easy, however, to find agreement on what it is about the Côte d'Or, and about Burgundian winemaking, that needs to be imitated. Is it the soil? Is it the climate? And if so, which elements of either? And if you wanted to make great Pinot more consistently than Burgundy manages to, what should you try to improve?

Might it not be a good idea to forget about Burgundy altogether and just follow your instincts? Vanya Cullen of Margaret River said she began making her best Pinot Noir when she gave up doing anything Burgundian and just let it stew in its own juice. An increasing number of New World Pinot-makers are indeed starting to follow their own instincts – and a number of New World Pinots are starting to taste very good indeed and not at all like Burgundy.

Climate

Burgundy is cooler than Bordeaux, with a colder winter and a slightly cooler summer, and a larger fluctuation between day- and nighttime temperatures. There is a risk of frost in spring, abundant rain in May and June, and more rain in October which, if it falls just before or during the harvest, can spoil the quality of the wine. Pinot Noir is an early-budding and early-ripening variety, and sensitive to rot: slow ripening in a cool climate suits it, because its berries are particularly sensitive to heat; yet, paradoxically, it needs quite a bit of warmth to ripen properly. Even with global warming, there are generally several years in each decade when Pinot Noir fails to ripen fully on the Côte d'Or, and many of the winemaking fashions there of the last few years have been aimed at extracting more colour and flavour from this not-very-strongly coloured and delicately scented and flavoured grape.

In California, UCD for many years used to advise growers that there was nowhere in California cool enough for it. Now it has become clear that Carneros, Santa Barbara, Russian River and the Sonoma Coast can produce smashing Pinot. And the growers of New Zealand's Central Otago point to a continental (indeed Burgundian) pattern of warm days and cool nights as being vital, rather than just a cool climate *per se*.

Climate change has transformed Germany's vineyards, and Pinot Noir has been the greatest beneficiary. From being a light-bodied, frail local delicacy, Pinot Noir from the new wave of producers covers the whole gamut, from mild and juicy, to world beaters. In Alsace, just to the north of the Pfalz, Pinot Noir is traditionally hardly more than a dark rosé, but bigger, burlier examples are now commonplace.

In the foreground is Gevrey-Chambertin's Lavaut-St-Jacques vineyard with gaps in the vines marking different ownerships of tiny plots. The wall marks the boundary of Clos des Varoilles within Les Varoilles vineyard, and the vines by the forest are Poissenot and la Romanée. All different, with various owners, and covering just a couple of hundred metres.

As Myron Redford of Amity Vineyards, Oregon, says, 'I still think the best Pinot Noir needs a cool climate, but my definition of "cool" has been expanded. Once you are in the right climate, however, *terroir* in the sense of soil, exposure, drainage and microclimate become the critical factors.'

Soil

Pinot Noir in the Côte d'Or is planted on the more limy soils, though ample supplies of the clay mineral montmorillonite are found in the mid-slopes where the best vineyards are. Montmorillonite has a high cation-exchange capacity, a cation being an atom or molecule with a positive electrical charge; cation exchange is how a plant extracts nutrients from the soil. In addition, the topsoil in the Côte d'Or is shallow, seldom much more than 1–1.5m (3–5ft) deep. This aids drainage and the fractured nature of the bedrock means that vines can generally find water even in drought conditions – but where the topsoil is thinnest there can be a greater concentration of virus-spreading nematodes. (It was the problem of viruses in Burgundy that led to the popularity of virus-free clones there; see below.) The high water table towards the foot of the slope usually defines the lower boundary of the Grands Crus, the exceptions being Clos de Vougeot and Bâtard-Montrachet, both of which fail to stop in time, and which include land that is too waterlogged for top quality.

Elsewhere in the world shallow, well-drained, low-fertility soils are generally favoured for Pinot Noir. Some focus on clay to give depth to the wine: Hamilton Russell in South Africa believes that people pay too much attention to the limestone in Burgundy, and not enough to its clay. But then Josh Jensen, maker of some of California's greatest Pinot Noirs at Calera, high in the Gavilan Mountains in San Benito County, says he couldn't do it without his limestone soils.

Cultivation

The usual density of planting on the Côte d'Or is 1m by 1m, which means 10,000 vines per hectare. Some vineyards are planted even tighter, at 1m by 0.9m, although in the Mâconnais the density can be 8000 vines per hectare. Most of the vines are Guyot pruned, though there are some trials of spur pruning and cordon; short pruning, it is acknowledged, is the main weapon in the battle to keep yields down and concentration up. However, if you prune too short the vine ends up as a bush and the fruit won't get enough sun. Dominique Lafon of Domaine des Comtes Lafon recommends leaving a longer cane and debudding one or two buds. 'It is better for the canopy and for aeration. You must think of the shape of the vine; shape is more important than the number of buds at pruning, because you need aeration to ward off botrytis.'

Crop thinning, generally done after *véraison* so that the green berries can be removed, is seen as a temporary solution to a problem, not a technique in itself. The ideal is to get the vine in balance, so that it is not overproducing in the first place; among the best growers, 35hl/ha is seen as the ideal yield for Pinot Noir. At Domaine de la Romanée-Conti average yields are even lower, at around 24hl/ha.

In New Zealand and Oregon around 2 tonnes/ha is a generally accepted figure for producing top-quality, ripe Pinot Noir. Compare that with yields for sparkling wine: in Champagne in the excellent 1990 vintage, yields averaged 110hl/ha. Houses that own their own vineyards generally use lower yields than growers who sell their grapes. In Chile 8–10 tonnes/ha is considered low, and 12–13 tonnes/ha is not uncommon.

Clones

Pinot Noir is notorious for its clonal variation. Some clones give pale-coloured, frail-tasting wine; others give robust, jammy flavours and dark colours; some give elegance, some chunkiness; some give large crops, some small. The 1970s were a bad time for Pinot Noir clones: high yields were the object, not high quality. New generations of clones, particularly the Dijon clones developed in Burgundy, have dramatically improved the picture as they move into vineyards across the world. But change doesn't happen overnight, and even on the Côte d'Or there is likely to be a higher proportion of the high-yielding Pinot Droit than many Burgundians care to admit. (Lower-yielding strains are Pinot Fin or Pinot Tordu.) On the Côte d'Or alone more than 1000 different clones have been noted.

But Dijon clones aren't a panacea. Some Burgundian growers, having planted new-generation clones, found them too high-yielding and have moved to massal selection from their own vineyards for greater complexity. Others believe that complexity comes from the *terroir*, and that greater vine age, plus good vineyard management, tends to eradicate differences between many clones. In the sunnier New World, many Dijon clones ripen and lose acid too fast.

The view that some Australians are now taking is that there can be more difference between Pinot from the top or the bottom of a slope than between clones. Some are even reappraising the old Australian workhorse clone, MV6, with exciting results.

At the winery

While good winemaking is absolutely essential for Pinot Noir, it is a curious fact that what makes a Pinot Noir wine interesting, rather than merely correct, is a series of low-level faults: aromas and flavours that are just slightly off squeaky clean. While Burgundians have generally reached this point by making their winemaking more correct in recent years, winemakers in some other countries perhaps have to become a little less correct if they are to make great Pinot Noir wine.

The norm in Burgundy is to destem all or most of the crop, and ferment a high proportion of whole berries, without a pre-fermentation crush. Only a few estates (including, it should be said, some top ones) leave the stems in. Chaptalization is accepted, and acidification not uncommon; you're not allowed to do both, but . . .

A pre-fermentation maceration or cold soak is often practised. It usually lasts just a few days, while the wild yeasts (cultured yeast is less common in Burgundy) slowly get moving. The point of it is that at 15°C (59°F) enzymes work on the cells of the skins to produce better aromas and tannins; it can be likened to tomatoes ripening on a windowsill. It gives more depth of flavour, and more detailed flavour. After a while, fermentation temperatures usually rise to 30–32°C (86–90°F), and the total time in tank, including any post-fermentation maceration, varies from two to three weeks.

Filtration

This does no great harm to the wine if done gently and properly, but an increasing number of producers in Burgundy and elsewhere don't filter their wine.

THE CABERNIZATION OF PINOT NOIR

How light should Pinot Noir be? How rich? Winemakers have been asking themselves this question ever since the early 1980s saw much lighter, perfumed, less weighty red Burgundies take over the market. Some consumers mourned the demise of the big, soupy wines of yore. (Go back far enough and the reason they were big and soupy was because they contained a good dose of red wine from elsewhere – usually the southern Rhône or North Africa.) Others relished the elegance and delicacy of the new wines, and saw them as authentic representatives of their *terroirs*.

But fashion in all other red wines currently focuses on big, rich flavours – like those of Cabernet Sauvignon. To be certain of selling in the international marketplace, wines must have huge but soft tannins, massive fruit, colossal intensity. Light reds are for wimps; they don't grab headlines, and they don't win blind tastings.

Accordingly more and more Pinot producers feel the pressure to make big, solid wines. And it's not just happening in California, where individual critics' opinions carry so much weight. In Burgundy too, Cabernization is a factor. Take the fad for 200 per cent new wood. Less than half new wood is usual in Burgundy, with the weightiest wines getting more than that, and the lighter wines less. About 18 months in barrel is normal. The 200 per centers age their wines for a year in new oak, and then rack them into more new barrels. Their object, they say, is not to produce hugely tannic, oaky wines: instead they aim to produce dark, voluptuous, rich wines from perfectly balanced, perfectly ripe grapes; wines that reflect their *terroir*. New oak, they believe, gives better oxidation, better interaction between the lees and the wine, and slows down the aging process. It also gives deeper colours and fixes the aromas. Sure. And adds massively to the tannins along with a great thud of vanilla and spice. Only the biggest wines actually get 200 per cent new oak, but really I always thought I was supposed to be looking for delicacy and beguiling scent in a Burgundy.

Will 200% oak, like Guy Accad's cold soak (a 'rediscovered' pre-fermentation maceration method), become 'a rediscovered traditional technique' and, in a modified form, enter the repertory? Will red Burgundy change its nature again and become a bigger, beefier wine? Only time will tell.

PINOT NOIR AROUND THE WORLD

Pinot noir may have a wide geographic spread, but its fussiness about soil and climate means that it is, in most places, in a minority. A considerable percentage of the grapes are used for sparkling wine, and because of its current popularity, not all the vines are planted in suitable, cool sites. But, bit by bit, new, high-quality Pinot producers are appearing.

Burgundy

The Côte d'Or is the heart of Burgundy as far as Pinot Noir is concerned; the surrounding regions produce what is often referred to as 'affordable Burgundy', which is at least a tacit admission that it is neither cheap nor usually particularly good value compared to other wines. Such bottles can be very attractive, but do not expect them to taste like the Pinots of the Côte d'Or.

In the Auxerrois to the north, Pinot is light and perfumed; in the Côte Chalonnaise to the south, it is, at its best, richly fruity with good structure, but not much complexity. In the Mâconnais Gamay is the main red grape, and Pinot Noir vines are often poor clones.

In the Côte d'Or, growers' reverence is reserved not for the grape variety but for the *terroir*. Burgundians do not look primarily for varietal character in their Pinot; indeed, some claim not to want varietal character at all. The vine is merely a conduit for the *terroir*.

But how do you express that *terroir*? Biodynamists in Burgundy (a growing number) believe that the overgenerous application of nitrogen and phosphorus fertilizers in the 1970s and beyond had the effect of masking the differences in *terroir* – a serious matter, since it is on *terroir* that the whole elaborate map of Premiers and Grands Crus is based. So it is not surprising that biodynamism is so influential here. It is helped by the small size of most Burgundian domaines, and by the high price of the wines, which means that the extra expense and manpower involved can be afforded and eventually recouped.

Ideally, then, each Côte d'Or vineyard should have its own recognizable flavour. To a large extent that is true – in Volnay, for example, Champans is more structured and dense than Santenots. But the large number of owners in each vineyard, with the Grand Cru of Clos de Vougeot, for example, having over 50 different growers owning vines and making wines that bear the vineyard name, means that the *terroir* is seen through the grower as well as through the grape. When the producer is good this adds another layer of complexity; but, inevitably, all producers are not of equal quality. The unevenness of Pinot Noir grape quality is mirrored, and indeed exacerbated, by unevenness among producers. However, standards of viticulture and winemaking have risen enormously in recent years.

The Côte de Nuits produces bigger, chewier, plummier, more solid reds than the Côte de Beaune. Even so, Marsannay reds are surprisingly scented though Fixin's solid reds are a bit clumsy. Gevrey-Chambertin is muscular, dense, sometimes black cherry scented; Morey St-Denis, with a clutch of top-class growers, is lean yet savoury and complex; Chambolle-Musigny is all roses and violets. Vougeot can be positively chubby. Vosne-Romanée is violets and cream, Nuits-St-Georges is black cherry and chocolate but Côte de Nuits-Villages is often a bit thin. In the Côte de Beaune, Aloxe-Corton is smooth, savoury and perfumed, Savigny-les-Beaune leaner, scented with strawberries; Beaune is round and soft. Pommard is chunkier, Volnay fragrant, and Chassagne-Montrachet solid and savoury.

Champagne

This cold region might not, at first glance, seem to be ideal for the early-budding Pinot Noir. The grape is horribly susceptible to spring frosts here, and clonal research in Champagne concentrates on frost resistance. Plantings are concentrated on the Montagne de Reims between Reims and Épernay – even, in this marginal climate, on certain north-facing slopes. Not very steep slopes, admittedly, but it is still surprising to find Grand Cru vineyards in such a position. Perhaps the chalk reflects enough heat to warm the air, and compensate?

Pinot Noir in Champagne never reaches what any other region would describe as ripeness. The warmest, south-facing sites are usually the best, but the undulating hills offer many different mesoclimates. The village of Ay, for example, is said by local growers never to get frost in spring – or perhaps hardly ever. The Marne valley broadens out on one side there, and the cold air can slip away – unlike at Épernay, a few kilometres away, where the valley retains the cold air. The river Marne runs right at the foot of the slope, too, which is the closest it gets to any of the Montagne vineyards: at Épernay, they say, there can be fog until 10 am when it's sunny in Ay. Picking is by hand, and Pinot Noir must be pressed carefully and quickly to avoid too much colour getting into the wine. The still wine usually has a pale pink tint, but this rarely shows in the finished wine: the colouring matter combines with the dead yeasts in the bottle after fermentation but before the wine is disgorged. And, of course, Pinot Noir is usually blended

Domaine de la Romanée-Conti
Just seeing the label makes my mouth water. This tiny vineyard produces Burgundy's most treasured red – scented, cerebral and profound.

Domaine Anne Gros
Le Grand Maupertuis refers to a single plot of vines in Clos-Vougeot. Anne Gros' style is for deep-coloured wines which are balanced and elegant.

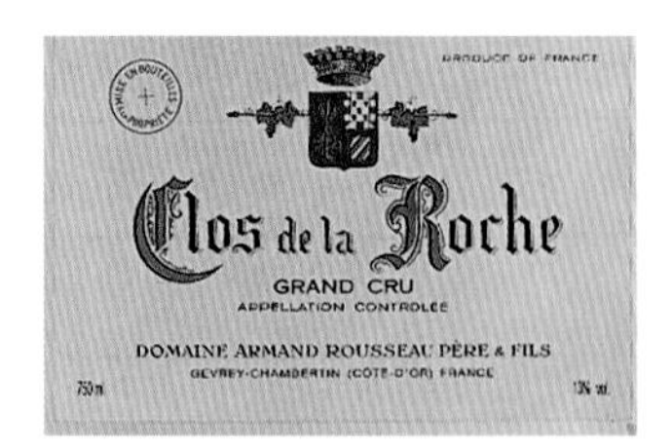

Domaine Armand Rousseau Père et Fils
Domaine Armand Rousseau is one of the great names in Burgundy. Clos de la Roche in Morey-Saint-Denis gives wines of silky texture and lingering scent.

Pinot Noir vines on the Montagne de Reims above the Grand Cru village of Verzenay in Champagne. Pinot Noir is successful here only because Champagne needs base wines of aroma and finesse, not richness, tannin and deep colour, so barely ripe, pale Pinot Noir fits the bill nicely.

with white Chardonnay. Only a few still red wines are produced, from Bouzy, Cumières or Ay, where the wines are ripest and have the darkest colour, but they need a very warm year to shed their tartness. Blanc de Noirs – white fizz made from black grapes – is thus Pinot Noir's best style of wine in Champagne.

Varietal character is no more the object here than it is in Burgundy – in fact probably even less so since the grapes are less ripe. Much of the flavour and style of Champagne comes from the aging period on the lees.

Other French Pinots

No other French regions get close to the ripeness and complexity of red Burgundy. There is some Pinot Noir in the Loire, most notably in Sancerre and Menetou-Salon, where in ripe years (these have been frequent since the 1990s) it can make wines with good fruit and structure. Most red Sancerre is, however, light and made to be drunk young; to make *vin de garde* here you have to dedicate your very best sites to Pinot Noir rather than to Sauvignon Blanc.

Alsace Pinot is generally light but plantings, often on light, fertile soils, have nearly doubled in the last 10 years. Some wines here are now made dark and unattractively oaky. Pinot Noir from the Savoie, Bugey and Puy de Dôme regions is light and aromatic, and high sites in the Languedoc can produce pleasant wines.

Germany

Amazingly, Germany is now the third-largest Pinot Noir grower, after France and the USA. There is a red wine boom in Germany, and Pinot growers have not been slow to catch on. The best Spätburgunders (as Pinot Noir can be known here) tread a line between the traditional German pale, sweetish style and the trap of overoaking and overextraction. But when they're good they're very good; and very expensive.

The most successful examples, some of them world class, come from Baden, Württemberg, the Pfalz and the Rheingau, but in the latter they have to compete for the best sites with Riesling. Rheingau producer Robert Weil has some Spätburgunder in a site too hot for Riesling, but usually a grower must choose between the two grapes. Some blend, particularly in Franken, with other grapes such as Blauer Portugieser. You don't get much Pinot character coming through, but they're nice, ripe, cherry- and raspberry-flavoured wines.

Austria

Blauer Burgunder (as Austria may call Pinot Noir) and other reds are on the increase in the warmer Burgenland, at the expense of less-fashionable whites like Welschriesling. Many reds are blends from Zweigelt, Blaufränkisch, Sankt Laurent and Cabernet Sauvignon. Overoaking and overextraction can be a problem, but hopefully only a temporary one as growers learn how to handle oak more deftly. Burgenland also has some oddities like Spätburgunder Trockenbeerenauslese: rich, white and sweet, they are rare and command high prices – but they're treacly, low acid oddities rather than vinous masterpieces.

Italy

Pinot Nero is hardly a new introduction here: it was first planted in Piedmont in 1825, though its pale colour and low acidity did not endear it to consumers, and its susceptibility to rot did

Domaine Michel Lafarge
Buying good Bourgogne Rouge can be just as tricky as buying Premier or Grand Cru Burgundy. This stylish example from Volnay has ripe, savoury fruit.

Producteurs de Mailly
This cooperative in Mailly on the Pinot Noir-dominated Montagne de Reims produces full-bodied Champagnes. This one is from black grapes only.

Karl-Heinz Johner
Johner is as passionate about new oak as he is about his grapes and so produces international-style Pinots from his vineyards in Baden, Germany.

little for its image among growers. Only in the last couple of decades have growers begun to look at better clones and better training systems (generally Guyot rather than pergola). It's very much a northern speciality, and quite a lot goes into sparkling wine, especially in Lombardy. On the still wine front, the west-facing slopes of Alto Adige east of the river seem to be particularly successful, helped by early morning shade and plenty of sun the rest of the day: one of the challenges has been finding spots with high fluctuations between day- and nighttime temperatures. In the north-west, blends with Barbera, such as Giacomo Bologna's Bacialè, are highly fashionable and often very good.

Rest of Europe

Spain has one or two respectable Pinot Noirs from Somontano, and there are moves to get it added to the list of approved grape varieties in Rioja. In Switzerland it is mostly blended with Gamay to make Dôle, though varietal Pinot Noir can be light, strawberryish and quite good here. In Eastern Europe it is planted widely though not well, and few examples from Romania, Hungary, Slovakia or Bulgaria really deliver. In England it grows very well for fizz, with a few successes as a still wine.

USA: California

Outside Burgundy the real Pinot Noir excitement is in the New World. So far California and New Zealand are ahead, with wines of richness and complexity, plus that elusive, fascinating fragrance and silkiness that are the most difficult to attain of all Pinot Noir's attributes. Lesser wines settle for charm and supple fruit. Most of California is too hot for Pinot Noir, but Carneros, where the temperature is 3–5 degrees cooler than the northern part of the Napa Valley, and where the lack of rich humus in the soil produces less vigorous vines and lower yields, seems to be able to do the business. Carneros was originally planted for sparkling wines, but replanting with superior clones has transformed the wines.

Carneros wines have wild strawberry fruit flavours, without a great deal of weight, although the best are marvellously balanced; other parts of Napa can give earthier, leathery flavours. In Sonoma County, Russian River has darker, blackberry flavours and Sonoma Coast is scented and sensuous. Further south, in Santa Barbara County, Santa Maria and Santa Rita Hill can be weightier and complex, with black cherry and plum fruit.

Pinot Noir is on the increase in California and new vineyards in the rugged Sonoma Coast, producing darker colour and brighter

CALIFORNIAN PINOT NOIR

Ted Lemon (right) is one of the most thoughtful and successful winemakers in California when it comes to this most capricious of grapes. California Pinot Noirs were initially derided in Europe for being too rich and heady. Some still are, but a new generation of producers, led by Ted Lemon, are working with growers to express the subtleties of different sites, convinced that places like Sonoma Coast or the Anderson Valley have terroir *characteristics just as valid as those of Burgundy. Where possible, organic or even biodynamic methods are pursued and earlier rather than later pruning is encouraged. The resulting wines are a delight, far more delicate than you might expect yet identifiably Californian, identifiably not Burgundy. Oregon was initially hailed as Pinot's American home given that it has a decidedly erratic climate veering towards the cool, damp and unpredictable – very much like Burgundy was the idea. California's claim is equally valid, but with far less of a nod towards Burgundy.*

Calera

In the early 1970s Josh Jensen planted Pinot Noir on limestone soil in California when the accepted wisdom was that it would never work there. He saw limestone as the key when everyone else there thought climate mattered most. He's been triumphantly vindicated.

fruit, are attracting attention. And with large companies planting Pinot hundreds of acres at a time, the potential for both quality and quantity is very exciting – though cult wineries may make Pinot Noir in smaller quantities even than Burgundy. But, helped by the film *Sideways*, all the wines sell.

Some of the improvement in quality in California can be put down to better clones: some of the early clones had been developed for less warm and fertile places. Exposed to the Californian sun they produced wines that lacked silkiness and finesse. New clones, especially Dijon clones, may be better adapted, though site, vine age and grower are probably more important to the final result since many Dijon clones ripen fast and lose acid too quickly. For sparkling wines the prime sites are cool ones with shallow soil and not too much sun: Russian River, Carneros and especially Anderson Valley all produce good sparklers.

Rest of North America

Charm and suppleness are generally the order of the day in Oregon, with a few producers making serious, structured, long-aging wine. It's still not a confident style: Oregon growers often try to imitate Burgundy, whereas Burgundy, with supreme confidence, just tries to make good wine. And quality is still erratic. Low yields are essential for quality, and Pinot Noir tends to get the best, warmest sites. Favoured soils include a clay loam called Jory, which is well drained and has high cation-exchange capacity, and gives wines with bright cherry fruit; Nekia soil, which is fairly similar and known as Aiken in California; and a clay loam called Willakenzie, which needs both drainage and irrigation. Nekia and Willakenzie give black cherry fruit and bigger, fuller tannins. There are about 300,000 acres/121,410ha of Nekia and Jory in the foothills of Oregon – obviously not all planted.

There's a bit of Pinot in wet western Washington and some in New York. Many feel that Canada should be a good stomping ground for Pinot Noir, but it hasn't been easy and British Columbia probably has had the most success.

Australia

Australia started off by planting Pinot Noir in sites that were too hot, where it produced baked, jammy wines. Greater vine age, better understanding of yields and cultivation and, as winemaker Michael Hill Smith puts it, 'lots of expensive bottles of Burgundy drunk in the middle of the night', are all paying off. The general quality level has improved, with a few wines of exceptional quality and complexity. One of the biggest changes has been in the clones planted: the two long-established clones, MV6 and G5V15, seem to have come from South Africa, and may have originated in the Lebanon. Dijon clones are proving better at producing the light, elegant wines which Australia craves. The areas around Melbourne are in the lead, especially Yarra and Mornington. Isolated producers elsewhere make good Pinot Noir. A lot is also used, blended with Chardonnay, for sparkling wine of high quality and often fuller fruit than Champagne.

New Zealand

There is no shortage of cool climate regions here, and Martinborough, in the Wairarapa region at the southern end of the North Island, and Central Otago, at the southern end of South Island, have carved out reputations as Pinot Noir centres in New Zealand. Martinborough makes elegant styles from old river terraces above the flood plain, while Central Otago produces higher-acid wines of thrilling intensity. Marlborough has developed an attractive mellow style, particularly in the Southern Valleys, where the soils retain water and ripening is slower and more even. Stony soils in much of Marlborough tend to advance ripening too quickly. Canterbury, especially around Waipara, can produce sensous, weighty examples, and Nelson's wines are silky and scented. It's starting to be a serious challenger to Burgundy – and yes, it will turn out to have grand cru sites, too.

Chile

Chile, partly by seeking Burgundian know-how and partly by hunting out the right climates, is capturing Pinot's elusive character at remarkably low prices. Flavour can be on the lighter side but the feeling of ripeness can be positively seductive. In terms of climate Casablanca is most producers' favourite Pinot region, but nearby Leyda, San Antonio and Chimbarongo and, way further south, Bío-Bío are staking a claim. Other parts of the coastal hills could also be good.

South Africa

South Africa is generally hotter for Pinot Noir than most competing New World wine regions, but there are some exciting coastal areas like Hemel-en-Aarde. The grape has also suffered from leafroll virus. The few producers who get it right make round, rich Pinots of considerable Burgundian stylishness.

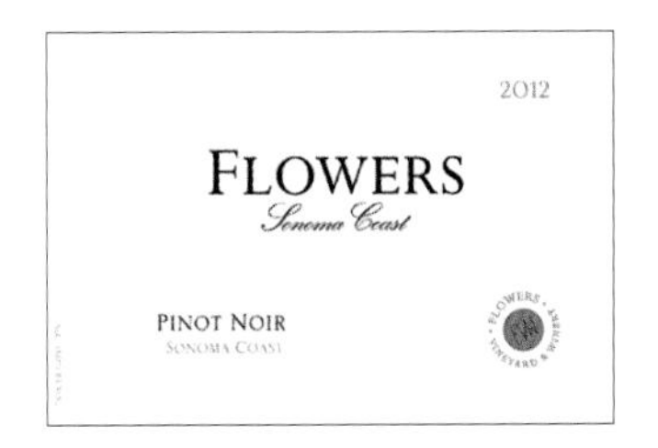

Flowers
One of the first producers to demonstrate that the scattered, remote vineyards of the rugged but cool Sonoma Coast just inland from the Pacific Ocean could produce California's most delicate, subtle, yet lingering Pinot Noirs.

Ten Minutes by Tractor
This single vineyard wine comes from the mould-breaking 10 Minutes by Tractor winery in Australia's cool but sunny Mornington Peninsula, where most of the top vineyards are now vinified separately.

Greywacke
The clay-rich soils of Marlborough's Southern Valleys are proving ideal to grow superb Pinot Noir. Kevin Judd at Greywacke makes a spicy, lush yet contemplative wine.

Viña Leyda
The rapid maturity achieved by Chile's Pinot Noir producers is demonstrated by Viña Leyda's production of single vineyard and single block wines in the cool, coastal Leyda region.

ENJOYING PINOT NOIR

How well does pinot noir age? It's a question that baffles many lovers of the grape. In Burgundy alone there are wines that will last up to a couple of decades and wines that seem hardly able to last the year. Admittedly, the two types do not usually come from the same producers, though sometimes they do come from the same communes.

Almost all Pinot Noir is delicious young; some is delicious mature. In between there is a period of adolescence, when the wine can be closed, even gawky, and can be as awkward at table as human adolescents. Watch out for that, and when it happens tuck your wine away for a couple of years: it is the only stage at which the grape does not give enormous pleasure. Otherwise you can catch its youthful fruit and perfume by opening it within months of bottling or you can wait until it is grown up and full of savoury spice.

Lesser vintages on the Côte d'Or should often be drunk within three to eight years: if a vintage is described as 'early drinking', then don't delay. Only the bigger, oakier wines are a little unfriendly in youth, as the oak hides the fruit.

Most Pinot Noirs from outside the Côte d'Or should be drunk young, within five years or so. The exceptions are the weightiest wines from regions like California's Santa Barbara, Sonoma Coast and Carneros regions, where they last longer: a decade should not be a problem for these.

Champagne is different again: Blancs de Noirs can be sturdy wines, though their aging ability depends partly on how much Pinot Meunier the blend contains. Most focus on weight and breadth over long ageability – and most are non-vintage. So keep them for another one to two years or so after you buy them, and then get the glasses out.

The taste of Pinot Noir

There's always something mouthwatering about good Pinot Noir; something that makes you want to go back to the glass again and again to pin down those elusive flavours because descriptions like strawberry, black cherry, game, leather, mushrooms, don't really tell you what it tastes like. Pinot is all these things, and none of them, sliding from one flavour to another. Words like 'complex', 'ethereal' or 'profound' may tell you even less about what it tastes like, but sometimes they are the only words that will do.

Simple Pinot wines are the easiest to describe. Less expensive fruit-first wines from the outlying parts of Burgundy, from Switzerland, from Carneros, from New Zealand, really do taste of strawberries. They are generally low in acidity, though the north of Burgundy can be an exception here, and low in tannin. Finer examples combine strawberries or black cherries with sensuous yet focused flavours, perhaps with a touch of incense and spice, sometimes with a pungent, gamy richness to balance the fruit.

Mature wines gain flavours of leather and woodsmoke, game and undergrowth, even a touch of rotting vegetables. Primary fruit is less important, though the wines should always taste fruity and perhaps slightly sweet. But labelling a mature Pinot with a particular type of fruit is usually quite impossible; every time you think you've caught the precise flavour, it's moved on to something else.

Robert Chevillon's highly rated Nuits-St-Georges domaine produces chewy, characterful wines from old vines. He owns vines in most of the commune's Premier Crus, and Les Vaucrains is east-facing and situated fairly high on the slopes (260–295m/850–970ft). The wines are quite austere but long-lasting. Felton Road's Pinot Noir is one of the most famous in the southern hemisphere, and comes from vineyards amid abandoned gold mines in Central Otago on New Zealand's South Island. It is proof that Pinot Noir can make great wine in places other than Burgundy.

MATCHING PINOT NOIR AND FOOD

The great grape of Burgundy has taken its food-friendly complexity all over the wine world. However, nothing can beat the marriage of great wine with sublime local food that is Burgundy's heritage, and it is Burgundian dishes that spring to mind as perfect partners for Pinot Noir: *coq au vin*, chicken with tarragon, rabbit with mustard, *jambon persillé*, *boeuf bourguignon*... the list is endless.

Pinot Noir's subtle flavours make it a natural choice for complex meat dishes but it is also excellent with plain grills and roasts, and with most dishes based on mushrooms. Richer examples are the ideal match for roast or casseroled game birds, and in its lighter manifestations from, say, the Loire or Oregon, Pinot Noir is a good match for salmon or salmon trout.

CONSUMER INFORMATION

Synonyms & local names

Numerous synonyms. Among others, Noirien in France, Savagnin Noir in Jura, Spätburgunder in Germany and Austria, Blauburgunder or Blauer Spätburgunder in Germany, Austria and Switzerland, Pinot Nero in Italy, Pinot Crni in Croatia and Serbia.

Best producers

FRANCE/Burgundy d'Angerville, Comte Armand, D Bachelet, G Barthod, A Bichot, J-M Boillot, Bouchard, Cathiard, Chandon Briailles, R Chevillon, Clair, J-J Confuron, Drouhin, C Dugat, Dugat-Py, Dujac, Faiveley, Grivot, Anne Gros, Jadot, Lafarge, Lafon, Lambrays, Dom. Leroy, Liger-Belair, H Lignier, Méo-Camuzet, de Montille, D Mortet, Mugneret-Gibourg, J-F Mugnier, Ponsot, Rion, Dom. de la Romanée-Conti, E Rouget, Roumier, Rousseau, de Vogüé, Vougeraie.
GERMANY Becker, Bercher, Fürst, Huber, Johner, Franz Keller, Kesseler, Knipser, Meyer-Näkel, Molitor, Rebholz, Schnaitmann, Stodden.
ITALY Ca' del Bosco, Franz Haas, Haderburg, Hofstätter, Nals-Margreid.
USA/California Ancien, Au Bon Climat, Calera, Dehlinger, Dutton Goldfield, Merry Edwards, Gary Farrell, Flowers, Kistler, Landmark, Littorai, Marcassin, Peter Michael, Morgan, Navarro, Patz & Hall, Rasmussen, J Rochioli, Saintsbury, Sanford, Sea Smoke, Siduri, Talley, Williams Selyem; **Oregon** Argyle, Beaux Frères, Bergström, Bethel Heights, Cristom, Domaine Drouhin, Domaine Serene, Ken Wright.
CANADA Blue Mountain, CedarCreek, Norman Hardie, Inniskillin, Mission Hill, Tawse.
AUSTRALIA Ashton Hills, Bannockburn, Bass Phillip, Bay of Fires, Bindi, By Farr, Castle Rock, Coldstream Hills, Cullen, Curly Flat, De Bortoli, Diamond Valley, William Downie, Freycinet, Gembrook Hill, Giaconda, Giant Steps, Grosset, Hurley, Knappstein, Kooyong, Stefano Lubiana, Marchand & Burch, Moorooduc, Oakridge, Paradigm Hill, Paringa, Stonier, Tamar Ridge, Tarrawarra, Ten Minutes by Tractor, Yabby Lake.
NEW ZEALAND Ata Rangi, Blind River, Cloudy Bay, Craggy Range, Dog Point, Dry River, Escarpment, Felton Road, Foxes Island, Fromm, Greywacke, Kusuda, Martinborough Vineyard, Neudorf, Palliser, Pegasus Bay, Peregrine, Quartz Reef, Rippon, Saint Clair, Schubert, Seresin, Te Kairanga, Valli, Vavasour.
SOUTH AFRICA Bouchard Finlayson, Chamonix, Paul Cluver, Crystallum, Hamilton Russell, Newton Johnson, The Winery of Good Hope.
CHILE Anakena, Casa Marin, Casas del Bosque, Cono Sur, Viña Leyda, Maycas del Limarí.

RECOMMENDED WINES TO TRY

Fifteen red Burgundies

Marquis d'Angerville Volnay Clos des Ducs
Albert Bichot Ch. Gris Nuits-St-Georges
Sylvain Cathiard Vosne-Romanée Malconsorts
Chandon de Briailles Corton
P Charlopin Marsannay En Montchevenoy
R Chevillon Nuits St-Georges Les St-Georges
Dugat-Py Charmes-Chambertin
Dujac Échézeaux
Faiveley Mercurey Domaine de la Croix Jacquelet
Vincent Girardin Maranges Clos des Loyères
Jean Grivot Vosne-Romanée les Beaux Monts
Henri et Paul Jacqueson Rully les Cloux
Louis Jadot Gevrey-Chambertin Clos St-Jacques
Denis Mortet Gevrey-Chambertin
Dom. de la Vougeraie Bonnes-Mares

Fifteen top New World Pinot Noirs

Ata Rangi Pinot Noir (New Zealand)
Bergström Temperance Hill (Oregon)
Bethel Heights Pinot Noir (Oregon)
Cloudy Bay Pinot Noir (New Zealand)
Domaine Drouhin Pinot Noir Laurène (Oregon)
Domaine Serene Pinot Noir (Oregon)
Escarpment Pinot Noir (New Zealand)
Felton Road Pinot Noir (New Zealand)
Giant Steps Pinot Noir (Australia)
Greywacke Pinot Noir (New Zealand)
Viña Leyda Cahuil (Chile)
Peter Michael Pinot Noir (California)
J Rochioli West Block Reserve Pinot Noir (California)
Te Kairanga Pinot Noir (New Zealand)
Yabby LakePinot Noir (Australia)

Five sparkling Blancs de Noirs

Ashton Hills Salmon Brut Vintage (Australia)
Edmond Barnaut Champagne Blanc de Noir Brut Non-vintage (France)
Billecart-Salmon Champagne Rosé Non-vintage (France)
Egly-Ouriet Champagne Blanc de Noirs Brut Vintage (France)
Schramsberg Blanc de Noirs Vintage (California)

Pinot Noir continues to tantalize growers worldwide. For many winemakers success with it represents the Holy Grail of winemaking, as it is one of the most challenging of all the international varieties.

Maturity charts

Most Pinot Noir from anywhere in the world other than the Côte d'Or follows the Wairarapa pattern of early drinkability, but that doesn't mean they don't age.

2010 Côte de Nuits red (Premier Cru)

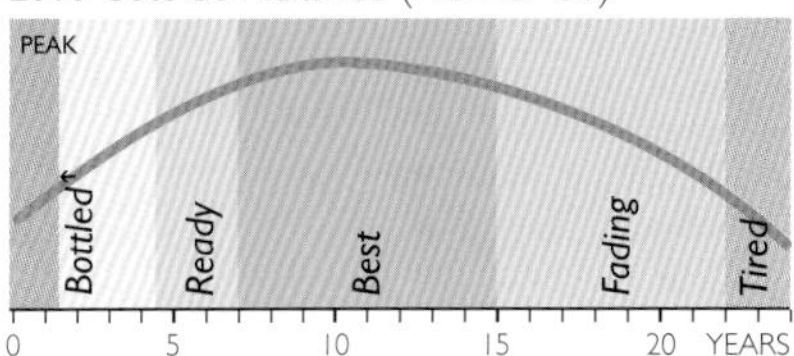

2010 was a small vintage, but the rather sunless summer produced wines of great intensity and character to drink young or, preferably, age.

2012 Top Santa Barbara Pinot Noir

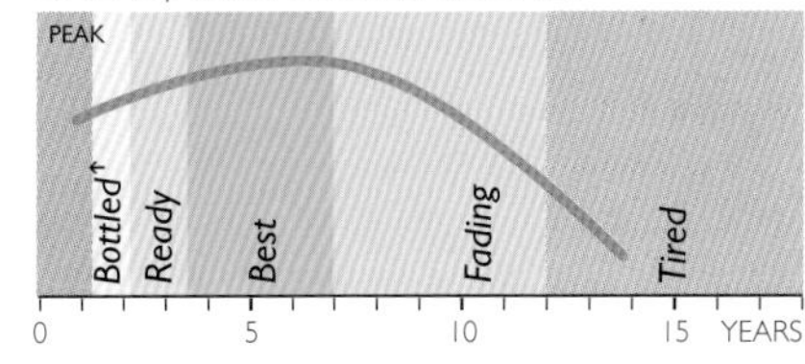

Santa Barbara Pinot Noir varies in longevity. Only the most concentrated wines will improve beyond five or six years; many should be drunk earlier.

2012 Top Wairarapa Pinot Noir

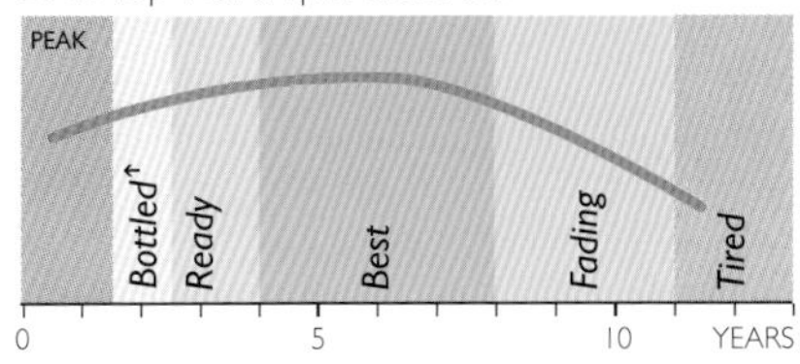

These are wines of good intensity and balance, drinkable early but great to age. Fruity and perfumed when young, the best have concentration and depth.

PINOTAGE

Dear, oh dear. I don't know any grape variety that arouses such fierce disagreement as Pinotage. It is potentially South Africa's national treasure – Professor Perold created it in 1925 at Stellenbosch University by crossing Pinot Noir and Cinsaut – and yet South Africans are some of its fiercest critics.

The South African wine establishment has long been obsessed with being as European as possible, even to the extent that they didn't like being described as 'New World' at all – 'We've been making wine since 1659' and all that stuff. And Pinotage is triumphantly different – capable of flavours no traditional European vine possesses. At its best it's a parade of deep, roaring flavours, oaked and unoaked, with an astonishing mixture of mulberry, damson and blackberry fruit, mixed with the November scents of marshmallows toasted over a Thanksgiving bonfire. That being said, it has proved a very difficult vine to manage. Low-yielding old bush vines can be excellent; and it needs very careful handling in the cellar to tame its wilder flavours and avoid intensive volatile esters. But nobody could agree on the best climate for it, far less the best soil.

Since 2000 quality has risen, as growers have become more focused on viticulture. They understand optimum ripeness much better now – this now goes for all grapes in South Africa. They reckon they can't get a decent Pinotage under 14 per cent alcohol, whereas it used to be 13 per cent, or even 12.5 per cent. And the new virus-free vines which went in about then, which give riper tannins, are starting to mature. They're getting better at handling oak, too. The American oak that often suits Pinotage is better quality these days, and better seasoned, though there is a vogue for mocha coffee-tasting oak, which creates – well, mocha coffee-tasting Pinotage.

At first Pinotage was trumpeted as offering the Holy Grail of Burgundy flavours from the Pinot Noir, and high yield and easy ripening from the Cinsaut (often called Hermitage in the Cape, hence the Pinotage name). In fact, it has never behaved or tasted remotely like Pinot Noir, or Cinsaut for that matter. But is it really the way forward for South Africa? Some say that it's best in blends – and not necessarily as the majority of the blend, either. Its least threatening incarnation may be as rosé. But bright and juicy and red is how I like it.

One or two other countries have had a go – New Zealand, Zimbabwe, Brazil, Canada, the USA, Australia, Israel and even Germany. Only New Zealand has had much luck so far.

THE TASTE OF PINOTAGE

Good Pinotage tastes and smells like no other wine – wonderful mulberry, blackberry and damson fruit, a flicker of lava and the deep, midwinter flavour of marshmallows toasted in front of the fire. There are two problems with Pinotage. One is that it is difficult to get the full flavours and yet control its rather aggressive tannins. And, secondly, those genuinely individualistic and exciting flavours do tread a knife edge and if the winemaking isn't really good, you can find a wine reeking of the spiritiness of gloss paint, the sweetness of candyfloss, and a volatile acidity tottering towards raspberry vinegar. But it also makes some delicious rosés, all flavoursome strawberry fruit and just the right weight.

Kanonkop has established itself as South Africa's leading exponent of Pinotage, resolutely sticking with it and refining it during long periods of unpopularity, particularly during its most unfashionable period in the late 1970s and 1980s. It has been estimated that some 60–70% of South Africa's Pinotage vines were uprooted during this period – including, inevitably, the sort of low-yielding, old bush vines that make the best wine. But some survived and unirrigated old vines are the key to this wine's concentration and depth.

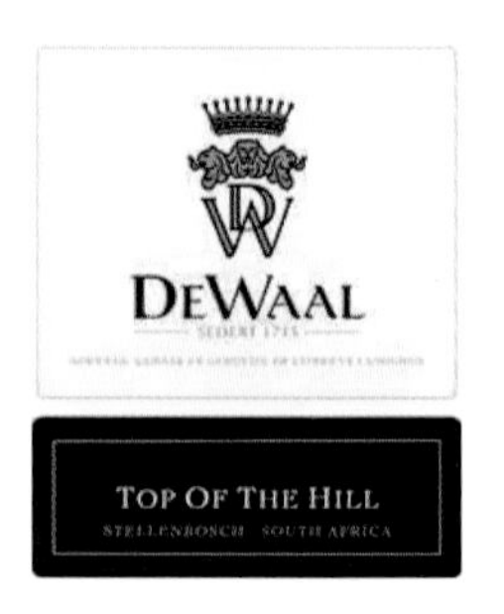

De Waal
CT de Waal was the first person ever to make a wine out of Pinotage – one barrel in 1941. The Top of the Hill block contains South Africa's oldest Pinotage, planted in 1950.

Spice Route
Charles Back of Fairview founded Spice Route winery in the 1990s to utilize the old vine fruit available in the Swartland region, in particular bush vine Pinotage.

Left: Picking Pinotage on the Warwick Estate in the Stellenbosch district. The grape is early-ripening, beating Cabernet Sauvignon to the winery by about a fortnight. ***Above:*** *Pinotage is high yielding – often too high for top quality. Many South African winemakers have yet to be convinced by it, and prefer Cabernet Sauvignon. But there is a Pinotage Association in South Africa, which is dedicated to improving quality, and new clones, without the viruses that plagued the vine in the past, produce much better flavours. It is higher in malic acid than other varieties, and vigorous conversion of malic acid to lactic acid can result in unbalanced and infection-prone wine. Picking ripe while still keeping the pH low, seems to be important to the success of the wine.*

CONSUMER INFORMATION

Synonyms & local names

The grape is known everywhere as Pinotage.

Best producers

SOUTH AFRICA Ashanti, Ashbourne, Backsberg, Graham Beck, Bellingham, Beyerskloof, Boplaas, Chamonix, Clos Malverne, DeWaal, Diemersfontein, Fairview, Grangehurst, Groot Constantia, Jacobsdal, Kaapzicht, Kanonkop, Laibach, L'Avenir, Longridge, Morgenhof, Newton Johnson, Robertson, Saxenburg, Simonsig, Southern Right/Hamilton Russell, Spice Route, Stellenzicht, Stony Brook, Swartland, Tukulu, Warwick, Wildekrans; **AUSTRALIA** Thick as Thieves, Topper's Mountain; **NEW ZEALAND** Muddy Water, Te Awa.

RECOMMENDED WINES TO TRY

Fifteen full-bodied Pinotage wines

Graham Beck The Old Road Pinotage
Claridge Trafalgar Bush Vine Pinotage
Clos Malverne Reserve Pinotage
DeWaal Top of the Hill Pinotage
Kaapzicht Steytler Pinotage
Kanonkop Estate Pinotage
L'Avenir Pinotage
Laibach Pinotage
Longridge Pinotage
Morgenhof Pinotage
Neethlingshof Lord Neethling Pinotage
Saxenburg Pinotage
Spice Route Flagship Pinotage
Stellenzicht Pinotage
Warwick Estate Old Bush Vine Pinotage

Ten lighter-style Pinotage wines

Ashanti Estate Pinotage
Avontuur Pinotage
Bellingham Pinotage
Beyerskloof Pinotage
Boplaas Pinotage
Fairview Pinotage
Groot Constantia Pinotage
Robertson Winery Pinotage
Simonsig Estate Pinotage
Swartland Winery Pinotage

Topper's Mountain

Co-fermenting Viognier and Shiraz is quite common but I'd never seen a Pinotage and Viognier label before (though some South African producers do a bit of co-fermenting). This wine comes from 900m (3000ft) high vineyards in the New England Tablelands of northern New South Wales, Australia.

PLAVAC MALI

This grape, found on the Dalmatian coast and on the Adriatic islands off Croatia, has close links with Zinfandel/Tribidrag, though the two varieties are not identical, as was once thought. Its similarities did, however, provide clues as to the origins of Zinfandel (see pages 300–309 and Primitivo, right). It may also be the case that several strains of Plavac Mali exist. Grown in torrid conditions, it achieves high alcohol (up to 17%) but is prone to low acidity and consequently, despite being able to age, a lot of bottles develop a rather fearsome rustic style. It is usually blended with other grapes, but is the only variety in such wines as Dingač and Postup. There are a number of varieties in the region called Plavac 'Something-or-other'; 'mali' means 'small' in Serbo-Croat.

PORTUGIESER

Common in both Germany and Austria for reds that are often too light, soft and sweetish to be of any interest except to the most red wine-deprived local. In Germany it is much planted in the Pfalz and Ahr regions. There's quite a bit in Austria where it goes by the name of Blauer Portugieser (see page 48).

In France it becomes called Portugais Bleu – Jura has a little; in Hungary and Romania it used to be called Kékoporto, which means the same thing, and in Croatia Portugizac Crni or Portugalkja. Another name is Oporto. In spite of all this determined association with Portugal, there is no proof that it came from that country, and many authorities reckon its origins are Austrian. Indeed, most Portuguese red varieties have considerably more character than Portugieser, so it's a bit of an affront. There is a pink (Grauer) Portugieser as well, and a green (i.e. white) one, which seem to be all part of the same family, not that anyone cares that much.

POULSARD

A fascinating Jura grape, early budding and early ripening and hence at danger from spring frosts, this gives pale, aromatic wines of great elegance. It's usually blended with Trousseau or Pinot Noir or used for rosé and sometimes labelled as Ploussard. Best producers: (France) Jean Bourdy, Jacques Puffeney.

PRESSAC

Malbec, usually known as Cot in Bordeaux, becomes known as Pressac when it is grown on Bordeaux's Right Bank (in the appellations of St-Émilion, Pomerol and Fronsac). The name is used to some extent in the nearby appellations of Bourg, Blaye and Entre-Deux-Mers, but the variety is declining in popularity. See Malbec pages 130–131.

PRETO MARTINHO

Portuguese grape found in the Tejo and Lisboa regions. It ripens early and gives deep-coloured wines with plenty of alcohol.

PRIÉ

A speciality (albeit on a small scale) of Italy's Valle d'Aosta, making both still and sparkling wine. The style is lean, clean and fresh. Best producer: (Italy) Vini Estremi.

PRIETO PICUDO

An interesting grape found in León in Spain. It has deep colour and an affinity for oak: it doesn't go solidly tannic in new oak, which Tempranillo can. It has slightly musky but not overt fruit. It now appears in several DO wines, but ought to be more cultivated since León red wines could do with beefing up. Best producer: (Spain) Bodegas de León-Ville.

PRIMITIVO

Fashionable southern Italian grape making big, burly, alcoholic wines and previously used mostly for blending. DNA fingerprinting has established that Primitivo is the same as California's flagship variety, Zinfandel (see pages 300–309), but more importantly Primitivo is the same as a very ancient Croatian vine called Crljenak Kasteljanski or Tribidrag, almost extinct and found only in a couple of minute Dalmatian plantings until its discovery in 2001 and 2002 by researchers desperate to pin down the true identity of Zinfandel.

Primitivo grows in Puglia, and if you know that map it's not far across the Adriatic to Croatia's Dalmatian coast. Indeed, a fair number of Dalmatians have settled in Puglia over the centuries, and some Puglian dialects sound more Dalmatian than Italian. So it's now pretty much accepted that Primitivo came to Italy from Dalmatia (see pages 300–309). It languished in obscurity in Puglia for many years, contributing colour and strength to blends and having a couple of DOCs of its own (Primitivo di Manduria is the main one), but seldom being seen outside the country. The popularity of Zinfandel has changed all that, and now Primitivo is popping up all over the place, including Australia and Chile, but it's usually called Zinfandel.

Felline

Felline has vines dating back to the 1940s on shallow red soil on top of a calcareous bedrock. These wines are rich and dense and aged in French and American oak.

PROCANICO

Synonym for Trebbiano (see page 284) grown in Umbria in central Italy. It has rather more character than most other Trebbiano sub-varieties.

PROSECCO

See Glera page 122.

PRUGNOLO GENTILE

One of Sangiovese's many Tuscan synonyms (see pages 222–231). This one is used in the Vino Nobile di Montepulciano zone.

PX

The usual abbreviation for Pedro Ximénez (see page 180).

RABIGATO

The name means 'Cat's Tail' ('Rabo di Gato') and other synonyms are Rabo de Asno (rabbit's) and Rabo de Carneiro (sheep's): all refer to its long clusters. It's a Douro grape used for table wines (and less so for Port) because of its citrus, orange blossom, mineral flavour and its ability to retain acidity. Its thin skin makes it susceptible to mildew and botrytis.

RABO DE OVELHA

From a cat to a ewe. The Ewe's Tail grape is unrelated to Rabigato but is often mistaken for it. It probably originated in Portugal's Alentejo, and gives large crops of large, thick-skinned grapes. The wine is light, pretty, won't keep long and is usually blended.

RABOSO

A highly tannic, highly acidic red grape found in northeastern Italy, where its austere wine lacks the charm of other acidic reds of the region such as Marzemino (see page 134). However, it makes a reasonable, rasping accompaniment

The grey skies and windbreaks may be horribly reminiscent of English beach holidays, but these are Ramisco vines for Colares wine grown on the coastal sand dunes at Azenhas do Mar, Portugal.

to pasta dishes. The grape has deep colour, not much sugar (so not much alcohol to add fat), and pretty lean fruit.

There are two unrelated vines called Raboso: one is Raboso Piave, which is also known as Raboso Friulara or Friularo, and the other is Raboso Veronese, which seems to be Raboso Piave's offspring. They take their name from the Raboso river, which is a tributary of the Piave. Both varieties are planted less than they once were. It is also found occasionally in Argentina where it may add a certain zip to some of their juicy, warm-climate red wines. Best producers: (Italy) Borletti, Cecchetto, Ivan Cescon.

RAMISCO

An oddity in today's wine world, Ramisco is the grape of Colares, on Portugal's Atlantic coast, where it grows in the sand dunes. Since this sand has never been infected with phylloxera, these Ramisco vines are ungrafted. The grapes have thick skins and colossal tannins and need time to soften. In time, good blackcurrant flavours are revealed but the tannis still prevail. But the Colares vineyard is shrinking and Ramisco has yet to catch the eye of winemakers from other regions. It would be nice if someone from somewhere else had a go at taming it. Australians, Chileans, South Africans . . . ? Anyone listening? Best producers: (Portugal) Adegas Beira Mar, Tavares & Rodrigues.

REFOSCO

A northeastern Italian red grape that gives highly acidic, deeply coloured wines often with somewhat rough, green tannins, but attractive black and blueberry fruit, and sometimes slightly grassy aromas. It is best young and unoaked but can develop good dark plum depth and black chocolate bitterness with a couple of years' aging. The grapes have good resistance to rot, but they ripen late, which is one reason for the high acidity and greenness. Despite all this talk of acidity and greenness, I'm quite fond of Refosco in its angular way.

There are various sorts of Refosco, and naturally, they're not related. The main one is Refosco dal Peduncolo Rosso, found mostly in Friuli-Venezia Giulia and in Slovenia, though anything called Refosk is Slovenia is probably Terrano, alias Refosco d'Istria. Friuli also grows some Refosco di Faedis, which has a bit of extra tannin and acidity.

REGENT

A 1967 crossing (Diana x Chambourcin; Diana is Silvaner x Müller-Thurgau) found especially in Germany but also in Switzerland and occasionally in England, Belgium and Sweden. It's extremely resistant to cold, as you might infer, and seems to be almost bomb-proof: downy and powdery mildews and botrytis can't do much to it, either. It makes full, supple, cherry-flavoured reds of some charm.

REICHENSTEINER

A modern (1939) German crossing of no great quality, Reichensteiner has Müller-Thurgau (see page 154) for one parent and a crossing of Madeleine Angevine (see page 129) and Early Calabrese, which sounds like a cabbage but is actually an Italian grape, for the other. It is resistant to rot and reaches high sugar levels but has dull flavours which, though not exactly resembling a Calabrese cabbage, have swathed themselves in the neutrality of the Müller-Thurgau, though with even less aroma. Most of Germany's Reichensteiner is in the Rheinhessen; there is also some in New Zealand and some in England – none of it is memorable. Best producers: (England) Chapel Down, Denbies, St George's, Stanlake Park.

RHEIN RIESLING

A synonym for Riesling (see pages 204–215). An alternative spelling is Rhine Riesling.

RHODITIS

An alternative spelling of Roditis (see page 217), a pink-skinned grape grown pretty much all over Greece.

RIBOLLA GIALLA

An attractive Friulian grape from northeast Italy that combines good acidity with a certain nutty flavour but lacks much aroma or fullness. Quality is generally good, although there is not much planted. It grows on hillsides in the Collio and Colli Orientali zones, as well as in Slovenia across the border, where it is called Rebula. Ribolla in Greece is unrelated, and it can give more lemony, flinty wines. Best producers: (Italy) Primosic, Matjaz Tercic, Le Vigne di Zamò.

RIBOLLA NERA

A synonym for Schioppettino (see page 243).

RIESLANER

A 1921 crossing of Silvaner and Riesling. It is mostly grown in Franken in Germany, and is actually one of the better modern crosses, although it's a bugger to grow. One massive plus is that it can make brilliant sweet wines when affected by botrytis, but since it finds it difficult to both rot and keep a crop, and anyway ripens very late and holds on to a fairly nippy acidity, plantings are not increasing.

RIESLING

How many years have I sat and listened to wine experts passionately promoting the claims of Riesling to be the world's greatest wine grape? Whatever that means. And for how many years have I thought, well, yes, up to a point, maybe, but, but . . . ? And how often have I thought OK, I'll replace a New Zealand Sauvignon with a Clare Valley Riesling in some public tasting I'm holding, or replace a Chablis with a dry Riesling from the Wachau in Austria or the Rhine in Germany – and then extol the wine's virtues, but I can see I'm not really getting through to half my audience. Oh, they're perfectly interested in the Riesling, but they'd rather be tasting a Sauvignon or a Chablis, and when the tasting is over and they charge off to get a glass of something to drink – it's far more likely that they'll be demanding the Sauvignon than the Riesling.

So does that mean the Sauvignon is a better grape than the Riesling? No. No more than a bunch of wine experts saying that the Riesling is far superior to the Sauvignon is necessarily valid. They grow in different places, favour different climatic conditions, and set out to produce different styles of wines. But actually, both can do a fair range, from intensely dry to intensely sweet. The Riesling is better at off-dry to fairly sweet styles, but then the Sauvignon is much happier at being fermented in a normal-sized oak barrel than the Riesling is.

On the banks of the River Mosel stands the town of Bernkastel, directly behind which climb steep Riesling vineyards, just like the ridged green stem of the traditional Mosel wine glass. The intricate ironwork shop signs typical of the Mosel region were surely originally inspired by coiling vine tendrils.

Riesling at its best is about purity, about a limpidity through which it may well be possible to almost taste the stones and the soil of the vineyard. This experience is very exciting, and would be wrecked by an attempt to ferment or age the wine in new oak. Which has meant that some of the enthusiasm for Riesling is engendered by those who don't like the flavour of new oak in their wines. Fair enough. The few new-oaked Rieslings I've had have been pretty weird and completely unrecognizable as Riesling, so I'll certainly support the view that new oak and Riesling is an abomination – all that tingling acidity and orchard-fresh fruit is lost. But the flavour of new oak on many grape varieties is not an abomination – it's a delight. Chardonnay, Sémillon, Chenin Blanc, Verdelho, Viura and many others take to oak-aging with gay abandon and generally excellent results. And the world loves them for it.

And here's another thing. Did you notice that word in the last paragraph? Chardonnay? Hmm. In some people's eyes, Chardonnay is the great Satan; the ruthless colonizer and destroyer of the world's vineyards and the world's palates. Riesling lovers, in particular, bridle at the runaway success of the easy-going, crowd-pleaser Chardonnay, and chunter among themselves that Chardonnay is a little slip of a thing, a flibbertygibbet with no depth and no complexity.

Well just as lovers of great art, great music, great literature don't expect the whole world to share their taste – in fact, they'd probably be mortified it did – it's probably time for Riesling lovers to stop objecting to the Andrew Lloyd Webbers and the John Grishams of the wine world and to become more smug in their Riesling devotion, to accept that Riesling isn't a flavour that everyone likes, to boast that it doesn't take kindly to dumbing down – as Chardonnay does. They should quietly be proud of the fact that in a wine world that spent the best part of a generation worshipping at the feet of rich, broad, super-ripe alcoholic wines, but which has, at least in white wines, begun to pine for something more elegant and restrained – they should be proud that Riesling has always stood for elegance, with its mouthwatering acidity, its icy cool perfume, and its array of mineral and citrous flavours.

And if the world suddenly wants more wines of a Riesling character, it won't be easy to provide them. Riesling has always been very choosy about where it can grow successfully. Wine companies have tried to plant it in easy conditions, to overcrop it and make it in industrial volumes. It hasn't worked. Even now, most of the Riesling grown in the world is dull and bland, resolutely refusing to be made into a crowd pleaser. Frankly, there's only just enough top-quality wine available for those of us who crave it as it is.

Riesling: From Grape to Glass
Geography and History page 206; Viticulture and Vinification page 208; Riesling around the World page 210; Enjoying Riesling page 214

GEOGRAPHY AND HISTORY

In the final decades of the 20th century Riesling suffered a fall from grace of epic proportions. One hundred years earlier the greatest German Rieslings had commanded higher prices than even the finest red Bordeaux; now just a glimpse of a German label was enough to deter many consumers even from taking the bottle from the shelf. Germany had, by the mass production of ever cheaper, ever poorer-quality wines, succeeded in getting the full force of wine snobbery turned on it. Riesling, Germany's finest grape was, in fact, an innocent victim, because the dismal parade of flaccid Liebfraumilchs and Niersteiner Gutes Domtals rarely had a single Riesling berry in them. Merely to be German was, in some markets, enough to damn the grape.

Riesling (and especially German Riesling) has, however, benefited from its years in the wilderness. In Germany Riesling has got drier, been through a painful stage of skeletal thinness, and emerged better balanced than before. The same is true of Austria.

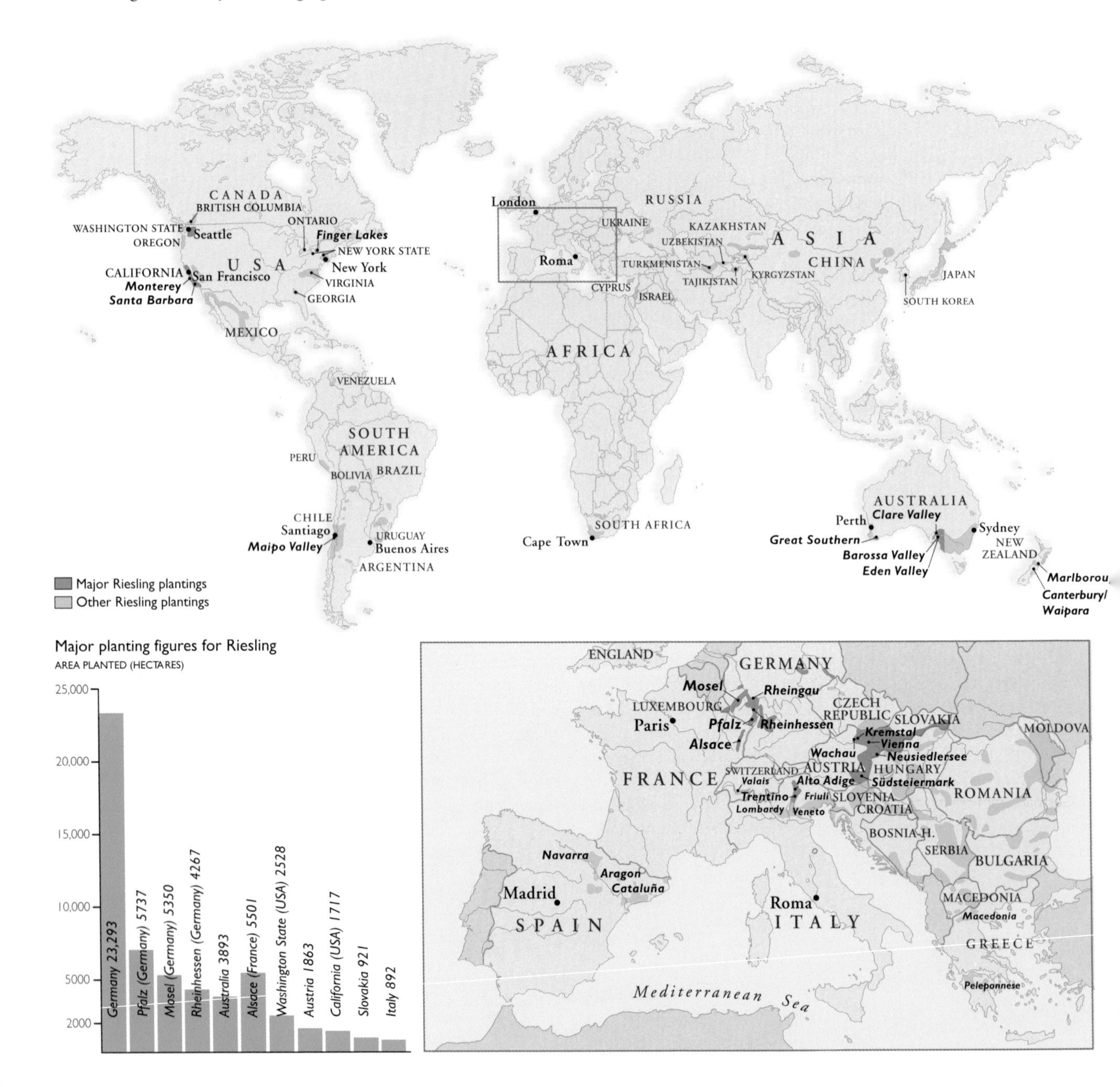

In Australia the profound wines of the 1950s and 1960s gave way to lighter, less interesting ones as attention shifted to Chardonnay and winemakers ceased to regard Riesling as a serious grape variety. That has changed again, and the Rieslings of the Eden and Clare Valleys rank as benchmark styles of the grape, along with those of Germany and Alsace.

The third benchmark style, that of Alsace, has both suffered and benefited from a lack of mass consumer interest. Styles have changed less here, which is great when it comes to those committed winemakers determined to produce fine wine. But cooperatives and merchant houses dominate proceedings here and too many of their wines are lean, lemony – and not all that cheap. They should take note of the competition.

Elsewhere in the world, Riesling has not created many waves – largely because so many New World leaders are basically hot countries with conditions too warm for successful Riesling. But Australia has worked out how to do it, so expect exciting things from other New World producers. To date, New Zealand, Washington, New York, Canada and Chile have produced some delightful, scented citrus styles.

HISTORICAL BACKGROUND

Riesling is probably the offspring of the Weisser Heunisch vine, which was widely grown in Germany in the Middle Ages, and which is also, remarkably, the parent of 75 other German vines as well as, under its French name Gouais Blanc, Chardonnay. A vine called 'Russelinge' is marked on an Alsace map of 1348, but more certain is a Rheingau invoice of 1435, when the grape is spelled 'riesslingen' – the modern spelling doesn't appear until 1552, in Hieronymus Bock's Latin herbal: it grew in the Mosel, the Rhine and around Worms. There is also a Rote, or Red Riesling, a pink-berried version of Riesling; some has recently been planted in Hessiche Bergstrasse and Rheingau.

Riesling's quality seems never to have been in dispute. But it was a luxury variety, grown by those who could afford its relatively low yields – mostly the Church and the aristocracy. Until the mid-18th century half the crop was taken in tax, making the high-yielding Elbling usually preferable; only after this tax was removed did Riesling begin its takeover of the Mosel. However, it had already begun to be planted as the sole variety in some vineyards – this was in itself a novel idea – starting with Schloss Johannisberg in the Rheingau in 1720–21.

In Alsace it was likewise admired but not widely grown. From 1870 to 1919, under German rule, Alsace was used for bulk wine and higher-yielding varieties were preferred, and Riesling plantings began to increase again only after 1919 with Alsace's return to France.

German immigrants took it to the New World. By the 1850s it was established in South Australia. Riesling made many of Australia's finest whites before being eclipsed by the Chardonnay surge of the 1980s. Now it's triumphantly back. By 1857 it was established in California and by 1871 in Washington State.

Autumnal Riesling vines in the Goldtröpfchen vineyard above Piesport in the Mosel Valley. Goldtröpfchen is a suntrap: the Mosel does one of its U-turns here, and provides a broad, sweeping slope where Riesling can ripen. However, the north-facing vineyards across the river have little hope of ripening their fruit and traditionally were usually planted to other crops.

Egon Müller makes stupendous wines from his Scharzhofberg vineyard, including some of Germany's greatest classic sweet wines. Here Egon is checking a bunch of shrivelled, nobly rotted Riesling.

The Nahe's reputation is as a producer of wines balancing the racy freshness of the Mosel with the grapier richness and scent of the Rheingau. These are the vineyards of Oberhausen and Schlossböckelheim.

VITICULTURE AND VINIFICATION

Despite being geographically diverse, Riesling is genetically rather stable. It does not mutate with the ease of the varieties in the Pinot family, and while there are some 60 different clones commercially available to growers in Germany (including the startlingly aromatic N90, grown by some innovative growers in the Pfalz and elsewhere) most of them are not hugely different from each other. The choice of clone is seldom a major factor in determining the style of the wine where Riesling is concerned.

Riesling wines can, however, exhibit a remarkable number of different flavours, from smoke to peach, from earthiness to petrol, from slate to spice to dried apricot. It reflects its vineyard more transparently than almost any other grape.

Terroir

The ability to translate the vineyard into the glass through the medium of winemaker and vine is what makes Riesling so endlessly fascinating to wine lovers. Growers in the Mosel region of Germany speak of the minerally flavour of wines from the village of Wehlen; of the blackcurranty note that comes from Piesport clay, of the steeliness of Traben-Trarbach's blue slate. In the Nahe region, wines from the Traiser Bastei vineyard at the foot of the Rotenfels cliff can, in good years, express the fieriness of Riesling grown in red slate. In Alsace, where the whole Grand Cru system of 50 top vineyards means anything only if there is a difference imparted to the wine by different *terroirs*, the soils which Riesling prefers are principally sandy clay and loam. However, it seems to be able to adapt to most soil types, providing they are well drained and offer the sheltered, sunny position it likes.

Harvesting Riesling in the Würzgarten vineyard at Urzig in the Mosel Valley. Riesling in the Mosel is trained in a particular and labour-intensive way, with one stake per vine. The stakes must be embedded in the solid rock beneath, itself no easy task. But then, standing without falling over the cliff isn't easy either.

To reflect its vineyard site to greatest advantage, and to produce wine with the best balance, the Riesling needs two things: a long, slow ripening season, and low yields.

Climate

Riesling is normally considered an early-ripening variety: only in Germany, where crossings like Müller-Thurgau can beat it with even greater speed, is it considered late-ripening. In a warm climate it ripens too early and quickly to have much interest: it acquires none of its characteristic complexity and tastes rather dull and flat. The vine's hard wood makes it highly resistant to winter cold, which makes it useful in New York State and Canada, as well as in the chillier parts of Germany, and it buds late, which gives it some resistance to late frosts. Its frost resistance means that winter pruning can begin early.

In the Mosel Riesling is nevertheless at its northern limit of cultivation in Europe, and must, if it is to ripen, have the sunniest, most sheltered sites; the angle to the sun of the steep vineyards is crucial if the vines are to benefit from every possible ray. In the Mosel Riesling can find itself too enveloped in mists coming off the river if it is planted at the foot of the slopes, and too cold if it is planted above about 200m (660ft). In South Australia the Clare and Eden Valleys, while they are on the face of it considerably warmer than the Mosel (Shiraz flourishes here, as well as Riesling), have such convoluted topography that many variations of soil, altitude and exposure are possible. Rainfall is low and nights are cool, and altitudes can rise as high as 400m (1320ft).

Yields

In Germany Riesling is considered low-yielding, simply because it appears so beside such prolific, cash-cow grapes as Müller-Thurgau. The latter can produce 300hl/ha; the turning point at which Riesling on the Mosel begins to produce much lower quality is somewhere between 120 and 150hl/ha. By most standards that is extremely high, and while Riesling yields are certainly higher in the Mosel than, say, in the Rheingau, top growers will seldom admit to making more than 50–70 hl/ha. There seems to be little difference analytically in the wine from high and low yields but the taste and texture are dramatically different.

Elsewhere, up to 60–70hl/ha is not uncommon, though winemakers in warmer climates may press less juice out of the grapes to avoid coarseness in the wine. Warmer climates give thicker skins, which can impart bitter-tasting tannins and phenols to the juice: Riesling from South Australia may have skins up to seven times as thick as Riesling from the Rheingau.

Riesling's compact clusters of small berries make it susceptible to *coulure* and grey rot, both of which are exacerbated by cold, wet, cloudy weather – and both of which can reduce yields.

Winemaking

The main discussion here is over the relative merits of wood and stainless steel. Stainless steel is favoured, both for fermentation and for aging, by those who seek a fresh, youthful briskness in their Riesling. When fermented and stored in steel Riesling has crystal clarity of fruit and can have an almost antiseptic cleanliness. When large, old wooden barrels are used the flavours are quite obviously different. The gentle oxygenation that occurs in old wooden barrels softens the edges of the wine and adds complexity though perhaps reducing perfume. The choice is a stylistic one, with the steely style generally being seen as more modern.

New oak versus old

Here there is less argument. A few producers use new oak barrels for fermenting and/or aging the wine, but only a minority. Riesling's acidity and floral perfume just don't go with new oak, and far from gaining complexity from the association, its beauty is smothered.

The only new-oak-aged Rieslings that are at all successful are those with plenty of weight. These tend to be from the Pfalz or from Baden, where warmer temperatures give greater alcohol and extract, and where the wines are likely to be fermented to dryness. (The more flowery style of Riesling made elsewhere in Germany loses everything in new oak.) In theory, Alsace Rieslings also fit the dry, weighty template, but while there have been a few experiments, Alsace growers do not generally favour new oak for any of their wines. Experiments with new oak elsewhere in the world tend to prove them right.

Varietal or blend?

Blended Rieslings are rare – not least because the tradition in both Germany and Alsace is of varietal wines. It is also true that Riesling is one of the few grape varieties that are complete in themselves, and need no improvement from others – to add another grape variety to Riesling even in small quantities means losing aroma, complexity and finesse.

Having said that, it can on occasion blend remarkably well, particularly with Pinot Blanc, for wines that combine finesse with weight. And it is in itself useful as a blender. Because of its high acidity, early picked Riesling can massively improve the balance of other grapes grown in warm conditions – Gewürztraminer or Muscat perhaps, or Semillon in the Barossa. It is also suitable for use as a sparkling wine, but has fallen out of favour partly because of the dominance of the Champagne blend – Chardonnay and Pinot Noir – in fizz production, but also because of the poor quality of much German Riesling-based Sekt. Carefully made from rather underripe, sharp fruit, sparkling Riesling can be delicious.

Harvesting Riesling for Icewine at Henry of Pelham estate in Ontario. Climate changes in Canada over the last decade mean that this harvest takes place about a month later than it used to.

RIESLING AS A SWEET WINE

This is one of the rare grapes that will make sweet wines which dazzle just as much as its dry ones. The reason is the grape's piercing acidity, which balances even the most intense sweetness so that a wine with 50 grams (1¾oz) of residual sugar can taste light, refreshing and even delicate.

Rieslings may be made off-dry or sweet by simply stopping the fermentation (by adding sulphur dioxide, centrifuging the wine or chilling it) or (where the law allows) by adding Süssreserve – sweet unfermented grape juice. The method chosen affects the flavour: the sugar in a ripe grape is composed of roughly equal amounts of glucose and fructose. The glucose ferments before the fructose, so if you stop the fermentation the residual sweetness is fructose – and fructose tastes more fruity and refreshing than glucose. If you ferment the wine to dryness and then add Süssreserve, you don't gain that extra dimension.

Fine sweet Rieslings are made with the help of *Botrytis cinerea*, or noble rot, the fungus that shrivels the grapes and concentrates their sweetness and acidity. Botrytis-affected wines usually come from Germany, Austria, Alsace, New Zealand and California. For German growers they are a flagship, though not always a profitable one. An estimate from the Rheingau puts the cost of producing a bottle of Trockenbeerenauslese (TBA) at €255 – around €50 more than its selling price. Another estimate is that it takes one picker one day to pick the grapes for a single bottle of Beerenauslese, and one picker one day to pick the grapes for half a bottle of Trockenbeerenauslese.

Yet another sweet style is possible where Riesling thrives in cold climates. For Eiswein/Icewine the grapes are picked frozen solid, when the night-time temperature plunges to –6°C (21°F) or lower. These low temperatures may not occur until January, but the wine nevertheless bears the vintage date of the previous year. When the grapes are pressed the water is left behind in the form of ice, and intensely sweet juice runs (very slowly) from the press. Some growers like to have some botrytized grapes in their Eiswein, while others prefer the purer flavours that come from having no botrytis-affected grapes.

Canada produces more Icewine than any other country – even more than Germany. By the late 1990s it made some 50,000 cases a year and brought Canada international attention. Oregon, Michigan and Luxembourg also make some.

RIESLING AROUND THE WORLD

Riesling producers around the world have two Old World prototypes to copy (or, alternatively, to rebel against). There is the flowery Germanic style, or the weightier, winier example from Alsace. So far only Australia has succeeded in establishing a home-grown style of equal stature, and its wines are more like those of Alsace than those of Germany. The Germanic style appears in watered-down versions elsewhere – showing, perhaps, how difficult it is to get right.

Germany

Rieslings from Germany depend on acidity as taut as piano wire, yet they can be as seemingly delicate as a butterfly. The fruit can be intense yet ethereal, the residual sugar (if there is any) must be integrated and honeyed, and if the alcohol is low, the extract must be sufficient to give the wine balance.

That, admittedly, is not the pattern all over Germany. In the South – in the Pfalz and Baden – the wines are drier and weightier, with more substance. But they should still have that taut, knife-edge balance. The further north you go, the more apparently ethereal the wines become. Wines from the Rheingau and Nahe have more weight than those from the Ahr or the Mosel; and within this latter region, wines from the Saar and, especially, the Ruwer, seem ever more delicate. The paradox of German Rieslings is that, like a consumptive operatic heroine, they appear to be ready to collapse in the first scene, yet are in fact quite capable of lasting the full three Acts and singing half a dozen physically demanding arias into the bargain. These frail-seeming Rieslings are fit, lithe athletes – and potentially some of the longest-lived wines in the world.

Concentration and yields

To fulfil their potential as the wine world's marathon winners, Rieslings need concentration. In the Rheingau the average yield is around 100–140hl/ha on the flatter land nearer the river, and perhaps 50–65hl/ha on the less fertile slopes. On a good soil, with good vineyard management, it is probably possible to raise the yield to 85–90hl/ha without quality problems – but that is in an ideal year, with sufficient sun and rain. The Rheingau doesn't always get enough sun. In the Mosel yields are generally higher, in spite of the exceptionally poor soil: 180hl/ha is not uncommon, and yields of this magnitude do not produce concentrated Riesling. Quality-conscious growers do not allow their vines to yield as generously as this: most quote an average yield of 50–70hl/ha, with the figure going down to 35hl/ha for very old vines. (The Rheingau equivalent might be an average yield of 45–50hl/ha from a top grower.)

There are few people with more energy and more desire to communicate their passion about wine than Ernie Loosen. He produces stunningly pure Riesling from Bernkastel, Graach, Wehlen, Ürzig and Erden.

More strictly selected wines like Auslese (made from selected clusters), Beerenauslese (from selected berries) and Trockenbeerenauslese (from selected berries that have been shrivelled by botrytis) are lower again.

Why should yields in the Mosel, where the soil in the best vineyards consists of nothing but flat shards of slate balanced on a steep hill, be so much higher? One reason is that Mosel Riesling sells for less in Germany than its counterpart from, say, the Pfalz, so growers are not prepared to sacrifice any quantity. Yet the Mosel demands greater effort: 1ha (2.5 acres) of vineyard in the Mosel requires 1200 hours of work per year. In the Pfalz, 800 hours is average. In the Saar and Ruwer, where the climate is even less promising than that of the Mosel, yields are lower. The regions are close together, but even half a degree centigrade can be the difference between losing part of your crop to frost and escaping, or between reaching, or failing to reach, an acceptable ripeness in your grapes.

Acidity

Acidity is the key to German Riesling. It is also the key to understanding the differences

Dr Loosen
Ernie Loosen makes his most memorable wines from the Erdener Prälat vineyard, with its red slate and probably the Mosel's hottest suntrap.

Gunderloch
In the Rheinhessen the Nackenheimer Rothenberg vineyard is also red slate; Gunderloch's TBA from here is one of the most thrilling from anywhere in Germany.

Leitz
Josef Leitz makes some of the Rheingau's purest Rieslings using large old oak casks for maturing his wines.

View over the town of Thann from the Clos St-Urbain vineyard on the hill of Rangen, Alsace's most southerly Grand Cru. Owned by Zind-Humbrecht, it is famous for Riesling. Here the grapes may be left on the vines until well into November. Thann is home to a local cult of St Urbain, whose litany prays for 'deliverance from the devastations of storm and frost'. The original chapel was built at the end of the 15th century, but destroyed in the revolution of 1789. The current chapel seen here was built in 1934.

between the major regions of Germany. Wines from Franken, where the summer is hotter, though shorter, are balanced at a much lower level of acidity than wines from the Mosel or Rheingau. Accordingly, they have less residual sugar.

In the warm Pfalz, residual sugar levels of as low as three grams per litre (very dry, in other words) are common, and acidity is likewise relatively low. In the Rheingau, a wine with 10g/l residual sugar, and acidity to balance, tastes only off-dry. That level of sugar in Franken would taste positively sweet. In the cold Saar, where acidity is higher, a wine with 40g/l sugar may taste as dry as a Rheingau with 10g/l.

Likewise, the type of acidity is important. Malic acid and tartaric acid taste totally different. Malic acid tastes raw and green like unripe cooking apples; tartaric tastes citrus, more intense, but riper. As a grape ripens, and its sugar content increases, so the total acidity in that grape decreases. But tartaric acid builds up while malic acid falls. At 40° Oechsle, the point at which the grapes begin to soften, Saar Riesling will have about 40 grams of acidity, all of which is malic. At 80° Oechsle – Kabinett level in the Saar – there will be about 10 grams of acidity, of which five will be tartaric. At 90° Oechsle – Saar Spätlese level – there will be eight grams of acidity, of which six will be tartaric. But to get this much riper acidity, you have to pick late: in cool years Saar growers may wait until November to pick. Picking earlier can produce wines with rawer acidity.

In the sunnier, more sheltered Mosel, the grapes are riper and the acidity is more likely to be the riper-tasting tartaric acid: nine grams of acidity in the Mosel can taste very different from nine grams of acidity in the Saar. Levels of the different types of acidity can determine the picking date of Riesling in Germany as much as sugar levels.

Clones

Although there is little clonal variation in Riesling compared to other vines, it is nevertheless true that the commercially available Riesling clones in the 1960s and 1970s were geared to high yields rather than high quality. The change to better, lower-yielding clones began in the early 1990s; by 2010 it was estimated that about one-third of German vineyards had been replanted to better clones.

Alsace

Riesling is on the increase here, but is choosy about where it is planted, preferring the hilliest, most sheltered sites. This might seem curious, given that Alsace is so much further south than much of Germany; but many of Alsace's vineyards are on the flat plains rather than on the steep, church-dotted hills beloved of photographers of the region. Soils on the plains are richer and more fertile, and while they should not be dismissed out of hand – good viticultural practice can produce expressive wines here – Riesling in less skilful hands can be thin and anaemic, or heavy and flat.

On the best hillside sites, particularly in the hillier Haut-Rhin in the south of Alsace, Riesling comes into its own. (There are 50 Grand Cru sites in Alsace; their names are a fair guide to the best sites, but not all the best wines are Grand Cru, and not all Grand Cru wines are worthy of the name.) The soils Riesling likes best here are sandy-clayey loams that warm up quickly in spring, although it is less fussy about soil than it is about aspect; key vineyards include Brand, Clos Ste-Hune, Elsbourg, Hengst, Kaefferkopf, Kastelberg,

Koehler-Ruprecht
This estate makes weighty, dry wines: this Auslese Trocken has 12.5% alcohol, and the structure and extract to go with both meat and fish dishes.

Domaine Albert Mann
The steep, south-facing Grand Cru Furstentum vineyard produces beautifully delineated yet ripe Riesling for Domaine Albert Mann.

Domaine Paul Blanck
Fastidious winemaking at this estate means that umpteen different Riesling cuvées are kept separate – with a detectable difference in taste.

Kirchberg (Ribeauvillé), Kitterlé, Osterberg, Rangen, Schneckelsbourg, Schoenenbourg, Sporen and Zahnacker. But the geology of Alsace is so convoluted that soil types often change within the same vineyard. Hence the habit of some top Alsace growers of picking such parcels separately and vinifying them as separate *cuvées*. Such *cuvées* may be blended later out of commercial necessity, but tasting them from barrel shows that Riesling in Alsace reflects its *terroir* just as clearly as it does across the Rhine in Germany – as long as the yields are not too high.

Permitted yields in Alsace are over-generous: 100hl/ha for the basic AC wine and 55hl/ha for Grand Cru. Wines cropped at that level have little character and no concentration; serious growers quote around 50hl/ha for AC wine and less for Grand Cru.

Why does Alsace Riesling taste so different from German Riesling – different even from the Rieslings of the Pfalz, only just to the north? One reason is the soil: the predominantly calcareous, clayey soils of Alsace give a fuller character than does the slate of, say, the Middle Mosel. Another reason is their higher alcohol: Rieslings here commonly have over 12.5 per cent alcohol, and may be chaptalized. They often spend longer in old barrels, too, which gives them greater roundness. But above all they are French wines, with the indefinable but recognizable French imprint. They are 'winier' wines than German Rieslings, lean and austere in youth but rich, honeyed and even petrolly in maturity.

They are also generally dry, although Alsace growers are notoriously cavalier about residual sugar levels. It is impossible to predict from a label whether an Alsace wine will be bone dry or medium. Vendange Tardive, or late harvest, wines may be dry or semi-sweet: Rieslings in this category must attain 95° Oechsle (220g/l sugar at picking). The much rarer Sélection de Grains Nobles wines are always sweet, and must have reached 110° Oechsle (256g/l sugar) at harvest.

Australia

Riesling is on a roll in Australia, although, since it only occupies a tiny percentage of total plantings, its current popularity may never amount to more than a mini-roll. Until Chardonnay overtook it in the early 1990s, Riesling was the most planted white grape in Australia, reflecting the belief current among many growers at the time that if you took a noble grape variety you could plant it anywhere regardless of climate or site and get great wine. It was (and still is) grown in many regions that were (and are) far too hot for such an early ripener; the resulting wines gave the variety a bad name.

Australia's best known Riesling region, the Clare Valley, has enjoyed substantial growth [illegible] Merlot and Shiraz between them have more than doubled over the same period. Investors in Clare, as elsewhere in Australia, are hypnotized by reds.

Clare is, on the face of it, a region far better suited to reds. Its climate is almost Mediterranean, but its topography makes it more adaptable than it would seem. Far from being a single valley, Clare is a series of gulleys, with hills rising to over 400m (1320ft) to the east and west. This already gives significant variations in temperature. In addition, nights are cool and rainfall is low, and the free-draining soils – which include red soil over limestone and shaley slate – allow for great variations of wine style. Watervale is traditionally the finest part of Clare for Riesling, though the Polish Hill River only a few miles north gives quite different wines with finesse and a certain mineral character. Clare's stylistic affinities are with Alsace.

The great 1970s Rieslings from Leo Buring set a standard in Australia which is still inspiring winemakers three decades later. Such wines are lean, even apparently rather simple in

ANOTHER NEW CLASSIC FROM AUSTRALIA

Jeffrey Grosset (right) has made a name for himself as the producer of some of the world's greatest cool, steely, yet scented Rieslings, from grapes grown in Australia's Clare Valley at Polish Hill and Watervale. Polish Hill Riesling has a minerally finesse due to the height of the vineyards – over 450m (1480ft) – and a slatey soil reminiscent of Germany's Mosel Valley.

Riesling was first planted in South Australia by the Silesians who settled in the Barossa Valley in the 1840s, bringing with them their brass bands, Würst and Lutheran churches, all of which still flourish there. There is some Riesling grown in the Barossa itself, but most has moved to the cooler hills of the Eden Valley, just above the Barossa to the east, or to the Clare Valley, a couple of hours' drive to the north.

Mount Horrocks
Mount Horrocks makes appetizingly limey Rieslings from terra rossa/limestone soil, and it ages to a toasty richness.

youth: many are drunk in their youth because their crispness and freshness are attractive then and seem to promise nothing more. But with time the acidity softens, the palate deepens, and they develop flavours of buttered toast and lime and a melting honey richness – and sometimes a whiff of kerosene; they emerge into maturity as some of the richest, most complex dry white wines in the world. And all without a hint of new oak.

The Australian style of Riesling varies from light and delicate to more powerful, but all share winemaking techniques of no oxidation, low-temperature fermentation in stainless steel, and early bottling. What you get is the unalloyed character of the grape.

In the Eden Valley the wines are more scented and immediately beguiling than in Clare; plantings have moved up here into the hills from their traditional place on the valley floor. Coonawarra and cool-climate Victoria makes some good examples. Tasmania's are fragile and delicious, and Great Southern and Frankland River in Western Australia are showing good, citrus form, particularly when made in a relatively low alcohol style. Some sparkling Riesling is also produced.

Austria

The finest sites for Riesling are the granite, gneiss or mica-schist terraces of the Wachau, where the climate is cool and the soil free-draining, and irrigation is both necessary and permitted. However, it accounts for only a small percentage of total Wachau plantings, and has only attained even that amount since the Second World War. It demands, and gets, the best spots: they include the Steinertal, Kellerberg, Schütt and Loibenberg in Loiben; the Tausendeimerberg, Singeriedl and Hochrain in Spitz; and Steinriegl, Achleiten and Klaus in Weissenkirchen. Alcohol levels are usually over 13 per cent – sometimes too high, so that the wines lose aroma. They often have a firm, minerally core and arresting, taut but spicy fruit and scent.

The styles, and soils, of the Wachau continue into the western part of Kremstal; elsewhere in Austria Riesling is more prevalent, and usually good, though less racy. The hilly region of South Styria, however, produces wine with good taut acidity, and Vienna itself makes some appetizing examples. Most Austrian Riesling is made dry: there is a little botrytis-affected Riesling made in that haven for botrytis, the shores of the Neusiedlersee, but it is vastly outplanted there by the totally different Welschriesling.

New Zealand

Riesling was introduced to the country in the early 1970s and is steadily on the increase. The cool climate is ideal for making wines of greater lightness and delicacy than are produced in Australia, and acidity levels are good, but all too often, except for a few thrilling and scented wines from South Island, the wines inexplicably lack excitement. That is not the case, however, with the late-harvest sweet wines, which have great concentration and raciness. Marlborough is a leading area for both sweet and dry styles, with Canterbury and Central Otago also good; Nelson also has a reputation for dry and late-harvest Riesling.

Canada

While dry Rieslings here are of good quality and weight, it is the sweet Icewines for which Canada, and in particular Ontario, is most famous. They have greater breadth than German Eiswein, and balance slightly less finesse and complexity with greater weight and an impressively direct sweetness.

USA

Riesling has been enthusiastically uprooted in California, though plantings still exist across the state, with good wines from Mendocino, Napa, Monterey and Santa Barbara. Oregon makes some pleasant examples, but Washington State has become a real Riesling expert in recent years, its reputation greatly helped by the involvement of German Riesling wizard Ernie Loosen in joint ventures. The same can be said of the Finger Lakes in New York State where sharp, citrussy, scented styles are probably the USA's finest. There are a few sweet botrytis examples in the USA.

Rest of the world

The vine is found widely in Europe, though west of Alsace only Spain has the occasional patch. Luxembourg makes dry, delicate versions and northeastern Italy produces good-quality wines that vary from light and aromatic in the Alto Adige to fuller but still dry in Friuli. There are substantial plantings in Slovenia, Croatia and the Czech Republic, also in Kazakhstan, Kyrgyzstan, Moldova, Russia, Tajikistan, Turkmenistan, the Ukraine and Uzbekistan. Riesling is planted in generally warm sites all over South America – but Chile is now having success in cool San Antonio and Bío-Bío. South Africa suffers from the same problem but lacks the option of exploring ever more southerly sites. China's Riesling has traditionally been quite good.

Bründlmayer
Bründlmayer makes some of Austria's most succulent Rieslings in Langenlois just north of the Danube. Heiligenstein is an exceptional south-west-facing site.

Steingarten
Steingarten is an isolated, exposed, sloping, stony soil in the hills above Australia's Barossa Valley. Its tiny yields of Riesling give wine of startling, impressive austerity.

Mission Hill
Canada's biggest claim to fame is Icewine. Riesling and Vidal are the two main grapes. Both the breezy Niagara Peninsula in Ontario and the steep lake side slopes of the Okanagan Valley in British Columbia are ideally suited to its production.

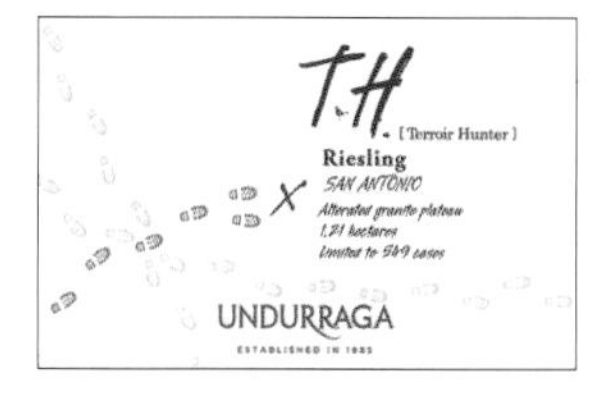

Undurraga
Tangy, citrussy Riesling, from a tiny windswept site near the Pacific Ocean in Chile, shows the potential of the grape in South America's coolest vineyard sites.

ENJOYING RIESLING

How long can Riesling age? Some Rieslings, it seems, can age almost indefinitely. It is possible to taste German Rieslings from the 1940s and 1950s that are still in fine fettle and, curiously, these are not always wines from the best vintages. Sometimes the high acidity levels that go hand-in-hand with cooler vintages have helped to preserve the wines into old age.

The question of aging Riesling is tied up with that of acidity. The acidity of many German Rieslings (in particular those of the cooler regions like the Saar) makes them taste lean and ungenerous in extreme youth. Taste them at this stage and you may well decide that Riesling is not for you. It is impossible to emphasize too strongly that good Rieslings should have bottle age: four to five years for a Kabinett, five to seven for a Spätlese, six to 10 for an Auslese, and 10 years plus for Beerenauslesen, Trockenbeerenauslesen and Eisweine. Simple QbA wines can be drunk a year or two after the vintage.

Dry (*trocken*) German Rieslings have a slightly different aging profile. They don't last as long as Rieslings with residual sugar, and become drinkable a little earlier.

Alsace Rieslings similarly need age – three or four years for simple AC wines, four or five years or more for Grands Crus, and at least five – ideally 10 or more – for Vendange Tardive wines. Austrian Rieslings are drunk young in Austria, though the best wines of the Wachau can improve for six or eight years. Top Australian Rieslings from the Clare or Eden Valleys will improve for perhaps 10 years, and last for 20.

The taste of Riesling

The adjectives that can be used to describe the flavour of this grape are as varied as the *terroirs* in which it grows. Slate soil gives it a characteristic smoky tang; in other soils it may taste minerally, steely, tarry, earthy, flowery or slightly spicy. Peaches and green apples are common descriptors; quince is also found, as is honey and citrus peel. Riper wines may taste of apricot or even pineapple. Australian Riesling often tastes of ripe limes and toast. Aged Riesling may acquire a characteristic smell of petrol or kerosene; this may not sound appetizing but is delicious. Look also for honey and marzipan and uncooked buttery shortbread, and buttered toast.

Botrytis cinerea, or noble rot, typically gives Riesling a dried apricot flavour, or one of honey, almonds or even raisins. Eiswein/Icewine, where it is made with unbotrytized grapes, has an icy smell literally like that of fresh snow; other aromas and flavours suggest lemons, peaches, apricots, passionfruit, pineapples or baked apples combined with high acidity.

Riesling is supposed to display the flavours of its particular vineyard site more than almost any other variety. Dr. Loosen agrees and makes finely textured, precisely focused wines from an array of great vineyard sites on the Mosel river between Graach and Erden. His Kabinett styles are low in alcohol, delightfully fruity, yet streaked with acidity and a positively rocky minerality. Jeffrey Grosset is based in South Australia's Clare Valley, and his Rieslings have equal intensity in a drier, limes-and-toast style that nevertheless give off a seductive perfume when young, and keeps their citrussy tang, when both young or old.

MATCHING RIESLING AND FOOD

Good German Spätlese Rieslings have the acidity to counteract the richness of, say, goose or duck, but the endless permutations of sweetness, dryness and weight in German wines mean that you do have to think out your food and wine combinations rather carefully.

For example, a well-aged Mosel Kabinett or Spätlese will be perfect with trout, or with smoked fish pâté. The Rheingau Halbtrocken equivalent will be good with fish in creamy sauces, though a traditionally sweet Rheingau Spätlese may well be too sweet for most dishes. But only the weightiest Auslesen and upwards should be attempted with desserts, and then only with desserts that are not oversweet. An Auslese from the Mosel will be too light for almost all desserts, and will be best drunk on its own.

Alsace Rieslings are far more food-friendly, and will partner everything from onion tart to spicy chicken dishes – and both Alsace and Australian Rieslings are perfect with Chinese and Thai food. Any dry Riesling with crisp apple or lime acidity will be a good match for salads.

CONSUMER INFORMATION

Synonyms & local names

Also known as Johannisberg Riesling, Rhine Riesling or White Riesling and in Italy as Riesling Renano. Don't confuse with Laski Rizling, Olasz Rizling, Riesling Italico or Welschriesling.

Best producers

GERMANY/dry Rieslings Bassermann-Jordan, Georg Breuer, Bürklin-Wolf, Busch, Christmann, Heymann-Löwenstein, Klaus Keller, Knipser, Koehler-Ruprecht, Künstler, Leitz, Rebholz, Sauer, Wittmann, J L Wolf; **non-dry Rieslings** Diel, Dönnhoff, Gunderloch, Haag, Haart, Heymann-Löwenstein, Karthäuserhof, von Kesselstatt, Kuhn, Künstler, Dr Loosen, Maximin Grünhaus, Molitor, Müller-Catoir, Müller-Scharzhof, J J Prum, St-Urbans-Hof, Schaefer, Schäfer-Fröhlich, Schloss Lieser, Selbach-Oster, Weil, Zilliken.

AUSTRIA Alzinger, Bründlmayer, Hiedler, Hirtzberger, J Högl, Knoll, Loime Malat, Nigl, Nikolaihof, F-X Pichler, Rudi Pichler, Prager, Schloss Gobelsburg, Schmelz.

FRANCE/Alsace J-B Adam, Beyer, P Blanck, Boxler, Deiss, Dirler-Cadé, Hugel, Josmeyer, Kientzler, A Mann, Muré, Ostertag, Schoffit, Trimbach, Weinbach, Zind-Humbrecht.

USA/New York Anthony Road, Fox Run, Dr Konstantin Frank, Ravines, Red Newt, Hermann J Wiemer; **Oregon** Chehalem; **Washington State** Chateau Ste Michelle (Eroica), Long Shadows.

CANADA Cave Spring, CedarCreek, Chateau des Charmes, Flat Rock, Quails' Gate, Tawse, Thirty Bench, Vineland Estates.

AUSTRALIA Tim Adams, Jim Barry, Bloodwood, Leo Buring, Castle Rock, Larry Cherubino, Crabtree, Eden Road, Forest Hill, Frankland Estate, Freycinet, Frogmore Creek, Gilberts, Grosset, Henschke, Houghton, Howard Park, Jacob's Creek (Steingarten), Kerrigan & Berry, Kilikanoon, Knappstein, KT, Peter Lehmann, Mesh, Mount Horrocks, Petaluma, Pewsey Vale, Seppelt, Three Drops.

NEW ZEALAND Auburn, Cloudy Bay, Dry River, Felton Road, Foxes Island, Framingham, Fromm, Mt Difficulty, Mount Edward, Neudorf, Pegasus Bay, Te Whare Ra, Villa Maria.

CHILE Casa Marín, Cono Sur, Viña Leyda.

SOUTH AFRICA Paul Cluver.

RECOMMENDED WINES TO TRY

Eight classic German non-dry Rieslings

Georg Breuer Berg Rottland Auslese
J J Christoffel Urziger Würzgarten Auslese
Dönnhoff Oberhäuser Brücke (Germany)
Gunderloch Nackenheimer Rothenberg Auslese Gold Capsule
Toni Jost Bacharacher Hahn Auslese
Karthäuserhof Eitelbacher Karthäuserhof Auslese Long Gold Capsule
Dr Loosen Wehlener Sonnenuhr Auslese
Willi Schaefer Graacher Domprobst Auslese

Ten dry European Rieslings

Paul Blanck Furstentum Vieilles Vignes (France)
Bründlmayer Zöbinger Heiligenstein Alte Reben (Austria)
Ch. Bela/Egon Müller Riesling Stúrovo (Slovakia)
Koehler-Ruprecht Kallstadter Saumagen Auslese Trocken (Germany)
Emmerich Knoll Dürnsteiner Kellerberg Smaragd (Austria)
Franz Künstler Hochheimer Hölle Auslese Trocken (Germany)
Leitz Rüdesheimer Berg Rottland (Germany)
Schoffit Rangen Clos St-Théobald Alsace (France)
Trimbach Clos Ste-Hune Alsace (France)
Weinbach Schlossberg Cuvée Ste-Catherine (France)

Twelve New World dry or off-dry Rieslings

Tim Adams Clare Valley Riesling (Australia)
Casa Marín Miramar Riesling (Chile)
Chehalem Riesling (Oregon)
Larry Cherubino Wall Flower Riesling (Australia)
Dry River Craighall Riesling (New Zealand)
Felton Road Riesling Dry (New Zealand)
Fireside Estate Estate Riesling (Canada)
Grosset Polish Hill Riesling (Australia)
Howard Park Riesling (Australia)
Long Shadows Poet's Leap (Washington State)
Neudorf Moutere Riesling (New Zealand)
Orlando Steingarten Riesling (Australia)
Thelema Riesling (South Africa)

Five sweet (dessert wine) Rieslings

Neethlingshof Noble Late Harvest (South Africa)
Ngatarawa Alwyn Reserve Noble Harvest Riesling (New Zealand)
Franz Prager Ried Achleiten Riesling TBA (Austria)
Reif Estate Riesling Icewine (Canada)
Horst Sauer Escherndorfer Lump Riesling TBA (Germany)

'You can't be a winemaker with Riesling,' says Mosel producer Johannes Selbach Oster. 'With Chardonnay you can have a recipe. With Riesling you decide what you are going to do and it turns out differently.'

Maturity charts

Riesling is one of the longest-lived of all white grapes. Some Australian examples from Clare and Eden Valleys are made to be cellared far longer than this chart shows.

2010 Mosel Riesling Auslese

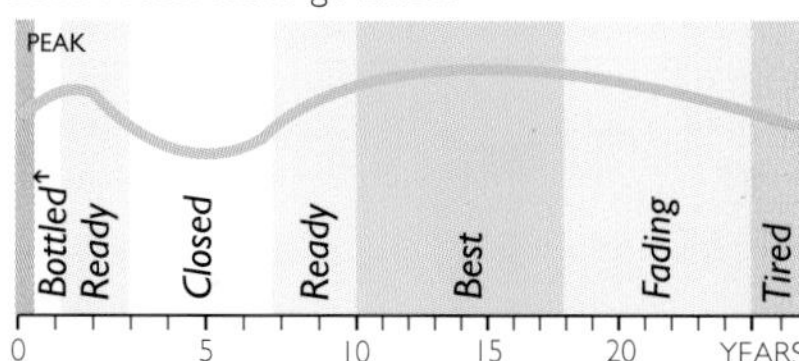

Germany has had a run of very good vintages. Riesling Ausleses go through a closed period before emerging into maturity.

2010 Alsace Riesling Grand Cru

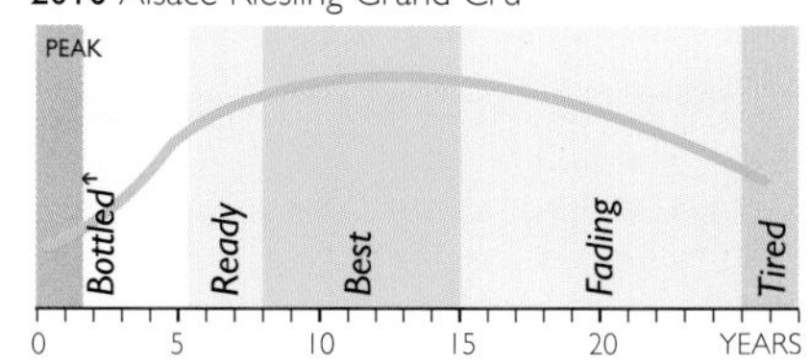

A long growing season, with a lot of cool weather, produced Rieslings of intense ripeness yet fantastic acids for long maturation.

2014 Top Clare Valley Riesling

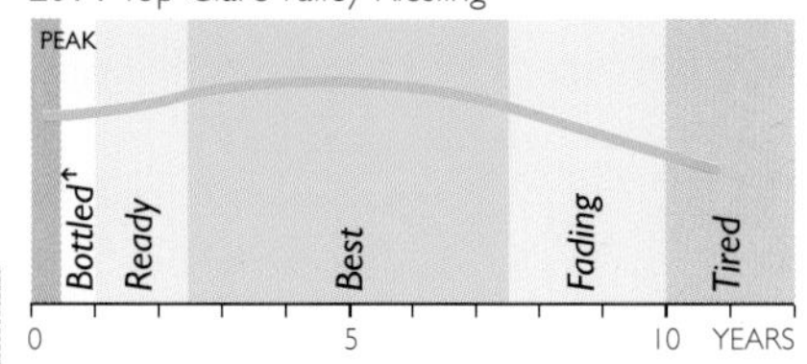

Most Clare Rieslings are at their peak within five years. A few, however, are meant for longer aging and will go on improving way beyond that time.

RIESLING ITALICO

A synonym for Welschriesling (see page 298), Hungary's Olasz Rizling and Croatia's Grasevina, a grape which is unrelated to the true Riesling of the Rhine and Mosel, and the sooner people start calling it by names like Grasevina, the better for its and for Riesling's reputation. But at the moment it is known as Riesling Italico in Italy, where it is far more widely planted than Riesling, and is found mostly in Friuli and the Veneto and other parts of the northeast. It has even spread further south to the Colli Bolognesi in Emilia-Romagna where the Vallania family at Vigneto delle Terre Rosse makes a delicious late-harvest wine as well as a flowery dry white from Riesling Italico blended with Riesling. Actually in the province of Treviso it is called Rismi.

Riesling Italico wine is light, flowery or nutty, and quite crisp. It's not on the same plane as true Riesling, but it has its attractions as an everyday wine with moderate perfume. Best producers: (Italy) Mazzolino, Pieropan, Vigneto delle Terre Rosse.

RIESLING RENANO

Italian synonym for Riesling (see pages 204–215) – the name is a translation of Rhein Riesling. It is outplanted in Italy by the lesser, and unrelated, Riesling Italico (see above) or Welschriesling (see page 298), and is mostly found, like Riesling Italico, in the northeast. The wines tend to be light and crisp, without much complexity. Best producers: (Italy) Paolo Caccese, Schiopetto, Le Vigne di San Petro, Villa Russiz.

RIESLING-SYLVANER

A synonym for Müller-Thurgau (see page 154), dating from the days when the parents of this undistinguished grape were believed to be Riesling and Silvaner/Sylvaner. They were, in fact, Riesling and Madeleine Royale.

Müller-Thurgau is known by this name in Switzerland, even though Dr Müller, who was responsible for the grape, came from the Swiss canton of Thurgau (the crossing was made in Germany). New Zealand has also sometimes preferred this name, and used to make what was probably the world's best Müller-Thurgau from it.

RIVANER

A synonym for Müller-Thurgau (see page 154) used in Luxembourg, where it is widely planted, and occasionally in England. It can produce pleasurable pale whites.

Every grape variety should have somewhere it can excel, even if its reputation is largely shabby. Rivaner, aka Müller-Thurgau, makes delightful, delicate wines on the banks of the Moselle River in Luxembourg.

RIZLING

Welschriesling (see page 298), under its various synonyms of Olasz Rizling and Laski Rizling, has been pressured to take this name to distinguish it from the superior Riesling.

RIZLING ZILVANI

Synonym for Riesling-Sylvaner (see left), and thus for Müller-Thurgau (see page 154).

RKATSITELI

Useful, all-purpose white grape found throughout the Russian Federation, especially in Georgia. Indeed, if we think of Georgia as one of the birthplaces of winemaking, there are unsubstantiated claims that Rkatsiteli seeds were found in clay jars going back to 3000 BC. If you're of a religious frame of mind, it is suggested that Rkatsiteli was the first vine Noah planted when the flood subsided, though I rather like the dry humour of world-famous wine geneticist Dr José Vouillamoz when he states, 'there is no botanical, archeological, historical nor genetic evidence for these assertions.' Faith? Now, that's another matter. Anyway, it is high-yielding, resistant to winter cold and high in acidity; the latter quality helps to make it (relatively) resistant to bad winemaking.

It is used for pretty well everything, including fortified wines and brandy, in most of the ex-Soviet states. There's loads in Ukraine, Bulgaria (it's the most planted white variety in both these countries), Romania and Moldova, though less nowadays in Russia itself. China, though, seems to like it, under the name of Baiyu. Visiting Western winemakers have seen it as having the potential for reasonably good quality, and it can produce wine with a quite powerful flavour of quince jelly or baked apple peel. My favourite examples are the baked apple peel examples from New York and Virginia. Best producer: (Moldova) Vitis Hincesti; (New York) Frank, (Virginia) Horton.

ROBOLA

High-acid Greek white grape which used to be thought to be the same as the Ribolla of north-eastern Italy (see page 203) and the Rebula of Slovenia, but which has now shown to be of a separate variety. In Greece it is found in the Ionian islands, and especially Cephalonia, where it grows ungrafted in limestone soils and gives lemony, almost chalky wines with plenty of character and weight, particularly when grown in mountain vineyards that can go as high as 300m (1000ft). Best producers: (Greece) Calligas, Gentilini.

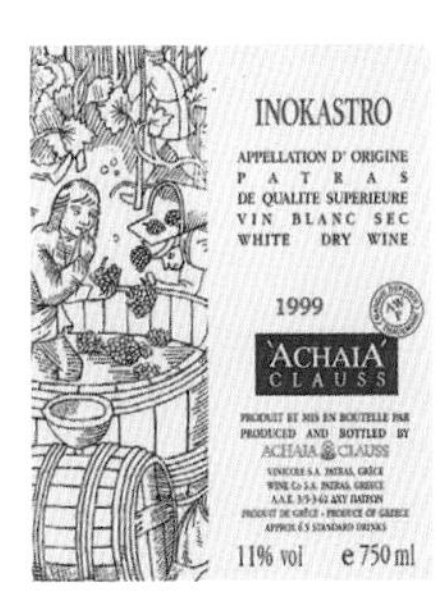

Gentilini

Gentilini makes a pale dry Robola with a fleeting perfume of orange blossom from ungrafted vines on the steep limestone-strewn slopes of Mount Ainos.

RODITIS

A pink-skinned, Greek grape with good acidity, especially when planted on cooler, north-facing slopes rather than on flat land. It often forms part of the Retsina blend, along with Savatiano (see page 243). There are numerous different strains, including some redder-skinned ones, which produce more complex, interesting wine. The wine is best young, and has good acidity even in warm spots. Also spelt as Rhoditis. Best producers: (Greece) Kir-Yianni, Kourtakis.

ROLLE

The same grape as Italy's Vermentino. As Rolle it's found in many blends in France's Provence, Rolle seems to mix happily with Roussanne, Marsanne, Viognier, Grenache Blanc and other local grapes. It makes characterful, aromatic wine with good acidity and adds freshness and bite to many. It is the main grape of Bellet, a wine that is hard to find in neighbouring Nice, never mind elsewhere. Rolle is also important in Corsica and Languedoc-Roussillon but doesn't seem to be the same variety as the Ligurian Rollo in northwest Italy. Best producers: (France) Bellet, Commanderie de Peyrassol, la Courtade, Crémat, Gavoty.

ROMORANTIN

White grape used for Cour-Cheverny, a Loire Valley appellation created in 1993 along with that of Cheverny. White Cheverny is made from Chenin Blanc, Chardonnay and Sauvignon Blanc, but Romorantin, which is found nowhere else, has its own AC. Plantings are on the decline, as the grape does not have any very distinctive character. When made traditionally it is a fierce, almost entirely unattractive wine, and when made in a modern, hygiene-conscious, stainless-steel style the examples I have so far tasted have seemed emasculated and forgettable. So if I had to choose between forgettably dull or memorably horrific, I think I'd choose the memorably horrific. Best producers: (France) le Chai des Vignerons, Huards, Philippe Tessier.

RONDINELLA

Rondinella is part of the Valpolicella blend. It lacks the elegance and aroma of Corvina, Valpolicella's main grape, but is well coloured and perfumed. It is disease-resistant and a reliably large cropper. It doesn't reach enormously high sugar levels, but it dries very well so is a useful part of the Amarone blend.

RONDO

Early-ripening *vinifera x amurensis* hybrid becoming very popular in Europe's chilly north – Denmark, Sweden, England et al. Deep-coloured, rather coarse wine, best blended with Regent.

RORIZ

Roriz, or Tinta Roriz, are the names by which Tempranillo (see pages 270–279) goes in Portugal's Douro Valley, where it is one of the most popular of the five varieties recommended by the port authorities. It is said to have been first planted at the Quinta de Roriz in the Douro in the 18th century. It produces large clusters of thick-skinned berries, and rich, tannic, intense wines with plenty of mulberry fruit and attractive floral fragrance, though fairly low acidity. In the Alentejo it is called Aragonez and produces fairly rich, mulberry-flavoured wines for drinking young. Also important in Dão, Tejo and Lisboa. Best producers: (Portugal) Quinta dos Carvalhais, Quinta do Côtto, Quinta do Crasto, Esporão, Casa Ferreirinha (Barca Velha), Niepoort, J P Ramos, Quinta dos Roques, Quinta de la Rosa, Casa Santos Lima, Sogrape, Quinta do Vale da Raposa.

ROSENMUSKATELLER

Rose-scented, deep pink Muscat grape occasionally found in Trentino-Alto Adige in northern Italy, where it is also called Moscato Rosa. Produces both sweet and dry wines. See Muscat pages 156–165.

ROSSESE

Probably Liguria's most characterful dark grape variety, Rossese is actually several varieties, all different. There's Rossesse di Dolceaqua, which in Provence becomes Tibouren and is rather good, and there's a raft of white or pink-skinned versions. Rossese Bianco, which is the same as southern Italy's Griollo, is also rather good. There's a different Rossese Bianco in Piedmont, another Rossese Bianco in Liguria, Rossese Bianco di San Biagio in Liguria and Rossese Bianco di Monforte in Piedmont. All unrelated, of course. Rossese di Dolceaqua is used mostly for rosé blends in France but for varietal light reds in Liguria, although the unrelated Rossese di Campochiesa can also be used. You'd be forgiven for mistaking one for the other in the glass: both have bright, crunchy raspberry and strawberry fruit with herbal notes. You'd also be forgiven for thinking 'life's too short...'

Anfosso

You don't see much Rossese di Dolceacqua outside the confines of Liguria in northwest Italy, and each example I've tasted has been different to the next.

ROSSOLA NERA

Red (rather than black) grape grown in Italy's Valtellina region in Lombardy, usually for blending with Nebbiolo. It has high acidity and ripens late.

ROTGIPFLER

Pink-skinned grape found almost only in Austria's Thermenregion south of Vienna. Its wines are full, even sturdy, with high alcohol, and it has a big spicy bouquet. It is generally made off-dry or sweet, and is blended with the slightly superior Zierfandler (see page 299) for the long-lived, spicy white wine called Gumpoldskirchner, named after the eponymous town. Best producers: (Austria) Karl Alphart, Johann Stadlmann, Harald Zierer.

ROUPEIRO

Portuguese grape grown primarily in the southern Alentejo region for gently aromatic, light whites but present under different names throughout Portugal. Pleasant when young, but seems to fall apart and oxidize quickly. It is also called Alva or Síria in northern Portugal, and is known as Códega in the Douro Valley. Over the border in Spain, it's called Doña Blanca, amongst other things. Best producers: (Portugal) Quinta do Carmo, Esporão.

ROUSSANNE

Sometimes it's not such a bad thing to be given up as a lost cause. Ever since I can remember, I've been reading that, of the northern Rhône twins, Roussanne has far more style and elegance and class than Marsanne, but that it was a tricky, finicky, inconsistent beast of a vine and so was being uprooted in favour of the gutsier Marsanne. And it was. So the obituaries continued to flow from every vinous pen – Roussanne is wonderful: Roussanne is doomed.

Well, maybe Roussanne was pushed to the brink, but we're now seeing a typical fashion-led revival of interest in Roussanne – interestingly, at the same time people decided that Marsanne is a pretty decent vine as well. And now that we can taste some examples of pure Roussanne, it becomes clear that it does possess an almost herbal minerally perfume and surprisingly elegant texture for a warm climate wine. Yet it is easy to see why, from the grower's point of view, Marsanne rather than Roussanne dominates in Hermitage, Crozes-Hermitage, St-Joseph and St-Péray – because Roussanne yields irregularly, is prone to powdery mildew and rot, and doesn't like strong winds. It also ripens late, and is prone to oxidation in the cellar.

Only a few producers, notably the Perrins of Château de Beaucastel and Jaboulet, still favour it, even though new clones have alleviated some of its worst problems: this is why Jaboulet's top white Hermitage and Crozes-Hermitage are so bright and scented, as is Beaucastel's Vieilles Vignes white Châteauneuf du Pape. Those who persevere love the finesse it adds to blends and its ability to age: it is excellent young, at three to four years, but then can enter a dumb phase, from which it emerges at seven or eight with greater depth and complexity. In Châteauneuf Marsanne isn't allowed, and Roussanne adds backbone and interest to Clairette, Bourboulenc and Grenache Blanc, but only occasionally is it a varietal wine. It likes a long growing season, and too much heat can send the alcohol up beyond 14 per cent and the wine out of balance. Some Languedoc-Roussillon and Provence examples are very exciting, particularly with a little judicious oaking.

In Savoie Roussanne (known there as Bergeron) can have an attractively glacial peppery, herby scent. Rhône Rangers in California like the grape. There is also some grown in Liguria, Tuscany, Australia and South Africa.

THE TASTE OF ROUSSANNE

The flavour of Roussanne is intense but nevertheless elusive and intriguing. It is reminiscent of pears or aromatic herbal tea, floral in youth and nutty and winey with age. If picked underripe it has high acidity. However, wines picked at less than full ripeness do not age as well; and fully ripe Roussanne manages to be both low in acidity and long-lived. Some of the white Hermitages of old have aged even better than the reds.

Château de Beaucastel's Châteauneuf-du-Pape Blanc Vieilles Vignes is, unusually for the appellation, made from 100% Roussanne. The normal bottling is about 80% Roussanne. Yields are low here at around 20hl/ha. The oak influence on this wine is extremely restrained – it is vinified half in stainless steel and half in one-year-old barrels and has superb aging potential.

Alban Vineyards

Based in the cool Arroyo Grande district of Edna Valley, John Alban is one of California's leading Rhône specialists. Roussanne is a difficult grape to grow and vinify but this wine is laden with honey notes.

D'Arenberg

A typically whimsical label from d'Arenberg – they didn't pick their first crop of Roussanne because the grapes were covered in tiny money spiders. This is nutty, melony and rich, but made without oak.

Left: *Tablas Creek is a joint venture established by the Perrin Family of Château de Beaucastel in France's Rhône Valley. The objective has always been to show how suitable the Rhône varieties are to California's relatively warm conditions, as epitomized by Paso Robles on California's Central Coast. Roussanne has been one of its great successes, usually blended in the classic Rhône manner, with Marsanne, Grenache or Picpoul but sometimes kept separate to show just how minerally yet scented its wine can be.* ***Above:*** *Plantings of Roussanne have risen almost twentyfold in France in the last 50 years.*

CONSUMER INFORMATION

Synonyms & local names

The grape grown in Provence as Roussanne du Var is unrelated, and although Roussanne is sometimes known as Roussette in the northern Rhône Valley, it is not the same variety as the Altesse (alias Roussette) of Savoie. Also known as Bergeron in the Vin de Savoie cru village of Chignin.

Best producers

FRANCE/Rhône Valley Beaucastel, Belle, Chave, Clape, Clos des Papes, Yves Cuilleron, Delas, Florentin, Font de Michelle, la Gardine, B Gripa, Guigal, Jaboulet, la Janasse, Jean Lionnet, Pradelle, Remizières, Marcel Richaud, Sorrel, Cave de Tain co-op; **Languedoc** Dom. Alquier, Cazeneuve, Chênes, Clavel, Estanilles, Lascaux, Mas de Bressades, Mas Bruguière, Nages, Prieuré de St-Jean de Bébian, La Truffière; **Provence** Borrely-Martin, Rabiéga, Trévallon; **Savoie** Quénard.
USA/California Alban, Andrew Murray Vineyards, Bonny Doon, Cass, Donelan, Fetzer, Quivira, V Sattui, Sobon Estate, Tablas Creek, Truchard; **Washington State** DeLille, McCrea Cellars.
AUSTRALIA D'Arenberg, Giaconda, Mitchelton, St Huberts, Seppelt, Torbreck.
SOUTH AFRICA Fairview, Ken Forrester, Rustenberg, Simonsig.

RECOMMENDED WINES TO TRY

Ten Rhône Valley wines

Ch. de Beaucastel Châteauneuf-du-Pape Blanc and Châteauneuf-du-Pape Blanc Vieilles Vignes
Domaine Belle Crozes-Hermitage Blanc
Chave Hermitage Blanc
Clos des Papes Châteauneuf-du-Pape Blanc
Yves Cuilleron St-Joseph Blanc Coteaux St-Pierre
Font de Michelle Châteauneuf-du-Pape Blanc Cuvée Etienne Gonnet
Ch. la Gardine Châteauneuf-du-Pape Blanc Vieilles Vignes
Domaine de la Janasse Châteauneuf-du-Pape Blanc
Marcel Richaud Côtes du Rhône les Garrigues Cairanne

Ten Languedoc-Roussillon and Provence wines

Domaine Alquier Roussanne/Marsanne
Ch. de Cazeneuve Coteaux du Languedoc Pic St-Loup Blanc and Coteaux du Languedoc Pic St-Loup Blanc Grande Cuvée
Domaine Clavel Coteaux du Languedoc Blanc
Ch. de Lascaux Coteaux du Languedoc Pierres d'Argent
Mas de Bressades Roussanne/Viognier
Mas Bruguière Coteaux du Languedoc Blanc
Ch. de Nages Costières de Nîmes Réserve du Château Blanc
Prieuré de St-Jean de Bébian Coteaux du Languedoc Blanc
Domaine de Trévallon Blanc

Seven other Roussanne-based wines

Alban Vineyards Edna Valley Roussanne (California)
Cass Roussanne (California)
Bonny Doon Le Sophiste (California)
D'Arenberg The Monkey Spider (Australia)
Mitchelton Viognier/Roussanne (Australia)
Sobon Estate Amador County Roussanne (California)
Tablas Creek Roussanne (California)

ROUSSETTE

See Altesse page 41.

[illegible]

A Bulgarian crossing of Nebbiolo and Syrah, this has low acidity, without much resemblance to either. It also has 50% more anthocyanins (colouring matter) than Cabernet, and is best drunk young.

RUBIRED

Widely grown in the hotter parts of California. It's a crossing of Alicante Ganzin with Tinto Cão, and the point of it is not flavour, of which it has hardly any, but deep and stable colour. It has dark purple juice and is widely used for concentrate, particularly for the Mega Purple brand which is used (perfectly legally) to darken more high-priced Napa reds than you might imagine.

RUBY CABERNET

A 1936 crossing of Carignane (Mazuelo) and Cabernet Sauvignon, produced in California. The idea was to combine the heavy cropping of the former with the elegance and complexity of the latter. In this it was only partly successful, since whatever Ruby Cabernet's qualities may be, elegance and complexity are not among them. It was intended for hot climates, and was heavily planted in the Central Valley, though the few examples produced in cooler spots seem more successful. It is mostly blended. In both Australia and South Africa its ability to produce high yields of dark wine with a rather earthy flavour yet decent black fruit has made it popular for fortifieds and for bulking up blends cheaply. Best producers: (South Africa) Daschbosch, Goudini, McGregor.

RUCHÉ

Seldom-seen red Piedmontese variety, also spelt as Rouchet, making interesting, florally aromatic pale reds with good tannin structure and sometimes penetrating acidity. It has its own DOC in Castagnole Monferrato south of Asti. Best producers: (Italy) Bava, Biletta.

RUFETE

A flavoursome Portuguese variety still found in older vineyards in the Douro, though it is not one of the approved five varieties. It's also found in Dão for rosé and sparkling, and rather more so in Beiras, where on the best sites, from low-yielding vines, it can produce remarkably good quality. Known as Tinta Pinheira in the Dão, and as Castellana over the border in Galicia.

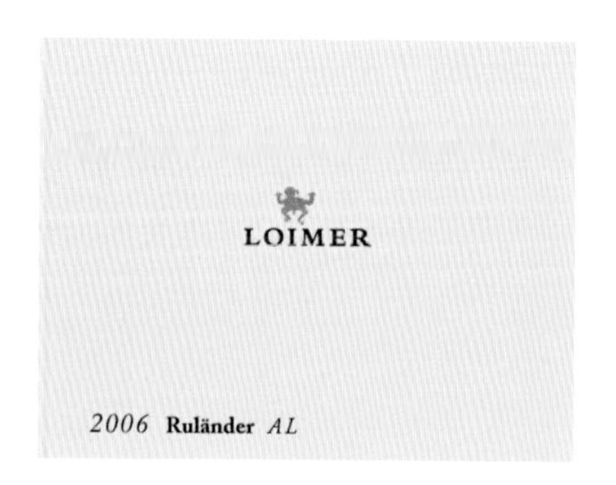

Loimer
Fred Loimer has built a remarkable, futuristic 'box' of a winery in the traditional Austrian region of the Kamptal. Conditions are warmer here than in the nearby Wachau and Rülander can be picked super-ripe. This label is for a 14% Halbtrocken Auslese.

Antonelli
Sagrantino can be a monumentally tannic red wine, but in the best examples the haunting dark flavour of cherries swirled in smoke makes the challenge worthwhile. There are a few sweet versions which are worth a tentative try.

RULÄNDER

Pinot Gris (see pages 186–187) in Germany may take either the name of Rülander or, increasingly, be called Grauburgunder (see page 123). The name Ruländer comes from Johann Seger Ruland, a wine merchant from the Pfalz region, who propagated the variety in the early 18th century. Like Pinot Blanc, it is genetically identical to Pinot Noir. Today in Germany it is grown most heavily in the warmer, more southerly wine regions, especially Baden, and also the Pfalz and Rheinhessen. In Baden it is usually known as Grauburgunder, and the wines are usually fermented dry. It produces quite weighty, fat wine; the aroma is mushroomy and earthy, with a modicum of runny honey. It's slightly spicy but less so than Alsace Pinot Gris.

In Austria, where it may also be called Pinot Gris, it makes rich, substantial wines in the Burgenland region and drier, lighter ones further south in Steiermark. In the more fashionable cellars of Austria and Germany it may be fermented in *barrique* to make big, ripe, dry honeyed wines. Best producers: (Austria) Loimer; (Germany) Salwey.

SACY

This French white grape is still just about found in the *département* of the Yonne in northern Burgundy, where it is grown for Crémant de Bourgogne, the local Champagne-method fizz, but you're a bit more likely to taste it in Saint-Pourçain, where it's called Tressallier. But I wouldn't go out of my way, because its virtues are high productivity and equally high acidity – fairly basic ones as virtues go.

That it is grown in the Yonne is a legacy of the days, in the 18th and 19th centuries, when, before the advent of the railways, the region was an important producer of wine for Paris. High-yielding varieties were the most prized by the growers, and lower-yielding Chardonnay was confined to the Chablis hills. Now Sacy is finally on its way out, having in the 21st century at last fallen victim to market forces.

SAGRANTINO

This Italian red grape, found around Perugia in Umbria, makes intense, tannic and strongly fruity wines with typically Italian cherry and smoke flavours, which used to be considered good for convalescents. It has its own DOCG in Montefalco, and its popularity is rising. The rarer *passito* wines from dried grapes are more powerful than the straight dry red versions. Some may be blended with Sangiovese in the DOC Montefalco. Sagrantino used to be written off as being austerely tannic, but better vinification has helped round out the fruit, and it is clear that this is a grape with a good deal of personality. Sagrantino has been planted in Australia, as the craze for Italian grapes continues, and there's every reason to believe it could work in California and Argentina. It should also be more widely planted in Italy. Best producers: (Italy) Adanti, Antonelli, Arnaldo Caprai, Colpetrone, Rocca di Fabri.

ST-ÉMILION

A synonym for Ugni Blanc (see page 285), especially in the Charente *département* in western France.

SÄMLING 88

Scheurebe (see page 243) is sometimes known by this name in Austria's Burgenland region, where it is used for high-quality sweet wines. Best producers: (Austria) Alois Kracher, Johann Münzenrieder.

SAMSÓ

Cultivated by Torres as an old Catalan vine as part of its research into rare varieties. It turns out to be either Cinsault or more likely, Mazuelo (Carignan). It's perfectly possible that the same name gets used for both vines, though Cinsault is barely planted in Spain. The Catalan authorities have settled on Mazuelo as the correct name, and DNA analysis one of these days will prove the matter one way or the other. It's a very late-ripening vine which can be very productive if grown on fertile soil: Torres grows it on slopes to curb its vigour. It's low in tannin and not very dark, but if the grapes are allowed to raisin on the vine it gives jammy, marmalady flavours; without this raisining process it seems to lack character.

SANGIOVESE

See pages 222–231.

SANGIOVETO A Tuscan synonym for Sangiovese (see pages 222–231).

SANKT LAURENT

The thing that excites everyone about Sankt Laurent is that it can produce velvety, richly fruity reds with a strong similarity to fully ripe Pinot Noir. The trouble, then, is that everyone tries to prove Sankt Laurent is some sort of earlier-flowering, earlier-ripening Pinot Noir. Leading Austrian producer Axel Stiegelmar even said it was a seedling of Pinot Noir – in other words both its parents were Pinot Noir, but, as happens with vine seedlings, it did not reproduce true to type. Well, genetically it seems to be a long way off Pinot Noir, but no one has yet come up with much more detail than that it probably originates in Austria – which, since most of the best Sankt Laurent I've tasted has been Austrian, sort of takes us back to where we started. But it is certainly like Pinot Noir – though perhaps even more like good Gamay – in its soft-centred, juicy cherry fruit.

The wine is best drunk young, and may not age well – but so what? It's a complete delight when it's young. It is particularly popular in Thermenregion of Austria and in southern Burgenland. With Blaufränkisch it is a parent of Zweigelt, Austria's most popular red grape.

Germany has quite a bit and it is one of the few red varieties to find any success in the cool climate of the Czech Republic and Slovakia. I suspect it would do well planted also in areas like southern England and Canada. Best producers: (Austria) Paul Achs, Gernot Heinrich, Juris (G Stiegelmar), Willi Opitz, Hans Pittnauer, Joseph Umathum.

Josef Umathum makes a speciality of Austria's own black grape varieties – Sankt Laurent, Zweigelt and Blaufränkisch – in his vineyards sloping down toward the warm, shallow Lake Neusiedl in Austria's Burgenland region. The lake never gets deeper than 2m (6ft) and the humid conditions also favour production of fine noble-rotted sweet white wines.

SAPERAVI

A dark-skinned, pink-fleshed, low-yielding variety originally from Georgia, and also very important in Moldova, Ukraine and parts of Russia, as well as Armenia and Bulgaria. It produces wines high in tannin, colour and acidity, which need time in bottle to soften to any degree of friendliness. It is late ripening and needs some warmth, or its acidity is too high for comfort; in such cases it is best blended. But I believe that as winemaking improves and they get their vineyards sorted out, we are going to find that Saperavi is a superb grape, indeed a classic grape just waiting for its moment in history to leap centre stage and do the jellyroll. Already I've seen 20- to 30-year-old examples from places like Moldova that had the style and perfume of a distinguished old Pauillac. Now that's high praise.

The Magarach research institute in the Crimea has crossed Saperavi with Cabernet Sauvignon to produce Rubinovy Magaracha, which seems to have some potential; it has also crossed Saperavi with Bastardo (from Portugal) and produced Magarach Bastardo. This latter is intended for fortified wine production. Which is all great fun. But I want to see Saperavi cherished and encouraged in all its own glory, and in the not too distant future, I'm sure I will. Best producers: (Australia) Gapsted; (Moldova) Vitis Hincesti.

IOVIS
F·R

SANGIOVESE

I'm a bit of a late convert to Sangiovese and its charms, but as any true 'born again' will tell you, the bug gets you far more potently if you come to it late. And I mean really late – like in the last 10 or 15 years. All through the 1970s and '80s when France was effortlessly the leading quality producer, and when the New World was still in its wine infancy, Italy as the globe's biggest wine producer did have a chance to wow the world with its most famous wine – Chianti, which is based on Sangiovese – but it failed to take the challenge seriously. Most people, me included, rarely came across Chianti except in raffia-covered flasks: you always ordered them in Italian *trattorie* and never anywhere else.

Although Chianti was supposed to be predominantly Sangiovese, antiquated wine laws meant that a substantial proportion was just as likely to be white Malvasia or Trebbiano. That didn't make for a very exciting red. There were some other local red grapes – Canaiolo is quite good – but the laws were so lax that much of the remaining blend was likely to be a great big soup of high strength, jammy this and that, trucked in from Puglia or Sicily to add – well, I suppose to add colour, alcohol and flavour, however rough. When you've got badly cared-for vineyards of a late-ripening variety like Sangiovese, diluted by high acid Trebbiano – well, you're looking at a rather unsavoury, harsh rosé if you're not careful, and a dollop of wine soup from the South could hardly make it any worse.

Sangiovese means literally the 'Blood of Jove'. Nice to think that this Roman god left off seducing mortals and dropping thunderbolts to give his name to this vine. Almost certainly of Tuscan origin, Sangiovese is still central Italy's most important grape. Florence, the heart of Tuscany, lies beyond and Keats's nightingale has strayed from Provence as this grape seems to give more truly 'a beaker full of the warm south'.

My first moment of enlightenment came on the side of a country road south of Florence one spring time. We'd stopped off in a village to buy some wine. 'Chianti?' the man asked. We nodded. 'Where's your bottle?' Sorry, didn't know we had to bring our own bottle. 'Okay, okay.' He found us a reasonably clean litre vessel, went out the back – and squirted a jet of bright red fluid into it, charged us almost nothing – and off we went. My first mouthful of this fantastic, slightly spritzy, sweet-sour red wine, cherries and cranberries, redcurrants and a splash of fresh thyme, was a revelation. *This* was Sangiovese? If nothing else, it made supreme picnic wine.

Then during the 1980s, a few leading lights in Tuscany surveyed the rampant success of Bordeaux and the abject failure of their own Tuscan wines whose names – Chianti, Brunello di Montalcino, Vino Nobile di Montepulciano – were widely known but equally widely despised, and set in train a massive movement for change. They were driven by high ambition and extreme seriousness. Bordeaux, with its glittering ranks of Classed Growths and international favourites, was the target, and for this reason a lot of Cabernet Sauvignon was planted in Tuscany. As it happens, Cabernet makes excellent wine blended with Sangiovese, but it usually dominates the blend, and the mood in Tuscany nowadays is to maximize the quality of Sangiovese without any help from the French interloper even though its help was crucial to kickstart the new era. So the lead grape has to be Tuscany's own Sangiovese. It needed a lot of thought and investment in the winery and work and investment in the vineyards as new quality clones and new trellissing systems and vine layouts have replaced the chaotic Tuscan sprawl.

It has been a remarkable success. Even the most basic Chianti has improved – yes, even the ones in the *fiaschi* or flasks, but at the top end, the Tuscans have embraced modern winemaking and extensive investment in things like new oak *barriques* and low-yielding vineyards and have produced a string of Italy's most challenging and ageworthy reds. Not remotely New World in softness or richness – but with real class, imbued with Florentine arrogance and an austere haughty beauty that makes no effort to seduce, demanding rather that you make the effort to understand. As producers ease off on the new wood, the flavours more and more reflect the earth, the character, the history of Tuscany. If you're one of those red wine lovers who like a little pain with their pleasure, Tuscan Sangiovese is increasingly the place for you.

Sangiovese: from Grape to Glass
Geography and History page 224; Viticulture and Vinification page 226; Sangiovese around the World page 228; Enjoying Sangiovese page 230

GEOGRAPHY AND HISTORY

Is the 21st century going to be the Italian century? Well, in grape terms it is possible. The 20th century saw the spread of all the great French grapes around the globe and the startling success of many of them. Cabernet Sauvignon, Pinot Noir, Syrah/Shiraz and Merlot have had many breathtaking successes. And yet the world is still thirsting for more red varieties and of the many possibilities, Sangiovese from Tuscany is probably the most famous. It is not, however, the most readily adapted to travel. It is a sensitive grape at the best of times, and needs far more attention to site, clone and yield than does Cabernet Sauvignon: plant it carelessly and you will be rewarded with wine that bears no resemblance to the great Tuscan reds, and very little resemblance to anything you'd really want to drink.

Yet the future could be bright. This is a grape that only started being taken seriously in its native land some 30 years ago: progress there has been dramatic. We are now well into

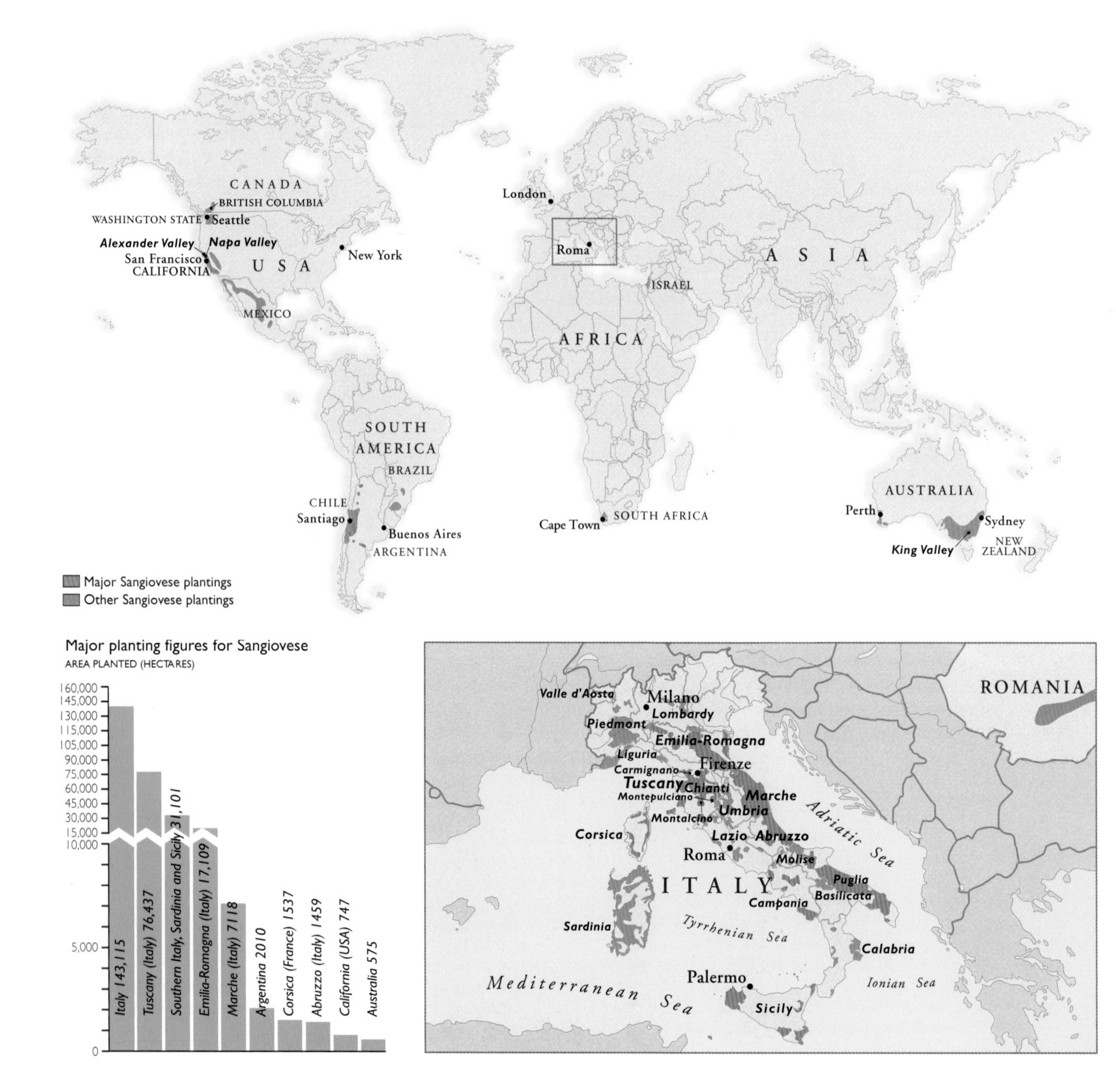

a new phase, and new clones and more care over the hows and wheres of planting have the potential to produce some fascinating wines in California and Australia in the future. However, it's not proving easy.

Sangiovese is not a grape for everyone, nor everywhere. It likes warm climates, but its vigour needs to be controlled; and if it is not treated with sufficient sensitivity the flavours in the finished wine will be harsh and unattractive. It has naturally high acidity, high tannins, but somewhat fugitive colour, and it is only now beginning to shine in friendly environments like Australia or Argentina. It is getting its chance to prove itself internationally because we are going through a period of enthusiasm for all things edible and drinkable from Italy. Slowly, the odd impassioned genius is unlocking the secret. If Sangiovese doesn't succeed, there's a queue of other red varieties – from Italy, but also from Portugal and Spain – awaiting their turn.

HISTORICAL BACKGROUND

We know Sangiovese's parents now: one is Calabrese di Montenuovo and the other is Ciliegiolo. The first, as its name suggests, is from Calabria, and is almost extinct now, and the second is an old Tuscan variety. Sangiovese itself is pretty old, and its first documented mention by name is in 1600. Its number of attested synonyms (29) indicates a long history and a wide spread: it's found, under different names, in Sicily, Corsica, Calabria, Puglia and Campania as well as Tuscany. Its origin might well lie in the South.

Sangiovese means 'Blood of Jove', and the vine is certainly as changeable as its namesake. It is genetically as varied as Pinot Noir, and so it is most useful to think of a broad spectrum of styles and qualities, with no clear dividing line between them.

The Sangiovese grown, as Brunello, in the Montalcino zone is not a separate clone: in fact around six different clones have been identified in Montalcino. (Montalcino cooled on the idea that its Sangiovese was unique when it was pointed out that in that case the clone could be planted elsewhere, and there could be a Brunello di Puglia.)

Sangiovese goes by different names in other parts of Tuscany, and its personality is difficult to pin down: neither the Prugnolo Gentile of Montepulciano nor the Morellino of Scansano have much uniformity. Nor does that of Emilia-Romagna: Tuscan growers in the 1980s blamed their problems on the Sangiovese di Romagna that they had planted in the 1970s, but there is enormous clonal variation in Emilia-Romagna as well. Indeed, some of the newest clones being selected and planted in Chianti originate there. But the difference is that now clones are being selected for colour, flavour and concentration of fruit, not high yields. Great vintages like 2010, but also less-favoured years, are showing the startling improvement already made with Sangiovese and the best is yet to come.

Vineyards are often interspersed with olive groves in Chianti, and here in the south of the region, in the commune of Castelnuovo Berardenga, the land is more gently sloping than in other parts of the Classico zone. The local climate is also a touch warmer and the style of wine consequently a little richer.

Renzo Cotarella has one of the most important wine jobs in Italy as the chief enologist for Antinori, whose empire spreads out from Tuscany to Puglia in the south and Piedmont and Franciacorta in the north.

Even the most innovative of producers may want to employ a consultant. Here the dynamic Federico Carletti (left) of Poliziano in Montepulciano shares his thoughts with enologist Carlo Ferrini.

VITICULTURE AND VINIFICATION

In some ways Sangiovese is a most obliging grape: it will produce light, juicy wines or big, complex ones according to where it is grown and how it is cultivated. But in other ways it is both demanding and inconsistent: it is early budding but late ripening, so likes a warm growing season, and in the marginal regions where it produces its finest quality it may do so only in four or five years out of 10. The Italians would like to think of Sangiovese as their equivalent to Cabernet Sauvignon. Up to a point it is. The top examples age really well, and love to be aged in oak. And just as most top Cabernet wines are blended with something else, so far for every great varietal Sangiovese there are probably more wines that are great blends. Work in the vineyard and winery is aimed at more focused fruit, increased intensity and colour, softer tannins and less aggressive acidity.

Climate

You've got to have a fair amount of warmth for this late-ripening variety, but too much warmth does not produce the best Sangiovese. In Italy it won't ripen well north of Emilia-Romagna – though this is because of rain, not because of temperature – and in Emilia-Romagna earlier-ripening clones are necessary to dodge the October rains. In Tuscany, where so far all the finest Sangiovese wines have been made, it is less reliable a ripener in Chianti than further south in Montalcino, where the nights are warmer and rainfall notably less. In Chianti it requires the best south- or southwest facing slopes, and altitudes of between 150m and 550m (500 and 1800ft), and the paucity of ideal sites means that only some 10 per cent of the region is actually given over to vineyards. September rains, too, can spoil a vintage, and they do so several times a decade.

In Montalcino it will even ripen on north-facing slopes; these certainly produce lighter, more elegant wines than sites on Montalcino's south- and southwest facing slopes, but the grapes ripen nonetheless.

In the south of Tuscany, the Maremma, a lot of Sangiovese is being planted for the rich, broad character it acquires in the hotter climate and shorter growing season. But here too much alcohol and too little aroma can be a problem.

Trying to duplicate the extremely varied climates of Tuscany in California and Australia is proving something of a headache. Tuscany's climates are more markedly continental than those of Australia, but likely regions include Langhorne Creek, Strathalbyn and Port Lincoln in South Australia, and Karridale and Margaret River in Western Australia: these are as warm as Tuscany, though not as continental. Better matches of climate could include Canberra and Young in New South Wales, the western parts of the Great Dividing Range and Heathcote in Victoria, and Stanthorpe in Queensland.

The variety has failed to take California by storm, though many winemakers feel that, so far, Sangiovese has simply been planted in the wrong places and in the wrong way. Exposing the clusters to too much sun seems to give poor results; Marchese Piero Antinori, who realized Sangiovese's full potential in Tuscany in the 1970s (see caption, opposite) and who bought Atlas Peak Vineyards in the foothills of the Napa Valley in 1993, points to the greater intensity of sunlight in California as a possible factor. Until California's great Sangiovese sites have been identified, Sangiovese wines are not going to rise above the 'interesting; must try harder' standard.

Soil

The soils of Tuscany are as varied as the climate and altitude. The heart of Chianti Classico is a highly desirable, friable, shaly clay called *galestro*; the lesser Chianti zones of Colli Senesi and Colli Aretini are clay; towards the coast the soils are lighter and sandier. Further south in Montalcino *galestro* alternates with

Vines near Tavernelle in the Brunello di Montalcino region of Tuscany with the hilltop town of Montalcino in the distance. Vines are grown up by the town at heights of 500m (1650ft) on galestro *soil and give the most scented and elegant wines, but they are also grown on clay soils in Val d'Orcia to the south at altitudes nearer 150m (500ft), giving heavier, more powerful wines. Although there is an increasing move towards single estate wines, many producers blend grapes from the different altitudes and soils.*

limestone *alberese*: these two soils, in Tuscany, produce the best wines, with good body and flavour.

In the New World insufficient attention has been paid to soil. Leaving aside the vexed question of *terroir*, Sangiovese is a vigorous variety which needs a lot of work to keep it in balance, so you don't want too fertile a soil. When planting is done on fertile land, the temptation might be to increase dramatically the density of planting in the belief that this will keep the vine in check, but this approach may result in a positive jungle of foliage and no sunlight getting to the grapes at all. The vigour of the soil is crucial in determining the density of planting Sangiovese.

When Marchese Piero Antinori launched Tignanello in the early 1970s, he set Tuscany on a new course. He blended Sangiovese with a significant percentage of Cabernet instead of the usual Chianti grapes, and all the wine was aged in small oak barrels.

Density and cultivation

In Tuscany planting densities have been rising steadily in recent years, with the Chianti 2000 research project advocating densities of 7000 vines per hectare or more. The traditional density in Chianti was 2700/ha: 'We planted vineyards for tractors, the opposite of what they did in France', says consultant winemaker Dr Alberto Antonini. 'If you go to 5000/ha you get an improvement in quality at the same yield per hectare. I have experimented with densities of 10,000/ha, but I find no improvement in quality from 5000/ha upwards. But you need very poor soil for high density, or you get problems of canopy congestion, and shaded fruit. For Sangiovese you need open canopies with good filtered light and fruit distribution, and reasonably low yield per vine. Density must depend on the soil, the vigour and the rainfall. You can't generalize.' The only generalization it is possible to make is that Sangiovese requires more attention in the vineyard – more cluster thinning, more selection, more careful canopy management – than Cabernet Sauvignon or Merlot of the same quality.

Yields

Here, too, it is difficult to make worldwide generalizations, except to say that reasonably low yields produce better quality than high ones. But what counts as reasonably low varies with the site, the vigour of the soil and the climate.

In Chianti, most estimates put 1.5kg per vine as the maximum for quality (that is equivalent to 7.5 tonnes per hectare at 5000 vines per hectare). In Tuscany Sangiovese is not considered very prolific, largely because recent work has been aimed at reducing fertilizers, restricting yields and planting less vigorous clones and rootstocks. In California it is said to grow like a weed, even on poor soils; in southern Italy and in Emilia-Romagna, too, it is vigorous. It is certainly less self-regulating than Cabernet Sauvignon, and if yields exceed 10–12 tonnes/ha quality is likely to suffer. But the vine is so varied, and the conditions in which it grows are so equally varied, that what is true for one spot may not be true for another. Certainly Argentina manages to get a very pleasant fruity red with a slight bitter twist from yields far higher than those tolerated in Tuscany.

At the winery

Modern Tuscan winemaking is aimed at softening the tannins of Sangiovese, and at getting those tannins ripe in the first place. Picking dates are 10 days to two weeks later than they used to be, which helps achieve better ripeness; and the length of the post-fermentation maceration on the skins, which had shortened to 7–12 days, has now lengthened again to three or four weeks. This gives greater polymerization of tannins – as does the (illegal) use of oak chips and the (legal) running of the wine into new oak *barriques* for the malolactic fermentation.

Barriques are not universal in Tuscany: the traditional aging of Sangiovese is in large oak *botti* of five or six hectolitres upwards in size. Most producers use a combination of different woods, sizes and ages of cask. Chestnut is often found in traditional cellars, but I've yet to see a new chestnut barrel. Sangiovese does seem to suck up the sweet vanilla of new oak with gay abandon, but the resulting wine mellows very attractively with age.

BLENDING SANGIOVESE

Varietal Sangiovese wines can be superb: witness Fontodi's Flaccianello della Pieve. It is more traditional, and probably still more common, however, for Tuscan Sangiovese to be blended with something else: Canaiolo Nero, Cabernet Sauvignon, Syrah, Merlot, what you will. Primitivo, Montepulciano and Nero d'Avola are also said to be added, though not legally, since these are not Tuscan grapes, and their addition would involve trucking wine up from the South. But why add other grapes at all? Why can't Sangiovese stand on its own?

The answer lies partly in the climate of Chianti, partly in the character of the grape, and partly in the long gradual decline of Italian viticulture, which has only been arrested and reversed in the past three decades.

Chianti is relatively cool, and rain is likely to descend just as the grapes are nicely ripe. In such circumstances Sangiovese will benefit from the extra colour and flesh and softness provided by another variety. Its colour is also a factor: it is a bit short on a group of colour-giving substances called acylated anthocyanins, so here, too, other darker grapes can help.

The reason that Cabernet Sauvignon entered the equation was because of the low reputation of Sangiovese both at home and abroad in the 1970s. To make world-class wines in Tuscany it seemed to be necessary to employ other grapes, and Cabernet Sauvignon had all the perfume, finesse and complexity that Sangiovese seemed to lack. It also had global renown as the great grape of the top red Bordeaux wines. In the 1980s a raft of super-Tuscan vini da tavola appeared, blending Sangiovese with Cabernet in every possible percentage.

At the same time Tuscans were busy studying their vineyards and their winemaking. What they learnt about Sangiovese convinced them that it could be a great grape in its own right, and it did not need to be dominated by Cabernet Sauvignon.

But while varietal Sangioveses will increase from the warmer parts of Tuscany, in Chianti the climate demands the option of blending. In that respect, it is just like Bordeaux.

SANGIOVESE AROUND THE WORLD

What should Sangiovese taste like when it is grown outside Italy? If New World producers shun the bitter cherry and tea scented styles of Tuscany, what will they replace them with to make the wine distinctly different from Cabernet and Merlot? Well, they're not really sure in most cases, so Sangiovese flavours worldwide are pretty haphazard and top Chianti flavours still seem to be the goal of many experimenters. But the most successful wines still keep the Tuscan graininess yet wrapped in ripe fruit.

Tuscany

Sangiovese in Tuscany has evolved over the past 35 years. For all the excitement about Cabernet Sauvignon, and the worry that Italian flavours were being drowned in a rush of Francophilia, Sangiovese has emerged more serious, better understood and more polished, with a distinctive Italian bitter-sweet savoriness to the fruit. It has also emerged confusingly varied in taste.

The most traditional styles emphasize the herb and bitter cherry flavours we have always associated with Chianti and other Sangiovese-based reds; the most international styles stress plum and mulberry flavours, and use new oak *barriques* for extra richness and spice. In between these extremes every possible permutation is being made.

The blend of grapes also varies: Chianti may now be made with 100 per cent Sangiovese, or it may include an admixture of other grapes, which may be Cabernet Sauvignon, Merlot, Syrah, Canaiolo Nero or other softening varieties. (Chianti has no fewer than seven subzones, of which Chianti Classico and Chianti Rufina produce the best, most substantial wines. Colli Fiorentini wines are lighter and fresher, Colli Senesi wines can be solid and rustic, and wines from the three remaining zones, Colli Aretini, Colline Pisane and Montalbano, have no distinctive character.) In Carmignano a percentage of Cabernet is called for in the DOCG regulations; for Vino Nobile di Montepulciano, where quality has only recently caught up with Chianti and Montalcino, 20 per cent of Canaiolo is supposed to be added, and perhaps sometimes even is. Only in Brunello di Montalcino must the DOCG wine be made entirely of Sangiovese – and usually is.

For the consumer, therefore, predicting the likely style of a Tuscan red is increasingly difficult. The raft of super-Tuscan vini da tavola which emerged in the 1980s did little to help the confusion: each had its own blend, its own fantasy name and its own style, even if that style was similar to that of several other super-Tuscans.

But it is unreasonable to complain of confusion when it is the mania for experimentation by producers that has given us the current massive leap in quality. The Chianti 2000 research project, which involved the universities of Pisa and Florence and the Chianti Classico *consorzio*, was very valuable, but it has also been outstripped by the work of individual estates, who have been doing their own work on clonal selection, planting density, cultivation methods, rootstocks and soil selection. And the improvement we have seen so far is primarily the result of better viticulture and better selection in old vineyards: good selected clones have only been available in the past 10 years: as Dr Alberto Antonini put it, 'The word "clone" didn't exist in Italy 20 years ago'. He believes that 50 per cent of the possible improvement in Tuscan Sangiovese is yet to come.

Rest of Italy

Most of the Sangiovese in Italy is concentrated in the centre; it is officially recommended in 53 provinces from Piedmont southwards, and authorized in a further 13, but it plays a progressively smaller part the further one strays from Emilia-Romagna and Tuscany.

In central Italy it is the workhorse red grape, producing everyday wines as well as world-class ones, and it may be made into *rosato* wines, sweet *passito* ones and even into Vin Santo. Umbria's finest examples include Torgiano, where Sangiovese may be blended with Canaiolo, and Montefalco Rosso, a blend with the local Sagrantino grape; in the Marche there is Rosso Piceno, blended with the Montepulciano grape. In Romagna all varietals come under the umbrella of Sangiovese di Romagna, which covers all conceivable qualities, up to and including some slick international *barrique*-aged versions. In the South it is mostly blended.

USA

It looked as though Sangiovese would take off in California on the back of the 'Cal-Ital' movement at the turn of the century, but it hasn't really got going. Some examples are OK, but more have been notable for their ambitious pricing than their quality to justify it.

Until the 1980s the only Sangiovese in the state was a small patch in Alexander Valley left over from pre-Prohibition days: the arrival of Tuscan innovator Piero Antinori (see caption, page 227) at Napa Valley's Atlas Peak in 1986

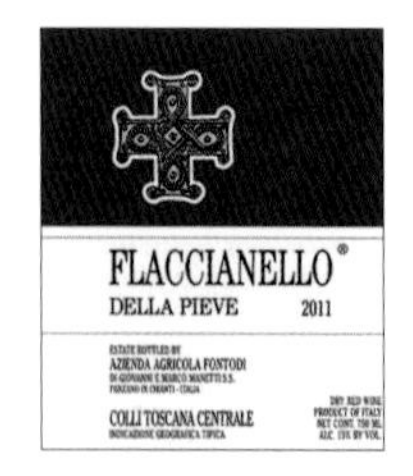

Fontodi
Fontodi makes exemplary Chianti Classico, but Flaccianello is 100% Sangiovese from a single vineyard of old vines, aged in new oak.

Antinori
The wine that ushered in the modern era in Tuscany, Tignanello is a single-vineyard, barrel-aged blend of 80% Sangiovese with Cabernets Franc and Sauvignon.

Biondi-Santi
Biondi-Santi is one of the most famous of old Italian wine names and its dense astonishingly long-lived reds first drew attention to Brunello di Montalcino.

It's tobacco we have to thank for Victoria's King Valley becoming the most important region for Italian varieties in Australia. Italian immigrants came over in the 1950s to grow tobacco – you can still see tobacco drying sheds in the valley. But as demand for tobacco dropped, many of them converted to grape growing. Pizzini are the leading growers and wine producers, and their Sangiovese vines give a wine with Australian juiciness matching Tuscan austerity. And if Sangiovese makes the perfect picnic wine, the Pizzini family seem to agree.

was the catalyst that set Sangiovese on its current somewhat rocky path. Antinori points out that Sangiovese is not easy to grow even in Tuscany; many Californian growers would ruefully agree that it requires far more work in the vineyard than they had bargained for and plantings since the early 2000s have been minimal. The total in 2013 was 1868 acres (755ha), most of it in Napa and Sonoma.

There is far from being a definitive style of California Sangiovese. Most of these wines have attractively bright fruit, cherryish and spicy, but they can be rustic or, alternatively, thin and lean. Cabernet or Merlot in the blend often helps the flavour, but too many California Sangioveses are already being made to taste too much like overoaked Cabernet.

Napa is generally hotter than Tuscany, and Tuscan-based consultant winemaker Dr Alberto Antonini believes that it is essential to extend the ripening period for as long as possible so that the tannins get fully ripe: sugar ripeness is no problem, and these wines often have 14.5 per cent alcohol compared to Tuscany's 12–12.5 per cent. Site selection and vineyard management will be crucial if we are to see any improvements. Washington State should also do well, but there are few good examples. Perhaps they're still struggling with site selection.

Australia

If interest in Sangiovese has failed to gain momentum in California, it is on the rise in Australia. Growers here seem to be taking a more planned approach to the grape – principally through better site selection and some decent clones – than was sometimes the case in California. Partly this is because Australia, as the world's driest continent, is accutely aware of the impact of climate change. Developing vineyards of warmer climate varieties like the Italian Sangiovese to begin replacing some of the cooler climate French varieties is a matter of increasing urgency. Both large and small wine producers are taking Sangiovese seriously, with some exciting results.

Central and South America

Mexico has some Sangiovese, though it hasn't produced any outstanding examples yet. Some Sangiovese is being planted in Chile, though producers there may be only at the beginning of a learning curve, starting with the question of which clone to plant and, even more importantly, where.

Argentina has a reasonable amount since waves of Italian immigrants often brought their Italian vine varieties with them. There haven't been any exciting top-level, oak-aged wines yet – largely because the quality revolution leaders here reckoned that other more international grapes would offer better rewards and recognition for less effort and headscratching. But some very nice juicy young reds are now appearing which manage to offer that characteristic bitter cherry kernel twist at the finish.

Poliziano
Poliziano is a leading light in the rejuvenated Vino Nobile zone in southern Tuscany. Asinone is its single-vineyard wine.

Seghesio
The Seghesio family are no newcomers to California: they've been growing grapes there for over a century. These Sangiovese vines were planted in 1910.

Chapel Hill
McLaren Vale is one of the most important areas of Australia experimenting with warm climate varieties and this was one of the first Sangioveses.

ENJOYING SANGIOVESE

Sangiovese can age, and age well, but most of its wines are intended to be drunk within a year or two of the harvest; they may develop an almost tomato-like leanness if left hanging around for too long. In fact, with so many different styles and qualities being made in Italy and with many of these being very recent innovations, it's pretty tricky to generalize about aging. In most regions there are a few producers trying to do something smart with Sangiovese that could merit aging, but even they themselves couldn't guarantee you a successful conclusion, especially when the use of new oak has been overdone.

The longest-lived wines are Brunello di Montalcino and the finest Sangiovese-based super-Tuscans. (Originally these all held the lowly status of vino da tavola, but now they have to move up the ranks of the Italian wine classification system, to DOC or IGT – Indicazione Geografica Tipica). In a good vintage these wines may well keep for 20 years, but most can start to be drunk after about five.

Vino Nobile di Montepulciano and the lighter Rosso di Montalcino can also be broached after five years, or sometimes less; all these should be drunk within eight to ten years. Carmignano has roughly the same lifespan.

Chianti is even more variable. Most basic Chianti should be drunk within three or four years of the vintage; Riserva may need a year or two more before it is ready. The very top wines, but only these, may last up to 15 years or more.

Other Italian Sangioveses should be drunk young unless they come from a producer who is specifically making a wine for aging. The same goes for New World Sangioveses: nearly all should be drunk within three or four years, and examples from Argentina are good within a year of the vintage.

The taste of Sangiovese

The traditional flavours of Tuscan Sangiovese are of bitter cherries and violets, with a certain tomatoey savouriness to the fruit, a definite rasp of herbs and a tea-like finish. Acidity is high and so is tannin: upfront fruit flavours are not the be-all and end-all of traditional Tuscan reds.

That has partly, though by no means entirely, changed. Those traditional flavours are likely nowadays to be richer and more concentrated, with better textures and finer tannins; the acidity is still there, but the greater concentration of flavour makes it less obvious.

The most international styles have a seasoning of vanilla and spice from new oak *barriques*, and the fruit leans more towards black cherry, plums and mulberries; where Cabernet Sauvignon joins the blend it is likely to be disproportionately dominant, with blackcurrant and plum flavours. These may become less noticeable as the wine ages.

Less ripe and concentrated Sangiovese can be stringy and rustic; that from warm climates can be heavier, broader and more alcoholic, tasting rather stewy and soupy and lacking the finesse of Tuscany's finest.

New World flavours vary from oaky, plummy Cabernet lookalikes to attractively bright, cherry-fruited bottles to those with high alcohol but unripe tannins. None so far really have the poise and finesse of good Montalcino. Not for the first time, emulators have found it surprisingly difficult to capture the quintessence of Italy.

Brunello di Montalcino is now the world's most famous and sought-after Sangiovese wine, but this success dates from as recently as the 1980s. It is well-deserved, since the best vineyards are fairly high, with infertile friable rocky soils. They are warmer than those of nearby Chianti, and also dryer, due to the looming Monte Amiata diverting most of the region's bad weather away from the vines. Constanti's 10ha (25 acres) of vines just below the hilltop town of Montalcino deliver consistently intense but elegant wines. Australia makes a richer, fruiter style, but Penfolds and others manage to keep the sweet sour Sangiovese character along with the juicier fruit.

MATCHING SANGIOVESE AND FOOD

Tuscany is where Sangiovese best expresses the qualities that can lead it, in the right circumstances, to be numbered among the great grapes of the world. And Tuscany is very much food-with-wine territory. Sangiovese-based wines such as Chianti, Rosso di Montalcino or Montepulciano, Vino Nobile and the biggest of them all, Brunello, positively demand to be drunk with food – such as *bistecca alla fiorentina* (succulent grilled T-bone steak), roast meats and game, calves' liver, *porcini* mushrooms, casseroles, pizza, hearty pasta dishes and almost anything in a tomato sauce (Sangiovese's acidity helps here), and tangy Pecorino cheese.

CONSUMER INFORMATION

Synonyms & local names

Also known (especially in Tuscany) as Sangioveto, Brunello or Prugnolo Gentile and Morellino. Corsica calls it Nielluccio.

Best producers

TUSCANY/Brunello di Montalcino Agostini Pieri, Altesino, Argiano, Biondi-Santi, Brunelli, Camigliano, Caparzo, Casanova di Neri, Casanuova delle Cerbaie, Case Basse, Castelgiocando/Frescobaldi, Centolani, Cerbaiona, Ciacci Piccolomini d'Aragona, Donatella Cinelli Colombini, Col d'Orcia, Costanti, Fuligni, La Gerla, Greppone Mazzi, Lambardi, Lisini, Mastrojanni, Siro Pacenti, Pian dell'Orino, Pian delle Vigne, Piancornello, Pieve Santa Restituta/Gaja, La Poderina, Poggio Antico, Pogio San Polo, le Potazzine, Salvioni, San Giuseppe, Livio Sassetti, Scopetone, Sesti, Talenti, Valdicava, Villa Le Prata; **Carmignano** Ambra, Arteminio, Capezzana, Piaggia, Pratesi, Villa di Trefiano; **Chianti Classico Riserva** Ama, Antinori, Badia a Coltibuono, Bibbiano, Il Borghetto, Brancaia, Cacchiano, Capaccia, Carpineto, Casaloste, Castellare, Castell'in Villa, Collelungo, Colombaio di Cencio, Casa Emma, Felsina, Fonterutoli, Fontodi, Isole e Olena, Il Mandorlo, Monsanto, Il Palazzino, Paneretta, Panzanello, Poggerino, Poggio al Sole, Poggiopiano, Querciabella, Rampolla, Ricasoli, Riecine, Rietine, Rignana, Rocca di Castagnoli, San Felice, San Giusto a Rentennano, San Polo in Rosso, Casa Solo, Terrabianca, Vecchie Terre di Montefili, Vignamaggio, Villa Cafaggio, Villa Calcinaia, Villa Rosa; **Chianti Rufina** Basciano, Bossi, Colognole, Frascole, Frescobaldi, Grati/Villa di Vetrice, Grignano, Lavacchio, Selvapiana, Trebbio; **Vino Nobile di Montepulciano** Avignonesi, Bindella, Boscarelli, La Braccesca/Antinori, Le Casalte, La Ciarlana, Dei, Del Cerro, Fassatti, Gracchiano, Il Macchione, Nottola, Palazzo Vecchio, Poliziano, Redi, Romeo, Salcheto, Sanguineto I e II, Trerose, Valdipiatta; **other central Italy** Boccadigabbia, La Carraia, Castelluccio, Lungarotti, Zerbina.

GREECE Karipidis.

USA/California Alexander Valley Vineyards, Altamura, Bonny Doon, Frey, Monte Volpe, Ortman, Pedroncelli, Seghesio, Sobon Estate; **New Mexico** Luna Rossa; **Washington State** Cavatappi. Kiona, Leonetti, Long Shadows.

AUSTRALIA Brown Brothers, Cardinham Estate, Chapel Hill, Coriole, Crittenden, Greenstone, Oatley, Penfolds, Pizzini, Scaffidi, Stonehaven.

ARGENTINA Norton.

SOUTH AFRICA Bouchard Finlayson, Kleine Zalze, Morgenster.

RECOMMENDED WINES TO TRY

20 classic Tuscan reds (Sangiovese and with other Tuscan varieties)

Biondi-Santi Sassoalloro
Castello di Brolio Chianti Classico
Castellare I Sodi di San Niccolò
Fattoria di Felsina Fontalloro
Castello di Fonterutoli Chianti Classico Riserva
Fontodi Flaccianello
Frescobaldi Chianti Rufina Montesodi
Isole e Olena Cepparello
Castello di Lilliano Anagallis
La Massa Chianti Classico Giorgio Prima
Monte Bernardi Sa'etta
Montevertine Le Pergole Torte
Castello della Paneretta Quattrocentenario
Poggio Scalette Il Carbonaione
Fattoria Petrolo Torrione
Riecine La Gioia
Rocca di Montegrossi Geremia
San Giusto a Rentennano Percarlo
Selvapiana Chianti Rufina Riserva Vigneto Bucerchiale
Castello di Volpaia Coltassala

Five Tuscan Sangiovese-Bordeaux blends

Antinori Tignanello
Castello di Ama Chianti Classico La Casuccia
Castello di Fonterutoli Siepi
Querciabella Camartina
Poggerino Primamateria

Nine New World Sangiovese wines

Bouchard Finlayson Hannibal (South Africa)
Coriole McLaren Vale Sangiovese (Australia)
Dalle Valle Napa Valley Pietre Rosse (California)
Long Shadows Saggi (Washington State)
Leonetti Walla Walla Valley Sangiovese (Washington State)
Ortman Paso Robles (California)
Penfolds Cellar Reserve (Australia)
Pizzini Sangiovese Shiraz (Australia)
Seghesio Omaggio (California)

Intensive research is pushing the quality of Sangiovese further and further forward. We have probably not yet seen the best it can do – and if that applies to Tuscany, which it does, it applies even more to the New World.

Maturity charts

Simple wines can and should be drunk within a few years; more concentrated, tannic versions need longer.

2009 Chianti Classico/Rufina (both Riserva)

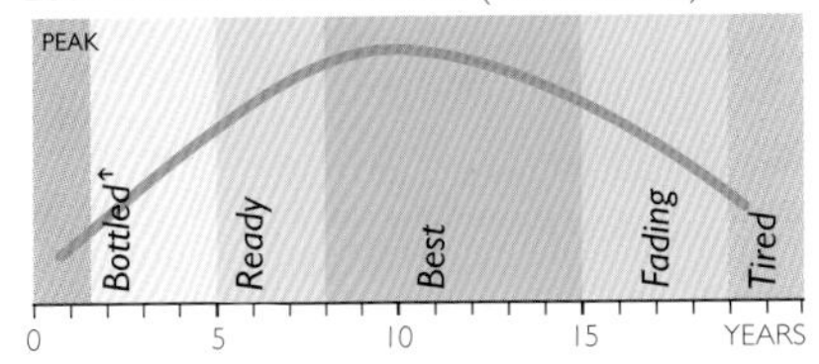

2009 was a hot, dry year in Tuscany, but the best Chiantis from the top estates are dark, ripe and long-lasting.

2008 Brunello di Montalcino

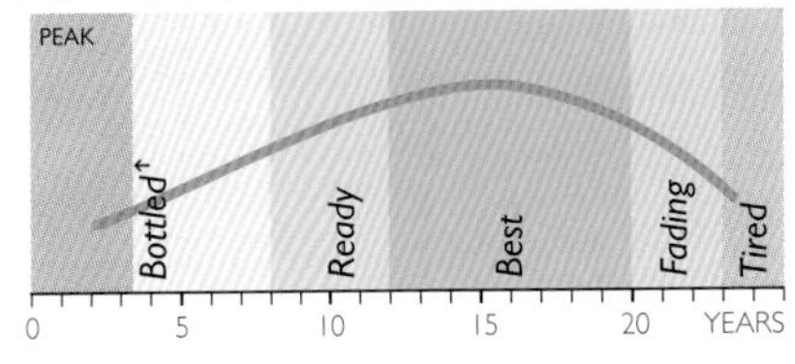

2008 was a hot year, but the heat wasn't extreme and nights were cool, creating deep, balanced wine which will age well.

2014 Chianti (Normale)

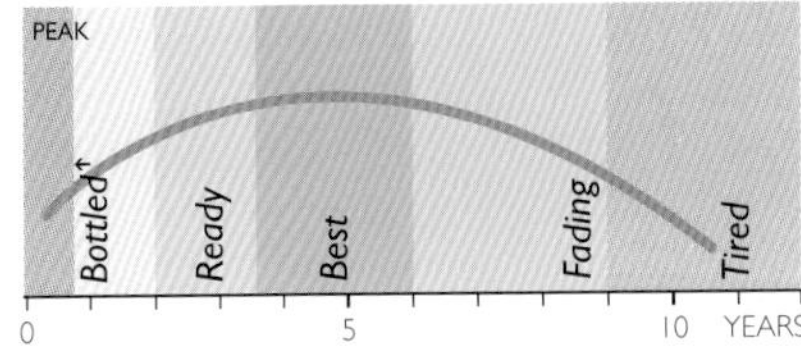

The 2011 vintage saw very difficult weather conditions in Tuscany; near drought in summer and blistering heat. Drink the wines young. .

SAUVIGNON BLANC

Why on earth does everyone make such a goddammed fuss about Sauvignon Blanc? Why can't they let it be? After all, it's such a simple grape, isn't it? The wine's not complex, it's not intellectually challenging, it's just a cracking good *drink*. You get a bottle, you whack it in the chiller, you whip off the screwcap, slosh it into everybody's glass and – hey – crisp, pure, tangy, thirst-quenching, yum. What a great drink! And I mean drink. A good glass of Sauvignon Blanc is like a good Gin & Tonic or a good chilled pint of golden ale. Just drink it. That's what it's there for.

The trouble is, winemakers can't leave well alone. They can't quite believe that what their audience love so much is Sauvignon's simplicity. Especially since a great part of Sauvignon's appeal is in its slightly underripe citrous, leafy flavours in a world gone mad for ripeness and even overripeness. One of the most worrying dogmas in wine today is a determination to achieve maximum phenolic ripeness in the wine grapes, and abhor anything with a sniff of underripeness about it. People who bang on about maximum ripeness – I wonder, have they ever tasted a perfect pear, a perfect Cox's apple or a perfect plum. Perfection is always at the 'just ripe' stage, never at the overripe stage.

A piece of the trellis from the potager garden at the château of Villandry illustrates the Loire Valley's long association with Sauvignon Blanc. The flavours of Sauvignon Blanc are linked to myriad fruit and vegetables but none more so than gooseberries. Behind the trellis stretch the vineyards and hills of the Marlborough region in New Zealand's South Island, the world's new classic area for Sauvignon Blanc.

The bigger argument with Sauvignon Blanc is between methoxypyrazine proponents and devotees of thiols. Sorry about the science. In bluntest terms, methoxypyrazines are the green element in Sauvignon's flavours, whereas thiols are the more exotic, tropical (if you're lucky) passion fruit flavours – if you're not lucky, armpit. I mean it. And not a washed one, either. I call Sauvignon a green grape – one that revels in the cool side of its personality, the green side, and one which is therefore almost always at its best coming from cool areas and cool vintages. The great classic cool area is Marlborough in New Zealand's South Island, followed by the cool coastal regions of South Africa and Chile. Grapes have a natural green quality from those areas. They also get decently ripe because the sun shines most of the time. And that makes it possible to keep crunchy, lime zesty, leafy freshness in the wine and add just a little exotic passion fruit and nectarine ripeness. Doesn't that make your mouth water? Well, that's probably the perfect style.

It's strange saying that the great classic areas for Sauvignon are places where grapes were hardly planted 30 years ago, sometimes hardly planted 10 years ago. But it's true. Sauvignon Blanc has been growing in Bordeaux and the Loire Valley for centuries. Bordeaux has made some great wines from it in Pessac-Léognan, but only as part of a blend with Semillon. The Loire made the most famous Sauvignons in pre-New Zealand history, but they had names like Sancerre or Pouilly Blanc Fumé – they never mentioned the grape's name on the label, so you didn't know you were drinking it.

It took the South Island of New Zealand, and the Marlborough region in particular, to show Sauvignon's astonishing flavours. This was an entirely new wine area with no *terroir*, no history, no reputation to protect. People realized that this brilliant, pungent, aggressively green yet exotically ripe style of wine was unlike anything the world had ever seen before. There had never been a wine with such outspoken, cut-glass purity of flavours whose whole purpose is to be refreshing and pleasurable.

I think it's fair to say that New Zealand Sauvignon Blanc changed the wine world by changing our ideas of what wine could be like, just as the Chardonnays of California and Australia did. And for a long time these two wines occupied seats in the opposite corners of the White Wine boxing ring: the pungent, green, aggressive Sauvignon, bare and unoaked, and the warm, round, soft, creamy, spicy, tropical Chardonnay with oak a key part of its attraction. Now, many other varieties are filling in the gaps between the two, but in the early years of the Wine Revolution, when we looked for leaders we found two whites – sexy Chardonnay, swathed in oak, and Sauvignon, naked as nature intended – and with attitude. Chardonnay has already started to reinvent itself in a perkier, less oaky style. Let's leave Sauvignon as it is.

Sauvignon Blanc: from Grape to Glass
Geography and History page 234; Viticulture and Vinification page 236; Sauvignon Blanc around the World page 238; Enjoying Sauvignon Blanc page 240

GEOGRAPHY AND HISTORY

I'm tempted to say there's no geography and no history of Sauvignon Blanc before 1973. That's when the first Sauvignon vines were planted in the Marlborough region of New Zealand's South Island. Within a few years they'd produced a wine of such shocking, tongue-tingling pungency that the world of wine was never the same again. Well, I'm tempted, but in fact Sauvignon Blanc had been around elsewhere for donkey's years, except that it never gave a wine with half so much excitement as New Zealand's offerings, and you virtually never saw the name Sauvignon on the label, so in any case you probably didn't know you were drinking it.

But look at the map – it's all over the place. New Zealand, sure, Australia, South Africa, Chile, northern Italy, Hungary, the Loire Valley, Bordeaux – almost everywhere you look you'll

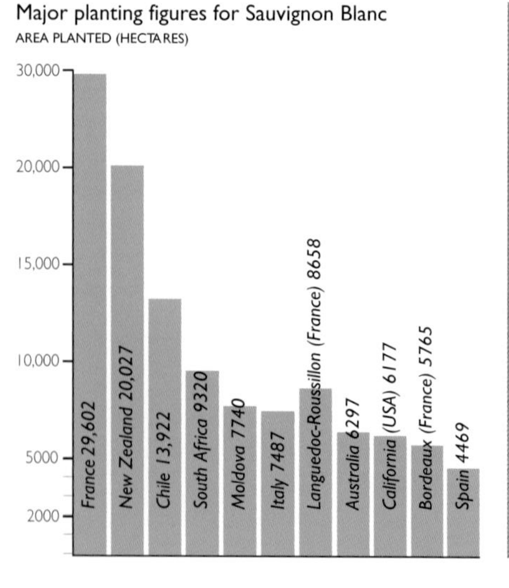

find Sauvignon planted. Which might imply it's a marvellously adaptable grape. But in fact it isn't. Sauvignon Blanc is, in fact, quite picky about its site, and what we're seeing at the moment is a sorting-out of styles, with only some places succeeding.

New Zealand, which first made us realize what Sauvignon could do, is suffering from a bit of an attack of confidence. Having shown the world how good Sauvignon could be, producers are now striving to find different experiences and flavours. I'd say: remember what made you famous in the first place; the audience still love it. By all means search for different sites and conditions which will naturally produce different flavours, but don't try to force the issue. Rant over.

Drier, sometimes tangy, sometimes more delicate, more subtle wines are appearing from South Africa in a variety of areas, both coastal and inland. Chile's coastal sites are providing real zing. Australia, with a few exceptions, is proving that Sauvignon grows better elsewhere. And California struggles gamely on. France just keeps on going much the same as usual.

HISTORICAL BACKGROUND

Both southwest France and the Loire Valley claim Sauvignon Blanc as an indigenous grape. The Loire looks more likely, but at some time in or before the 18th century, and presumably in Bordeaux, it got together with Cabernet Franc to produce the seedling that became known as Cabernet Sauvignon, and if it had done nothing else in its existence, wine lovers would have to thank it for that.

Its current fame in the southwest, however, is of quite recent date. Until the late 1980s it languished behind Sémillon and Ugni Blanc in terms of the numbers of hectares planted and its wine was generally rather raw and earthy. If it did indeed originate in the southwest and spread from there to the Loire (the opposite seems more likely), then we would have a rare instance of the sort of happy accident by which a vine that produces generally indifferent wine at home (and frankly 'indifferent' is high praise for most of the Bordeaux Sauvignon that was creeping out into the critical spotlight until the 1990s) suddenly excels elsewhere.

The Sauvignon Blanc planted in Bordeaux in the 19th century must have been mixed up with Sauvignon Vert, alias Sauvigonasse, a pretty uninspired poor relation of the real Sauvignon. Since Chile got its Sauvignon from Bordeaux cuttings before phylloxera and the two were mixed up there, it is reasonable to assume its field blend was inherited.

There is also a pink mutation of Sauvignon Blanc, known as Sauvignon Gris: Chile and Bordeaux both have some of this. It gives 20 per cent lower yields than Sauvignon Blanc, one degree more alcohol and a less pungent but spicier aroma. The berries are more deeply coloured than those of Sauvignon Blanc. However, Sauvignon Gris seems not to be the same as the Sauvignon Rouge mutation found in small quantities in the Loire. And modern history – well, that started in New Zealand, in 1973.

Vineyards in the rolling hills above Bué and Venoize in the Sancerre appellation in the eastern Loire Valley. There are 14 different communes in the appellation, so wine styles are necessarily heterogeneous. In fact, the much-vaunted differences between Sancerre and Pouilly-Fumé are far less than the differences to be found within the Sancerre appellation, with its varying soil types and its hills that rise to 400m (1320ft).

Pale chalky soils – these are at Bué – are crucial to the freshness and minerality of the best Sancerres, and their alkaline nature gives finely balanced, sharply focused flavours.

The late Didier Dagueneau was always the Pouilly-Fumé producer most determined to push the boundaries. Here his son Benjamin shows off his new egg-shaped oak fermenter.

VITICULTURE AND VINIFICATION

The ideal style of Sauvignon Blanc seems to be up for grabs at the moment. There are far fewer really raw, green, underripe flavours around. The good greenness you'll find is that of nettles, blackcurrant leaves, lime zest and green apple skins – thrillingly refreshing, when you can find it. But there is a trend to greater ripeness, and that brings a risk of low acidity which some producers correct by throwing buckets of tartaric acid into the vat. It sort of works – but you can taste the acidity separately if they've been too enthusiastic. Overripe Sauvignon goes oily, sweaty and over-rich, which is even less nice. So warmer regions just don't suit it. Many producers in Australia, New Zealand and South Africa deal with the problem by picking at different levels of ripeness – some underripe, some perfectly ripe, and some overripe. Done cleverly, this can produce palate-tingling wines of seamless balance. But why not plant the vines in a cool but sunny spot in the first place?

Climate

To get that too-often-elusive balance between sugar ripeness, acidity and aroma, the right climate is crucial. Once you've got it then the soil can certainly influence the flavour of Sauvignon; but the climate must be the first priority for anyone thinking of planting Sauvignon Blanc.

It's both a late budder and an early ripener: it doesn't therefore need enormous heat. In France it flourishes in both the maritime climate of Bordeaux and the more continental climate of Sancerre in the Loire Valley; New Zealand's climates, too, are maritime. Slow ripening gives better flavour development, but optimum aroma occurs just before optimum sugar ripeness. This is important. The greatest intensity of aroma is found just before the ideal balance of sugar and acidity; choosing the picking date means a slight compromise in one direction or the other. Ideally, it should, of course, be as slight as possible. Part of the improvement in quality in Sancerre in recent years has been produced by better-judged picking dates; and while logically these should be later, since the wines clearly taste riper, better balanced and more interesting than they did, most growers, when questioned, claim they pick earlier. Global warming, they say, has advanced the maturity of the grapes. I'd say awareness of the success of Sauvignon from New Zealand's South Island is an equally likely answer.

In New Zealand, different systems of canopy management have been aimed at producing riper grapes in a cool but sunny climate and with soils of high potential vigour. Fatter, more tropical wines have become common, but with raised levels, too, of residual sugar. There's a bit of a battle going on between the green, zesty champions and the tropical fruit boys. Since New Zealand set a world standard for tangy, citrus styles that other wine countries desperately try to copy, the move to a fatter, riper, indeed sometimes sweaty style is bonkers.

Kevin Judd was the original winemaker at mould-breaking Cloudy Bay in Marlborough, New Zealand, whose startling Sauvignon Blanc changed our wine world for ever. Now he has a new winery, Greywacke, also in Marlborough. Here he is taking a bit of a nap on some old barrels containing Wild Sauvignon. The dog looks thirsty.

In warmer climates growers generally pick early to keep acidity in their grapes, but in so doing seldom get the best aromas. California is a classic example: few of its Sauvignon Blanc wines have any character, so much so that the belief that Americans don't like Sauvignon Blanc becomes a self-fulfilling prophecy.

In Chile, many winemakers pick Sauvignon Blanc at several different times: unripe for grapes with high malic acid; riper for red and green pepper flavours; at perfect ripeness, and at overripeness. Where did they learn such habits? Why, the South Island of New Zealand, and as some New Zealand producers try to soften up their wines they should remember the original, brilliant formula.

Soil

The question of soil and Sauvignon Blanc is really confined to France, with a glance at New Zealand; no other countries pay such attention to the issue.

In the Loire Valley soils in Sancerre and neighbouring Pouilly vary from chalk over Kimmeridgean marl – this produces the best balanced wines, with richness and complexity – to the compact chalk, or *caillotte*, found at the base of the hills – this gives finesse and perfume – to flint, or silex, which gives wines with a certain gunflint sparkiness and vigour. There are also warm terraces of sandy or gravelly soil near the river which give spicy, floral flavours to wines which are the earliest maturing of all. Marl gives wines that age better, but the longest-lasting wines come from silex.

In Bordeaux the soils usually allotted to Sauvignon are more alluvial and produce high yields, which accounts for the lesser pungency of the wine here.

In New Zealand's Marlborough region the soils are particularly varied. Stony or sandy soils over shingle with poor fertility and good drainage compete with heavier, fertile clays and fertile silt. All produce good but different styles of Sauvignon. Further south in the Awatere Valley a mixture of silt loams, gravels and sands produce exciting results, and there is even limestone in the Ure Valley.

In the Sauvignon Blanc heartland of the flood plain of the Wairau Valley soil types, both fertile and infertile, run in bands that go from east to west. This means that if vines are planted from north to south, then a single row can contain vigorous, late-ripening vines with large canopies and weaker, early-ripening ones with small canopies. Interestingly, this mix of weak and strong, of tropical and citrus, unripe and ripe, can produce really interesting, pungent stuff. The heavier soils tend to be

Sauvignon Blanc vines at Montana's Brancott Estate near Blenheim, Marlborough – the original Sauvignon vineyard in the South Island. Blenheim usually gets more sunshine hours than any other town in New Zealand, thanks to the Southern Alps, which provide a handy rain shadow. Yet it's still quite cool in viticultural terms, so Sauvignon Blanc doesn't race to overripeness and flabbiness.

later ripening, and give more herbaceous flavours; stony soils are warmer and thus earlier ripening, and give riper, lusher flavours. Mix them cleverly and you've got pungency plus ripeness.

Yields

If what you're after with Sauvignon Blanc is aromatic, crisp, fresh wine for early drinking, then yields are not an enormous problem. The maximum yield in Sancerre is 68hl/ha, including the *plafond limite de classement*, by which growers are allowed an increase in yield in prolific years, and actually Sauvignon Blanc tolerates these sort of levels reasonably well. If you want more serious, fatter, weightier wines, ones that will last and improve for several years, then 50hl/ha is enough. At 40hl/ha the wine may well be very serious, with extract and staying power, but it will not be most people's idea of Sancerre. Yields are lower than this at good Graves and Pessac-Léognan estates in Bordeaux: under 40hl/ha at a few estates, notably Château Pape-Clément, but more usually between 40hl/ha and 55hl/ha. Since most top Bordeaux whites are barrel-fermented, these lower yields are crucial for a deep, balanced, ageworthy wine.

In the New World yields are higher again: at least six tons per acre (108hl/ha) in California, and around 8–12 tonnes per hectare (58–87hl/ha) even in Chile's relatively low-yielding Casablanca Valley. In the Central Valley yields can top 15 tonnes per hectare (109hl/ha).

At the winery

Fermentation temperatures are a major point of difference between producing Sauvignon in the Loire Valley and the New World. In the Loire, fermentation (either in steel or wood) is at 16–18°C (61–64°F), in order to avoid the tropical fruit aromas that the New World seeks with its cooler temperatures. These relatively warm fermentations produce wines that are minerally rather than exuberantly fruity: Loire winemakers prefer their wines to reflect their *terroir* rather than the grape variety.

Denis Dubourdieu, Professor of Enology at Bordeaux University and high priest of Sauvignon, whose work has lifted Bordeaux Sauvignon to a level inconceivable 20 years ago, points out that it is the peaks of temperature during barrel fermentation – these peaks can touch 25°C (77°F) – which give richness and varietal aroma to the wine. Fermentation in stainless steel, with tightly controlled temperatures, gives wines with fewer *terroir* characteristics but the potential for explosive varietal fruit.

But there are far more ways of tinkering with Sauvignon in the winery than merely adjusting fermentation temperatures. Pre-fermentation skin contact is used by some producers for more expressive fruit flavours.

When South Island New Zealand Sauvignon Blanc first hurled itself on to our unsuspecting palate with thrilling flavours of gooseberry, green pepper, asparagus and passionfruit, one reason for the wine's intensity was an unavoidable period of skin contact. Why? Well: there were no wineries at the time in the whole of South Island and grapes had to travel by truck and ferry all the way to Auckland up through North Island. At the height of harvest time that could mean grapes and juice sloshing around together for up to 24 hours. A wine of particular pungency could always be blamed on a traffic jam at the ferry.

The gooseberry/green pepper/lime zest flavours come from a group of flavour compounds called methoxypyrazines, which can be created, even in warmer areas, by shading the grapes from the sun. Many producers now eschew methoxypyrazines in favour of the more tropical-tasting, lusher thiols, but these carry with them an ever-present threat of armpit sweat. A skilful viticulturalist will try to balance the canopy shading to maximize the good points of both.

Aging on the fine lees works well: the lees protect against oxidation, and if the wine is in barrel they prevent it from becoming too oaky. *Bâtonnage* is used to increase weight; there is even some malolactic fermentation and new oak creeping into the Loire, though the style is atypical and I don't find the addition of creaminess and the reduction in tangy acidity that malolactic brings to be a particularly attractive objective. New oak is more usually thought of as belonging to Bordeaux and the New World: the term Fumé Blanc, which has no legal meaning, may be applied to New World Sauvignons (or indeed wines from other white grapes) with new oak aging. Most of the examples come from California where the term originated. I suppose the idea for the name came from Pouilly-Blanc Fumé in the Loire, but the flavours of these wines have nothing whatever to do with traditional Loire Sauvignon Blanc.

SAUVIGNON BLANC AROUND THE WORLD

Frankly, I generally prefer my Sauvignon Blanc as a pure thrilling blast of all the grape has to offer – unashamed, unblended, love it or loathe it. But some faint hearts prefer to temper the wine's fierceness by blending – and Sémillon is usually the grape of choice. Indeed, in Pessac-Léognan, even I actually prefer the Sémillon-Sauvignon Blanc blend.

France: Loire Valley

The finest appellations here for Sauvignon Blanc are Sancerre and its neighbour, Pouilly-Fumé; and in spite of the tradition that Pouilly-Fumé is marked by a characteristic whiff of gunflint, the difference between the two wines is less than differences found between Sancerre's different soils and villages. The 'Fumé' was appended to Pouilly to distinguish the area's Sauvignon from its Chasselas, which has the AC of Pouilly-sur-Loire.

At the moment Sancerre is a more go-ahead AC than Pouilly-Fumé. The older style of grassy, gooseberryish fruit is being replaced by richer, more peach and melon notes. The growers of the Loire have certainly been spurred into doing better by the success of New Zealand's South Island. New Zealand Sauvignon is not made in a style they admire. They want to achieve something different. Hmm. That's fair enough. But I would suggest that their first step towards making the best wine they can might be to respect New Zealand's brilliance, rather than turn their noses up at it.

Lesser, but cheaper, and attractive, distinctive wines are produced nearby in Quincy, Reuilly and Menetou-Salon. These are snappy, fresh wines, but lack the depth of a good Sancerre or Pouilly-Fumé: Quincy is the most intense and gooseberryish; Reuilly and Menetou-Salon have some of the Sancerre snap and nettly tang; Touraine Sauvignon is satisfyingly green and crunchy.

France: Bordeaux

Denis Dubourdieu, Professor of Enology at Bordeaux university and highly respected enologist, has been the greatest single influence behind the enormous improvement in quality of Sauvignon Blanc at all levels. Winemaking, viticulture and clones have all improved, particularly in the last 10 years; even basic Bordeaux Blanc (which may be 100 per cent Sauvignon or Sémillon, or a blend of Sauvignon and Sémillon with or without a little Muscadelle) is now pretty reliably fresh. The scented dry white wines of the Graves and Pessac-Léognan are often fermented and aged in new oak.

In the sweet wine appellations of Sauternes, Barsac, Loupiac, Cérons, Cadillac and Ste-Croix-du-Mont, the proportion of Sauvignon Blanc in the vineyards varies between 10 and 40 per cent. Its thin skin makes it highly susceptible to botrytis, and its acidity adds freshness to the blend.

Rest of France

Sauvignon Blanc is widely planted throughout the southwest, and is particularly successful in Gascony and the Dordogne. In the Languedoc it is fairly popular for IGP wines, usually made in a slightly fat, fruity style. However, the warm climate and high yields combine to make it difficult to produce tangy Sauvignon with Loire-style aroma and freshness though there are one or two standouts. Burgundy has one tiny outpost at St-Bris near Chablis.

Rest of Europe

Spain's most notable Sauvignon Blanc comes from the Rueda DO in Castilla y León, where it was introduced in the early 1980s. Styles are ripe, but the peachy fruit is balanced by a nettly acidity, usually achieved by picking the grapes as early as August. Some new oak may be used. It is also an authorized variety for the La Mancha and Rioja DOs.

Austrian versions of Sauvignon Blanc often have classic nettly, blackcurrant-leaves fruit, and restrained, understated, often excellent wines come from the Sudsteiermark region.

It is not known when Sauvignon Blanc was first planted in Italy, but it seems, in its early days at least, to have been grown alongside Sauvignonasse. Its first port of call was in Piedmont, although such Sauvignon Blanc as is made there today (there isn't much, and Gaja's is the best) was planted in the 1980s and 1990s. It is a grape of the Italian North and produces its most typical varietal aromas further east, in Collio, Friuli and Alto Adige. Italian producers may make Sauvignon as a varietal, or they may blend it with anything and everything: Chardonnay, Müller-Thurgau, Ribolla, Picolit, Vermentino, Inzolia, Tocai, Malvazia Istriana, Pinot Bianco and Erbamatt (a very rare white from Lake Garda).

The grape has potential for good quality in central and eastern Europe; there are large plantings in Romania and Moldova; the Czech Republic, Slovenia and, especially, Hungary are making appetizing wines.

Alphonse Mellot
Wine from 87-year-old vines in the La Moussière vineyard, fermented and aged in 900-litre vats to mark the 19th generation of winemaking Mellots.

Château Couhins-Lurton
This property produces a wonderfully rich and intense barrel-fermented Sauvignon Blanc which, unlike most Pessac-Léognan wines, contains no Sémillon.

Cloudy Bay
The most famous New Zealand Sauvignon Blanc of them all, Cloudy Bay achieved instant cult status with its first vintage in 1985.

New Zealand

The first Sauvignon Blanc was not planted in Marlborough, in the South Island, until the 1970s, when Montana made the inspired decision to plant a trial plot. The main attraction of this then little-known region was its cheapness, and its possible suitability for Müller-Thurgau, then a far more important vine in New Zealand. Montana's first Marlborough Sauvignon was made in 1980; it is now the leading region for the grape. The typical Marlborough style of Sauvignon Blanc – the climate is cool, dry and sunny – is distinctly tangy and citrussy, though some producers now seek more tropical fruit. The Awatere subzone to the south produces Marlborough's sharpest, most mouthwatering Sauvignons, lime zest fresh with tomato leaf and green capsicum crunch. Elsewhere in the South Island, Nelson's wines are softer, but citrous and Central Otago's are bitingly fresh. In the North Island, Hawkes Bay wines are mellow and some Wairarapa examples are tangy and earthy.

Wine was first made at Steenberg in Constantia in 1695. Vines were replanted in the 1990s as it became clear that the vineyards were ideally suited to cool climate varieties, with False Bay (seen in the distance) chilling the air and slowing ripeness. These vines produce one of South Africa's best Sauvignon Blancs.

Chile and South America

Much of Chile's original Sauvignon Blanc is in fact Sauvignonasse, though just how much is not clear: a distinction between the two varieties was made in the early 1990s. Plantings after about 1995 are of Sauvignon Blanc proper. The leading region though not the largest, is the cool Casablanca Valley, where plantings began in about 1990, but far more is planted further south in the Central Valley in much hotter conditions. Casablanca, with its Mediterranean climate, has similar daily temperature variation to Marlborough. The new regions, where everybody is scrambling to find land and grapes, are nearby San Antonio and Leyda, Limarí, coastal Aconcagua and several other new, chilly coastal locations, as well as Bío-Bío and further south. The wines are light and crisp but intense, with flavours ranging from gooseberry, mint and tomato leaf to light green melon and crunchy apple flesh and greengage plums. Sauvignon Blanc is also found in Mexico, Argentina, Uruguay and Bolivia. In Brazil, according to Galet, the vine called Sauvignon Blanc is really Seyval Blanc.

Australia

Australia isn't a Sauvignon paradise. There are good examples from Adelaide Hills, Padthaway, Orange and Tasmania, and Margaret River, in Western Australia, makes some excellent blends with Semillon. But on the whole one has to wonder if Australia is really suited to it. Most wines are simple and may taste slightly confected.

USA

Sauvignon Blanc is widely planted in California and even though it is now the third most planted white grape, it still lies way behind Chardonnay and French Colombard. Oaked versions of the wine often use the name Fumé Blanc. There is some in Washington State and Oregon, but acreage is declining. It is also planted on Long Island, New York State, and in Virginia.

South Africa

Vivid flavours of nettles, herbs, lime zest and gooseberries are common here. In the last decade, South Africa has become one of the world's most stylish Sauvignon producers, by making a considerable effort in identifying numerous different sites which give a variety of fascinating flavours, sometimes green but gentle, sometimes really crunchy. The West Coast north of Durbanville, and the far south near Cape Aghulas, produce the tangier styles, but the gentler delights come from Elgin, Constantia, Robertson and Stellenbosch.

Casa Marin
Chile has outstanding growing conditions for Sauvignon near the coast, where stiff Pacific breezes keep temperatures down and freshness up.

Henschke
Henschke are better known for their profound Shiraz reds, but they also take advantage of the cool Adelaide Hills to produce crisp Sauvignon.

The Berrio
Wonderfully snappy Sauvignon from Elim, on the tip of Africa, where breezes from the Antarctic Benguela Current provide chilly conditions.

ENJOYING SAUVIGNON BLANC

Generally speaking, Sauvignon Blanc is not a wine made to last. Its attraction is its youthful freshness and zest, and the fact that it can be drunk immediately, in the spring following the harvest. Most Sauvignon Blancs, if kept longer than a couple of years, fade rapidly and lose their aroma.

Those that will improve in bottle come not just from particular areas but from particular growers who take a decision to make a wine for keeping. This means, first and foremost, restricting yields: serious, ageworthy wines do not come from the same generous crops as light, early-drinking ones.

The best Marlborough Sauvignon Blancs intensify their flavours for between five and 10 years, depending on the vintage. Top Sancerres and Pouilly-Fumés can develop in bottle for five to eight years, developing flavours of honey and toast to replace those of nettly fruit, but always keeping a mineral streak; top white Graves and Pessac-Léognan change dramatically with age. They often start out with nettly acidity, bright nectarine fruit and gentle custardy oak and change over 10 to 15 years to magnificently deep nutty, creamy wines. Domaine de Chevalier, Smith-Haut-Lafitte, Malartic-Lagravière, Haut-Brion and Laville-Haut-Brion can last even longer, and may well stay in excellent shape for 20 years or even longer.

Sweet wines are made from blends of Sauvignon and Sémillon in Sauternes and Barsac, and sometimes from pure Sauvignon in California and New Zealand. Classed growth Sauternes reach maturity after a decade or so, but are nevertheless delicious younger; California and New Zealand sweet versions vary in their ageability, but most will improve for up to five years. Top Australian examples are beautiful at two to three years, but should age for a decade or so.

The taste of Sauvignon Blanc

When I'm trying to describe why I absolutely love the flavour of Sauvignon Blanc, I have to accept it's one of those grapes some wine people simply can't stand. That's okay. They don't have to drink it. All I ask is that they don't try to change it into something bland and soft or they'll find me challenging them to a furious fistfight. I love its taste of gooseberries, its taste of green peppers sliced with a silver knife, passion fruit and kiwi scattered with lime zest, nettles crushed up with blackcurrant leaves. These are the kind of flavours that make Sauvignon for me irresistibly refreshing.

If you like riper tastes, well, Sauvignon develops a spectrum of white peach, nectarine and melon masking any excess acidity; wines with a touch of botrytis may have a whiff of apricot. Sancerre and Pouilly-Fumé often have a minerally streak, particularly if they are grown on flinty silex soil: generally clay gives more richness, chalk lightness and perfume.

Lower fermentation temperatures produce a range of tropical fruit flavours: pineapple, banana and guava – dangerous unless balanced by good acidity.

New oak aging will give the wines a vanilla sheen; malolactic fermentation may add butter to the palate. With bottle age Sauvignon takes on tastes of honey and toast and quince, less obviously fruity, but rich and complex.

Botrytized sweet wines have flavours of pineapple and marzipan, oranges and apricots, with often piercing acidity to cut through the richness.

Although Sauvignon Blanc originates in Western France, its modern fame rests squarely on New Zealand's shoulders. New Zealand Sauvignon hit the world in the 1980s with tingling, mouthwatering green fruit and a citrussy attack that virtually ushered in an entire new era of modern white wine. Astrolabe continue the tradition in the coolest part of Marlborough on the South Island – Awatere Valley, dry, challenging and open to the cold southern winds. Domaine de Chevalier makes one of France's classier, barrel-fermented whites by blending 30% Semillon in with the Sauvignon Blanc – a practice also copied in other parts of the world like California, South Africa, Western Australia and New Zealand.

MATCHING SAUVIGNON BLANC AND FOOD

This grape makes wines with enough bite and sharpness to accompany quite rich fish dishes as well as being an obvious choice for seafood and for Thai dishes. The characteristic acid intensity makes a brilliant match with dishes made with tomato, but the best match of all is white Sancerre or Pouilly-Fumé with the local sharp *crottin* goats' cheese of the Upper Loire Valley. With their strong, gooseberry-fresh taste, Sauvignons make good thirst-quenching apéritifs.

CONSUMER INFORMATION

Synonyms & local names

Sometimes called Blanc Fumé in the central Loire; there are variations called Jaune, Noir, Rose or Gris and Violet according to the berry colour; it is called Muskat-Silvaner or Muskat-Sylvaner in Germany and Austria (though Steiermark often uses the name Sauvignon Blanc); oaked versions are also known as Fumé Blanc in California and Australia. The variety Sauvignon Vert or Sauvignonasse is unrelated.

Best producers

FRANCE/Pouilly-Fumé J-C Chatelain, Didier Dagueneau, Ladoucette, Masson-Blondelet, de Tracy; **Sancerre** G Boulay, H Bourgeois, F Cotat, F Crochet, Alphonse Mellot, Vacheron; **Bordeaux/Pessac-Léognan** Brown, Dom. de Chevalier, Couhins-Lurton, Fieuzal, Haut-Brion, la Louvière, Malartic-Lagravière, Smith-Haut-Lafitte; **Graves** Clos Floridène.
ITALY Colterenzio co-op, Peter Dipoli, Edi Kante, Lageder, Vie di Romans, Villa Russiz.
AUSTRIA Gross, Lackner-Tinnacher, Neumeister, Polz, E Sabathi, Sattlerhof, E & M Tement.
SPAIN Hermanos Lurton, Palacio de Bornos, Marqués de Riscal, Javier Sanz, Sitios de Bodega, Torres.
USA/California Araujo, Brander, Coquerel, Dry Creek, Flora Springs, Grgich Hills, Heitz, Honig, Kenwood, Mondavi, St Supery, Spottswoode.
AUSTRALIA Angullong, Bannockburn, Bird in Hand, Brookland Valley, Larry Cherubino, De Bortoli, Hanging Rock, Houghton, Karribindi, Katnook, Lenton Brae, Logan, Longview, Nepenthe, S C Pannell, Philip Shaw, Shaw & Smith, Stella Bella, Tamar Ridge, Word of Mouth.
NEW ZEALAND Astrolabe, Blind River, Brancott, Cloudy Bay, Dog Point, Gladstone, Greywacke, Lawson's Dry Hills, Man O'War, Matua Valley, Neudorf, Palliser, Pegasus Bay, Sacred Hill, Saint Clair, Stoneleigh, Te Kairanga, Te Mata, TerraVin, Vavasour, Villa Maria, Yealands.
CHILE Casa Marín, Casas del Bosque, Concha y Toro, Cono Sur (20 Barrels), Errázuriz, O Fournier, Viña Leyda, Luis Felipe Edwards, Montes, San Pedro (Castillo de Molina), Santa Rita (Floresta), Undurraga.
SOUTH AFRICA Graham Beck, Cape Point, Cedarberg Ghost Corner, Constantia Glen, Neil Ellis, Flagstone, Fleur du Cap, Fryer's Cove, Hermanuspietersfontein, Iona, Klein Constantia, Mulderbosch, Oak Valley, Quoin Rock, Springfield, Steenberg, Thelema, Vergelegen.

RECOMMENDED WINES TO TRY

Ten New Zealand Sauvignon wines

Astrolabe Marlborough
Blind River Marlborough
Cloudy Bay Marlborough
Dog Point Marlborough
Te Kairanga Wairarapa
Greywacke Marlborough
Saint Clair Marlborough
Vavasour Marlborough Single Vineyard
Villa Maria Marlborough Clifford Bay Reserve
Yealands Marlborough

Five classic Loire Sauvignon wines

J-C Chatelain Pouilly-Fumé
Francis & Paul Cotat Sancerre Chavignol la Grande Côte
Lucien Crochet Sancerre Cuvée Prestige
Didier Dagueneau Pouilly-Fumé Pur Sang
Alphonse Mellot Sancerre Cuvée Edmond

Eight classic Sauvignon-dominated dry white Bordeaux wines

Ch. Brown Pessac-Léognan
Domaine de Chevalier Pessac-Léognan
Ch. Couhins-Lurton Pessac-Léognan
Ch. la Louvière Pessac-Léognan
Ch. Malartic-Lagravière Pessac-Léognan
Ch. Margaux Bordeaux Pavillon Blanc
Ch. Pape-Clément Pessac-Léognan
Ch. Smith-Haut-Lafitte Pessac-Léognan

Eleven other New World Sauvignon wines

Casa Marín Cipreses (Chile)
Cederberg Ghost Corner Sauvignon Blanc (South Africa)
Concha y Toro Terrunyo (Chile)
Errázuriz Aconcagua Coastal (Chile)
O Fournier Centauri (Chile)
Luis Felipe Edwards Marea de Leyda (Chile)
Montes Outer Limits (Chile)
Viña Leyda Sauvignon Blanc (Garuma) (Chile)
Shaw & Smith Sauvignon Blanc (Australia)
Steenberg Black Swan (South Africa)
Vergelegen Sauvignon Blanc (South Africa)

The exciting thing about Sauvignon Blanc is its wonderful, unabashed fruit salad bowlful of flavours, all tumbling over one another. These tastes are most obvious when the wine is young, but, especially when barrel-fermented and blended with Sémillon, Sauvignon can produce deep, complex, long-lasting wines.

Maturity charts

Sauvignon is usually made for early drinking, though some Sancerres and Pouilly-Fumés will keep and improve for much longer.

2012 Sancerre/Pouilly-Fumé

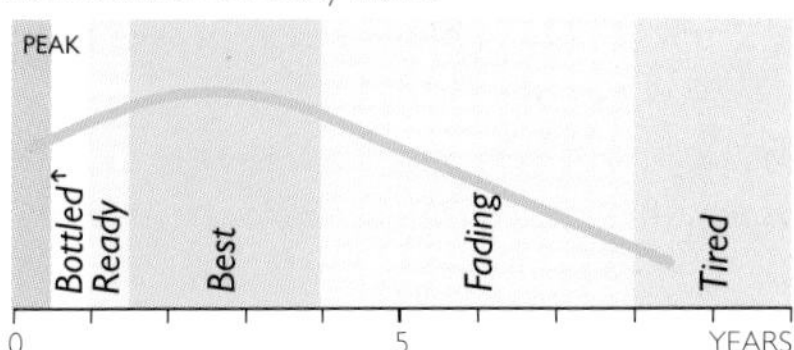

A few producers make Sancerre that will age in bottle, but generally it is a light, fresh wine intended for early drinking.

2012 Pessac-Léognan Cru Classé

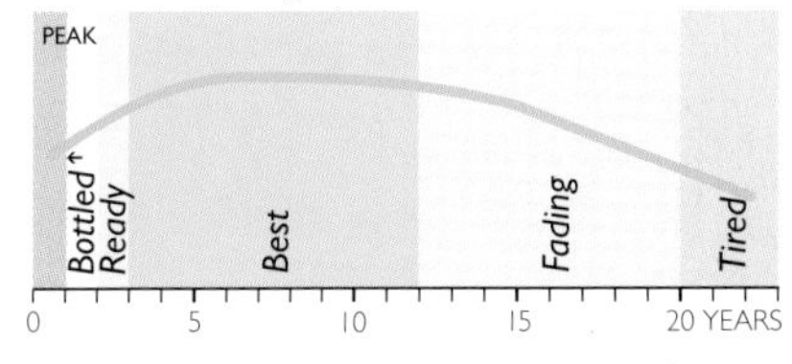

Recent vintages, even of slow developers like Dom. de Chevalier, are brilliant young, but also age beautifully.

2014 New Zealand Marlborough Sauvignon Blanc

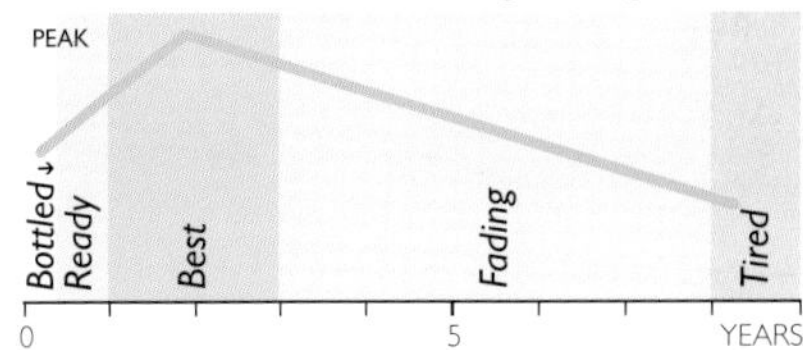

A couple of years in bottle is generally enough for even the top wines; more everyday bottles are best drunk within the year of the vintage.

SAUVIGNONASSE

It was only in the early 1990s that a distinction was made in Chilean vineyards between Sauvignon Blanc, which was what producers thought they had, and Sauvignonasse, which was what they actually did have, much of the time. The two varieties look similar: it is an easy mistake to make. And since the original cuttings of 'Sauvignon Blanc' that populated Chile's vineyards came from Bordeaux in the 19th century, the Chilean field mix merely reflected the 19th century Bordelais field mix.

Sauvignonasse is also known as Sauvignon Vert, but unlike Sauvignon Gris (also found in small amounts in Chile) it is not a mutation of the more pungent Sauvignon Blanc. Sauvignonasse did share the vineyards of Bordeaux with Sauvignon Blanc, but it lacks the assertive nose, the acidity and the staying power of the latter. Until it's about three months old it can have quite good aroma and flavour, but it is an aroma of green apples rather than the blackcurrant leaf and gooseberries of Sauvignon Blanc. It has little character if picked underripe, and reaches high levels of alcohol – up to 14.5 per cent – very easily. Acidity, however, drops rapidly, and at more than 13 per cent alcohol the wine can taste dull and featureless.

It's not clear just how much Sauvignonasse lingers in Chilean vineyards – the authorities recognize only Sauvignon Blanc, so that's how it appears in the records – but it may be as little as 200ha (500 acres). New plantings in the north and on the coast are all Sauvignon Blanc.

Sauvignonasse used to be known as Tocai Friulano in Italy, where it was imported in the 19th century, but its official name now is simply Friulano to prevent confusion with Hungary's Tokaji. It would be interesting to see if Chile could produce versions as impressive as those of the best growers of Collio and Colli Orientali del Friuli by dramatically lowering yields and taking their winemaking a little more seriously. See also Sauvignon Blanc pages 232–241.

SAUVIGNON GRIS

An alternative name for Sauvignon Rosé, which is a pink-skinned version of Sauvignon Blanc (see pages 232–241). It is much less aromatic than Sauvignon Blanc, but makes powerful, rather interesting wines. Chile has some, as does Bordeaux: Château Smith-Haut-Lafitte in Pessac-Léognan, for example, sometimes adds around 5 per cent Sauvignon Gris to its dry white which is otherwise entirely Sauvignon Blanc. Best producers: (France) Carsin, Courteillac, la Ragotière.

Savagnin vines growing on the limestone and marl soils of the Château Chalon appellation, in the Jura region of eastern France. The Vin Jaune from here is one of the very few flor-growing wines in the world. I'm not convinced that it's as good as good sherry, but I'm beginning to understand its hidden charms.

SAUVIGNON VERT

An alternative name for Sauvignonasse (see left) and its official title in Chile.

SAVAGNIN

A very old variety that probably comes from northeast France (where its main stronghold is still the Jura) or southern Germany. There is a possibility that it is directly descended from wild vines; but on the other hand, Pinot might be one of its parents. The only thing that is certain is that Savagnin has an enormous number of mutations. Gewürztraminer, Traminer and Heida (Païen) are all, in fact, Savagnin, even though all are discussed separately in this book. And they are very different.

For Savagnin Rose and its aromatic form Gewürztraminer, see Gewürztraminer pages 112–121. Here we'll look at Savagnin Blanc, which seems to be its earliest form.

Savagnin is found in the Jura region, and is best known for the local speciality Vin Jaune, though it also makes straight table wines of startling structure, pungency and minerality. For Vin Jaune, the Savagnin ripens late, being picked at 13 to 15 per cent potential alcohol in November or even December. The *flor*-like yeast covering, here called *voile* or veil, grows more slowly and more thinly than the *flor* of Jerez (see Palomino Fino page 178), and dies earlier. There is no solera system used in the Jura; in Jerez and the other sherry towns it is the constant refreshing of wine in solera that keeps the *flor* alive. In addition, the much cooler temperatures of the Jura do not encourage such lavish growth. The wine is left in cask for six years and three months, by the end of which time it has developed a pungent, oxidized, nutty flavour with piercing acidity. Once bottled it is said to be able to last 50 years or more. I'm still keeping my one bottle.

Savagnin is grown throughout the Jura region, and while it is a permitted addition to any white wine there, it is usually kept for Vin Jaune. To recap, there is also a Savagnin Rose which is the same as Traminer (see page 284). The *musqué* form of Savagnin Rose is the more aromatic Gewürztraminer (see pages 112–121). Australia makes good examples tasting of green apple core and lemony acid, although the growers thought they were planting Albariño. In Switzerland, as Heida or Païen it is full-bodied and fairly spicy. It's a speciality of the Valais, and is grown high up (1100m/3610ft high up) in Visperterminen. Best producers: (France) Arlay, Jean Bourdy, Hubert Clavelin, Jean-Marie Courbet, Durand-Perron, l'Étoile, Henri Maire, Montbourgeau.

SAVAGNIN NOIR

A name occasionally used in the Jura and parts of Switzerland for Pinot Noir (see pages 188–199). It should not be confused with the local grape Savagnin (see left).

SAVATIANO

Greece's workhorse grape, usually neutral and low in acidity, and used for inexpensive branded wines and for retsina. More acidic grapes, especially Assyrtiko and Roditis, are often added to retsina to give balance. The pine resin is added in pieces to the must, and removed only when the wine is racked. If grown on good sites and picked slightly earlier, Savatiano can produce surprisingly well-structured wines. Best producers: (Greece) Achaia-Clauss, Kourtakis, Semeli Winery, Skouras, Strofilia.

SCHEUREBE

Scheurebe was long thought to be a crossing of Riesling with Silvaner, aimed at finding an improved version of Riesling – or possibly a more perfumed version of Silvaner, depending on who you talked to. However, while one of its parents is certainly Riesling, the other is unknown.

Its wine lacks the taut elegance of good Riesling and even at its best tends to be clumsier, but it is complex and rich at high Prädikat levels, making powerful sweet wines with a fantastic flavour of ripe pink grapefruit swathed in honey which age quite well, though not for as long as equivalent Rieslings. It ripens to higher sugar levels, is high-yielding, and seems to produce its most exciting wines in the Pfalz. When made dry, there is a danger of catty, white grapefruit flavours if the grapes are underripe; Scheurebe like this can be raw and aggressive. It is, however, far and away the most successful of the modern German crossings, and the only one highly regarded by serious winemakers.

In Austria it is known as Sämling 88 (see page 220) – seedling number 88 was the seedling selected from all those propagated by Scheu in 1916. Best producers: (Germany) Andreas Laible, Lingenfelder, Müller-Catoir, Hans Wirsching, Wolff-Metternich; (Austria) Alois Kracher.

SCHIAVA

This Italian red grape (or grapes, since the name covers several more or less similar but genetically different vines) produces perfectly pleasant everyday wines in Trentino-Alto Adige, but doesn't seem capable of anything of real excitement. It yields generously, and gives wines of light smoky strawberry fruit and a mildly creamy texture. Concentration, depth and complexity, however, are generally lacking. It is declining as growers see the greater commercial opportunities of weightier reds, but still covers a substantial area.

Its Italian name means 'little slave' – the variety is very amenable, though its Süd-Tirol German name, Vernatsch, means 'of local origin', suggesting that it has long been thought of there as being a local variety. The name it is given in Germany itself, Trollinger, also suggests a link with the Tyrol. Records of the vine in Trentino-Alto Adige go back to the 13th century. (See Vernatsch page 287 and Trollinger page 284.) The range of Schiavas includes Schiava Grigia, or Grauvernatsch; Schiava Gentile, or Edelvernatsch; and Schiava Grossa, or Grossvernatsch; and Schiava Lombarda. Schiava Grossa is the least distinguished, but particularly high yielding, so inevitably it is the most planted.

It is the main grape in DOC Santa Maddalena, where it may be given more character by Lagrein or some other red grape and I have to say I've had some really lovely gentle, fresh picnic wines sitting among the vines high above the city of Bolzano. It is also found in numerous other DOC and non-DOC wines in the area. Declining Schiava may be, but in the high mountains of Alto Adige in early summer, well, find me a more delightful red than one of these. Best producers: (Italy) Cornaiano co-op, Cortaccia co-op, Franz Gojer, Gries co-op, Lageder, Josephus Mayr, Thomas Mayr, Niedermayr, Georg Ramoser, Hans Rottensteiner, Heinrich Rottensteiner, San Michele Appiano co-op, Santa Maddalena co-op, Termeno co-op.

SCHIOPPETTINO

A fairly characterful northeastern Italian variety recently rescued from terminal decline. It is native to Friuli, and is also known there as Ribolla Nera. Its flavour is peppery and raspberryish, fairly light in body and alcohol, and with high acidity. There is also a local young and fizzy version. Best producers: (Italy) Dorigo, Davide Moschioni, Petrussa, Ronchi di Cialla, Ronco del Gnemiz, La Viarte.

SCHÖNBURGER

A German crossing (Pinot Noir with a crossing of Chasselas Rose and Muscat Hamburg) now more grown in England than in Germany, where it is concentrated in the Rheinhessen and Pfalz. It ripens easily, yields well and is disease-resistant. Its berries are pink but it is used for making white wine. Its perfume is heavy and somewhat Muscatty which is very attractive in a light English wine grown in England's cool climate, but can be rather cloying in wine from warmer climes. Best producers: (Canada) Gehringer; (England) Carr Taylor, Danebury.

SCHWARZRIESLING

The German name for Pinot Meunier (see page 183); it is found mostly in the Württemberg. region. Best producers: (Germany) Dautel, Drautz-Able, Fürst zu Hohenlohe-Öhringen, von Neipperg.

SCIACARELLO

A grape once thought to be unique to Corsica but now revealed as the Tuscan Mammolo. However, since it's probably been growing in Corsica since the 11th century, you can see why the Corsicans claimed it as their own. It is at its best in the southwest of the island, around Sartène and Ajaccio, and while its wines are light in colour and not particularly tannic, they have a lively herby pepperiness. With age they develop hints of woodsmoke and tobacco. The name means 'the grape that bursts under the teeth' and – guess what – the grapes have tough skins and lots of juice. Best producers: (France) Albertini Frères, Clos Capitoro, Clos Laudry, Martini, Peraldi, Torraccia.

SCUPPERNONG

A vine found in the southwestern states of the USA and in Mexico, Scuppernong is a *Vitis rotundifolia* vine, and belongs to the genus Muscadinia. The thick-skinned berries grow in small clusters and are low in sugar; chaptalization is normal. Pressing can be difficult, too, because of the thick, fleshy pulp. The flavour is strong and musky and the wines are usually made sweet.

Virginia Dare, a North Carolina wine that enjoyed great popularity in the early years of the 20th century, was made from Scuppernong and named after the first child born in the American colonies to English settlers. It seems bizarre now, with the dominance of California as a USA grape grower, but, largely due to Scuppernong, North Carolina was, for a time in the 19th century, the USA's most prolific grape grower, and Scuppernong is still the official state fruit.

SEIBEL

A group of French hybrids produced by Albert Seibel (1844–1936). Seibel 4643, also known by the somewhat ambitious name of Roi des Noirs, used to be widely planted in western France, though is so no longer: its wine was rustic and dark. Other Seibels include 7053, otherwise known as Chancellor; and 5279, or Aurore, both early ripeners planted here and there in North America.

SÉMILLON

You can't say that Sémillon hasn't been given a chance. Half the wine countries in the world have given it a chance, and almost as many have decided that it can't seem to produce anything remotely interesting, so let's rip most of it out again. That happened in South Africa – where it was very popular largely because Semillon rewarded the very favourable, benign vineyard conditions there with oceans of dull, tasteless juice that just happened to be fine for distilling into brandy if little else. Luckily brandy was what most South Africans drank at the time. Chile was overrun with it for much of the 20th century. Argentina and the rest of South America also gave it its head, and Sémillon thanked them by making wines that were a byword for dullness and which revelled in attracting sulphur like carrion attracts crows.

And, of course, there's Bordeaux in southwest France. Bordeaux has the biggest plantings in the world, but even these are a fraction of what they used to be. Barely 50 years ago, half of Bordeaux's production was white, mostly from the Sémillon grape. Now Bordeaux has 121,270ha (299,665 acres) of vines. Only 7021ha (17,350 acres) of these are Sémillon, and the most important of these are to the southeast of the city of Bordeaux, in Graves, Pessac–Léognan and Sauternes. And it's here that Sémillon can excel, though not without help. In the first case, human intervention has discovered that if you ferment Sémillon in barrels it takes on a wonderful waxy, creamy quality. Blend it with Sauvignon, which adds crispness, leafiness, citrous tang, and you have memorable dry white wine, as good as white Burgundy. A few people round the world have succeeded with this formula, mostly in South Africa and Western Australia.

Seen here in the honeyed autumnal light evocative of its precious golden wine, Château d'Yquem is the supreme example of the majestic sweet wines of Sauternes. Pickers will go through the vineyards up to a dozen times during the harvest, picking only the most 'nobly rotted' grapes in a constant attempt to make the best sweet wine in the world.

But it's greatest triumph is due to nature's helping hand. In the Sauternes region, nature creates perfect conditions for the grapes to rot in the vineyard. Now, rot is the curse of the red wine grower. Yet rot is the making of the crop in Sauternes. But not any old rot. Noble rot, so called because it concentrates the sugar in the grapes without turning the juice sour and undrinkable, needs very special conditions to flourish, and in Sauternes' little patch of land it gets them. The local rivers Ciron and Garonne get very foggy in the autumn mornings. If that fog is compounded by rain, we're in trouble. All kinds of rot will threaten, none of them noble. But if the sun fills the sky and burns off the morning fog, the whole vineyard becomes muggy and humid – and hot. And in these conditions noble rot sets to work thinning the skins of the Sémillon grape and intensifying the sugar to such an extent that memorable sweet wine is created.

There's one other place where terrible weather has conspired to make great Semillon wine – Australia's Hunter Valley, north of Sydney. Nature knows that no sane person would try to grow grapes there, and ever since the first magnificent obsessives decided to have a go in the 1830s nature has done her best to flout their efforts. She's washed away the decent soils with a never-ending succession of tropical storms. She's arranged subtropical heat during the summer, brackish bore water unfit for irrigation, and frequent winter droughts, just in case you were thinking you might build a few dams to store water to help your vines survive. And to make quite sure you got the message that grape-growing is doomed in the Hunter, she arranges cyclones to sweep down the coast and batter the valley just before the hapless grapes are ripe – just so you know who's really in charge. And without me attempting to seek refuge in rhyme or reason, it is precisely these woeful conditions that have created the classic Hunter Valley Semillon.

In the occasional perfect summer, with the grapes fully ripe and the harvest safely in, Hunter Semillon is full and fat, a bit blowzy – good grog, but quick to flower and fade. But in the years when nature does her worst, when the grape's alcohol level sometimes barely reaches 10 per cent alcohol, the result – if you wait 10 years for it to mature – is one of the world's great classic whites. Unsurprisingly, completely unlike any other white wine in the world.

Sémillon: from Grape to Glass

Geography and History page 246; Viticulture and Vinification page 248; Sémillon around the World page 250; Enjoying Sémillon page 252

GEOGRAPHY AND HISTORY

Where has all the Sémillon gone? Look at the map and it's just isolated patches: southwest France has the most, and there's some in Australia, some in Chile and other South American countries, and a bit in South Africa.

Yet 30 years ago some three-quarters of Chile's white vines were Semillon (it's usually seen without the accent in New World winemaking countries). It smothered South Africa – in 1822 it covered 93 per cent of the vineyard area, and was referred to with perfect logic simply as Wyndruif, or wine grape.

Chile now has much less. South Africa has much, much less. First of all, Sémillon gave way to Chardonnay; now white grapes are in their turn giving way to red. And Sémillon, planted all over the world for its disease resistance and its ability to produce large quantities of grapes, might have had its day, were it not for two facts. One is that it produces outstanding sweet wine in Sauternes and Barsac on the left bank of the Garonne in Bordeaux. The other is that it produces outstanding dry wine in Australia's Hunter Valley north of Sydney, and blended with

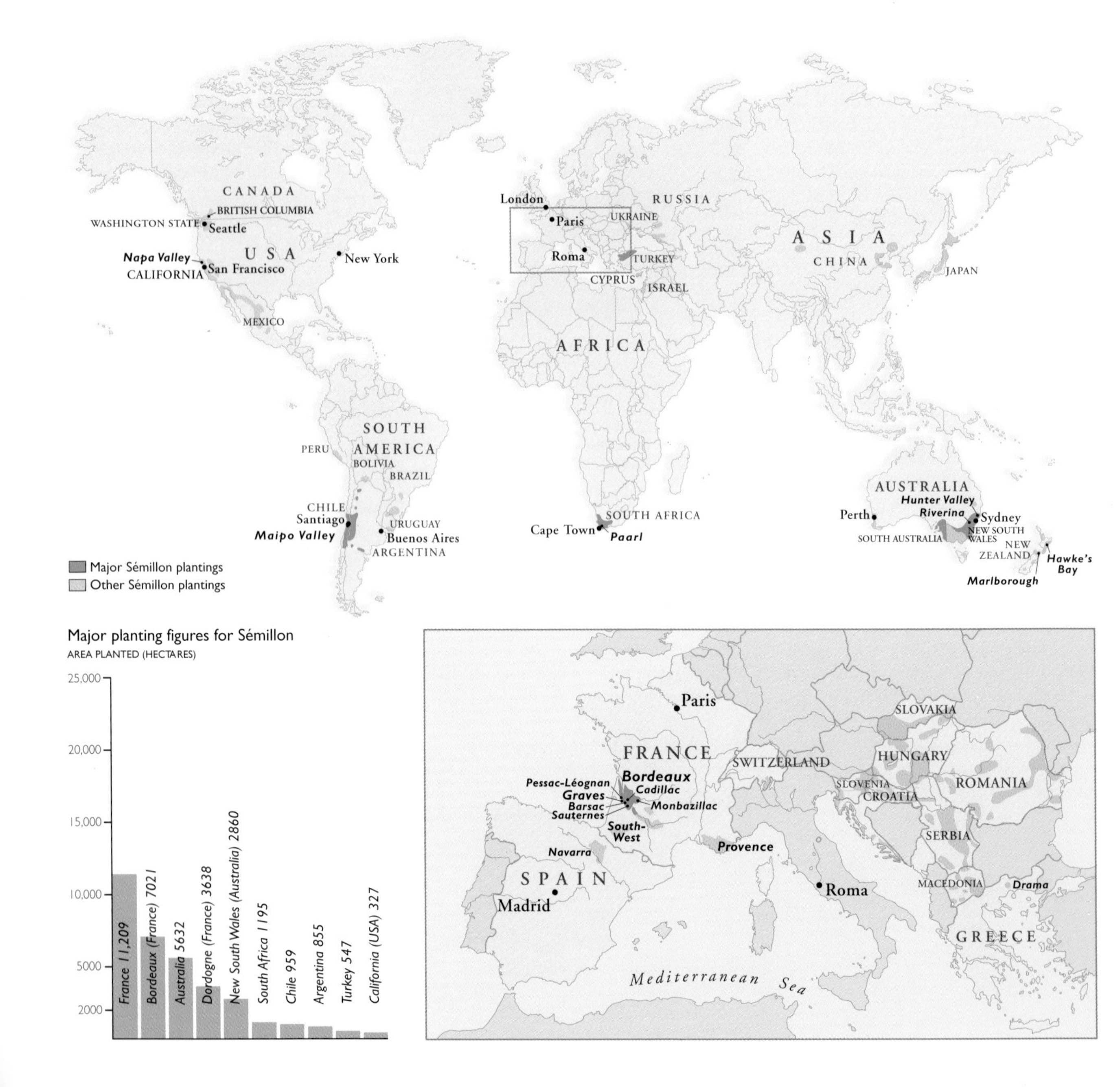

Sauvignon, in Western Australia's Margaret River, a few spots in South Africa, and, above all, in Pessac-Léognan and Graves, next door to Sauternes. And that's just enough of a CV to keep Sémillon being planted around the world.

The thing is, Sémillon was planted as an all-purpose grape – you could make dry or sweet wine from it, sherry, brandy, whatever. Yet it isn't an all-purpose grape. At high yields it is dilute and thin; when underripe it is green and stringy. Even its sweet wines usually need a touch of Sauvignon Blanc to brighten them up. And then, in misty, humid Sauternes and the subtropical Hunter Valley where the conditions are hardly suitable for grape growing at all, it produces world-class wine. So why does it make so few great wines in much easier conditions? Let's have a look.

HISTORICAL BACKGROUND

The ampelographer Galet thinks that Sémillon probably originated in Sauternes and spread from there to the rest of the Gironde. It was found in St-Émilion by the 18th century, and indeed St-Émilion used to be one of its local synonyms; nowadays this is a synonym for the Ugni Blanc variety, which Sémillon does not remotely resemble. It was still planted in St-Émilion on a small scale, along with Sauvignon Blanc and Muscadelle, for white wines well into the 20th century, and even now there are a few fugitive vines there.

When was Sauternes first made sweet? The usual date given, for want of concrete evidence of an earlier date, is the mid-19th century. To suppose that it was not made sweet before, however, requires a suspension of disbelief: Tokaj had been famous for its botrytis-affected sweet wines since the late 17th century, so the technique of making sweet wines from botrytized grapes was well known; and *Botrytis cinerea* occurred in Sauternes and Barsac then just as it does now. Picking in the region in the 18th century was not until late November; in Cadillac – on the right bank of the Garonne – the Abbé Bellet, who kept records of every vintage from 1717 to 1736, confirms that by October the grapes were affected by noble rot, and were picked in selective *tries*. He does not confirm that only the rotten ones were used, but if the non-rotten ones had been the most desired, and had been picked in the first *trie*, it is hard to see the point of further pickings. Clearly the selective trips through the rows of vines were to pick out the grapes affected by noble rot. And he would have found it pretty hard to make ordinary dry wine from such grapes.

If you judged a grape solely on its appearance, it's hard to see why anyone would make wine from ugly, squishy, noble-rotted grapes. But squeeze the gooey syrup from the grapes and lick your fingers and the fabulous rich flavour will persuade you in a flash.

A golden autumn day at Château Suduiraut at Preignac in the Sauternes appellation. This is just the sort of weather that favours the development of noble rot: humid, foggy nights are essential, but if dampness persists throughout the day, the rot will turn grey and ignoble and there will be little chance of rich, sweet wine.

A carved coat of arms at Château d'Yquem, Sauternes. The Sauvage family owned d'Yquem until 1785, when the Lur-Saluces family acquired it. They held it until 1999, when it was bought by LVMH.

A gooey mess of nobly rotten Sémillon grapes. Grapes in this state may look hideous but they are highly desirable for their super-concentrated sugars. They are difficult to press and slow to ferment.

VITICULTURE AND VINIFICATION

I can see why Sémillon was such a popular variety in the old days. It will grow just about anywhere and will recklessly produce gigantic crops of grapes. Which taste of...? Er, nothing really. Just vatsful of juice of no discernible personality but enough sugar to ferment into whatever kind of concoction you fancy. From dry white to sweet, from sherry to brandy. But in one or two little corners of the globe, if you really restrict its crop and vinify it skilfully, it has for centuries produced world-class dry and sweet whites. It buds slightly later than Sauvignon Blanc but ripens earlier, and is actually less subject to noble rot than Sauvignon. It is resistant to most other diseases, but rainy years can produce attacks of grey rot. This is caused by the same fungus as noble rot, but has different, more harmful results – most importantly it tastes horrible.

Climate

It is an unarguable fact that if you were to match your vine varieties with your sites on a rigid degree-day basis you would never plant Semillon in the Hunter Valley. It is, by this reckoning, far too warm there to produce dry Semillon, and indeed far too humid, especially when the humidity takes the form of tropical downpours at harvest time. But Semillon thrives there, and even more curiously, it is rainy years that often produce the best Hunter Semillon, years which are so poor in accepted terms that the alcohol level hardly reaches 10 per cent. There's got to be a reason. Up to a point, there is. So here's the science. It still won't all add up, because Hunter Semillon is one of the wine world's great enigmas. But here goes.

That the region produces such good Semillon is partly due to its cloud cover: the temperature may rise to 42°C (108°F) during the day, but the sunlight is muted. The Hunter has a relatively low ratio of hours of bright sunshine to effective day degrees, and this helps to keep sugar levels down. (It also helps to mitigate tannins and astringency in the red wines of the region.) The humidity also helps: high afternoon relative humidity is associated with higher acidity levels in grapes, and low-acidity Semillon needs all the help it can get in this department.

The climate in the Hunter Valley could hardly be less like that of Bordeaux, except that autumn humidity here is essential for the growth of *Botrytis cinerea* in Sauternes. Early morning mists, caused by the confluence of the ice-cold river Ciron and the warmer Garonne, spread back up the Ciron valley and encourage the growth of botrytis in the neighbouring Sauternes vineyards. The botrytis has the effect of concentrating both sugar and acidity, again giving the grape's low acidity levels a helping hand.

Interestingly, Bordeaux's relative humidity and sunshine hours compare with those of the Upper Hunter, though of course the latter is much, much hotter. The Lower Hunter, however, has less relative humidity than Bordeaux, and slightly fewer sunshine hours.

Soil

In the Hunter Valley the usual soils for growing vines are light, sandy ones, simply because heavy soils become quagmires after heavy rain. In a way, suitable doesn't come into it in the Hunter. There are some soils that will support vines – whatever variety – and there are poor clays that won't support anything at all. Period. The best-drained soils in the Lower Hunter Valley include the friable loam and friable red soils. In the Upper Hunter black silty loams over dark clay loam are successful; the red-brown duplex soils, which resemble those of the Lower Hunter, are better drained, which is not automatically an advantage in this hotter, drier region where irrigation is essential.

The soil in Sauternes is sandy gravel in varying depths over calcareous clay; Barsac is much flatter, and lacks the gravel of Sauternes. Barsac has well-drained, sandy, limy soil, and vine stress can be a problem in very dry years. In Sauternes, by contrast, the clay subsoil can be poorly drained where the topsoil is very thin:

Picking grapes is never easy work – it's backbreaking at the best of times. But when you have to search each cluster for berries of the right degree of noble rot, as is happening here at Château d'Yquem in Sauternes, then it requires great concentration. Only skilled pickers can be trusted with such a task.

Château d'Yquem put in some 10km (6 miles) of drainage in the 19th century to correct this. Because Sémillon ripens earlier than Sauvignon Blanc, it may be planted on more clayey soil in Sauternes; it is sometimes suggested that clay can favour the development of botrytis. Sauvignon Blanc may be planted on the gravel. In the Graves, however, where a different style of wine is desirable, Sémillon gets the warmer soils and the better-exposed sites. The AC rules state that the vines must be eight years old for Sauternes; some châteaux maintain that they don't get good levels of botrytis until they are 10. Root depth seems to be a factor in determining whether rot turns noble or grey; shallow-rooted vines tend to get grey rot.

Cultivation and yields

In Bordeaux growers may leave a long cane in order to have some spare buds in case of spring frost; what may happen is that the buds at the end of the cane grow vigorously and impede the development of those further down. The cane may be trained in an *arcure* to balance this.

Yields must be low if quality is the aim. Sauternes is the lowest: the legal limit is 25hl/ha, and most leading properties make much less. Yquem famously makes just one glass of wine per vine, or 9hl/ha. In Monbazillac legal maximum yields have been cut from an over-generous 40hl/ha to a sensible 27hl/ha. In the lesser regions of Bordeaux yields may reach 80–100hl/ha. Australian yields are around 3.5 tons per acre or 8 to 9 tonnes per hectare (roughly 60hl/ha); in New Zealand Semillon produces thinner, grassier wine at 10–17 tonnes/ha.

At the winery

The big question with Sémillon is to oak or not to oak? It certainly has a great affinity with oak, in particular with new oak, and the increased proportion of new oak used by the Sauternes châteaux since the mid-1980s has been a factor in the improvement of the region's wines. In the Graves, too, it is commonly fermented and aged in new oak; in youth these blends of Sémillon and Sauvignon Blanc can seem too intensely oaky but they age remarkably well. New oak here is being used with a more delicate hand than it was a few years ago, as is sulphur in Sauternes: Sémillon oxidizes easily, and one of the effects of noble rot is to make it require extra sulphur to protect it against oxidation. It takes a strong nerve for a grower to hold back. A few Sauternes are still noticeably sulphurous in youth, though this should fade with age.

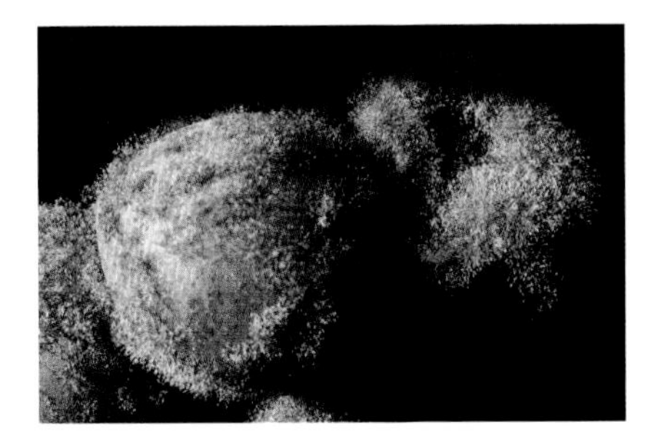

They look terrible, don't they – awful, soggy, covered with furry rot. But this is noble rot at work, and the worse the grapes look, the sweeter their juice will be and the more luscious the wine.

The alternative style in Bordeaux and the southwest is the stainless steel-fermented one of crisp fruit and youthful acidity. Such wines are not intended to age, and a large proportion of Sauvignon Blanc is essential if the wine is to have sufficient flavour; young, unoaked Sémillon can taste lemony and grassy, but that's about all. If it doesn't have Sauvignon to help it along, then it needs oak. Sometimes, as in the great Sémillon-based dry whites of the Graves and Pessac-Léognan, it gets both.

In Australia, unoaked Semillon is a classic style of the Hunter Valley north of Sydney, where unpredictable, subtropical conditions often mean the grapes have to be picked very early or they rot on the vine. These wines, neutral – indeed positively acidic and tart – in youth, mature (after a decade or so) into rich, honeyed toastiness – tasting for all the world as though they had spent their infancy in new oak barrels, although they haven't. Australian wine buffs love to bemuse visiting Brits in blind tastings with these unoaked beauties that we unerringly pronounce to be mature French Burgundy. The Hunter has toyed with oak-aging and now mostly rejected it; in South Australia, too, producers have mostly pulled back on the oak, though some is still successfully used in Western Australia. For the consumer, and perhaps for the producer, oak-aging means that the wine has more complexity at a young age: unoaked Hunter ones are beloved by those who know them, but they are about as far removed from wines for instant gratification as it is possible for a white wine to get. It's not a difficult wine to make. As Bruce Tyrrell says, 'just chuck it in a tank and leave it'.

SÉMILLON AND BOTRYTIS

What makes Sémillon and *Botrytis cinerea* so suited to each other? And what makes some rot noble and delicious and other rot merely grey and unpleasant-tasting?

Noble rot and grey bunch rot are the same fungus: both are *Botrytis cinerea*. The fungus is now thought to infect the berry at fruit set, and can develop in either direction, depending on circumstances; and it is by no means unknown to have grey rot and noble rot together, even on the same cluster.

For botrytis to turn noble, fluctuating humidity is necessary: just the sort of damp, foggy nights and early mornings, followed by warm sunny days, that are typical of a Sauternais autumn. Under these conditions fungal growth will be limited, and its metabolism modified. Constantly high humidity favours grey rot and heavy rain encourages berry-splitting; you need a team of experienced pickers to tell one from the other.

The flavour of botrytized Sémillon is not solely due to the concentration of sugars and acidity, though the drying and shrivelling of the grapes (which helps to protect against invasion by other bacteria that could produce off-flavours) is an important part. Noble rot metabolizes grape acids, especially tartaric, which falls dramatically in infected grapes. The concentration of the remaining acidity, however, means that the pH of botrytized Sémillon grapes is slightly, but only slightly, higher than that of healthy ones. It also metabolizes sugar, and total sugar content drops by 35–45 per cent; this is compensated by water loss as the grapes shrivel, and the remaining sugar becomes more concentrated. It produces glycerol, which contributes to the silky mouthfeel of botrytized wines, and it both degrades the esters that give white wines their fruitiness, and destroys the terpenes that give varietal aroma. Sémillon has so little varietal aroma and so few fruity flavours to start with that this is no great loss; and in their place the wine gains great aromatic complexity. (Muscat, by contrast, loses more than it gains, which is why most sweet Muscats are made sweet by the fortification method or by shrivelling.)

Sweet Sémillon's youthful aromas come from over 20 aromatic compounds synthesized by noble rot. Sotolon is one: in conjunction with others, it helps to give nobly rotten wines their distinctive honeyed aroma.

SÉMILLON AROUND THE WORLD

Sémillon's classic styles – Sauternes, white Pessac-Léognan and Graves, and the Hunter Valley – are so unlike each other that if growers elsewhere wanted to make serious Sémillon they would be forgiven for not knowing where to aim. Sémillon also seems to be its own worst enemy; unless severely regulated it produces huge amounts of insipid wine.

Sweet white Bordeaux

Sémillon is the main grape throughout Sauternes, Barsac, Monbazillac and the lesser sweet and semisweet white regions of Cadillac, Ste-Croix-du-Mont, Loupiac and Cérons. It accounts for between 60 and 90 per cent of most vineyards, the balance being Sauvignon Blanc (of which yields are slightly higher, so the percentage of Sauvignon in the wine may be greater than in the vineyard); and sometimes a few per cent of Muscadelle. Sauvignon adds freshness, lightness and acidity and Muscadelle aroma.

All these regions follow, more or less closely, the practices of Sauternes and Barsac, and the great improvements in quality in the latter regions since the 1980s have spread, with encouraging results. Monbazillac, for example, used to permit mechanical harvesters – a complete contradiction of the style of the wine, since botrytized wines depend on selection, and the hand and eye of the picker. Since 1994 these harvesters have been phased out, and minimum must weights have been increased from 13 per cent potential alcohol to 14.5 per cent. In fact, careful selection here can regularly produce potential alcohol levels of 18–19 per cent, and more in top years.

The rules in Sauternes are stricter. The alcohol in the finished wine must be at least 13 per cent, and is usually 14 per cent; the residual sugar usually amounts to another four to seven per cent potential alcohol. Château d'Yquem, for example, picks at between 20 and 22 per cent potential alcohol. It might be necessary to send the pickers through the vineyards for up to 10 separate passages to attain these levels, though three or four times is more usual. Much depends on the year: noble rot arrives more years than not, but it is unpredictable and often patchy. Years in which it blankets the vineyards, like 1990, are few and far between. Balance can also be a problem in Sauternes: Yquem sends constant instructions to its pickers to select more botrytized grapes or more healthy ones, as the balance of the must requires.

These four Sémillon grapes show different levels of botrytis infection – barely infected on the right through to shrivelled and furry, but intensely sweet, on the left. Most Sauternes will contain some of each level and only the best châteaux in the best years will manage nothing but the two on the left.

Chaptalization is permitted, though the best properties claim only to use it in poor years; cryoextraction is also permitted. This is an expensive technique which involves leaving the picked grapes for about 20 hours in a cold room to freeze their water content so that just the sweet juice runs from the press. It can be useful in a rainy year, but it is no substitute for careful selection. There are, inevitably, years in which noble rot fails to appear. In these years wine will be made from overripe and shrivelled grapes only: it may be sweet and concentrated and pretty tasty, but will lack the characteristic flavours of Sauternes.

Dry white Bordeaux

The lighter, sandier soils of Pessac-Léognan (where reds outnumber whites four to one) are the ones usually given over to white wines, of which at least 25 per cent of the blend must be Sauvignon Blanc. Usually Sauvignon's figure is much higher, and may even be 100 per cent, though Sémillon is valued for the richness it brings to the blend, and for its affinity for the new oak in which the wine is usually fermented and aged. Flavours can be remarkably exotic, with apricot, nectarine and mango, nuts, custard and buttered toast complementing the leafy greeness of the Sauvignon Blanc, and the wines can age for many years. Dry whites from the rest of Bordeaux and the southwest are seldom intended for aging. The blend may range from 100 per cent to no Sémillon. It is being outplanted not just by Sauvignon, but by more fashionable red varieties.

Château d'Yquem

This most famous of all Sauternes properties is owned by the multinational company LVMH, which also makes luggage, perfume and champagne.

Château Climens

Climens produces one of the richest wines in Barsac, where the rules allow wines to be labelled either as Barsac or as Sauternes.

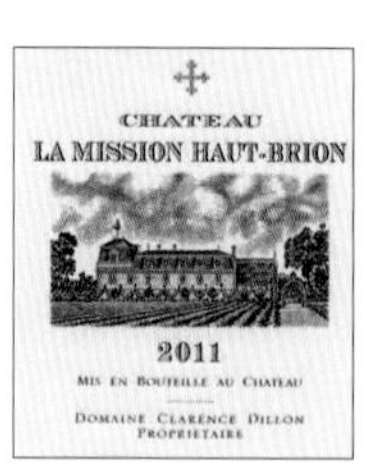

Château La Mission Haut-Brion

This wine used to be called Laville-Haut-Brion. Slow to open out, good vintages become deeper and more complex over 20 to 30 years.

The Lindemans Ben Ean winery in the Lower Hunter Valley. This is one of the great historic wineries of an historic region: vines have been grown in the Lower Hunter since the 19th century. The unoaked Semillons for which it is known are a unique world wine style and predate by many years the birth of modern Australian wine. Difficult to appreciate when young, they were nearly washed away by a tide of fashionable Chardonnay – nearly but luckily not quite – and now they're back, better than ever.

Australia

Hunter producer Bruce Tyrrell's comment that 'Semillon has a structure and acidity like Riesling; it's like Riesling in everything except flavour' might come as a shock to anyone comparing a high-acid Saar Riesling with a low-acid Bordeaux Sémillon. But Semillon in the Hunter Valley does have much higher acidity than it manages in France. The ripeness at which it is picked is a factor: for unoaked Semillon, which needs acidity to age for a decade or more in bottle, picking is often at 10.5–11 Baumé. For Semillon destined for oak-aging, more ripeness and substance is required, and picking is at 12–12.5 Baumé, to prevent the wine being swamped by the wood. If you pick at that ripeness for unoaked Semillon, the fruit loses its freshness and gets broad and blowzy after five or six years. The best vintages for unoaked Hunter Semillon are often the rainy ones: in 1971, says Tyrrell, it rained for three months, 'and the wine is still going strong'. A bit of botrytis – say 10–20 per cent – is also useful for the complexity and aroma it adds.

Semillon from the rest of Australia is usually oaked; only that from Hilltops in New South Wales seems to share the Hunter structure. The warm irrigated regions produce large amounts of fairly basic but pleasant dry white that is often blended with Chardonnay and seasoned with oak chips. The Riverina area of New South Wales gets natural botrytis infection and has produced some astonishing Sauternes-like sweet wines.

USA

Plantings in California are falling rapidly, and what little there is is usually blended with Sauvignon Blanc. Clones and yields may be at fault, but the main problem is probably that few care enough to treat the vine seriously. Clos du Val makes a good varietal, though it is not as long lived as a Graves or Hunter version; Semillon in California is often more successful as a sweet late-harvest wine. There were experiments in the 1950s and 1960s by Myron and Alice Nightingale with spraying picked grapes with botrytis spores; Beringer made some wines commercially this way into the 1990s. The results were astonishingly intense, but lacked finesse. Results from botrytis infection in the vineyard have generally been better – occasionally, in Napa, excellent. Both Oregon and Washington State make a little, generally grassy, Sauvignon-ish wine, though Washington can excel in a nutty style.

New Zealand

Marlborough, Gisborne and Hawkes Bay are the main regions for Semillon in New Zealand but plantings in 2013 accounted for less than 1 per cent of the overall total. The usual clone here – UCD2 – has loose knit clusters like the Semillon in California, and unlike the compact clusters of Bordeaux. It is often blended with Sauvignon Blanc, and can improve the longevity of the latter. One or two interesting sweet wines have been made from Semillon.

Rest of the world

Chile had 959ha (2370 acres) in 2011 – which shows how great the decline has been since 1985's figure of 6195ha (15,308 acres). A few producers, notably Morandé and Casa Silva, see it as having potential, and there are some old vines capable of interesting quality. Argentina has some. In South Africa, it is usually seen as a bulk variety except in Stellenbosch and Constantia, which makes fine barrel-fermented Sémillon–Sauvignon blends, and in Franschhoek and Swartland, which possess precious, ancient bushes. There is some in Croatia and other parts of Eastern Europe.

L'Ecole No 41
Washington State's Sémillon plantings are in decline, but L'Ecole No 41 continues to make a world-class example, helped by a small addition of Sauvignon.

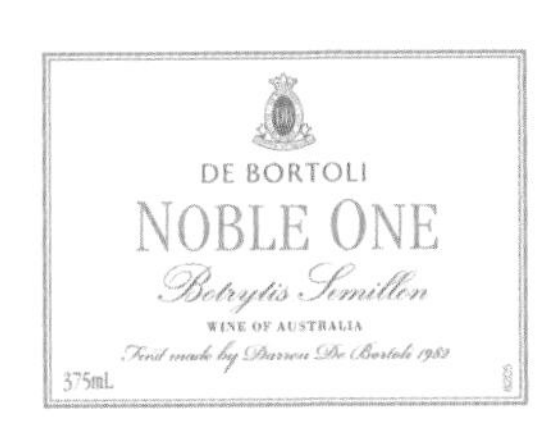

De Bortoli
The wine that has shown that Australia, too, can produce outstanding nobly rotten Semillon. Intense and sweet, it compares favourably with top Sauternes.

Ghost Corner
The extremely cool, windy conditions at Elim – on the southern tip of Africa – give leafy, crunchy flavours similar to Sauvignon Blanc.

ENJOYING SÉMILLON

Ask most of the world's winemakers how Sémillon ages, and you'll be told that it doesn't. In most countries it's a bland, bulk wine for drinking young.

Australia and France know differently. This is where Sémillon from a serious producer can last 20 years – and if botrytis is brought into the equation, wines from top châteaux can still be drinking a century later. Most Australian and French Sémillons, happily for those of us with neither the cellar nor the expected lifespan to keep wines for a century, are less extreme in their longevity.

Simple Bordeaux Blanc should be drunk within a year or two of the vintage; even decent quality white Pessac-Léognan and Graves should mostly be drunk within five years. Classed Growth wines can last longer, from 10 to 20 years in a good vintage, and are good at almost any age; the exceptions, and the longest-lived wines, are the magically rich yet dry Haut-Brion and Laville-Haut-Brion, which are seldom at their best before their fifth birthday but which positively demand 20 years in a good vintage.

Much is written about Sauternes needing time to mature. Of course it does improve in bottle, but it is also utterly delicious young, and while Classed Growth wines should probably not be touched for at least 10 years, lesser wines, and those from Monbazillac, can be drunk with great enjoyment earlier.

Unoaked Hunter Semillon is, however, less expressive in youth. It needs five or six years to show much character, but the rainiest, most acidic vintages (the best ones, in other words) can last 20 years or more. Upper Hunter Semillon develops faster than Lower Hunter wine. Oaked Semillon can sometimes last three or four years, but the point of the oak is to make the grape more complex, and therefore drinkable, at a younger age. It is always drinkable immediately upon release, though many improve with a little extra aging.

The taste of Sémillon

'Battery acid' is how winemaker Michael Hill Smith succinctly describes the flavour of young, unoaked Hunter Valley Semillon. The same could be said of much German Riesling; and like Riesling, Hunter Semillon develops astonishing flavours of honey and toast. After six or 10 years in bottle it does not have clear fruit flavours, although in youth it is citrous and fresh; maturity subsumes its fruit into a rich, silky wininess.

Oaked Australian Semillon is different: being picked riper it has richer fruit flavours, of greengages and apricots and mangoes, all mixed with the custardy vanilla of the wood. It is richer, broader and fuller than good white Graves or Pessac-Léognans, which usually have a substantial addition of Sauvignon Blanc for acidity and freshness; they are generally tighter and more subtle, and elegance, finesse and a complex nuttiness are more to the fore, particularly with maturity. In young Graves wines, if they are not dominated by new oak, the Sémillon adds creaminess, sometimes almost like egg custard, and nectarines to the Sauvignon, which in any case is less grassy in Bordeaux than it is in somewhere like New Zealand.

The flavours of Sauternes are of marzipan, apricots, mangoes, honey, nuts, toast, pineapple, peach, orange, honeysuckle, beeswax, barley sugar and coconut, all wrapped up in a creamy, silky, unctuous texture, enthralling and rich. If that seems like an awful lot of flavours, get yourself a mature bottle of top Sauternes and sip it reflectively – don't hurry. You may end up needing more words, not fewer.

Tyrrell's Vat 1 Semillon comes from vines dating back a century and more in the Hunter Valley. The wine is usually fairly low in alcohol and is bottled without any oak aging after only six months' aging. It starts out life lean, lemony and leathery, but takes on an amazing, honeyed, waxy, toasty richness with age. Château Rieussec is one of the most unctuous and compelling of Sauternes wines, immensely rich, almost syrupy, when young but able to age to a majestic mix of honey, barley sugar, pineapple and marmalade after 10–20 years.

MATCHING SÉMILLON AND FOOD

Dry Bordeaux Blanc wines are excellent with fish and shellfish; fuller, riper New World Semillons are equal to spicy food and rich sauces, often going even better with meat than with fish. Sweet Sémillons can partner many puddings, especially rich, creamy ones. Sémillon also goes well with many cheeses, and Sauternes with Roquefort or other blue cheese is a classic combination.

CONSUMER INFORMATION

Synonyms & local names
Known in 19th-century South Africa as Wyndruif ('winegrape') or 'groen' ('green') grape and in Australia's Hunter Valley as Hunter Valley Riesling. Called Boal in the Douro Valley, Portugal.

Best producers
FRANCE/Bordeaux/Graves and Pessac-Léognan Ardennes, Brondelle, Chantegrive, Dom. de Chevalier, Clos Floridène, Fieuzal, Domaine la Grave, Haut-Brion, Landiras, Laville-Haut-Brion, Latour-Martillac, Pape-Clément, Rahoul, Respide-Médeville, Roquetaillade-la-Grange, Seuil, Vieux Château Gaubert, Villa Bel Air; **Sauternes/Barsac** Bastor-Lamontagne, Climens, Clos Haut-Peyraguey, Coutet, Doisy-Daëne, Doisy-Védrines, Fargues, Filhot, Gilette, Guiraud, Haut-Bergeron, les Justices, Lafaurie-Peyraguey, Lamothe-Guignard, Laville, Malle, Myrat, Nairac, Piada, Rabaud-Promis, Raymond-Lafon, Rayne-Vigneau, Rieussec, Sigalas-Rabaud, Suau, Suduiraut, la Tour-Blanche, Yquem; **other Bordeaux** Birot, Carsin, Cayla, Chantegrive, Cros, Fayau, Grand Enclos du Château de Cérons, Haura, Lagarosse, Loubens, Loupiac-Gaudiet, les Miaudoux, Noble, Pavillon, la Rame, Reynon, Ricaud, Seuil, Sours, Toutigeac, Turcaud; **Bergerac** le Raz, Tour des Gendres; **Monbazillac/Saussignac** l'Ancienne Cure, Bélingard, Bellevue, la Borderie, Clos d'Yvigne, Grande-Maison, les Hébras, la Maurigne, Miaudoux, Richard, Theulet, Tirecul-la-Gravière, Treuil-de-Nailhac, Verdots.
USA/California Amador Foothill, Chalk Hill, Far Niente, Peter Michael, Simi, Spottswoode, St Supéry, Swanson; **Washington State** Amavi, Chateau Ste Michelle, Columbia, L'Ecole No 41, Woodward Canyon.
AUSTRALIA Tim Adams, Brokenwood, Leo Buring, Cape Mentelle, Chain of Ponds, Cullen, D'Arenberg, De Bortoli, De Iuliis, Fermoy Estate, Huntington Estate, Peter Lehmann, Lindeman's, McWilliams, Moss Wood, Mount Horrocks, Nepenthe, St Hallett, Stella Bella, Torbreck, Tyrrell's, Vasse Felix, Westend, Yalumba.
NEW ZEALAND Alpha Domus, Pegasus Bay, Selaks, Seresin, Sileni.
SOUTH AFRICA Boekenhoutskloof, Cape Point, Cederberg Ghost Corner, Fairview, Mullineux, Neethlingshof, Sadie, Steenberg, Stellenzicht, Vergelegen.

RECOMMENDED WINES TO TRY

Classic sweet white Bordeaux
See Best producers for Sauternes/Barsac left.

Ten other French sweet white wines
Ch. d'Arche Sauternes
Ch. Bastor-Lamontagne Sauternes
Ch. Barréjats Sauternes
Grand Enclos du Château de Cérons Cérons
Ch. Loubens Ste-Croix-du-Mont
Ch. les Miaudoux Saussignac
Ch. du Noble Loupiac
Ch. la Rame Ste-Croix-du-Mont Réserve du Château
Ch. Theulet Monbazillac Cuvée Prestige
Ch. Tirecul-la-Gravière Monbazillac

Ten French Sémillon-dry blends
Ch. Bauduc Les Trois Hectares, Bordeaux Blanc
Ch. Carsin Bordeaux Blanc Cuvée Prestige
Ch. de Chantegrive Graves Cuvée Caroline
Clos Floridène Graves
Domaine la Grave Graves
Ch. Haut-Brion Pessac-Léognan
Ch. Laville-Haut-Brion Pessac-Léognan
Ch. le Raz Montravel Cuvée Grande Chêne
Ch. Reynon Bordeaux Blanc Vieilles Vignes
Ch. Tour des Gendres Bergerac Cuvée des Conti

Ten dry New World Semillon wines
Tim Adams Clare Valley (Australia)
Bethany Barossa Valley Wood Aged (Australia)
Boekenhoutskloof Franschhoek (South Africa)
Cederberg Ghost Corner Semillon (South Africa)
L'Ecole No 41 Columbia Valley Barrel-Fermented (Washington)
McWilliams Hunter Valley Mount Pleasant Lovedale (Australia)
Moss Wood Margaret River (Australia)
Sileni Semillon (New Zealand)
Stellenzicht Stellenbosch Reserve (South Africa)
Tyrrell's Hunter Valley Vat 1 (Australia)

Five sweet New World Semillon wines
Tim Adams Clare Valley Botrytis (Australia)
Chateau Xanadu Margaret River Noble Semillon (Australia)
De Bortoli Noble One (Australia)
Swanson Crepuscule (California)
Yalumba Eden Valley Botrytis (Australia)

Sémillon ready to be picked. The green grapes have turned to gold and a few have even begun to speckle. Leave them and they will nobly rot for sweet wine if the weather permits. Pick them now and they may lack freshness but will add wax and lanolin texture.

Maturity charts
Sauternes and Pessac-Léognan can be drunk young, but greatly improve with age while unoaked Hunter Semillon demands age.

2009 Sauternes Premier Cru Classé

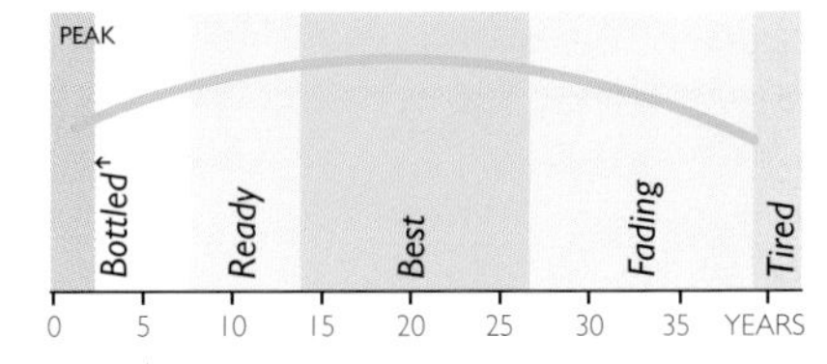

2009 Sauternes are outstanding, supremely sweet, wonderfully fresh, impressively structured. Lovely now, they will last and last.

2010 Pessac-Léognan Cru Classé

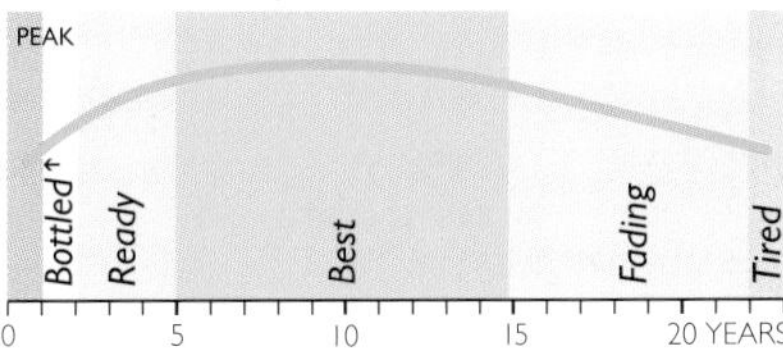

Modern Pessac-Léognan whites are delicious at bottling but gain fascinating depth and compexity with aging between 5 and 15 years.

2011 Hunter Valley Semillon (premium unoaked)

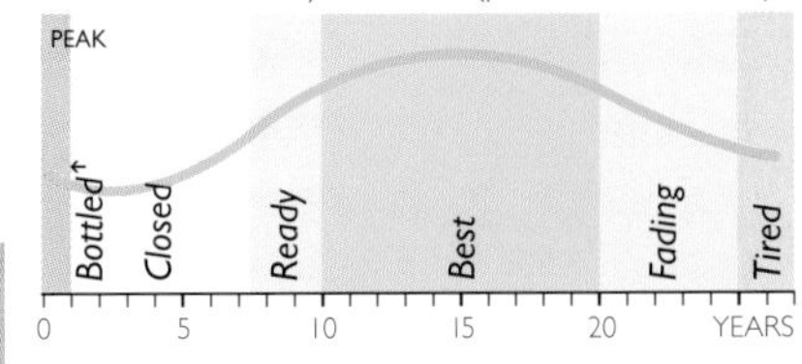

A warm, early vintage of waxy wines. The acidity starts high, but the wines emerge eventually into an amazing, toasty maturity..

SERCIAL

One of the noble grape varieties of Madeira, now much reduced in area and really only found around Câmara de Lobos in the south of the island, and around Porto Moniz in the northwest. It is often grown at high altitude, and being in any case late ripening, it is the last variety to be picked.

It is high in acidity and makes the driest and lightest of all Madeira styles. Since 1993 any Madeira with the word 'Sercial' on the label must contain at least 85 per cent of that grape. Modernizing efforts by the island's largest producer, the Madeira Wine Company, mean that its Sercial Madeiras are slowly becoming less austere in style, and retaining more fruit flavour. This is partly achieved by maturing the wine not by the short, sharp shock of a warm *estufa* (which nowadays is used only for the cheaper wines of Madeira), but by the gentler *canteiro* method, in vats stored under the roof. Since the brilliant, shocking tang of Sercial, and its astonishing longevity were two of the reasons for Sercial wine's greatness, I hope modernization doesn't mean 'dumbing down'.

Sercial is the same grape as the Esgana Cão, or 'Dog strangler', of the Portuguese mainland (see page 98). Best producers: (Portugal) Barros e Sousa, Blandy, Cossart Gordon, Henriques & Henriques, Leacock.

The northern coast of Madeira near the village of Seixal is one of the best spots for growing Sercial, but as always in Madeira, other vines – here, notably, the hybrid Jacquet – outnumber the noble variety.

SEYVAL BLANC

A French hybrid vine that was produced by crossing two Seibel hybrids – Seibel (see page 243) being the name of a group of French hybrids. Seyval Blanc is a so-called Seyve-Villard hybrid, one of a group of about 100 French hybrids bred by Bertille Seyve and Victor Villard.

As it is not pure *vinifera*, it is forbidden by EU law from being used in Quality Wine. Ironically, it is one of the most successful varieties of all in England, producing well-structured wines which, at their best, have a strong, Chablis-like austerity to start with and attain a complex honeyed richness with bottle age. Seyval Blanc may be coarse and dull in warmer France, but every grape has a day of glory, and low-cropped Seyval in cool England can make a smashing dry white. The fact that the EU would like to stop England using it for quality wine is merely proof, for some, of a French-inspired conspiracy against poor old Blighty. The big investments in English wine, however, are directed at sparkling; so the spotlight has moved away from Seyval Blanc of its own accord. Elsewhere it is found in Canada, and the eastern USA. Best producers: (England) Breaky Bottom, Camel Valley, Chapel Down, Three Choirs; (USA) Benmarl, Wagner.

SHIRAZ

The Australian name for Syrah (see pages 258–269) But it's a lot more than that – to many drinkers, the very word Shiraz evokes images of rich, dark, spicy red wines that perfume the brain and fill the nostrils with billowing waves of flavours. Many of these drinkers would not know what Syrah was, but Australian Shiraz is their nirvana.

Nowadays you can often tell what style a winemaker is after by whether he or she labels the wine Syrah – for a more austere, less ebullient style, or Shiraz – for a richer, riper, more hedonistic experience. And consequently you may find the names Shiraz or Syrah in countries from Australia to Chile, South Africa and even France. To generalize, the Australian Shiraz style is richer than Rhône Syrah, with broader, sweeter fruit, and more flavours of chocolate and berries instead of the smoke and minerals of the northern Rhône. Syrah is generally thought to be the original name of the grape – but is it? The town of Shiraz in ancient Persia is one of the first recorded centres of winemaking and the Phoenicians went there and traded their vines around the Mediterranean. The vine, though, is now thought to have originated in the Rhône Valley.

SILVANER

See pages 256–257.

SÍRIA

This is what we should really be calling the grapes we know as Spain's Doña Branca and the Alentejo's Roupeiro; it has a good few other synonyms as well, but the Dona Branca of Portugal is a separate vine. It's an old variety that seems to have originated in northwest Iberia; it makes intensely aromatic wines – think of lime blossom and peach, with a lush texture – but it doesn't keep its aromas for long because it oxidizes so easily and so early. Producers who take it seriously and address these problems can produce some lovely wines.

SOUSÃO

Portuguese black grape found in the Douro Valley. It is not one of the five varieties recommended for planting by the port authorities, but plays a part in one of the greatest of all ports, Quinta do Noval's Nacional, as well as in many others. It is normally thought of as a rather rustic grape that contributes intense colour and a certain raisiny fruit to the blend, but which does not age particularly well. It is the same grape as Vinhão (see page 287), which is the main grape grown for red Vinho Verde. It appears in Galicia as Sousón. In South Africa it is regarded as one of the better fortified wine varieties for its high sugar levels and deep colour. There is a little in Australia and California, too. Best producers: (Portugal) Quinta do Noval; (South Africa) Boplaas, Die Krans, KWV, Overgaauw.

SPANNA

Nebbiolo is known by this name in the Vercelli and Novara hills of Piedmont. It is found under this name in the DOCs of Gattinara, Ghemme and several others, usually blended with other grapes. The wines have a reputation for being somewhat tough and stringy, but improved winemaking is yielding better results in the form of rounder fruit and riper tannins. See Nebbiolo pages 166–175.

SPÄTBURGUNDER

German name for Pinot Noir (see pages 174–185). Styles of this wine have changed out of all recognition in Germany, with the pale, sweet wines of yore being superseded by dark, dry, sometimes overoaked international versions. The best wines come from the Pfalz and Baden regions. Best producers: (Germany) Bercher, Rudolf Fürst, Huber, Karl-Heinz Johner, Meyer-Näkel, Rebholz.

STEEN

Traditional South African name for Chenin Blanc (see pages 82–91), although the latter name is now generally used. On the decline, it now covers less than 18 per cent of the total vineyard area there, but it is still easily the most planted variety in the country. It is South Africa's national workhorse grape, producing everything from dry table wine, through sweet and sparkling wines, to fortified wines and brandy. The growing band of producers determined to improve the quality and image of the grape in South Africa, and already producing some of South Africa's most interesting whites, prefer the name of Chenin Blanc. Best producers: (South Africa) Cederberg, Glen Carlou, Hartenberg, L'Avenir, Mullineux, Stellenzicht, Villiera.

SUMOLL

You're most likely to come across this in the Canaries, though it's also found occasionally in Catalonia. It makes fairly unremarkable but attractive wine: nicely fruity with good acidity and freshness. Sumoll Blanc, also found in Catalonia, is unrelated.

SILVANER

Silvaner (see pages 256–257) is spelled with a 'y' in all countries except Germany.

SYRAH

See pages 258–269.

SZÜRKEBARÁT

Hungarian name for Pinot Gris (see pages 186–187). At their best the wines are lively, with good earthy, spicy fruit, but winemaking and viticulture are of variable standards and a vaguely grubby flat flavour is still unfortunately more common than the bright, minerally, honeyed style it can achieve.

TAMÎIOASA

The Romanian name for Muscat. Tamîioasa Alba or Tamîioasa Românească are names for Muscat Blanc à Petits Grains, while Tamîioasa Ottonel is Muscat Ottonel and Tamîioasa Hamburg or Neagra is Muscat of Hamburg.

Tamîioasa Românească is used, unusually, for nobly rotten wines in Romania (sweet Muscats are not usually achieved through noble rot), and the name may appear on the label as the only grape in such wines, although in the historic sweet wine region of Cotnari in the northeast of the country a blend of several grapes, including Grasa, Fetească Alba and Frîncusa as well as Tamîioasa Românească is supposed to be traditional. Noble-rotted versions of Tamîioasa lack the characteristic Muscat Blanc à Petits Grains aroma, and have instead a more typically botrytis spice and marzipan note. These are difficult times for Romanian wines, but there are still occasionally lovely examples of Tamîioasa, both young and old wines, to make you hope that, one day, Romania will begin to fulfil its massive potential once more. Best producers: (Romania) Cotnari Cellars, Dealul Mare Winery.

TAMINGA

Recent Australian cross bred to withstand warm conditions. It yields generously, keeps its acidity, and gives aromatic, grapy-tasting wine, which makes it very useful for sprucing up otherwise flavourless blends.

TANNAT

A grape of the Basque region of southwest France, seldom seen on wine labels there, but now making its mark as one of the more interesting grapes grown in South America.

In France its wines are most notable for their high tannin levels, and in Madiran the grape benefits from being blended with Cabernets Sauvignon and Franc as well as Fer, to make it more approachable. It also plays a part in Côtes de St-Mont, Irouléguy, Tursan and Béarn. All these reds have been or are being improved by modern winemaking methods, which emphasize fruit and soften tannins, but all can be tough to the point of astringency if not made well. The method of tannin-softening known as micro-oxygenation – bubbling small amounts of oxygen through the wine – was developed in Madiran for Tannat. A case of necessity being the mother of invention. The grape gives deep colours and takes well to oak aging. But it is generally in decline in France. This may be reversed by research showing that Tannat has record high levels of procyanidins, the heart-friendly chemical in red wine. Ninety-year-old men are surprisingly common – and healthy – in the deep southwest.

In Uruguay Tannat is increasing in quality every year, and ideally is marked by its fine, ripe tannins and elegant blackberry fruit. These wines seem far more European in style than most from South America, and it's a matter of opinion whether they could actually use more of the juicy, ripe fruit that is so typically South American, or whether they should jealously guard their rather more severe European style. Plantings at Maldonado out on the Atlantic Coast may provide the perfect compromise.

Growers in Uruguay report a difference in character between their old vines, which are descendants of the original cuttings brought from southwest France, and the newest clones imported from France. The newer ones tend to give more powerful but more simple wines, with an extra degree of alcohol – 13–13.5 instead of 12–12.5 per cent – but less acidity. The best solution, if complexity and depth are the aims, might be to produce virus-free clones from the old vine stocks – but, of course, the removal of viruses can itself change the character of the wine. A few good examples in Argentina, California, Brazil and Virginia. Best producers: (France) Aydie, Barréjat, Berthoumieu, Brana, Cayrou, la Chapelle Lenclos, Crampilh, Montus, Producteurs Plaimont; (Uruguay) Establecimiento Juanico, Pisano, Castel Pujol, Hector Stagnari.

TARRANGO

A modern Australian crossing, bred in 1965 from Touriga x Sultana to give wines of fair colour and acidity but low tannin – Beaujolais-style, in other words. Quality is pretty good and prices are low. Tarrango needs hot climates to ripen properly and is at its best in the torrid, irrigated Riverland region where producing a bright, breezy, juicy red in such semi-desert conditions is quite an achievement. Best producer: (Australia) Brown Brothers.

SILVANER

I used to look forward to Silvaner (or Sylvaner as the French and most other people spell it) wines in blind tasting competitions for the simple reason that it seemed to have almost no taste of its own, except for a certain green apple peel acidity when it came from a decent Alsace producer; and consequently, you could always taste the *terroir*. Well, the vineyard anyway. Actually what I really mean is the mud, the earth, and, if the wine was mature, it developed a strange taste of tomatoes. So I suppose I'm saying it never exactly tasted that clean, but I could sure as hell spot it in a blind line-up.

But that doesn't mean it was necessarily unattractive because, in either Alsace or Germany, surrounded by wines that were either highly aromatic or definitely on the sweet side, this broody, earthy quality actually made it rather appetizing. However, having earthiness as your main taste characteristic is not going to endear you greatly to producers in new wine regions looking for new vines to plant. Despite having once been widespread throughout Central Europe, Silvaner is now in decline and has made virtually no mark at all on the New World, though California has a few vines, and one or two Australians have had a sniff of it.

It seems to have originated in Austria – Savagnin getting together with Osterreichisch Weiss. Savagnin's alter ego is Traminer, and Osterreichisch Weiss can still be found very occasionally near Vienna. It was introduced to Germany in the replanting that followed the Thirty Years War.

It's still Franken's main speciality. It has that dry (usually less than [illegible] grams per litre of residual sugar), slightly earthy fruit, good apple peel acidity and a lightness that can veer towards the insubstantial, but at low yields it reflects the subtleties of its *terroir* and can last well for a few years. In the Rheinhessen it also occupies some good sites. I still find that earthy undertow in the wines, but it can make good apple-crisp dry whites as well as surprisingly long-lived, fat, sweet styles – Spätlese and Auslese. The old East German vineyards in Sachsen and Saale-Unstrut also produce good examples.

In Alsace most growers have turned from Sylvaner to Pinot Gris and Pinot Blanc, which is a shame, because from a serious grower it can be a weighty, structured delight. You can still find a little Sylvaner in Central and Eastern Europe – Hungary, Slovenia, Russia and the Czech Republic have some, and Austria has a few plots. But the only other places where it performs remotely memorably are Italy's Alto Adige and Switzerland (it's called Johannisberg there) where in some of the warmer sites in the Valais it can produce surprisingly deep, golden, minerally wine.

THE TASTE OF SILVANER

Even at its best, Silvaner seems to have more of a style than an actual flavour. In Germany its wines are dry, light and gently earthy, but never pungently so, and never at all aromatic. In Alsace it takes on the region's prevailing smoky spice, together with greater breadth, but still has little in the way of identifiable flavour. Italy's Alto Adige makes a fairly bright acid style and Switzerland can add a little earthy honey. But most examples won't age and will start to taste of tomatoes if you do try to cellar them.

The Franken region, spread along the banks of the River Main to the east of Frankfurt, is unusual in Germany, in that the best wines are made from Silvaner rather than Riesling grapes. This is due to the area's continental climate, which used to mean that winter could suddenly close in with its murderous frosts before the Riesling was fully ripe, yet the earlier-ripening Silvaner could have been picked at full maturity. So a lot of the best vineyard sites are also given over to Silvaner, and its dry, earthy, savoury make Franken Silvaner the best in the world. The Fürst family have been making wine at Bürgstadt since 1638, and the vines are from a steep, south-facing slope of warm, iron-rich weathered sandstone soils.

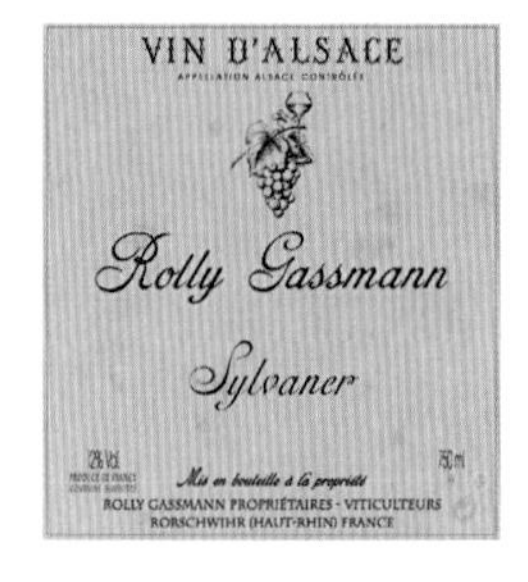

Domaine Rolly Gassmann

Sylvaner has been in decline in Alsace for a long time, but since 2005 some sites can now make Grand Cru Sylvaner and its earthy, savoury merits are starting to be appreciated.

Horst Sauer

The Lump vineyard at Eschernderf in Franken has excellent southeast exposure: Silvaner gets very ripe here. Generally speaking, the vine doesn't like the very hottest sites, but it's not complaining chez Herr Sauer.

Left: As is so often the case in Germany, the village – in this case, Escherndorf in Franken – is on the flat land beside the river, while vines are planted on the steep slopes above, benefiting from the sunny aspect. Here, the bends in the meandering River Main create several excellent vineyard sites. ***Above:*** *Growers in Germany like Silvaner for its early ripening and consistently high yields – higher than Riesling. In Franken it likes fairly warm sites with deep, rich soil and with good moisture retention. Its character is subtle rather than strong, but it can make wines of all levels of sweetness, up to Trockenbeerenauslese.*

CONSUMER INFORMATION

Synonyms & local names

The grape is spelt Silvaner in Germany and Sylvaner in France and Austria. The Swiss have many synonyms for it, including Johannisberg in the Valais.

Best producers

FRANCE/Alsace Auther, P Blanck, Bott-Geyl, Albert Boxler, Bursin, Dirler-Cadé, Kientzler, Seppi Landmann, René Muré, Ostertag, Pfaffenheim co-op, Rolly Gassmann, Martin Schaetzel, Bruno Sorg, Marc Tempé, Weinbach.
GERMANY Bürgerspital, Rudolf Fürst, Fürstlich Castell'sches Domänenamt, Freiherr Heyl zu Herrnsheim, Staatliche Hofkeller, Juliusspital, U Lützkendorf, Horst Sauer, Schmitt's Kinder, Störrlein.
ITALY Abbazia di Novacella co-op, Köfererhof, Peter Pliger/Kuenhof, Valle Isarco co-op.
SWITZERLAND Robert Gilliard, Caves Imesch.
AUSTRALIA Ballandean Estate.
SOUTH AFRICA Overgaauw.

RECOMMENDED WINES TO TRY

Ten German wines

Rudolf Fürst Bürgstadter Centgrafenberg Silvaner Kabinett
Fürstlich Castell'sches Domänenamt Casteller Hohnart Silvaner Kabinett Trocken
Freiherr Heyl zu Herrnsheim Niersteiner Rosenberg Silvaner QbA Trocken
Staatliche Hofkeller Würzburger Stein Silvaner Spätlese Trocken
Weingut Juliusspital Iphofer Julius-Echter-Berg Silvaner Spätlese Trocken and Würzburger Stein Silvaner Spätlese Trocken
U Lützkendorf Pfortener Köppelberg Silvaner Trocken
Horst Sauer Eschendorfer Lump Silvaner Auslese and Eschendorfer Lump Silvaner Spätlese Trocken
Schmitt's Kinder Randsackerer Sonnenstuhl Silvaner Spätlese Trocken

Ten Alsace wines

Blanck Sylvaner Vieilles Vignes
Dirler Sylvaner Vieilles Vignes
Kientzler Sylvaner
René Muré Sylvaner Clos St-Landelin Cuvée Oscar
Domaine Ostertag Sylvaner Vieilles Vignes
Pfaffenheim co-op Sylvaner Vieilles Vignes
Rolly Gassmann Sylvaner Vallée Noble
Martin Schaetzel Sylvaner Vieilles Vignes
Bruno Sorg Sylvaner
Weinbach Sylvaner Réserve

Eight other Sylvaner wines

Abbazia di Novacella co-op Alto Adige Valle Isarco Sylvaner (Italy)
Ballandean Estate Late Harvest Sylvaner (Australia)
Robert Gilliard Johannisberg du Valais Porte de Novembre (Switzerland)
Caves Imesch Johannisberg du Valais Sylvaner (Switzerland)
Köfererhof Alto Adige Valle Isarco Sylvaner (Italy)
Overgaauw Stellenbosch Sylvaner (South Africa)
Peter Pliger/Kuenhof Alto Adige Valle Isarco Sylvaner (Italy)
Valle Isarco co-op Alto Adige Valle Isarco Sylvaner Dominus (Italy)

SYRAH/SHIRAZ

Most of the wine tastings I conduct each year are not for wine connoisseurs. But they are always for enthusiasts – and I want them to go away happy that they've learned a bit about this wine tasting lark. So I have to use wines with loads of flavour. For instance, I start out with Sauvignon Blanc from New Zealand because you'd have to have the palate of a lamp-post not to appreciate and recognize some of the powerful fruit flavours there. And after roaming through the world of white and red, I need a grand finale: I need a great big walloping wine packed with personality and passion to finish off with. I need a wine to put a smile on people's faces and start a jitterbug in their hearts. And I always choose Australian Shiraz, as Syrah is called there – from the Barossa Valley or McLaren Vale. But as the cult of Shiraz spreads, I can see myself using Californian, South African, Argentinian, Chilean or New Zealand examples. Or, of course, French – where they'd call it Syrah, if they put the name of the grape on the label at all, that is. And when I ask people what their favourite wine is – they chorus, 'Shiraz'. And if I'd asked them what the greatest wine grape in the world was – they'd probably scream, 'Shiraz'. And perhaps they're right. Perhaps it is. You wouldn't have found many supporters for that view a generation ago. The name Syrah was largely unknown to anyone who wasn't a red Rhône fanatic. In Australia the name Shiraz was applied to a style of rich, ripe, ebullient red wine that was generally thought of as an old-fashioned alternative to Cabernet Sauvignon.

Whether it originated in the northern Rhône, its European heartland, in Syracuse in Sicily or was brought back from Shiraz in Persia, there is something exotic about this grape. It's a nice idea but probably untrue that Syrah reached Europe at roughly the same time as the damask rose, and along the same route. In the far distance is the chapel on the Hermitage hill high above the river Rhône and in the foreground some of the wild violets that seem to perfume the wine itself. The parrot is an Adelaide Rosella whose habitat includes the Barossa Valley where Shiraz excels.

Ah, yes. Cabernet Sauvignon. The grape that has enjoyed a century and more revelling in its reputation as the world's great red wine grape. And until the 1980s, there wasn't a suggestion of any rival to its position. Bordeaux reds were thought of as the world's greatest. Anyone in the New World wishing to be taken seriously would plant their best land with Cabernet Sauvignon and pray the wine turned out something like Bordeaux.

But why wasn't anyone doing the same with Syrah? Well, the fact is that there was a large amount of good-quality Bordeaux red available and it had first been exported to northern Europe and then around the world more than 800 years ago. The great red wines of the Rhône, however, were very scarce, hardly written about, rarely exported and consequently few of the New World pioneers had ever tasted them, so why on earth would they even think of planting Syrah? Only Australia, pursuing its own merry path thousands of miles away in the southern Pacific, had significant plantings – and they were mostly for making port.

Well, luckily the Rhône began to get its share of the limelight during the 1990s – just at a time when Bordeaux was starting to look very expensive and, after a string of poor vintages, not very good value. And, what a surprise, Australia suddenly realized what a fantastic grape Shiraz was and how lucky they were to have so much of it. The 1990s was also a decade when warm climate flavours began to be appreciated more than cool climate ones, and if you have to look for France's greatest warm-climate variety, Syrah is unquestionably the star. And if you have to decide on the most opinionated, bumptious, most irrepressibly self-confident of the warm climate nations, Australia takes the crown without breaking sweat. So the Aussies blew the Shiraz trumpet long and stridently, and the wine world listened and planted the vines furiously. But the 21st century has seen a quality breakout in the more restrained 'Syrah' style of the Rhône Valley, and the wine world listened again. Two great red wine originals at opposite ends of the taste spectrum, from one great grape variety, and if anything, the cooler, more scented 'Syrah' styles are beginning to overtake the richer, riper 'Shiraz' styles in popularity as growers seek out less sun-baked slopes to plant their vines. So what's it to be? Shiraz or Syrah? Which one is fighting for the title of 'greatest wine grape'? I'd say they both are.

Syrah/Shiraz: from Grape to Glass

Geography and History page 260; Viticulture and Vinification page 262; Syrah/Shiraz around the World page 264; Enjoying Syrah/Shiraz page 268

GEOGRAPHY AND HISTORY

A map showing the spread of Syrah (or Shiraz, the name it takes in the New World) would have appeared very different 30 years ago. Then it would have looked like a grape in decline. And in any case, there would be only two patches on the map with any significant plantings at all – Australia and southern France. In Australia its vineyards were shrinking under the pressure to plant fashionable Cabernet Sauvignon, and indeed Chardonnay. Only in its homeland of the Rhône Valley – in particular Hermitage – and in the Midi, where it has long been regarded as an 'improver' vine, ideal for bolstering the aroma and flavour of otherwise tough reds, was it recovering from a long period of drab stagnation.

In the first half of the 19th century the picture would have looked different again. Bordeaux would have been highlighted on the map. There were small but significant amounts planted here, at châteaux as eminent as Cos d'Estournel, Lafite and Latour.

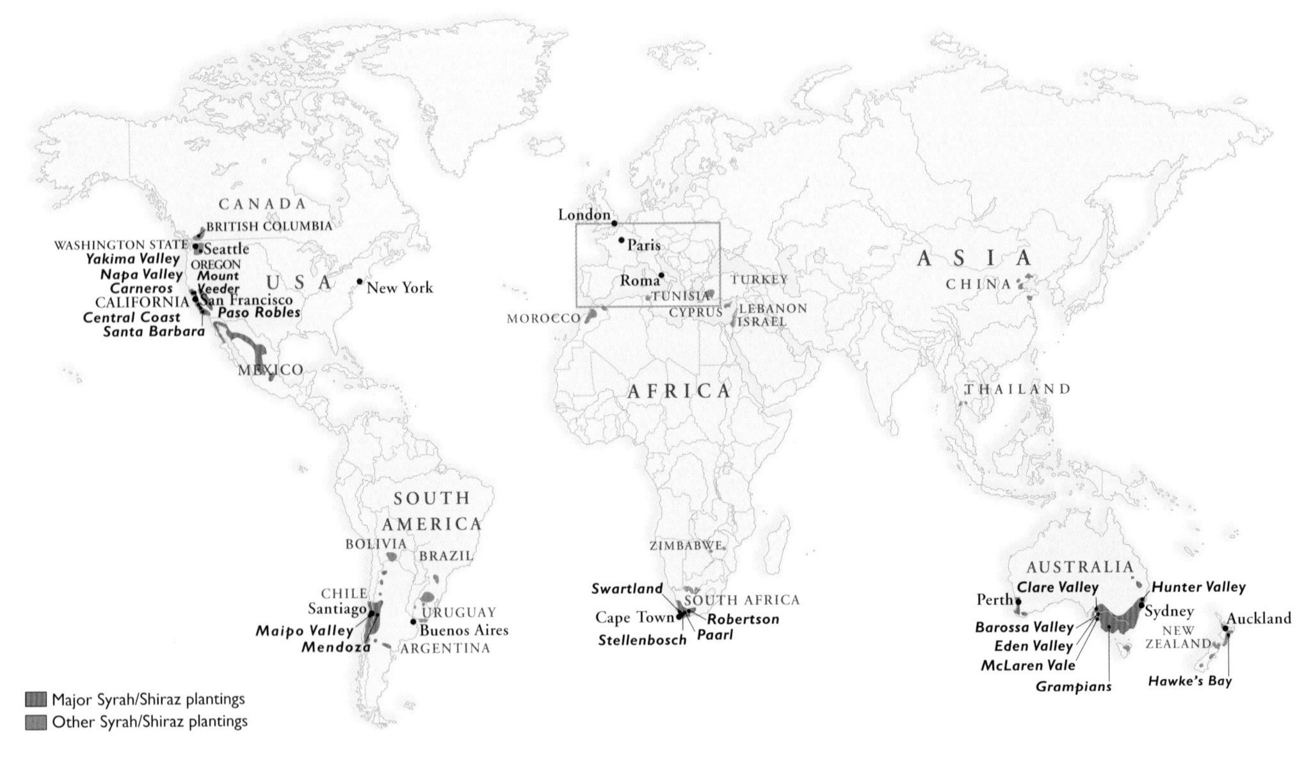

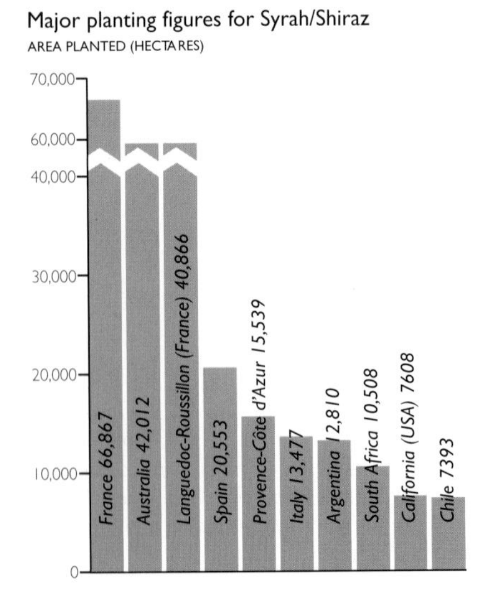

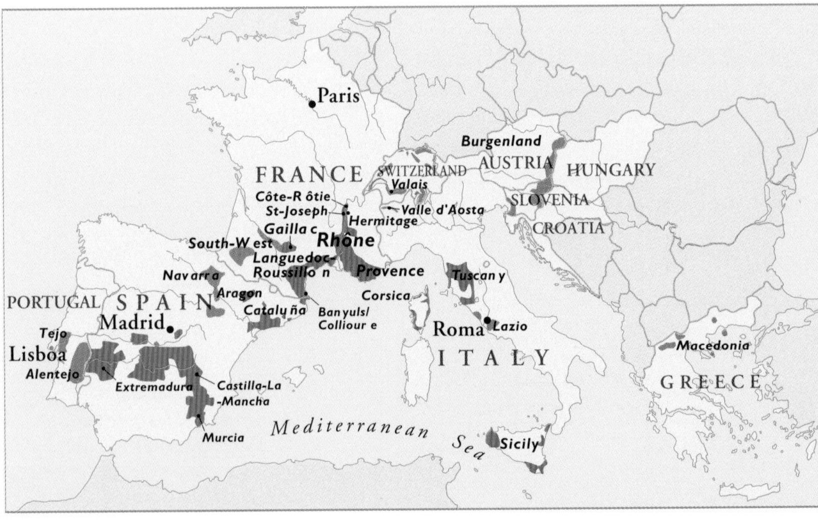

Syrah was also grown and vinified in the Rhône, to be blended with top-level red Bordeaux to add colour and structure; the resulting 'improved' wines were described as 'Hermitagé'. As merchant Nathaniel Johnston wrote to his partner Monsieur Guestier in Bordeaux in the early 19th century, 'The Lafitte of 1795, which was made up with Hermitage, was the best liked wine of any of that year'.

But while Syrah blends like a dream with other varieties and also makes some of the finest, most exotically aromatic of red wines on its own, it is quite fussy about where it grows. Now that it is fashionable again, it has spread rapidly across the world – into California, Spain, Switzerland and Chile, Argentina, South Africa and New Zealand, and even, more cautiously, into the vineyards of Austria, Italy, Bulgaria and Portugal. Growers are now more prepared to admit that in the right relatively cool conditions it can be more successful than Cabernet or Merlot. But its choosiness means that it is unlikely ever to become as widespread in every wine region of the world as Cabernet Sauvignon.

HISTORICAL BACKGROUND

It is somehow appropriate that a grape so startlingly perfumed should come laden with legend. Were the first vines on the hill of Hermitage in the Rhône planted by St Patrick as he made his way to the monastery of Lérins by the sea? Were they planted by a returning Crusader, Gaspard de Stérimberg? Did the Syrah vine originate in Persia and take one of its names from the city of Shiraz? Did it originate in Egypt, and travel to the Rhône via Syracuse in Sicily, gaining its name on the way? Or did it simply originate in the northern Rhône and stay there?

Sadly for romantics, it looks as though the last theory is the correct one, though mere reality won't stop me preferring all the other stories. A vine known as Allobrogica was being cultivated in the northern Rhône by the Gallic tribe of that name, the Allobroges, during the Roman Empire, and seems to have been selected from vines growing wild in the locality. The wine rapidly gained a reputation for fine quality and an unusual tarry flavour.

Did this at some point develop into the vine we know as Syrah? DNA fingerprinting at the University of California at Davis has confirmed that Syrah's parents are red Dureza, from the Ardèche, to the west of the Rhone, and Mondeuse Blanche, an old Savoie vine to the east. (It also turns out to be related to Pinot Noir, possibly as a descendant.) Syrah in the northern Rhône does, nevertheless, have enormous genetic variation, which would be expected if this was indeed its birthplace.

Syrah's introduction to Australia is less of a mystery. Scottish-born James Busby (1801–71), often described as the father of Australian viticulture, settled in New South Wales in 1824. In 1832, during a four-month visit to Europe, he collected more than 400 vine cuttings; Syrah was one of the most successful at adapting to the hotter, drier conditions of its new homeland.

A little chapel dedicated to St Christopher sits on the peak of the hill of Hermitage in the Rhône. The hill is said to have got its name from Gaspard de Stérimberg who lived there as a hermit after his return from the 13th-century Crusades, though the name of Hermitage was not used for wine until the 16th century.

Vines have been grown on the hill of Hermitage (seen here across the Rhône) at least since the Romans occupied the area. The town at the foot of the hill is Tain, which the Romans knew as Tegna: Pliny and Martial both mention its wines by name.

Vines on the Côte Blonde sector of Côte-Rotie. The traditional method of training vines here is up chestnut stakes arranged in a wigwam, to protect the vines from damage by the Mistral, the cold wind that blows down the Rhône Valley.

VITICULTURE AND VINIFICATION

Syrah's reputation is for wines that seem positively to ooze sunshine and warmth, especially if you take the marvellous monsters of South Australia as your starting point. But if you start from the perfumed delicacy of, say, a Côte-Rôtie from the northern Rhône, you would think you were dealing with a much cooler climate vine. Syrah/Shiraz will lose its heavenly floral fragrance if the climate gets too hot – though if it gets too cold, as in a poor vintage on the Rhône, the wine will taste too much like root vegetables for comfort. And in the heat of Australia's Barossa Valley or McLaren Vale, it can go from ripe to overripe in as little as a day.

It grows vigorously and gives its finest, most fascinating flavours on poor soils; on fertile soils the canopy must be open and well spread out. You can't overcrop it and produce excellent wine, though pleasant light wine is possible at high yields. And as for aging it in new oak – well, it has so much personality it doesn't need much new oak. Too much and it will taste like Barossa Shiraz – wherever it comes from.

Climate

Syrah has a low tolerance both of too much heat and too little. It buds late and ripens early to mid-season, and in too hot a climate will rush to overripeness. It is naturally resistant to most disease, though grey rot, bunch rot and certain viruses can be a problem.

In terms of climate the Rhône Valley is marginal. The vine needs sheltered, south-southeast- or southwest-facing sites here, where it is protected from the full strength of the Mistral, which originates further east and rips down the valley at speeds of up to 145km/h (90mph). The only virtue of such winds for the vine grower is that they help to dry the grapes after often violent rainstorms.

Hermitage has a mean temperature a couple of degrees higher than that of Côte-Rôtie, and Cornas can be hotter again, set as it is in a south-facing amphitheatre, protected from the wind. Growers in the northern Rhône seek the warmest spots for Syrah, and thus the greatest ripeness; in Australia, where Shiraz is frequently planted in regions considerably hotter than the Rhône, the last few decades have seen first a crisis of confidence, and then renewed enthusiasm for the heavyweight styles that result.

In the Barossa, source of Australia's benchmark Shiraz, drought is a problem on the hot valley floor. The best Shiraz comes from old, old vines whose roots are deep enough to find what little water there is. Much top Shiraz from the Barossa is dry-farmed; irrigated vines here can require 5 litres of water per vine per day during the summer.

In Australia in the late 1980s a lot of effort went into seeking out cooler climates, in an attempt to mimic the climate of the Rhône. First results produced a lot of rather stringy wines, but good growers have persevered, and now the scented cool climate styles, coming especially from parts of Victoria, Canberra and Western Australia, are some of the most sought-after, several of them giving wines with something of the Barossa's ripeness and something of Côte-Rôtie's perfume.

Soil

This vigorous vine needs shallow, rocky, well-drained soils if it is to produce its most intense flavours. The best Rhône sites have soils derived from primary rock, especially granite, which retains the heat. Côte-Rôtie's Côte Blonde vineyard has some gneiss (a form of granite), and the Côte Brune has mica schist (a heat-retaining schistose rock that is rich in potassium, magnesium and iron; mica weathers to clay); the granite of the southern slopes of Côte-Rôtie gives softer, more aromatic wines, while the schistose of the northern slopes gives greater tannic structure. Parts of Hermitage, Crozes-Hermitage and St-Joseph are planted on alluvial terraces, and Hermitage also has granitic soil in the west.

In the New World site selection has, as we've seen, traditionally been by climate rather than by soil, and many different soils

Breakfast time – on the flat road, not on the precipitous slopes – during the harvest at Côte-Rôtie. Picking has to be by hand on these slopes: much of Côte-Rôtie is terraced, making any sort of mechanization impossible. Spraying against rot, however, is often done by helicopter, which saves many man-hours.

are planted with Shiraz. In Australia's McLaren Vale alone, major soil types include thin, shaly soil over limestone in the north, with next to no underground water; deep sandy soils to the southeast; clay on limestone, sand on marly limestone clay and grey loam on clay to the west and southwest; grey clay over limestone and red earth over limestone in Willunga Flats near the sea; and heavy red loam and shale in the foothills of the Southern Mount Lofty Ranges. These five major *terroirs* produce, respectively, concentrated, spicy, bold wines; fleshy, soft ones; wines with peppery, spicy, dark plum characters; wines with firm tannic structure; and potentially heavily cropped wines with a danger of lack of intensity.

Cultivation

When I think of the Rhône, I immediately think of stubby bush vines, gobelet-trained with one stake per vine, as in Hermitage, or Guyot simple-trained with four stakes per vine, as in Côte-Rôtie, for maximum protection against the wind. However, even within the Rhône Valley Syrah is subject to many variations in training, with training on wires increasingly popular on flatter ground where some mechanization is possible. It grows floppily, and where it is particularly vigorous it needs careful trellising to spread out its canopy: training methods elsewhere vary from bush, as in the oldest vines of the Barossa, to new systems of canopy management like Lyre and Smart-Dyson.

Yields

As usual, concentration goes hand in hand with excitement. The usual maximum permitted yield in the appellations of the northern Rhône is 40hl/ha, and while average yields have sometimes risen, particularly in Cornas, the steep hillside vineyards usually get less than this. Rhône producers maintain that at over 45hl/ha quality suffers; Australian viticulturalist Richard Smart, however, believes that it is possible to increase the crop to 10 tons/acre – that's between 150 and 180hl/ha – with no loss of quality, through a combination of canopy management, correct pruning and water stress. (I'm still waiting to taste the results, Richard.) Barossa producer Charles Melton, by contrast, puts the limit for quality at 2 tons/acre (36hl/ha). (I've tasted his. Yummy.)

At the winery

The buzzword here is 'co-fermentation'. Well, it's not the buzzword in the Rhône, because there they've been doing it for aeons. In Australia, however, co-fermenting your Shiraz with a spot of Viognier, Côte-Rôtie-style, is now commonplace: it gives not just perfume but some extra silkiness of texture, but it can easily be overdone. Some producers reckon 2 per cent Viognier is enough; few would go higher than 5 per cent.

A grape as rich in colour-producing anthocyanins as Syrah can take hot fermentation temperatures of 30–35°C (86–95°F), and a maceration of anything from a few days to three weeks to maximize extraction. It used to be traditional in the Rhône to include all the stems in the fermentation vats, which made for some relentlessly tannic wines; now total or partial destemming is more common, with some growers even taking up an extreme anti-stem position. Everything we want from the grape is in the skins, they say: add the stems and you're adding harsh tannins. Not if the stalks are ripe; and not if you vinify with care, then the stems add a haunting snappy perfume. It's horses for courses, but a judicious addition of stems really does change the flavour and the texture.

You can still find old open fermenting tuns made of wood in the northern Rhône, but now closed vats and stainless steel are more common. In Australia some growers are returning to shallow, open fermenters with excellent results.

Later picking of riper grapes in France means that the wines are better suited to longer macerations (it also means, incidentally, that the stems, if used, are likely to be riper and less green), and to the small new oak barrels that are increasingly used for aging the wine. Côte-Rôtie has the biggest concentration of small new oak barrels – too many, I'd say; I'm longing for them to get older; Hermitage has fewer. Cornas, under the influence of enologist Jean-Luc Colombo, has been looking more kindly on new oak in recent years but I'm not convinced northern Rhônes benefit from much new oak. The oak here is French, more subtle in its flavours than American. Australian Shiraz is often run into new small American oak barrels before the fermentation is complete. Finishing the fermentation in new oak gives better integration of oak flavours, and helps to fix the colour. But new oak must, however, be handled with care: it can easily muffle the excitement of Syrah. Grange, the archetypal Australian Shiraz, has the concentration to withstand lashings of American oak; far too many of would-be Granges don't.

SYRAH AS A BLEND

We've become so accustomed to varietal Syrahs and Shirazes (from the Rhône and the south of France, California and – in the form of Shiraz – from Australia) that it's easy to forget that this is also a superlative blending grape. That was for long its role in the south of France, where it was being added as a cépage *améliorateur* ('improver' variety) to the appellation wines of Provence and Languedoc-Roussillon long before Australian-trained winemakers starting bottling it as a varietal vin de pays. In the southern Rhône, too, it plays its part in the blend along with such grapes as Grenache and Mourvèdre.

Even in the northern Rhône it has commonly been blended – and, more remarkably, with white grapes. Up to 20 per cent Viognier may be added in Côte-Rôtie – though with only five or six per cent of Viognier grown in the whole appellation, the figure is much less than that, and often zero. In Hermitage, the blending grapes are Marsanne and Roussanne: 15 per cent may be added, though again, it is increasingly rare to add any. The reasoning was that white grapes helped to soften Syrah's high tannins; a function performed nowadays by picking at greater ripeness and better control of fermentation and maceration. New oak, too, can paradoxically help to polymerize and thus soften tannins, although one might think its only role would be to add more. So-called Rhône Ranger producers in California – lovers of all things Rhône-ish – produce some impressive blends of Syrah with the likes of Grenache and Mourvèdre; in the Napa Valley and Paso Robles, temperatures can be on the hot side for Syrah, making it less successful as a varietal. Areas like Carneros, Sonoma County and Santa Barbara further south are more successful.

Until the late 1990s the typical Australian red blend was Shiraz and Cabernet Sauvignon, mimicking an old 19th-century Bordeaux style. Cabernet/Merlot blends are now more popular, and most Shiraz is being planted for varietal wines. But blends with Viognier are flavour of the month here, too, for extra aroma and complexity. In Tuscany, Syrah blends well with Sangiovese, and in Sicily with Nero d'Avola; in Spain with Garnacha, Cabernet Sauvignon, Cariñena and Tempranillo, even all together.

SYRAH/SHIRAZ AROUND THE WORLD

You can tell a winemaker's allegiance from the name he chooses for the grape. In France's Rhône Valley it's always Syrah, but then this is the archetype of the minerally, smoky herbs style. In the Languedoc it may sometimes change its name to Shiraz, reflecting brand loyalty to the Australian style of sweeter, more chocolaty fruit. In Australia it's almost always Shiraz. In other countries like Spain, South Africa, Argentina and the USA you get both names, according to what role model the winemaker wants to adopt – French Hermitage or Australian Grange. If anything, I'd say the Syrah producers are making more progress.

France: Rhône Valley

Until the 1970s, the Syrah vineyards of the northern Rhône were struggling for survival. The terraces on the granite slopes overlooking the river had been replanted after phylloxera, but what was the point in working them? The wine sold for a pittance. Most wasn't domaine bottled: it was either sold by the *pichet* in local bars, or taken away by the *négociant* houses – in fact, it is thanks to the latter that many of these vineyards survived at all. What changed was not the wine or the grape, but the market. People at home and abroad woke up to these wines: and the decades since have seen rapid improvements in winemaking and the expansion, not always for the best, of the vineyards.

Terroir matters here for Syrah just as much as it does for Pinot Noir in Burgundy, and the steepness of the slopes (up to 55 degrees in Côte-Rôtie) gives maximum exposure to the sun, which is crucial in a climate that is marginal for the grape. Unlike Pinot Noir in Burgundy, however, the best northern Rhône Syrahs are nearly always blends of several different terroirs.

Syrah in the northern Rhône tends to be peppery at lower ripeness levels (11–12 per cent potential alcohol), but fruity and perfumed at 12.5–13.5 per cent potential alcohol. Côte-Rôtie can be paler in colour than Hermitage and more aromatic, with floral and roasted characters on the nose. Hermitage is firm, minerally and tannic; Cornas is the darkest and most robust of the lot, but lacks the thrill of Hermitage. Crozes-Hermitages and St-Joseph are lighter versions, more peppery, sometimes coffeeish, sometimes floral, increasingly good. The reasons there are such differences in flavour between Syrahs grown close to each other are to do with aspect, temperature and soil. The Hermitage hill's daunting slopes shouldn't and don't give the same flavours as the flatter vineyards of Crozes-Hermitage.

Let's take the two main appellations of the northern Rhône as examples. The Côtes Brune and Blonde in Côte-Rôtie are both larger sectors and specific *lieux-dits* (named plots of land) within those sectors. The Côte Brune, the broad northern sector, has clay- and iron-rich soil and produces the denser, more long-lived wine; but Viognier, the aromatic white grape which is permitted to be blended with Syrah here, does better on the more gneiss and limestone soils of the Côte Blonde. However, since it ripens before Syrah yet must be picked at the same time (the grapes must be added at fermentation), it is generally picked overripe and quantities used in the wine, if any, are smaller than its 5 per cent share of the total vineyard would suggest.

Other key *lieux-dits* in Côte-Rôtie include La Mouline, which makes rich, balanced wines that fade slightly earlier than some others; La Landonne, that makes solid, backward wines which take longer to come round and which last for decades. La Turque is firm, concentrated and elegant. These are generalizations because the chief producer is Guigal whose wines are a) very oaky, and b) very, very expensive. Consequently, comparisons are rare.

At Hermitage, the granitic les Bessards vineyard makes dark, tough, concentrated wines that are the backbone of many blends. The loess soil of l'Hermite makes softer, more mellow wine; les Beaumes, with limestone and ferruginous clay, gives scented, complex wines with fairly low tannins; chalky le Méal gives supple wines; the brown limestone of les Greffieux also gives supple styles, but is even better for the inclusion of white grapes, Marsanne and Roussanne.

It is not just the soil which gives flavour differences in Hermitage. Jean-Louis Chave, arguably Hermitage's greatest producer, believes that the remarkable smoky bacon depth of traditional Hermitage was due to an old Syrah type called Serine – with oval-shaped berries and now very rare. Certainly, Serine's dense, savoury flavours seem completely different from the scented lush fruit displayed by many modern clones. He's on a mission to preserve whatever's left. If he's in New Zealand, he might take a look at the ancient Busby clone there.

In the southern Rhône Syrah loses its monopoly of the red vineyards. The North is part of what the French label the Septentrionale zone; the South is in the Méridionale. The North's climate is Continental, with cool

Guigal
The grapes for La Turque come from 1 ha (2.5 acres) on the Côte Brune, just at the boundary with the Côte Blonde. The wine contains 5–7% Viognier.

Chapoutier
Chapoutier's Ermitage Le Pavillon uses Syrah vines that are on average 70 years old, giving a concentrated, perfumed wine.

Isole e Olena
Paolo de Marchi first planted Syrah in Chianti to beef up his Sangiovese. Now it's bottled alone and is fantastic – Syrah in fruit flavour but still Tuscan.

winters and warm summers; the South is hotter and drier. Too hot, often, for Syrah, which can race to overripeness, and good producers give it the cooler, north-facing sites to slow it down a little. Grenache gets first refusal of the best sites in the South, though there are some high-altitude (up to 550m/1800ft) spots where the harvest can be two weeks later than the norm; it will be interesting to see if the vogue for Syrah produces some new cooler plantations here.

France: the South

Syrah plays a very important part both in Provence and in the Midi. It is mandatory in some appellations and optional in others, and along with Mourvèdre it is one of the main *cépages améliorateurs*. For varietal Syrah here one must usually look to the IPG category of wines, and in particular to those from the Languedoc, where it may even be rechristened Shiraz in deference to the national allegiance of the winemaker.

These Shiraz vines, in Henschke's Hill of Grace vineyard, are over 100 years old. That makes them older than just about all the Syrah in the northern Rhône. The Henschkes call them 'the grandfathers', and the Barossa vines from which they were taken as cuttings some time in the 19th century were themselves brought to Australia from Hermitage before phylloxera struck the French vineyards. Hill of Grace Shiraz is now one of Australia's most expensive wines.

Australia

If the Australian settlers who first planted Shiraz in their new, untried country had been searching for a place that mimicked Hermitage, they would never have touched the Barossa or Hunter Valleys.

But of course they weren't looking for ideal sites; they had a hot climate and wanted to plant a grape that could cope. For them Shiraz was a dream come true. It was first planted in the Hunter in the 1830s and in the Barossa in the 1840s, and from the 1860s until the 1970s and 1980s, when the vogue for Cabernet Sauvignon took over, Shiraz was the national workhorse grape. It could do anything: it could produce light, soft, jammy flavours and big, beefy reds. It could make sparkling wines (one producer used to strip the colour out with carbon) and it could make fortified wines. Not surprisingly, it was taken less than seriously and it was easy to presume that Cabernet Sauvignon seemed to be the grape of the future.

In 1986 a government vine-pull scheme was the opportunity growers had been waiting for: they could get rid of their Shiraz, and be paid for doing so. Those who hung on could find themselves having to sell their grapes for Shiraz raisin muffins. Some of the biggest companies, however, did hang on, though not always enthusiastically. Penfolds, Lindemans, Wynns, Tyrrells and other major names continued to grow Shiraz, often blending it with the more fashionable Cabernet. What might have happened if the supply of virus-free Cabernet vines had been able to satisfy demand doesn't bear thinking about; we should just be grateful that Australians came round in the nick of time to appreciating what they had in their own backyard.

Even after the vine-pull scheme had done its worst, there remained far more old, ungrafted Shiraz vines – often planted in the middle of the 19th century – in the Barossa and Hunter Valleys than anywhere else in the world. These vines had originally come as cuttings from the Rhône, and so, along with ancient Grenache and Mourvèdre, form a sort of living museum of pre-phylloxera France.

The knowledge of how to handle Shiraz had come from French expertise at around the same time, and had been handed down the generations even during the years when cheap fortified Shiraz was the rule. By the time Shiraz was resurrected in public esteem the knowledge still existed, though growers had forgotten its source. Barossa producer

Jim Barry
Jim Barry's The Armagh is about as big as Australian Shiraz can get – and that's pretty big. American oak is part of the secret, great fruit the rest of it.

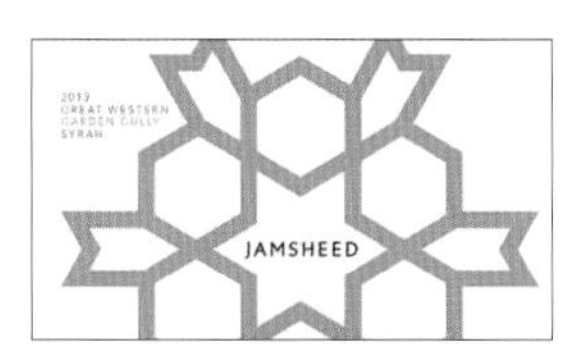

Jamsheed
A leader in the new generation of Australians making cool climate Syrah. Jamsheed grapes are from ancient vines in the Great Western region of Western Victoria.

Clonakilla
Tim Kirk triumphantly pioneered the Côte Rôtie method of co-fermenting Syrah with Viognier in Australia from his vineyard near Canberra.

Robert O'Callaghan remembers meeting Rhône producers like Chapoutier and Jaboulet and being astounded that they knew the same things that he did.

Shiraz is now Australia's most significant red wine – it makes up one-quarter of the country's total plantings and has easily surpassed the old favourite, Chardonnay. It gives spectacularly good results when taken seriously – it has become, in many respects, Australia's premium varietal. After some ill-starred forays into too-cool climates, styles seem pretty established. Barossa makes dark chocolate-tasting wines to McLaren Vale's milk chocolate; Eden Valley wines have more focused black fruit. Clare Valley's fruit is more dark red. In Victoria the Grampians produces wines that have a touch of Rhône spice. Heathcote and Beechworth, with their aromatics, finesse and supple richness, could end up among the best. Geelong and Sunbury can have problems ripening, but make elegant, restrained styles, and the Hunter makes surprisingly elegant, gentle, balanced wines for such a subtropical place that used to be famous for the pong of the 'sweaty saddle'. Canberra's cool Shiraz is scented, Western Australia's dark and peppery.

Fashionability has not put an end to Shiraz's versatility in Australia. It continues to make port styles of high quality, and sparkling reds that have attained a small but devoted following abroad. There are two styles of these: those made with young base wines are simple and sweetly blackberryish; those made with aged base wine are remarkable – they taste like serious Shiraz – but with bubbles. Wicked! Shiraz could also have a future as a base grape for a fuller, richer, white sparkling wine with very different flavours to those of Champagne.

Much Shiraz, especially in South Australia, is descended from the 19th-century small-berried originals. Clonal variation in the Barossa, and in any vineyards planted with massal selections from old vineyards, reflects that in the traditional Rhône. The commercially available modern clones, are brighter, higher-yielding and sometimes beguilingly scented, but they have less potential.

USA: California

The first Syrah was planted here in the 1970s but it took a loose knit group of winemakers known as Rhône Rangers to show that it could be an alternative to the ubiquitous Cabernet: deeply coloured but less tannic, well-flavoured and food-friendly.Fashion has not been kind, though. In the 1990s, the Californians planted Shiraz like mad, but the American consumers never seemed to quite get it, and when they wanted an alternative to Cabernet and Merlot they preferred Pinot Noir. At one time, there was eight times as much Syrah being grown as was being drunk. A pity, because much Californian Syrah is delightful, particularly from

THE GRANGE STORY

It was 1949 when the late Max Schubert (right) fell in love with the great reds of Bordeaux; he returned to Australia determined to recreate them in the Penfolds winery where he worked as winemaker. But there was hardly any Cabernet in Australia, which is why Grange is made from Shiraz – there was loads of that planted, though it was generally used to make sweet 'port'. Schubert named his wine Grange Hermitage in honour of the famous Rhône red; in the 1980s the European Union insisted that Hermitage be dropped. The first commercial vintage was 1952; and in 1957 Schubert was told to stop making the wine. Nobody liked it; nobody wanted it. Did he obey? Well, no. And just as well: come 1960, those first vintages were tasting so superb that Penfolds revoked its order. Moral: never obey orders.

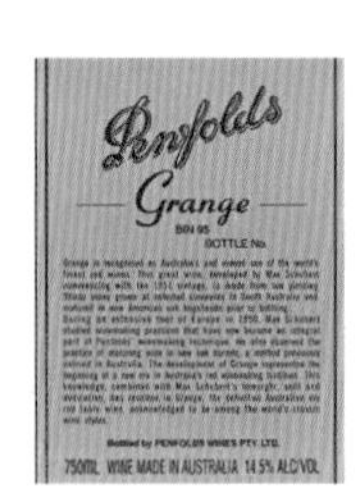

Penfolds

Grange (then known as Grange Hermitage) first found public acclaim in Australia in 1962 – after a decade of criticism. These days you'll pay more for a single bottle of Grange than a whole ton of Shiraz grapes cost back in the wine's early days.

Carneros, old plantings in Napa Valley, San Luis Obispo and Santa Barbara. And there's French interest, too: the Perrins of Château de Beaucastel in Châteauneuf-du-Pape have a joint venture with their US importer, Robert Haas, called Tablas Creek Vineyards, growing Syrah west of Paso Robles.

As in Australia, there is a search for climate and style. The hottest parts of the Napa and the warmer sites in Paso Robles can be too hot, and Syrah is then best blended with Mourvèdre and Grenache. Here, as elsewhere, Syrah is a most obliging blender. Varietal Syrahs declare their allegiance on the label, with those calling themselves by the grape's French name being generally spicier and meatier than the in-your-face fruit-first styles that go under the name of Shiraz. These 'Shiraz' are a more recent development in California, and they generally denote some Australian or New Zealand influence in the winery.

Syrah in California (which is usually larger berried than in Australia, either because it is a different clone or because of the climate) should not be confused with Petite Sirah. What California calls Petite Sirah is Durif; according to Dr Carole Meredith of UCD, 'Durif/Petite Sirah is a seedling of Syrah. It was known that Durif was "derived" from Peloursin, but how so was not known. What we know now, from DNA research, is that Durif grew from a seedling resulting from a cross between Peloursin (mother) and Syrah (father). We think this cross was not deliberate because Dr Durif would certainly have reported the pollen source as Syrah if he had known it. It was probably just a chance pollination in his research plot.'

USA: Washington State

The first Syrah plantings were by David Lake MW of Columbia Winery in 1985 and since the mid-1990s there has been a rush of new plantings, and here consumers do seem to like what they taste: in 1999 for the first time new plantings overtook those of Merlot, the state's second red grape. The hottest, earliest ripening vineyards give blackberry, cassis and plum flavours; cooler sites in the Yakima Valley are more mulberry and black cherry with some bacon fat. McCrea Cellars, for one, blends a little Viognier into its best *cuvée*.

South America

Argentina has the most, largely in Mendoza. Growers like it for its big crops of 12–14 tonnes/ha, but this cropping level stops it excelling. Even at this level, quality is respectable if canopy management is good, but the best producers are now working on yields only half as high with much improved results. Its favoured climates tend to be like the southern Rhône; in hotter spots it loses structure and acidity.

In Chile, where plantings are increasing fast, the grape seems to be perfectly at home, with more potential, some think, than either Merlot or Carmenère. A battle for bragging rights is developing between oaky, dense, warm-climate styles and thrillingly scented, cool-climate examples. Colchagua is making impressive warm styles led by Montes Folly, while the newer regions of San Antonio, Limari and Elqui are showing massive potential for richly fruited, scented reds of remarkable originality. I think I prefer these newer styles because they really are bringing an entirely new style to the Syrah table.

Rest of the world

As Shiraz it is being planted widely in South Africa, helped by the availability of better clones. Swartland is the most exciting area, producing some wines of remarkable scent and style and generally calling its wines Syrah. Paarl and Stellenbosch are also major producers with some fine examples when the oak has been kept in check.

In New Zealand it needs the warmest sites. The Gimblett Gravels area in Hawkes Bay, Martinborough and Waiheke Island have produced some beautiful wines, but expect to see exciting things from Marlborough in the future. In Austria's Burgenland it is being planted in small quantities at the expense of lesser white varieties. In Switzerland's Valais, where it has been planted in small quantities since 1920, it can make good, deep-coloured wines if the yields are kept down to 40hl/ha and the cap of skins and pips is well punched down during fermentation to maximize extraction. It is the first vine to bud in the spring, and one of the latest to ripen. Work is currently being done on selecting the best clones for the Valais.

Syrah is scattered in small quantities all over Italy, and blends well with local varieties like Sangiovese and Nero d'Avola as well as being made as a varietal. In Rioja it has experimental status only, but has made some interesting wines elsewhere in Spain, in Portugal and in Greece. It plays a small part in the blend at Chateau Musar in the Lebanon, and there is some, rather less distinguished, in Morocco and Tunisia. As with most of the major varieties Chinese Syrah will be coming your way soon.

Kongsgaard
Memorable Syrahs aren't common in California, but Kongsgaard uses fruit from volcanic soils in the cool Carneros region to brilliant, savoury, meaty effect, moderating its California power.

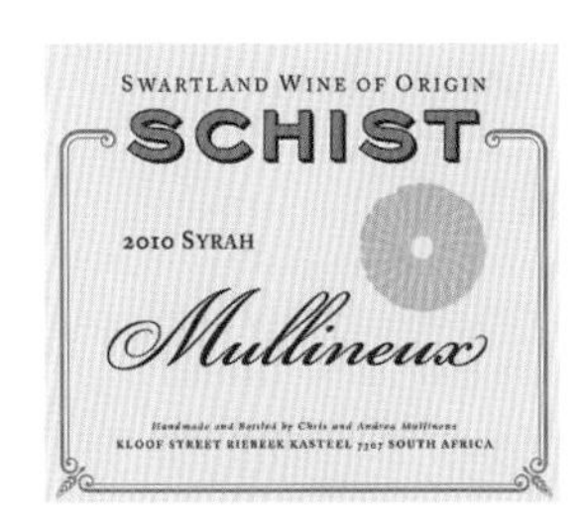

Mullineux
The Mullineux family are leaders of the new wave in South Africa producing wonderfully scented, amazingly smooth-textured Syrahs. Swartland is in the centre of this movement and even new vines give exciting wines.

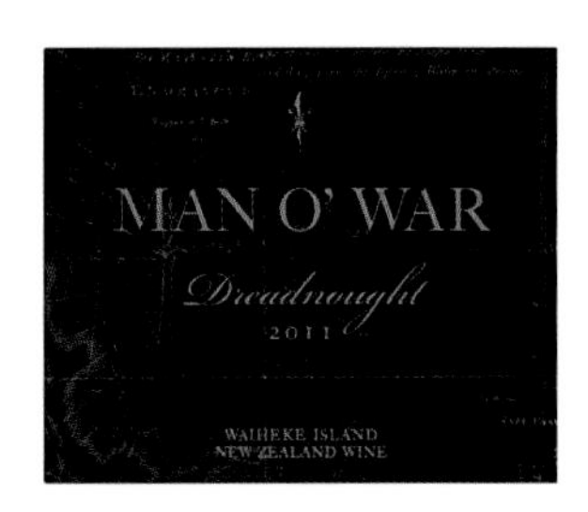

Man O'War
Syrah from dizzingly steep hillsides on Waiheke Island out in the bay opposite New Zealand's capital Auckland. The wine is rich with blackberry fruit, but smoky, peppery and scented with violets.

Matetic
Matetic expertly straddles the warm and the cool in Chile's San Antonio Valley a few miles in from the sea. The granite soils suit Syrah perfectly and the wine is rich, but fresh and scented.

ENJOYING SYRAH/SHIRAZ

Our knowledge of how Syrah ages might, at first glance, seem rather limited. Until the 1960s Cornas was seldom domaine-bottled. Hermitage and Côte-Rôtie were domaine-bottled earlier than that. But it was common until the 1970s to bottle wines on demand rather than when the development of the wine in cask called for it. Wines could vary dramatically according to whether they were bottled early or late.

Hermitage is the longest-living and slowest-maturing wine in the Rhône. Good vintages need 10 or more years to loosen up, though the palate can remain tight and tannic even after the bouquet has developed plummy, spicy characters. Lesser vintages may be ready in five or six years. Top Côte-Rôties take nearly as long as Hermitage, though the Côte Blonde is earlier maturing than the Côte Brune, and top Cornas can need nearly a decade, going through a dumb phase between two and six years. Each vintage sees more good Crozes-Hermitage and St-Joseph, and they are often delicious after about three years, though they can age well.

Syrah's influence in Châteauneuf-du-Pape is waning because, with climate change, it's difficult to stop Syrah overripening. Even so, most Châteauneuf is ready to drink by five years old but lasts much longer. Languedoc Syrahs are getting richer, but are usually good from two years old.

Most Australian Shirazes can be drunk within five years, though top old-vine versions, and especially Grange, may need up to a decade and will keep easily for 20 years or more. Cool-climate Chilean, South African and New Zealand Syrahs are excellent on release and will age a decade. eeee

The taste of Syrah/Shiraz

Young Syrah often smells, surprisingly for a red wine, of exotically scented flowers like carnations and violets. To those flavours add woodsmoke, perhaps with a few sprigs of rosemary thrown on to the fire, plus raspberries, blackberries and blackcurrants. Côte-Rôtie is more perfumed than Hermitage, which in turn is more blackcurranty, and both share a whiff of wood smoke. Hermitage, too, has a centre of tremendous succulence and richness – though if you open a bottle too young, you may wonder where this is – and it keeps this richness longer than any comparable French red wine. With age it gains gamy, leathery smells, an almost chocolatey character, and a whiff of violets and tobacco.

In Australia, flavours are creamier and more chocolatey, with blackcurrant and black cherry fruit. Hunter Valley Shiraz is leather-scented and mellow; Barossa versions are all black cherry, chocolate and spice. Clare wines have purer fruit and fine structure and Coonawarra fruit is bright and peppery. The Victoria style mixes peppery fruit with floral scents and increasingly lush texture. Nearly all Aussie Shiraz gains a thick dark taste of liquorice, prunes and black chocolate as it ages, but a few stay perfumed and blackcurranty.

The chemical compound that gives pepperiness to the grape is alpha-ylangene, and it's found in peppercorns, as well as in other grape varieties. It's present in Shiraz at just a few parts per billion. But if you can't smell it, don't worry: 20 per cent of us can't.

The ancient, Cambrian soils of Heathcote are so red even the sheep are crimson. Honest! I've seen them. But these deep, red-stained, gravelly loam soils grow brilliant Shiraz grapes for Jasper Hill. At a height of 300m (1000ft), with the Mount Camel Range funnelling cold winds onto the vines, ripening is slow and the wines have impenetrable colour yet palate-soothing tannins. Gerin's Côte-Rôtie is made from 17 different plots of land shared between the granitic Côte Blonde and the schistous Côte Brune. Between 5 and 10% Viognier is included and, unusually, a small proportion of American as well as French oak is used.

MATCHING SYRAH/SHIRAZ AND FOOD

Whether from France (in the northern Rhône), Australia, California or South Africa, this grape needs food with plenty of flavour. Northern Rhône Syrah goes well with game, beef and venison; the same goes for the more concentrated, leathery or smoky Australian Shiraz. Australian and South African versions are perfect for picking up the smoky flavours of barbecued food.

Lighter, more berryish wines are good with turkey, guinea fowl, lamb casseroles and even chicken, providing the chicken has some flavour. Liver is a good partner, too. Light Shiraz can be a good match for the subtle spices of Indian dishes, but Chinese or Thai flavours are seldom successful with this grape. Hard English cheeses like Cheddar are good.

CONSUMER INFORMATION

Synonyms & local names

It's called Syrah in France and Shiraz in Australia. Most producers in other regions use either name depending on the style of wine being made. Also called Petite Syrah in the northern Rhône. However, it is not to be confused with Petite Sirah, a grape grown mainly in California and Central and South America.

Best producers

FRANCE/Rhône Valley Allemand, F Balthazar, G Barge, A Belle, Chapoutier, J-L Chave, Y Chave, Clape, Clusel-Roch, Dom. du Colombier, Courbis, Coursodon, Yves Cuilleron, Delas Frères, Duclaux, E & J Durand, B Faurie, Ferraton, Pierre Gaillard, J-M Gérin, Gonon, A Graillot, Gripa, Guigal, Jamet, P Jasmin, S Ogier, V Paris, Rostaing, M Sorrel, Tardieu-Laurent, G Vernay, F Villard; **Languedoc-Roussillon** J-M Alquier, Estanilles, Gauby.
ITALY Stefano Amerighi, Bertelli, Cottanera, D'Alessandro, Fontodi, Isole e Olena, Le Macchiole, Planeta.
SPAIN Abadía Retuerta, Casa Castillo, Finca Sandoval, Enrique Mendoza, Pago de Vallegarcia.
PORTUGAL Cortes de Cima, Quinta do Monte d'Oiro.
USA/California Alban, Araujo, Beckmen, Dehlinger, Dutton-Goldfield, Kongsgaard, Krupp Brothers, Lagier Meredith, Andrew Murray, Pax, Qupé, Ramey, Sean Thackrey, Truchard; **Washington State** Betz, Cayuse, Gramercy, Hedges, Syncline.
CANADA Church & State, Colaneri, Creekside, Jackson-Triggs, Moon Curser, Painted Rock.
AUSTRALIA Tim Adams, Jim Barry, Best's, Rolf Binder, Brokenwood, Grant Burge, Chapel Hill, Clonakilla, Craiglee, Dalwhinnie, D'Arenberg, De Bortoli, Dutschke, John Duval, Gemtree, Hardys, Henschke, Hewitson, Jamsheed, Jasper Hill, Peter Lehmann, Majella, Charles Melton, Mount Langi Ghiran, S C Pannell, Penfolds, Rockford, Seppelt, Shaw & Smith, Torbreck, Turkey Flat, Tyrrell's, Wendouree, The Willows, Wirra Wirra, Yalumba, Yering Station.
NEW ZEALAND Bilancia, Craggy Range, Dry River, Elephant Hill, Esk Valley, Fromm, Man O'War, Millton, Passage Rock, Stonecroft, Te Awa, Te Whare Ra, Trinity Hill, Vidal, Villa Maria.
CHILE Casa Marín, De Martino, Errázuriz, Falernia, Lapostolle, Viña Leyda, Loma Larga, Matetic, Maycas del Limarí, Montes, Tabalí, Tamaya, Undurraga.
SOUTH AFRICA Boekenhoutskloof, De Trafford, Eagles' Nest, Fairview, The Foundry, Hartenberg, Mullineux, Porseleinberg, Sadie Family, Saxenburg.

RECOMMENDED WINES TO TRY

Twelve classic northern Rhône Syrah

Chapoutier Ermitage le Pavillon
Jean-Louis Chave Hermitage
Auguste Clape Cornas
Clusel-Roch Côte-Rôtie les Grandes Places
Dom. du Colombier Hermitage
Delas Frères Hermitage les Bessards
Ferraton Crozes-Hermitage le Grand Courtil
Alain Graillot Crozes-Hermitage la Guiraude
Jamet Côte-Rôtie
Dom. du Mortier St-Joseph
Réne Rostaing Côte-Rôtie Côte Blonde
Marc Sorrel Hermitage

Eighteen other classic Shiraz/Syrah wines

Bilancia La Collina Syrah (New Zealand)
Brokenwood Graveyard Vineyard Shiraz (Australia)
Clonakilla Shiraz (Australia)
Elephant Hill Syrah (New Zealand)
Fromm Syrah (New Zealand)
Hedges Bel Villa (Washington State)
Henschke Hill of Grace Shiraz (Australia)
Jamsheed Garden Gully Syrah (Australia)
Kongsgaard Syrah (California)
Viña Leyda Syrah Reserva (Chile)
Man O'War Dreadnought Syrah (New Zealand)
Matetic Syrah (Chile)
Moon Curser Contraband Syrah (Canada)
Mullineux Schist Syrah (South Africa)
Penfolds RWT (Australia)
Porseleinberg Syrah (South Africa)
Sean Thackrey Orion Old Vines Rossi Vineyard (California)
Trinity Hill Homage Syrah (New Zealand)

Five sparkling Shiraz wines

Fox Creek Vixen Sparkling Shiraz Non-vintage (Australia)
Peter Lehmann Black Queen (Australia)
Charles Melton Sparkling Shiraz (Australia)
Rockford Black Shiraz Non-vintage (Australia)
Seppelt Show Sparkling Shiraz Vintage (Australia)

Shiraz is the comeback kid: one moment growers were pulling it up as fast as they could and the next the whole world was crying out for old-vine Shiraz. Or failing that, any Shiraz at all. And there's a massive difference between old-vine Shiraz and young, high-yielding Shiraz.

Maturity charts

This grape is still new to most regions except the Rhône Valley and Australia, so identifying national styles is tricky.

2010 Hermitage (Red)

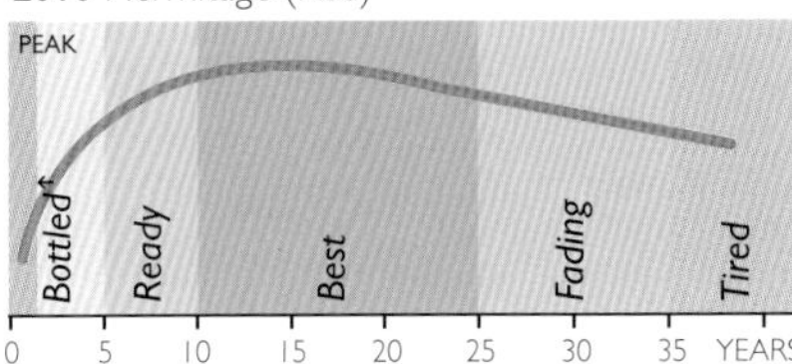

After the intensely rich 2009s, 2010 was a more typical Rhône vintage and the wines are dark, dry, rich but beautifully balanced for aging.

2010 Barossa Valley Old Vine Shiraz

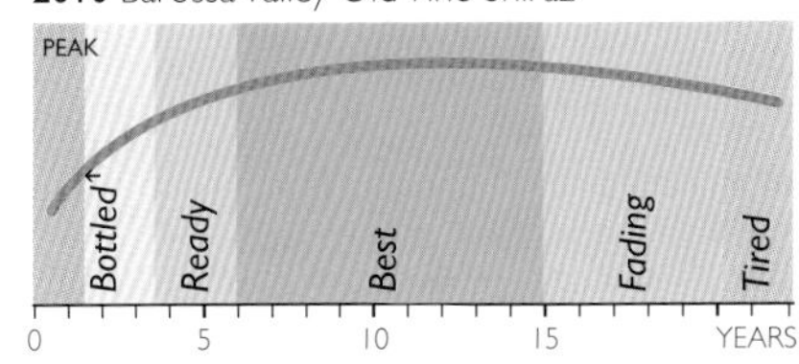

Styles here vary, but the most intense wines can combine early drinkability with the ability to age for a couple of decades or more.

2014 California Syrah

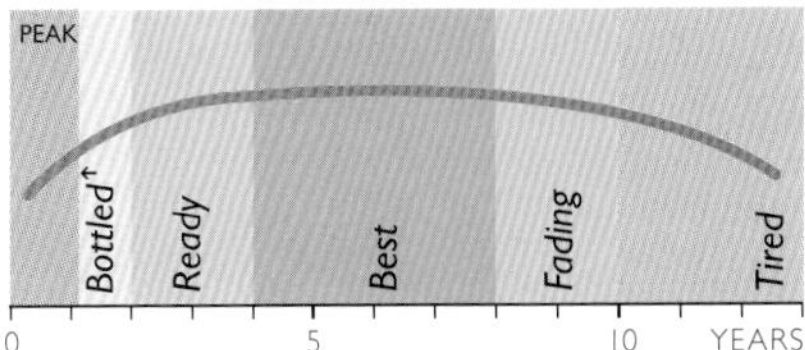

A typical California Syrah or Shiraz does not yet exist. Most wines are made to be drunk well within 3–4 years; gutsier ones are good young but will last longer.

E·R

TEMPRANILLO

It must be tremendous to be the dominant grape variety in one of the world's greatest wine countries. In the new age of wine becoming a global village, it must be delightful to watch other countries in both hemispheres taste with delight your handiwork in a myriad of famous wines, and you need only sit and wait for them to head back home clutching cuttings, determined to see whether they can match the mighty original.

What a pity that this is just a dream inside your head, because hardly anyone knows your name, or what wines you're responsible for. It must make things even worse when fine bottles are broached and people coo about your qualities – and then they call you something completely different.

And that's the problem with being Tempranillo. An awful lot of people drink your wine and think it's smashing. But they don't know you provided the grapes. Yet the great majority of top red wines in Spain do come from the Tempranillo in Spain, where there are about three times as many plantings as there now are of Garnacha, traditionally the dominant grape, much of whose vineyard space has been taken over by Tempranillo, even where Garnacha is much more suited (as in Rioja Baja). So why isn't it better known?

The history and grandeur of Tempranillo is conveyed by references to two Spanish artists of the 17th century: first, Diego Velázquez, the great Spanish court painter whose paintings of the royal family, including the Infanta Maria Teresa depicted here, were much admired at the time, and second, Juan Sánchez Cotán, who is best known for his still lifes with their detailed realism.

Well, in Rioja it is well known. Tempranillo is called Tempranillo there, but it's only in relatively recent times that Rioja has been accorded any great fame and respect outside Spain. And remember that when the Californians and Australians and South Africans and the rest were looking for grape varieties to transplant back to their distant lands, they were obsessed by the French classics – they wanted to make Bordeaux lookalikes and never cast a glance south of the Pyrenees. But if they had glanced at Ribera del Duero – and they could have done since Vega Sicilia has been Spain's most famous and expensive red wine for a century – they'd have discovered that the main grape was called – well, Tinto Fino, actually. Where's this taking me? Ah, the grape is actually Tempranillo, but they don't use the name. Convenient. What about port then, over the border in Portugal? Ah, you must be thinking of Tinta Roriz. Or Aragonez. Again, what's the relevance here? Roriz and Aragonez are Tempranillo. Yet again, they don't use the name.

And we're not finished yet. Back in Spain, Tempranillo is the major grape in most of the other top red wines. But Toro's best grape is Tinto de Toro. Valdepeñas and La Mancha boast about their Cencibel. Penedès swears by the Ull de Llebre – and so it goes on, through province after province. Always an excellent red grape, always called something different. But always actually Tempranillo. I can't think of any red grape that so completely dominates a country's quality wines as Tempranillo does Spain's – and which hides under so many aliases, with Portugal willingly adding a few of its own.

Since the New World winemakers are the ones who have made grape varieties famous by putting their names on the labels, you have to attract these trailblazers if your grape is to achieve any renown. Tempranillo was in the wrong place – nobody wanted to copy Spanish wines. And Tempranillo was never on the label, anyway. Spain and Portugal followed the French appellation idea of naming their wines after areas, not grapes. And since table winemaking in Iberia was dreadfully old-fashioned, who would care to spend time and money finding out if their grapes were actually any good?

It was only when a few 'flying winemakers' turned up in Iberia that the vineyards actually got a close look. Such winemakers have turned Iberia's wine into some of the most exciting in Europe. And time and again, they've discovered that the local grape which they've transformed into the best grog was – the Tempranillo. Which is why Australian wineries such as Pondalowie and Gemtree and New Zealanders like Trinity Hill and Elephant Hill, having made delicious experimental wines, are taking a punt that of all the emerging new grape varieties in their countries, Tempranillo is the one to back.

Tempranillo: from Grape to Glass

Geography and History page 272; Viticulture and Vinification page 274; Tempranillo around the World page 276; Enjoying Tempranillo page 278

GEOGRAPHY AND HISTORY

Suddenly, everyone's talking Tempranillo. If you're one of those who's not, let me explain. Tempranillo is the great Spanish grape that is at the heart of Rioja and Ribera del Duero, Spain's most famous red wines. It turns up almost everywhere else in Spain under a variety of names like Cencibel, Tinta de Toro and the rest, and could rightly be said to express the fundamental character of Spanish red wines from the cool high vineyards of the northeast to the arid plains way south of Madrid. As such, we've all drunk Tempranillo wines, even if we've not seen the name on the label. But Spain's explosion of quality in the last few years has meant that its own grape varieties have suddenly come under examination by growers worldwide keen to find commercial alternatives to Cabernet Sauvignon and Merlot.

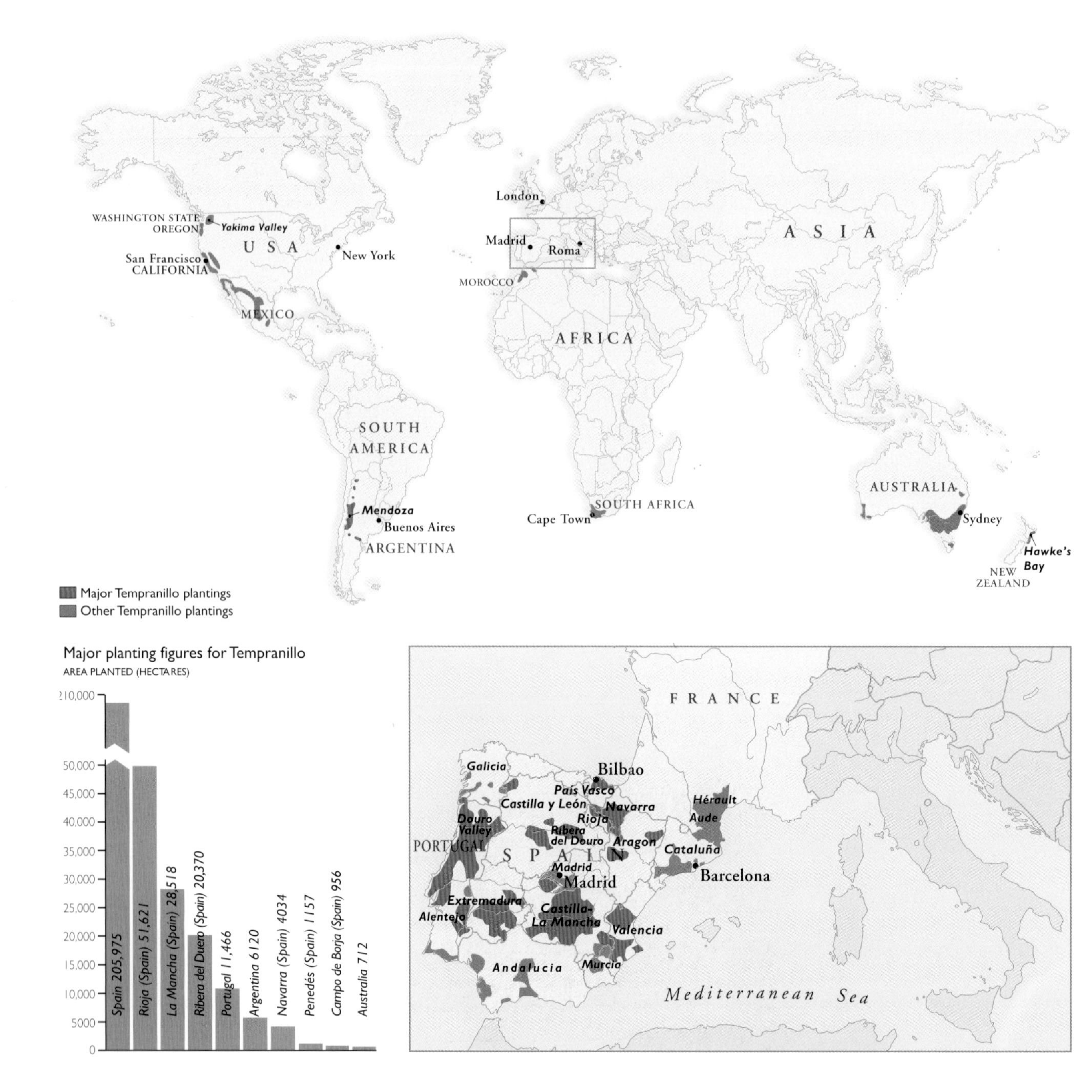

Southern France has had experimental plantings for quite a while. Portugal has a good deal under the names Tinta Roriz and Aragonez, and Argentina has old vines. Italy has some, but usually calls it Malvasia Nera. But it is the current wave of interest from such places as Australia, California, Washington State, and possibly New Zealand that just might make the name Tempranillo internationally famous. What's the attraction? The world is mad for red wine – but it wants it to be both reliable and different. It wants lush textures and appealing fruit, low tannin and not too much acidity. And it wants a grape that tastes recognizable, no matter what the geographical source. Tempranillo can do all this. It has the right exuberant flavour, and what it lacks in perfume can be supplied by other grapes. It takes to new oak like a dream. It lacks acidity, but that can be rectified by canny blending. It will grow obligingly in warm climates, and it will produce anything from light young wines to older, oak-aged ones. It may not easily reflect *terroir*, or be that complex, but it will taste of itself and be utterly, utterly seductive.

HISTORICAL BACKGROUND

Tempranillo probably originated in Rioja or Navarra, but it's very difficult to guess when. The earliest reliable mention is in 1807. It could be a lot older than that, or just a bit older; we don't know.

From Spain it spread to Portugal and to Argentina and Mexico – and, more surprisingly, to Italy, where the Malvasia Nera of Tuscany and Basilicata is none other than Tempranillo. Its arrival in the Douro Valley seems to date from the 18th century, when it was planted at Quinta de Roriz by the quinta's Scottish owner, Robert Archibald and obligingly called Roriz. Archibald is also thought to have introduced Tinta Francisca to the Douro.

In Portugal, its Spanish origins are reflected in its local name of Aragonez. In the Setúbal Peninsula it is known as Tinto de Santiago, again pointing to a Spanish source.

In Spain, local names include the Catalan Ull de Llebre (Ojo de Liebre in Spanish) in Penedès; Tinto Fino in Ribera del Duero; Cencibel in Valdepeñas; and Tinto de Madrid, Tinto de la Rioja, Tinta del País and Tinta de Toro in other parts. All these regions tend to claim that their Tempranillo is a local clone that gives a unique character to their wines; the truth is that no fewer than 552 different clones have been identified. Cencibel and Tinto Fino are not local clones; they are merely local names for a vine which may well behave slightly differently in different conditions. Variations include Tempranillo Peludo, or Hairy Tempranillo, which has more down on the back of the leaf, and Tempranillo Blanco, a chance mutation found some years ago and now authorized for use in the Rioja blend.

The name comes from the Spanish word *temprano*, meaning early, and reflects its early ripening nature; it also buds late, and needs only quite a short growing season. Which is why it clearly suits the cool conditions of much of Rioja, but it makes me question whether it should really be so widely planted in warmer parts of Spain.

The house of the Rodriguez family on the beautiful Remelluri estate up in the hills above Labastida in Rioja Alavesa. For many years Remelluri, a single-estate wine, was a rarity in Rioja, a region where bodegas have traditionally bought grapes from many different spots. Now there is an increasing focus on individual terroirs.

A traditionally bush-trained Tempranillo vine in Rioja: the local name for this form of training is en vaso. More and more growers, succumbing to the delights of mechanization, are training their vines on wires.

Oak barrel-aging is crucial to the character of Rioja and nowhere more so than at Bodegas Muga, where every step of red winemaking is carried out in oak, and they retain their own cooper.

VITICULTURE AND VINIFICATION

Climate and winemaking seem to be the main influences on the flavour of Tempranillo – in fact, until quite recently the flavour most associated with it was the vanilla softness of American oak. This was because the only widely known Tempranillo-based wine was Rioja, whose reputation during the 1970s and 1980s had been built almost entirely on the felicitous marriage of vanilla-scented oak and strawberry-flavoured Tempranillo. There is no long tradition of single-vineyard wines in Spain, and little emphasis until recently on the importance of different *terroirs* to the styles of individual wines. Most Tempranillo was and is a blend of different soils and climates and, especially in Rioja and Ribera del Duero, different grapes; more emphasis on individuality might lead to greater understanding of how the vine reacts to different *terroirs*. The most likely place for this to occur is Ribera del Duero, though single-vineyard Riojas are now beginning to make their presence felt.

Climate

To get elegance and acidity out of Tempranillo, you need a cool climate. But to get high sugar levels and the thick skins that give deep colour you need heat. In Spain these two opposites are best reconciled in the continental climate but high altitude of Ribera del Duero. Here, at up to 850m (2800ft), summer daytime temperatures may hit 40°C (104°F), which gives sugar and colour, but fall dramatically by up to 20°C (68°F) at night, thus conserving acidity. Summers are short here and winters long and hard: there may be as few as a hundred frost-free days a year.

Nearby Toro also has a short, hot ripening season, though at 230 days it is significantly longer than Ribera del Duero's 171 to 198 days. Vines are planted here at between 600 and 750m (2000 and 2500ft). Rioja Alta is appreciably lower at 500–600m (1650–2000ft), and Rioja in general is less continental in climate: producers point to the temperate, Atlantic-influenced west wind that blows across their vineyards, while the Sierra de Cantabria mountains give shelter from the colder north wind. The result is lower alcohol in Rioja – typically 12.5–13 per cent, compared to 13.5–14 per cent in Toro – and less pronounced structure and fruit than in Toro and Ribera del Duero. In Portugal's part of the Douro Valley, vineyards are planted at almost all altitudes, with the highest ones giving wines with the greatest acidity, and ripening two to three weeks after the lowest ones, but even late-ripening sites can usually produce grapes with pronounced perfume and fruit.

Soil

Soils in Rioja are relatively homogeneous compared to some regions: in the north they are mostly clay-based, and in Rioja Alta and Rioja Alavesa there are patches of chalky or iron-rich subsoil. The best wines, as is so often the case, come from the chalky clays: chalk gives acidity and elegance, clay gives body. But the ferruginous clays are full of trace elements that can add complexity.

In Toro most of the soils are alluvial, and there is some limestone in the subzone of Morales; an admixture of clay, because it holds water, enables the vines to survive the hot summers.

Both in Ribera del Duero and in the Douro Valley further west, the river Duero/Douro flows through schist. At the Spanish end there is also a lot of chalk: limestone and chalk comprise a third of the soils in the west of Ribera del Duero, and over half the soils in the east. Limestone and chalk, here again, give acidity. Aragonez from the Alentejo in south-east Portugal lacks the acidity of Tinta Roriz from the Douro, but can be a delicious, juicy mouthful.

Telmo Rodríguez has been behaving like a one-man revolution in Spanish wine since the late 1980s. Based in Rioja, but active in Valdeorras, Rueda, Ribera del Duero, Toro, Alicante, Cigales and Malaga – phew! – he and his partner Pablo Eguzkiza fight to save old vineyards, preserve and propagate ancient genetic material, and re-create the traditional field blends of the different areas. They also make inspiring wines mixing the best of tradition with modern imagination.

Cultivation

The traditional pruning method for Tempranillo in Spain is the same as for so many Spanish grapes: *en vaso*, or *gobelet*. Three or four branches are left, with about 20 fruiting buds in all. Mechanization is not possible with this system, and so training on wires, usually with double cordon, is on the increase. Indeed, it seems to suit the vigour and upright habit of the vine. Densities for mechanical harvesting are lower: about 2200 vines per hectare, compared to the more traditional figure of 2500–3000 per ha. In the Douro Valley training on wires is the norm.

Tempranillo crops generously: sometimes too generously for quality because, like Pinot Noir, it is considered a very yield-sensitive vine. In the Alentejo, 1.5 to 2 kilos per vine is about the limit for quality: go to twice that and the wine will be dilute. In the Douro, seven tonnes per hectare (equivalent to about 49hl/ha) is considered quite high; in Argentina, however, 12 tonnes per hectare is considered quite low. In Argentina's warmer, most heavily irrigated spots the vine will give 30 tonnes per hectare (or about 200hl/ha) of dilute wine, but as a simple soft glugger sold at a low price – well, it still has enough flavour to perform that task rather better than most other varieties.

In Rioja the legal limit is 45hl/ha (49hl/ha in Ribera del Duero), but while many vineyards produce less than this, there's no doubt much of the region overproduces – particularly on newer valley-floor vineyards. Indeed, dilution through overcropping has been one of Rioja's major problems in recent years. Training on wires increases yields, and in addition to this, irrigation was introduced in the late 1990s.

At the winery

Ribera del Duero is leading the way in Spain, producing the sort of dark, rich wines that fashion requires. Rioja is following, with longer maceration times, shorter oak aging and more use of French oak in place of the American oak that used to be the norm; some of the newest and priciest Riojas are unrecognizable as traditional Rioja. But that is because Rioja's traditional style is one that comes from long oak aging rather than from *terroir* or particular fruit characters. Now the flavour profile of Tempranillo is shifting more towards plums and black cherries, and away from strawberries, toffee and spice.

Winemakers are just beginning to understand what Tempranillo can do, and so far what it seems to be best at is wines in the full but soft Merlot mould. This might point to a bright future in California and Australia. It's prone to reduction if fermented in closed containers. Maceration times vary enormously, though too much extraction can give oily, rancid flavours, since the pips are less hard than those of Cabernet Sauvignon, and need to be handled gently. For the more modern wines, the malolactic fermentation is done in barrel to fix the colour and tannins. But the crucial difference between old-style and modern Rioja winemaking is in the length of wood aging. Tempranillo is particularly resistant to oxidation, so is able to take a lot more wood aging than many varieties – even when the crop is high and the colour and structure are light. But such wines, even if they're technically healthy, aren't a great mouthful because the creamy oak will completely dominate the mild fruit. A lot of Rioja is still made in this style, particularly at Reserva level, but more modern wines are usually vinified to maximize fruit and bottled earlier, for freshness.

There are few wine regions except, perhaps, California's Napa Valley, where architects have been given such a free rein as in Rioja. There are some simply stunning modern buildings, and there are elegant, subtle buildings like Torres' new eco-friendly Labastida winery, doing all it can to blend in with the landscape rather than dominate it.

Carbonic maceration is also very popular in many regions, either for a part of or the whole of the blend, for early-drinking wines, and the Viño Joven – young wine – made from Tempranillo is very successful: deep, soft and bursting with fruit.

Much Tempranillo is blended. It may only need a seasoning of other varieties – Garnacha, Mazuelo, Graciano, Cabernet Sauvignon, Merlot, Syrah and so forth – to add perfume, acidity, flesh, or whatever the particular lack might be. But that seasoning can make all the difference. Although Tempranillo resists oxidation successfully, many unblended Tempranillo wines seem to gain little with extended aging. A seasoning with other grapes adds the spark which makes for a complex, interesting maturity.

TEMPRANILLO AND OAK

The taste of oak, which has long been associated with Rioja and thus with Tempranillo, traditionally comes from 225 litre barricas bordelesas of American oak. These were introduced by French merchants who came prospecting for wine during France's phylloxera crisis in the 1860s and '70s. (Phylloxera did not hit Rioja until slightly later, in 1901.)

Tempranillo and American oak get on like a house on fire: the rich vanilla flavour of the oak suits the ebullient fruit of the grape. But new oak flavours are not the ones generally found in Rioja: instead the wine gets its character from long aging in old barrels. The wine acquires a mature character over the years from all its exposure to air, and needs no further aging in bottle, although it nevertheless seems to be able to survive in bottle – albeit in what can sometimes seem a mummified form – for many, many more years.

These are the rules in Rioja: Crianza and Reserva reds must spend at least a year in oak; Gran Reservas must spend at least two years. Joven wines are unoaked. There are also legal minima for the time wines must spend in bottle or tank before release. Traditional Rioja may be given much longer in old oak barrels than the law demands. Not surprisingly, there are an awful lot of barrels in Rioja: more than 600,000 at the last count.

However, the trend in Spain is towards shorter oak aging, and a greater use of new oak – and French oak at that. The lead was given by Ribera del Duero, which pioneered a style of more youthful fruit and more pronounced new oak. More and more Rioja bodegas are now busy making wines in the same style.

But alongside the move towards new oak – seen not just in these classic regions, but in every part of Spain where Tempranillo is grown – there is an equal move towards unoaked, juicily fruited Tempranillo, even in such oak-obsessed areas as Rioja and Ribera del Duero. Given the amount of Tempranillo growing all over Spain, even in less well-known regions, this is an encouraging trend. Wines aged for many years in old oak are no longer the norm: in fact, they are rapidly becoming a niche market – even, just possibly, an endangered one.

TEMPRANILLO AROUND THE WORLD

Is Tempranillo a great grape? The jury is still out on this one. Just as Italy's Nebbiolo has excelled only in Piedmont thus far, so it's only Spain that has produced world-class Tempranillo – as yet. But Portugal's new wave winemakers have made some lovely examples, and since it is basically a forgiving grape I'd expect it quickly to find fans among New World producers.

Rioja

The region of Rioja is flourishing. Prices for these wines have risen and the Spanish market seems to have an insatiable demand for Alta Expresion wines: super-premium bottles sold for high prices pulled out of the air. Even the producers admit that such prices have no relationship to costs. Not surprisingly, producers are eyeing every patch of half-decent land not yet planted with grapes. Obtaining planting rights is a problem, particularly for bodegas – the authorities prefer to give planting rights to growers – but the price of such potential vineyard land continues to rise. Rioja has been in this foment of investment and exploitation for a while now. It's up to the authorities to try to balance demand for planting rights with the need to protect quality.

To be honest, almost all the region's top vineyards were in production 30 years ago. You could argue that if more planting rights are not granted, then growers will simply overproduce even more to meet demand. Depressing, but true. And already production has shot up: in 1970, around 40,000ha (98,900 acres) were under vine in Rioja. By 2013 it had risen to 61,840ha (152,810 acres) – but production had more than doubled.

In Rioja Alavesa, the only part of the Rioja region located in the Basque country, and thus under different local government, there is a policy of encouraging growers to become bodegas. In the long run this must surely produce greater individuality in Rioja. Single-vineyard wines, along with special *cuvées*, are becoming more popular, and are very much in line with fashionable thinking, but Tempranillo does not seem to reflect its *terroir* as clearly as some other grapes. So a single-vineyard wine will not necessarily be more interesting than a carefully created blend.

There are, however, differences between the three parts of Rioja. Rioja Alavesa produces the most delicate, scented wines; those from Rioja Alta, where 66 per cent of the region's Tempranillo is planted, are firmer, darker, richer. Overall more than 86 per cent of Rioja's vineyards are planted with Tempranillo, but further east in the hotter Rioja Baja Garnacha takes over as the main grape, and makes the best wine, though too much Tempranillo is planted and its wines are frequently rather thick and stewy.

Ribera del Duero

This region, where Tempranillo covers some 80 per cent of the vineyard, owes its fame in the first place to Vega Sicilia, which was making world-class wine here more than 100 years ago. More recently a raft of new bodegas has been established: in the late 1990s there were fewer than 60, but by 2013 there were nearly 300. The vineyard area has not, however, increased by the same amount; grape prices have shot up.

Many of the wines are sold as Jovenes, full of blackberry and mulberry fruit; Crianzas and Reservas have one year's oak aging, and Gran Reservas have two years. Reservas and Gran Reservas must have additional aging in tank or bottle before release. At least 75 per cent of the wines must be Tinto Fino; the rest may be Cabernet Sauvignon, Garnacha, Malbec, Merlot or Albillo. The most complex wines are generally the blends. Although it is easy to blame Rioja for too much oak and over-extraction in the winery, Rioja is actually on the mend, with gob-smacker black-hearted reds far less common now than a few years ago. Yet excessive use of oak and dense textures are still a problem in Ribera del Duero – such a pity when the local Tempranillo is capable of some of the most beautiful blackcurranty fruit and savoury acid balance in the whole of Spain.

Other Spanish Tempranillos

There are few places in Spain where Tempranillo does not pop up: in fact, Spanish growers are planting more Tempranillo than Cabernet Sauvignon, Merlot and Syrah put together. In Navarra it can be silky and voluptuous; in La Mancha anything from light, pale, fresh and blended with white grapes, to surprisingly intense and juicy; in Cataluña red-fruited and generally blended; in Somontano perhaps a little green; in Toro, increasingly fashionable and often massively oaky, dark, savoury, solid, low in acidity and high in alcohol; in Costers del Segre well balanced and savoury.

Portugal

In the Douro Valley, Tinta Roriz is one of the five official grape varieties recommended for

El Pison

Exceptional dense but scented and exciting 'special cuvée'. Many producers do these 'High Expression' wines; only a few, like Artadi's, are successful.

Marqués de Murrieta

Murrieta's Gran Reserva Especial has been known to spend up up to 216 months in barrel – and still taste wonderful. This 2005 made do with 30 months.

Bodegas Vega Sicilia

Vega Sicilia's legendary Unico is traditionally Spain's most famous red wine. Tinto Fino (Tempranillo) makes up 65–80% of the blend,.

It was a Scot who first planted Tempranillo in Portugal at his Quinta de Roriz in the Douro Valley – hence the Tinta Roriz name – and, as so often in the Douro, the view simply takes one's breath away. Douro properties always grow a number of varieties, but Tinta Roriz is the second most planted after Touriga Franca.

port, and, with Tinta Franca, is often favoured by growers over the others. It flourishes in all parts of the region, but prefers soils that are rich in minerals. The wine can be almost astringently tannic but can have a heavenly scent, so new wave winemakers are finding ways of taming this harshness and emphasizing its raspberry and mulberry fruit and floral fragrance. It has less colour than some other port grapes, but its ability to resist oxidation means that it keeps its colour well; this lighter tint also makes it highly suitable in the blends for tawny ports.

Further south, in the Alentejo where it is known as Aragonez, it has less tannin and acidity than it does in the Douro and this has in the past consigned it to a subsidiary role since it quickly loses colour and freshness. But new wave winemakers are now finding ways to capitalize on its lower tannin and acidity and we are now seeing gorgeous, juicy, plummy, spicy reds, ready to guzzle as soon as they are bottled.

Will these wines age? That's not the point. In Dão the giant Sogrape company is persuading its growers to graft over to Aragonez, in an attempt to add some perfume and fruit to what is traditionally a rather lean wine and plantings in Portugal have increased rapidly since the mid-1990s.

Australia

This is one of Australia's fastest growing varieties in percentage terms. The big wine companies like Tempranillo because it has the natural acidity to deal with hot regions like Riverland; smaller producers are starting to express frustration with its high yields. They are the main reason why Australia has yet to produce any really impressive examples. You may want 5 tons per acre, but the wretched vines want to give you 15. But then even in Rioja the clones weren't designed for low yields. There's nothing wrong with either climate or winemaking, though, and juicy examples are starting to appear.

South America

Tempranilla, as the grape is sometimes called in Argentina, is used as a workhorse grape, and until recently was never thought of for good wine. But that's largely because the Argentinians grossly overproduce it. With lower yields, companies like Finca El Retiro and O J Fournier have made some really serious wines, but the toughness of the tannins shows the variety is wrestling with the warm conditions.

Rest of the world

I've had quite nice examples from Mexico and it has been grown in the south of France, particularly in the Aude department, for over 20 years. It's being tried in Bulgaria, too.

In California it is probably the same as the variety known as Valdepeñas, a grape in decline and little regarded. But only a couple of Napa producers have had a go with it. Expect young tyros to take it more seriously in the next few years, expecially since it can make wines similar in softness and texture to Merlot. Globetrotting viticulturist Richard Smart feels it has potential in Oregon, either in the warmer south or in the cooler Willamette Valley; and there have been some successful attempts in Washington and Texas.

Bodegas Alejandro Fernandez
Alejandro Fernandez is known as the 'King of Tempranillo', which he calls the 'Queen of Grapes' – Pesquera is 100% Tempranillo.

O. Fournier
Argentina's Uco Valley has some exceptional old Tempranillo vines, but it took Spanish wine producer O. Fournier to give them some respect. .

Gemtree
McLaren Vale is one of Australia's most experimental wine regions. Gemtree excel with their Tempranillo grown on limestone soils cooled by sea breezes.

ENJOYING TEMPRANILLO

Straight Tempranillo can be absolutely gorgeous young – crunchy, juicy, herb-scented, irresistible. But its reputation, established in places like Rioja and Ribera del Duero, is as a wine that ages. Well, pure Tempranillo rarely evolves much with age; it cries out for a slug of something else – Graciano, Mazuelo, Cabernet Sauvignon – and in wines destined for aging it usually gets it.

Although Rioja was the first region to make a name for itself, Ribera del Duero has grabbed the mantle of making Tempranillo's most age-worthy wines – and it's interesting that the soil (a lot of limestone) and the climate (massive temperature differences between day and night) encourage acid retention and development of perfume and fruit. If only they'd use less oak. Here the finest Gran Reservas can last up to perhaps 30 years – I just had a positively virile 1953 – and even lesser wines can last a decade quite happily. But wines labelled Joven or Crianza, here as everywhere in Spain, are not intended to be aged: Joven wines should be drunk within a year or so, and most Crianza wines within a couple of years. The best Riojas are only slightly less age-worthy – a couple of beguiling 1964 Gran Reservas this year have proved the point. But traditionally made Rioja, which has been aged for several years in old oak barrels, follows a different aging pattern to more modern wines that are bottled earlier. Traditional Rioja seems to age relatively fast in its first few years in bottle, before reaching a plateau which continues for a couple of decades before gradually fading. Is this a form of mummification? The wines seem to evolve little in this period; they couldn't be said to improve. But they certainly last. Tempranillo and Tempranillo blends from elsewhere in Spain age according to the nature of the blend: five to 10 years is probably the limit.

In Portugal, new-wave Roriz from the Douro Valley and Aragonez from the Alentejo should be drunk young, but port lasts for donkey's years and Tempranillo/Roriz plays an important part in the blend. For more details see Touriga Nacional pages 282–283.

The taste of Tempranillo

Think of a cross between Cabernet Sauvignon and Pinot Noir, and you have the flavour of Tempranillo. Well, sort of. It has the deep colour and rich flavour of the one, plus the strawberry fruit of the other – yet the complexity of neither at its best. But complexity is not the point of Tempranillo: its attractions are its lush texture and its supple, exuberant fruit, all blackberries and black cherries, mulberries and raspberries. In Ribera del Duero and Toro these flavours have a sensational savoury butter and blackcurrant slant when young and move towards tobacco, plums, prunes and cocoa with age. In lightweight Tempranillo intended for early drinking, the taste is more of strawberries and plum jam. Overripe Tempranillo is figgy and sweet; with long oak aging the flavours become savoury and strawberryish, with a touch of coffee bean and dried fruits.

Acidity can vary from low to quite high, and tannins are generally soft and ripe, though Toro and Ribera del Duero are sometimes a bit tough, as can be Roriz in the Douro. In lesser Ribera del Duero wines the acidity and oak between them can outpace the fruit in a poor year. Both tannins and acidity are, however, essential if the wine is to improve for more than a year or two.

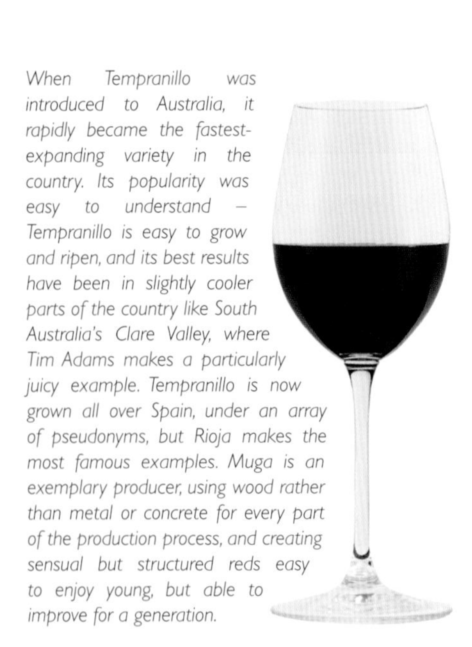

When Tempranillo was introduced to Australia, it rapidly became the fastest-expanding variety in the country. Its popularity was easy to understand – Tempranillo is easy to grow and ripen, and its best results have been in slightly cooler parts of the country like South Australia's Clare Valley, where Tim Adams makes a particularly juicy example. Tempranillo is now grown all over Spain, under an array of pseudonyms, but Rioja makes the most famous examples. Muga is an exemplary producer, using wood rather than metal or concrete for every part of the production process, and creating sensual but structured reds easy to enjoy young, but able to improve for a generation.

MATCHING TEMPRANILLO AND FOOD

Spain's best red native grape makes aromatic wines for drinking young, and matures well to a rich and usually oaky flavour. Northern Spanish cuisine is well suited to the gentle oaky flavours of Navarra and Rioja, as well as the more intense fruit of Ribera del Duero. Tempranillo is good with game, local cured smoked hams and sausages, especially spicy *chorizo*, casseroles and meat grilled with herbs; it is particularly good with roast lamb. It can partner some Indian dishes and goes well with soft cheeses such as ripe Brie.

CONSUMER INFORMATION

Synonyms & local names

Tempranillo has many synonyms in Spain – the Ribera del Duero uses Tinto Fino and the La Mancha region, especially Valdepeñas, uses the name Cencibel. Other synonyms include Tinto del País, Tinto de Toro, Tinto de Madrid and (in Penedès) Ull de Llebre (Catalan) and Ojo de Liebre (Spanish); in Portugal it is known as (Tinta) Roriz or (Tinta) Aragonez/Aragonês. Sometimes called Tempranilla in Argentina.

Best producers

SPAIN/Castilla-La Mancha Dehesa del Carrizal, Uribes Madero, Manuel Manzaneque; **Rioja** Allende, Altos de Lanzaga/Telmo Rodriguez, Artadi, Baron de Ley, Bilbaínas, Campillo, Campo Viejo, Contador, Contino, CVNE, DSG Vineyards, Faustino, Viña Ijalba, López de Heredia, Marqués de Cáceras, Marques de Murrieta, Marqués de Riscal, Marqués de Vargas, Martinez Bujanda, Abel Mendoza, Montecillo, Muga, Viñedos de Páganos, Palacios Remondo, Remelluri, Fernando Remírez de Ganuza, la Rioja Alta, Riojanas, Olivier Rivière, Roda, Señorio de San Vicente, Sierra Cantabria, Torres, Valdemar, Valenciso, Valpiedra, Valsacro; **Ribera del Duero**, Aalto, Alión, Alonso del Yerro, Arzuaga, Aster, Los Astrales, Balbás, Hijos de Antonio Barceló, Briego, Casajús, Cillar de Silos, Convento San Francisco, O Fournier, Goyo García Viadero, Hacienda Monasterio, Hacienda Solano, Matarromera, Montecastro, Emilio Moro, Pago de los Capellanes, Pago de Carraovejas, Pedrosa/Pérez Pascuas, Pesquera, Pingus, Protos, Rodero, Hermanos Sastre, Tarsus, Torres, Valduero, Valtravieso, Vega Sicilia; **Navarra** Chivite, Guelbenzu, Ochoa, Orvalaiz; **other Spanish wines** Albet I Noya, Pirineos, Romero Almonazar, San Isidro, Schenk, Viñas del Vero.
PORTUGAL Quinta dos Carvalhais, Cortes de Cima, Esporão, João Portugal Ramos, Quinta dos Roques, Quinta de la Rosa, Quinta do Vale da Raposa.
USA/California Bokisch, Justin, Twisted Oak, Viader, Kenneth Volk; **Oregon** Abacela; **Washington State** Cayuse.
AUSTRALIA Tim Adams, Crittenden, Delatite, Gemtree, La Linea, Mayford, Nepenthe, Pondalowie, Sanguine, Tar & Roses, Willunga 100, Yalumba/Running with Bulls.
NEW ZEALAND Elephant Hill, Trinity Hill.
ARGENTINA Anubis, Finca El Retiro, O Fournier, Salentein, Zuccardi.

RECOMMENDED WINES TO TRY

Ten classic Ribera del Duero (or equivalent) wines

Abadía Retuerta El Campanario
Bodegas Alión
Ismael Arroyo Val Sotillo Gran Reserva
Alejandro Fernández Pesquera Janus
Condado de Haza Alenza
Bodegas Mauro Vendimia Seleccionada
Pedrosa Gran Reserva
Dominio de Pingus Pingus
Teófilo Reyes Crianza
Vega Sicilia Unico

Ten classic Rioja wines

Finca Allende Rioja Aurus
Artadi Rioja Reserva Pagos Viejos
Bodegas Bretón Rioja Alba de Bretón
Contino Viña del Olivos
Marqués de Griñon Rioja Coleccion Personal
Marqués de Murrieta Rioja Reserva Dalmau
Marqués de Vargas Rioja Reserva Privada
Muga Rioja Reserva Especial Torre Muga
Remelluri Rioja Gran Reserva
Bodegas Roda Rioja Reserva Roda II
Señorío de San Vicente Rioja Tempranillo

Ten other Spanish Tempranillo wines

Albet I Noya Penedès Tempranillo Collecció
Chivite Navarra Gran Feudo Viñas Viejas
Guelbenzu Navarra Tinto
Ochoa Navarra Tempranillo Crianza
Orvalaiz Navarra Crianza
Pirineos Señorio de Lazán Reserva
Romero Almonazar Ribera del Guadiana
San Isidro Castillo de Maluenda Calatayud
Schenk Las Lomas Utiel-Requena
Viñas del Vero Somontano Tempranillo

Ten non-Spanish Tempranillo wines

Anubis Tempranillo (Argentina)
Quinta dos Carvalhais Dão Tinta Roriz (Portugal)
Cortes de Cima Alentejo Aragonez (Portugal)
Finca El Retiro Tempranillo (Argentina)
O Fournier Tempranillo (Argentina)
Gemtree Tempranillo (Australia)
Quinta dos Roques Dão Tinta Roriz (Portugal)
Quinta de la Rosa Tinta Roriz (Portugal)
Trinity Hill Tempranillo (New Zealand)
Quinta do Vale da Raposa Douro Tinta Roriz (Portugal)

Can we expect a flood of Tempranillo from all corners of the earth in the next few years? Australian growers are planting it furiously, and it could be terrific in South America. Washington Tempranillo is pretty smart.

Maturity charts

Most Tempranillo can be drunk quite early. Only a few top Ribera del Dueros and Riojas need extended aging.

2009 Ribera del Duero Crianza

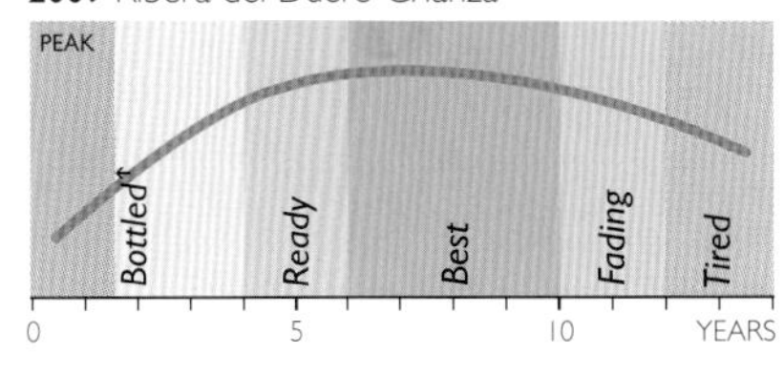

A hot dry year saved by September rains produced outstanding Ribera del Deuros, drinkable young, but capable of long aging.

2010 Rioja Reserva

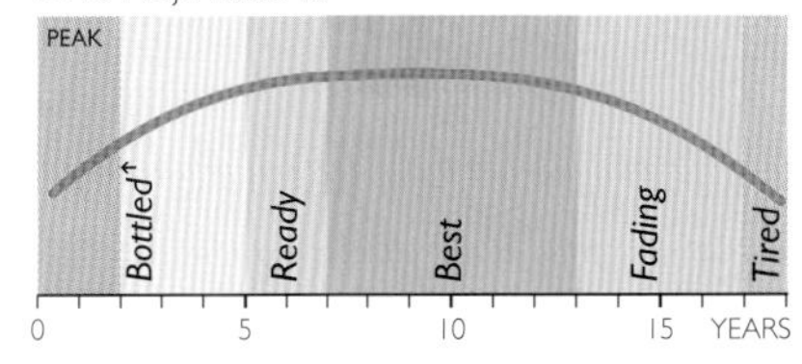

A tremendous vintage in Rioja, with a little more restraint and style than the richly impressives 2009s. Good for aging.

2012 Toro

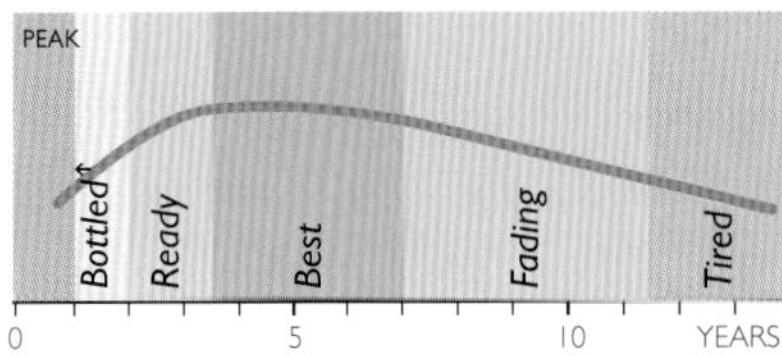

Toro seldom gains complexity in bottle, and is at its best within four or five years of the vintage. More modern styles are still chunky and chewy but easier to drink young.

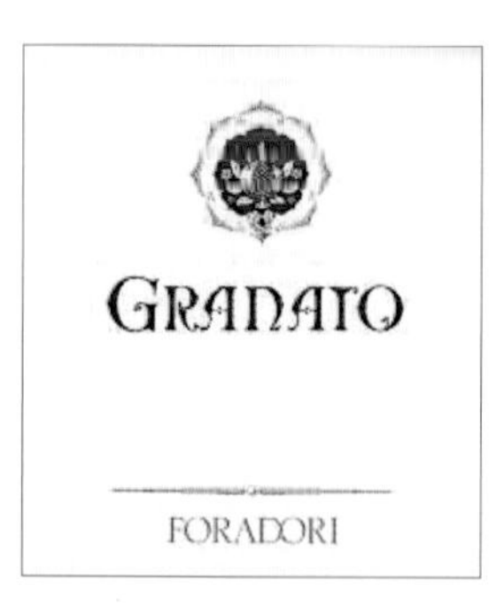

Foradori
Elisabetta Foradori has made strenuous efforts to rediscover the old genetic material of Teroldego after a period since the 1970s when clones were developed mainly for high yields. Granato is triumphant proof of her success.

TEROLDEGO

Teroldego may at the moment be a relatively little-known northeastern Italian grape, but it has a devoted fan club, and could have great potential in parts of the New World – New Zealand and Brazil, for example. Californian producer Jim Clendenen of Au Bon Climat is on the record as saying that California should have planted Teroldego and Barbera instead of Merlot – there is a little in California – and that if Teroldego replaced Syrah on the slopes of Crozes-Hermitage in the northern Rhône the wine would be greatly improved. Not everybody would go that far, but there is no doubt that it is an interesting grape. And not surprisingly: DNA fingerprinting has revealed that it is a full sibling of Dureza, which is one of the parents of Syrah. And it is closely related to both Lagrein and Marzemino.

In Trentino, which is almost its only home, it is often overcropped, and produces perfectly pleasant red with a rough-cut leafy earthiness that is best drunk young. But if the yield is restricted, the tannins allowed to get really ripe, and a bit of *barrique* aging added, the wine gains enormously in complexity and depth. It retains all its typically Italian bitter-cherry fruit but adds smoke and plums and mulberries; the acidity is in balance, though high (this is Italy, after all) and the tannins give a good firm backbone. Its only DOC is Teroldego Rotaliano. Best producers: (Italy) Barone de Cles, Marco Donati, Dorigati, Endrizzi, Foradori, Conti Martini, Mezzacorona, Cantina Rotaliana, G Sebastiani, A & R Zeni.

TERRANO

A grape of northeastern Italy, Slovenia and Croatia – and what's more, it's been there since the Middle Ages. What one expects of red grapes from round here is lots of tannin, lots of acidity, and black-fruit flavours, and Terrano delivers all these very gracefully.

TERRANTEZ

Almost extinct Madeira grape of very good quality. It is very low yielding – like, nothing at all in some vintages. Tastings of old bottlings reveal a wine of great complexity and length. Best producers: (Portugal) Barbeito, Barros e Sousa, Blandy.

TERRET

It is tempting to see Terret – or the Terrets, since it is a grape that comes in different colours – as a relic of the old wine world. Terret Gris and Terret Blanc, both of which make white wine, used to be standbys of the vermouth industry in the south of France; as that industry fades, there is less and less need for grapes that will produce up to 150hl/ha of wine that is fresh, light and dry but without much else to say for itself.

Terret Noir has a bit more to say for itself, but it too is in terminal decline. It is part of the permitted blend in Châteauneuf-du-Pape, Corbières, Minervois and other wines of the South, but it is light in colour and body. It adds perfume to the blend and marries quite well with sturdier wines like Mourvèdre and Grenache. That, though, is not enough to justify its presence in most modern vineyards. It is so prone to mutation that it is possible to find the same plant giving black grapes and white grapes. Best producer: (France) Mas Jullien.

TIBOUREN

The same grape as Liguria's Rossese (see page 217). In France it's found in the Var and around St-Tropez, where it contributes to the character of the local rosé.

TIMAROSSO

Piedmont variety, almost extinct but now being rescued. And it's worth it: think of honey, nuts and ripe citrus, held together with minerality and good acidity. Best producers: Walter Massa, Luigi Boveri.

TINTA AMARELA

See Trincadeira page 284.

TINTA ARAGONEZ

Like Aragonez, a Portuguese synonym for Tempranillo (see pages 270–279). It takes these names in the Alentejo. It gives wines with attractive blackberry aromas, but yields must be kept severely in check. Best producers: (Portugal) Quinta do Carmo, Cortes de Cima, Esporão, J M da Fonseca, J P Ramos, Reguengos de Monsaraz co-op.

TINTA BARROCA

A useful part of the port blend, favoured in the Douro Valley because it will produce good deep colours even on north-facing slopes and high up, and is unworried by drought. In fact, it even prefers north-facing sites, because it ripens early, and its thin-skinned berries do not like too much strong sunlight. Hotter sites do, however, give it the tannin that it can lack; a soft, even slack texture despite being fairly full-bodied is its main drawback. It does not always age well. It is seldom successful as a varietal, since it lacks the tannins to fix the colour, although its high sugar gives it a blackberry flavour and some roundness on the palate.

It is highly regarded for fortified wines in South Africa, where it is spelt Barocca, but plantings are small. It also makes sturdy, earthy table wines. Best producers: (Portugal) Quinta da Estação, Ferreira, Ramos Pinto.

TINTA CAIADA

If you want the best wines from this widespread vine, you'll have to go to Portugal's Alentejo region east of Lisbon, where producers are taking it seriously, succeeding in getting it properly ripe and making rather good fresh reds for early drinking. But it has plenty of other names in Portugal, and a good few more in Spain, in Sardinia and in Corsica. Its name in Somontano, where it probably originated, is Parraleta.

TINTA FRANCISCA

A port grape, though not one of the five officially recommended varieties. Its wines are high in sugar but light in body and flavour.

TINTA MIÚDA

Portuguese name for Graciano (see page 122), found mainly in Lisboa and Tejo. Yields are low and the wine is sturdy and full-bodied with a flowery aroma. Best producers: (Portugal) Arruda co-op, Casa Santos Lima.

TINTA NEGRA MOLE

See Negramoll page 176.

TINTA RORIZ

A northern Portuguese synonym for Tempranillo (see pages 270–279). It is more often known just as Roriz (see page 217).

TINTO CÃO

High-quality but low-yielding port variety grown in the Douro Valley. The name means 'Red dog', in keeping with the Portuguese habit of naming vines after animals whenever possible. In what way Tinto Cão resembles a dog of any colour is, however, not clear.

It is believed to have appeared in Portugal in the 17th century, and possibly originated in the Douro Valley. Its colour is not particularly deep, but it has good aroma – spicy when grown in warm sites, floral in the cooler sites to which it is best suited – and when young seems inferior to the other four recommended port varieties. It needs age, and after five years it develops considerable finesse. It can also give attractively aromatic red wines in Dao. It is attracting some attention in California and Australia. Best producers: (Portugal) Quinta do Vale da Raposa; (USA) Quady, St Amant.

TINTO FINO

The name given to Tempranillo in Spain's Ribera del Duero region (see pages 270–279). It is sometimes claimed to be a particular clone of Tempranillo, but this seems not to be the case: instead, it is simply an instance of a vine producing better balanced and structured wines of more complexity in a different climate and in different soils.

The vines are grown at about 750m (2500ft) above sea level, which makes it Spain's highest wine region, and summers are hot, winters cold and springs and autumns short. Nights are cool: in August and September there can be a difference of 20°C (68°F) between daytime and night-time temperatures. This produces more intensely flavoured wines with good acidity and powerful blackcurrant fruit. There is a rather short vegetative period between the last spring frost and the first autumn frost, too, so it is very important that no ripening time is lost through drought or rain Best producers: (Spain) see Tempranillo page 279.

TINTO DEL PAÍS

Spanish synonym for Tempranillo (see pages 270–279).

TOCAI FRIULANO

Tocai Friulano is the same grape as Chile's Sauvignonasse and arrived in Italy from France during the 19th century (see page 242). In retrospect it is hard to see how the Chileans could have confused the flavour of this light, delicately floral or appley wine with the pungent gooseberry or white peach-flavoured Sauvignon Blanc, but the plants do look very similar. In Italy's Friuli, where it is the main white grape, it can make wine of high though subtle quality, with good structure and balance and some depth. It can also turn out very large quantities of unexciting everyday wine rather as it is prone to do in Chile. Either way, it is best drunk young. There is some Tocai Friulano (or at least, a vine of that name) in Argentina. It is probably the same as the Tocai Italico of the Veneto region (but not of the Breganze DOC), but has nothing whatever to do with Tokay d'Alsace, a now-redundant name for Pinot Gris in Alsace.

Producers of Hungary's legendary Tokaj dessert wine complained about the use of the names Tokay d'Alsace and Tocai Friulano and so the EU decreed that both names had to go. For the time being the Friuli producers seem to have settled on the name 'Friulano'. Best producers: (Italy) Rocco Bernarda, Borgo Conventi, Borgo San Daniele, Borgo del Tiglio, Drius, Livio Felluga, Marco Felluga, Jermann, Edi Keber, Miani, P Pecorari, Princic, Paolo Rodaro, Ronchi di Manzano, Russiz Superiore, Russolo, Schiopetto, La Viarte, Vie di Romans, Villa Russiz, Volpe Pasini, Le Vigne di Zamò.

TOKAY D'ALSACE

For many years this was Pinot Gris' name in Alsace, but the EU bureaucrats decreed that it should be phased out from 2007 to avoid any confusion, however unlikely, with the Tokaj wine of Hungary. Tokay-Pinot Gris was a compromise name. All Alsace labels now use Pinot Gris (see pages 186–187).

TORBATO

A characterful Italian white grape found on Sardinia, where it was heading for extinction until rescued by the company of Sella & Mosca in the last century. It produces well-structured dry wines with good body and buttery richness. It used to be grown in the south of France and is now being reintroduced under its local name of Tourbat; it's blended in small quantities into Vins Doux Naturels such as Banyuls, Maury and Rivesaltes. Best producer: (Italy) Sella & Mosca.

TORRONTÉS

Argentina's white speciality, Torrontés is, in fact, several varieties, not all of which have the full Muscat-like aroma for which this Muscat-Criolla cross is known. The aroma of Torrontés can also be reminiscent of Gewürztraminer; it also brings to mind the smell of air freshener. It is floral, soapy and sometimes lightly spicy; when not overcropped and when carefully vinified, it can be positively heady and beautifully refreshing. I've never come across a Torrontés wine that has aged well beyond a couple of years. Less than a year is usually best.

Argentina's most planted and most aromatic Torrontés is Torrontés Riojano, named after the Argentine province of La Rioja. The less aromatic and less widely planted Torrontés Sanjuanino likewise takes its name from San Juan province. The least aromatic variety is Torrontés Mendocino or Mendozino, which is found further south, in Rio Negro at the northern end of Patagonia. Cafayate is turning out to be the best region for Torrontés Riojano, where it's being planted on terraces at an altitude of up to 2300m (7500ft) – there is one vineyard nearer 3000m (10,000ft). There is some Torrontés in Chile – mostly Torrontés Riojano, which is used for pisco, the local brandy.

Spain also has a vine called Torrontés and logically it should be the same as at least one of Argentina's versions, but this doesn't seem to be so. Spain's various Torrontés are found in Ribeiro, around Madrid, in Montilla-Moriles and in the Canaries, and the name also crops up in Portugal – again, for different grape varieties. Best producers: (Argentina) Colomé, Etchart, Norton, Michel Torino; (Spain) Bodega Alanis, Viña Meín, Vitivinícola del Ribeiro, Emilio Roja, Vilerma.

TOURIGA

More of an abbreviation than a synonym, this is what Australians call Touriga Nacional (see pages 282–283), though the name seems to have been used erroneously in the past for several other beefy red vines as well. California's Touriga can be either Touriga Nacional or Franca (see below). Best producers: (Australia) Brown Brothers, St Hallett.

TOURIGA FRANCA

Called Touriga Francesa until 2001, this is a robustly flavoured port grape, one of the five official varieties, with good colour and tannins, though without the quality of Touriga Nacional (see pages 282–283) or Tinta Roriz (see facing page). Its great advantage is its powerful aroma of mulberries and roses, which adds an exotic note to the blend. It yields well but needs hot sites to ripen properly. It is a crossing of Touriga Nacional and Marufo, an old and much-synonymed vine found in Spain (mostly Castilla-La Mancha) and Portugal. It is the most widely planted vine in the Douro, and is spreading to other parts of Portugal. Best producer: (Portugal) Casa Santos Lima.

TOURIGA NACIONAL

Portugal's wine revolution is so recent and the changes for the better in winemaking and wine styles are so dramatic that it is difficult to say which of a marvellous array of individuals is Portugal's best grape variety. But it may be, just may be, Touriga Nacional. I've no doubt that this variety can produce thrilling flavours – dark, damsony fruit, violet and new leather perfumes – but it also has a fairly powerful tannic attack which means it often does best when blended with other varieties – and luckily Portugal has a whole fistful just waiting for the call: Touriga Franca, Tinta Roriz (a.k.a. Tempranillo), Tinto Cão, Periquita, and a heap of others. In this, Touriga Nacional is like Cabernet Sauvignon – a grape of tremendous personality but with aggressive tendencies, which is often much better when softened by blending with something else.

However, Touriga Nacional is also the star grape in the great fortified port wines of the Douro Valley, even though it may make up only three or four per cent of the vineyard plantings. Even that, however, is up from 20 years ago. Low yields are the problem – they are half or even less than those of the other top grapes. But new clones are much better yielders, and there seems to be little reason now why plantings shouldn't increase. There are also hundreds of mutations of Touriga Nacional, producing anything from dense, dark, heady reds in small quantities to large crops of pale, insipid juice. At best, its berries are tiny; this is, of course, the source of its intense flavour and its powerful tannins.

In Dão, where it originates, it used to cover 90 per cent of the vineyards until phylloxera; as so often, post-phylloxera replanting favoured more productive varieties, like the burly Baga, and it is only now once more on the increase; the regulations demand a 20 per cent minimum in the blend and the quality of Dão has soared, surely not coincidentally, since the new emphasis on Touriga Nacional. It is also increasing in Bairrada, Estremadura, Beiras and the Alentejo, though the great heat of Portugal's far south may be a bit much for it.

In the New World, only Australia, with its great 'port' tradition, has shown much interest in Touriga Nacional, but producers there have so far found it tricky to isolate clones that balance productivity with quality. Certainly, it should be looked at by red table wine producers in such warm places as Chile, Argentina, California, South Africa and even neighbour Spain when they want to expand their range.

THE TASTE OF TOURIGA NACIONAL

Young Touriga Nacional has a deep and intense aroma that is reminiscent of young Cabernet Sauvignon, even to the extent of mixing dark, sweet fruit with a leafy freshness and hint of violets. As it matures as a port, it gains wonderfully rich mulberry and blackberry richness without losing its black peppery attack and its seductive scent of flowers. As a table wine, it takes a while to lose its tannin, unless blended with other varieties, but the fruit gets sweeter and richer with age.

Quinta do Crasto's revival as a wine estate dates from the 1980s, when its current owner returned from a sojourn in Brazil – he'd gone there in 1975 after the Portuguese revolution. Now Crasto makes both port and table wine from the same grapes. Like most port estates, it has both old mixed plantings and newer ones planted by variety; it also has a fantastic situation, perched right on top of a mountain overlooking the river Douro. As so often in Europe, the Romans were here first: the name comes from the Latin word castrum, *or fort.*

Taylor's
Quinta de Vargellas is the backbone of Taylor's vintage port, and also makes a wonderfully scented, deep-flavoured single-quinta wine. Apart from Touriga Nacional, the blend for this includes Touriga Franca and Tintas Barroca, Roriz and Cão.

Quinta dos Roques
Varietal Touriga Nacional wines like this one from Quinta dos Roques are becoming more common in Dão, which is good news because Quinta dos Roques shows the wine can have dark fruit, floral scent and a burly, smoky intensity.

Left: *Terraces at Quinta do Noval in the Douro Valley with parcels of the ungrafted Nacional vines on the left. Nacional is a field blend, but as individual vines die they are usually replaced with Touriga Nacional.* *Above:* *Touriga Nacional is important in both the Douro and Dão, but to judge from the amount of clonal variation it probably originates in Dão. There is a village called Tourigo in the Dão, and nearby is another village called Mortágua, which is another of Touriga Nacional's synonyms.*

CONSUMER INFORMATION

Synonyms & local names

Sometimes called Preto Mortágua in Portugal. Known as Touriga in Australia (see page 281). Touriga in California is probably Touriga Franca.

Best producers

PORTUGAL/Dão Boãs Quintas, Carvalhais, Quinta de Lemos, Pellada, Roques; **Douro** Aliança, Cachão, Calem, Churchill, Cockburn's, Côtto, Crasto, Croft, Ferreira, Fonseca, Gaivosa, Graham, Niepoort, Noval, Ramos Pinto, Sandeman, Sogrape, Taylor, Vale da Raposa, Warre; **rest of Portugal** Cortes de Cima, Cortezia, Esporão, Pancas, Primavera, Casa Santos Lima.

SOUTH AFRICA Boplaas, De Krans.

RECOMMENDED WINES TO TRY

Ten premium ports containing Touriga Nacional

Churchill Vintage Port
Cockburn Vintage Port
Croft Quinta da Roêda Vintage Port
Fonseca Guimaraens Vintage Port
Graham Malvedos Vintage Port
Niepoort Vintage Port
Quinta do Noval Vintage Port
Sandeman Vau Vintage Port
Taylor Quinta de Vargellas Vintage Port
Warre Traditional Late-Bottled Vintage Port

Ten Douro reds made wholly or mainly from Touriga Nacional

Aliança Foral Grande Escolha
Calem Lagarde Sá Touriga Nacional
Quinta do Côtto Grande Escolha
Quinta do Crasto Touriga Nacional
Ferreira Quinta da Leda Touriga Nacional
Quinta da Gaivosa Vinha de Lordelo
Quinta do Noval Corucho
Ramos Pinto Duas Quintas Reserva
Sogrape Reserva
Quinta do Vale da Raposa Touriga Nacional

Eight other Portuguese reds

Boãs Quintas Dão Touriga Nacional
Quinta dos Carvalhais Dão Touriga Nacional
Quinta da Cortezia Estremadura Touriga Nacional
Quinta de Lemos Dão Touriga Nacional
Quinta da Pellada Dão Touriga Nacional
Caves Primavera Beiras Touriga Nacional
Quinta dos Roques Dão Touriga Nacional
Casa Santos Lima Estremadura Touriga Nacional

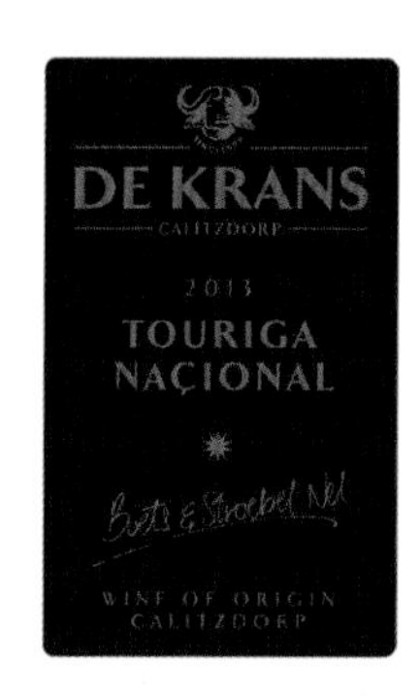

De Krans

Based at Calitzdorp in the boiling hot Klein Karoo region, De Krans makes some of South Africa's best 'port' styles from traditional Portuguese varieties, but also makes dry reds from these grapes, led by a dark, spicy Touriga Nacional..

First Drop

Touriga Nacional should work brilliantly in South Australia, but a huge variety of clones has led to wines ranging from insipid to mighty. First Drop, in McLaren Vale, seems to be getting it right.

TRAJADURA

One of the grapes grown in northern Portugal for Vinho Verde, Trajadura is less aromatic than Loureiro or Alvarinho, and tends to be blended with one of them, usually the former. It is early ripening and needs to be picked before full ripeness if it is to keep its acidity. It contributes lemony, peppery fruit to the blend. Over the border in Spain it becomes Treixadura (see right). Best producers: (Portugal) Quinta da Aveleda, Quinta da Franqueira, Casa de Sezim.

TRAMINER

This name is a synonym of Savagnin Blanc (see page 242); Gewürztraminer (see pages 112–121) is a form of Savagnin Rose, itself a mutation of Savagnin Blanc. Often it's not easy to tell them apart, particularly in somewhere like Italy's Alto Adige. In practice, the two names are often used interchangeably. In Visp in Switzerland's Valais, Traminer is called Heida. Traminer is found throughout central and eastern Europe, in Austria, Germany and Romania, Italy, and in Australia where the name is usually used for Gewürztraminer. Best producers: (Canada) Hainle; (Germany) Andreas Laible, U Lützkendorf, Fritz Salomon, Wolff-Metternich, Klaus Zimmerling.

TREBBIANO

Trebbiano is Italy's least avoidable grape: there are very few regions where you do not run the risk of bumping into it. It is permitted in around 80 DOCs and probably produces about a third of all Italy's DOC white wine. Total monotony is avoided by the fact that Trebbiano is a collection of vines rather than one single one; nevertheless, boredom is the state most often induced by Trebbiano.

All Trebbiano is distinguished, if that is the word, by its high acidity and neutral flavour. The better sorts can add some leafy fruit and even some waxy depth, and just now and then Trebbiano comes up with some pretty pleasant wines.

The best Trebbiano is Trebbiano di Soave, which is sometimes known as Trebbiano di Lugana or Trebbiano di Verona. This isn't surprising because Trebbiano di Soave is, in fact, Verdicchio, a pretty decent white from the Italian East Coast. Even so, the boring Trebbiano is far more widely planted, even in Soave. Trebbiano di Soave/di Lugana is, however, the only grape used for Lugana, from the southern end of Lake Garda, a wine of significantly more weight and character than most Soave. Umbria's Procanico, alias Trebbiano Toscano, has some character, though most Trebbiano Toscano (which takes the equally undistinguished name of Ugni Blanc in France) has zero personality. Unfortunately, it's the most widely planted Trebbiano of all.

It is followed in terms of area planted by Trebbiano Romagnolo, Trebbiano d'Abruzzo, which might or might not be the same as Bombino Bianco, also often grown for DOC Trebbiano d'Abruzzo, and Trebbiano Giallo. There is also Trebbiano Modenese, much favoured by the balsamic vinegar industry, and Trebbiano Spoletino, which is actually rather good, and being rescued from obscurity. There are other Trebbianos too, though these seem to be the main ones.

Trebbiano was much blamed in the past for being part of the (red) Chianti blend, though now we know more about the benefits of co-fermentation we should be less quick to condemn: if co-fermented (which it might have been originally, but certainly wasn't latterly) it might have helped to fix the always elusive colour of Sangiovese. The addition of Trebbiano to the blend was mandatory, even if, as time went on, fewer and fewer producers put any in. After a lot of wrangling and arm twisting by the more progressive producers, Trebbiano is no longer required in Chianti or, indeed, allowed in Chianti Classico. Wines like Galestro sprang up in order to mop up the surplus Trebbiano; however, with plantings of Trebbiano generally in decline now in Tuscany, production of Galestro is already becoming a bit of an irrelevance.

Trebbiano is present, under one name or another, in Bulgaria, Greece, Russia, Portugal (as Thalia), Mexico (for brandy), Brazil, Argentina, Uruguay, South Africa, California (a bit) and Australia, where it is sometimes added to red to up the acidity. Best producers: (Italy) Antinori, Barberani, Ca' dei Frati, Falesco, Ottella, Provenza, Valentini, Visconti, Zenato; (USA) Iván Tamás, Viansa.

TREIXADURA

The Trajadura of Portugal (see left) takes this name when it crosses the border into Spain. In Galicia it performs much the same role as it does in Vinho Verde, adding light lemony fruit to the blend. In Ribeiro it is the main grape, and is blended with Torrontés and Lado; in Rías Baixas, Albariño and Loureira are its partners. Best producers: (Spain) Gargalo, Viña Meín, Vitivinícola del Ribeiro, Emilio Rojo, Vilerma.

TREPAT

Red-skinned Catalan grape much used for pink Cava, and also used for still wines in Conca de Barberà and Costers del Segre. The wines taste fresh, red-berryish and quite light.

TRESALLIER

A synonym for Sacy, this grape is better known now under this name, given to it in St-Pourçain, though 'better known' is perhaps an exaggeration for such an obscure wine. These days no more than half the blend for St-Pourçain may be Tresallier, with the balance being Chardonnay and/or Sauvignon Blanc. Tresallier gives light, acidic wines, and is declining. Best producer: (France) St-Pourçain co-op.

TRIBIDRAG

This, believe it or not, is the original identity of Zinfandel, aka Primitivo. It's an indigenous Croatian variety that goes back at least to the 15th century. From virtual extinction it is now spreading in the Dalmatian region. It can still be found in Montenegro under the name of Kratosija and occasionally in the Republic of Macedonia.

TRINCADEIRA

A Portuguese variety grown for port in the Douro valley and for table wines further south, especially in the Dão and Alentejo. In the Douro it's known as Tinta Amarela, and is less favoured than it once was because of its susceptibility to rot; it also has to be picked at exactly the right moment, and the window of opportunity between underripeness and overripeness is very short. On the plus side it has an intriguing tea-like perfume and good colour. It also yields well. In more southerly parts of Portugal it gives good quality reds of intensity and depth. The biggest planting outside Portugal is apparently at Fairview in South Africa, where it was bought in the belief that it was Tempranillo. Best producers: (Portugal) Quinta das Maias, Valle Pradinhos.

TROLLINGER

The German name for Italy's Schiava Grossa or Grossvernatsch (see page 243). The name seems to point to Tyrolean origin, and in Germany it is grown almost exclusively in Württemberg. German producers also prefer the higher-yielding Grossvernatsch or Schiava Grossa clone – almost inevitably the one with the least quality – and use it to satisfy the local taste for sweetish, reddish wine. The style is thus very different to the light, dry, mildly acidic wine of Trentino-Alto Adige. The vine may also be known as Blauer Trollinger; it also doubles as a table grape.

TROUSSEAU

A now little-planted grape in the Jura region of eastern France which has trouble ripening unless planted on the warmest gravel soils. It makes sturdy, dark wine but yields irregularly, and Pinot Noir has tended to replace it in many vineyards. It is often blended with Poulsard, which contributes finesse. It is the same as Spain and Portugal's Bastardo. There is also a paler mutation, Trousseau Gris, which may be the same as the rare variety called Gray Riesling in California. Best producers: (France) Jacques Forêt, Frédéric Lornet, Henri Maire, la Renardière, Rolet, Tissot, Voorhuis-Henquet.

UGNI BLANC

This neutral-tasting, acidic variety is the same as the Trebbiano Toscano of Italy (see opposite), and probably arrived in France along with the establishment of the Papal court at Avignon in the 14th century, a kind of vinous Trojan Horse from the Vatican since, despite its mediocrity, it is now France's most widely planted white grape. It is still found in Provence, and in the Midi, where its wines tend to be less acid, but its main French stronghold is in the Cognac and Armagnac brandy regions of the South-West. Here, at the limit of its cultivation, it produces wines of much lower alcohol – 9 per cent or lower – than in the Midi, where it easily reaches 11 or 12 per cent, but its acidity is searingly high – good for making brandy, but requiring considerable skill to make it into attractive table wine.

Although good Charente white is hard to find, Côtes de Gascogne wine from the Armagnac region has been a considerable success, especially when the Ugni is mixed with a bit of Colombard. Yields are high everywhere: up to 150hl/ha. Ugni features in many Midi blends, though usually in a minor (and decreasing) way. Other French synonyms are Clairette Ronde (Provence), Rossola (Corsica) and St-Émilion (Charentes). It is grown in Mexico, Brazil, Argentina and Uruguay; South Africa, California and Australia also have some, under either one name or another, as do Bulgaria, Romania and Portugal, where it is called Thalia. Best producers: (France) Meste-Duran, Producteurs Plaimont, le Puts, Tariquet.

ULL DE LLEBRE

The name given to Tempranillo in Cataluña in northeast Spain. See pages 270–279.

UVA RARA

Uva Rara is often called Bonarda in the Novara/Vercelli region of Piedmont in northwest Italy, but Bonarda is a name they play loose and wild with in northern Italy. Bonarda and Uva Rara also swap names in Lombardy, where there is a lot of real Uva Rara grown. They also share a soft scented character, to confuse matters further, and are both used to soften Nebbiolo's ferocious tannins. Uva Rara isn't the same grape as Bonarda, but sometimes you'd never know. Best producers: (Italy) Antichi Vigneti di Cantalupo, Antoniolo, Nervi, Travaglini.

Picking Verdejo grapes in Rueda, Castilla-León. Verdejo is one of Spain's best white vines, and is seldom improved by an admixture of dull Viura. The best Rueda has little or none of the latter.

UVA DI TROIA

See Nero di Troia page 176.

VALDIGUIÉ

Now hardly grown in France, Valdiguié was once a workhorse grape of the southwest, producing huge quantities of poor wine. It can also occasionally be found still in Languedoc and in Provence.

In 1980 the French ampelographer Galet identified the variety known in California as Napa Gamay as being merely Valdiguié. It had already been exported from California to Australia, Brazil and Uruguay under the name of Napa Gamay. Beringer and J Lohr in California make a varietal under the name of Valdiguié, and the wine shows that throwing the full force of New World winemaking at the grape can certainly bring out lots of juicy fruit and a quite attractive Beaujolais style. Best producers: (USA) Hop Kiln, J Lohr.

VERDEJO

A good-quality Spanish grape making greengage- and pear-flavoured wine that is one of the mainstays of Rueda, one of Spain's top DOs for white wines, whose blend must include at least 50 per cent Verdejo. The other native grape in the Rueda blend is the rather monochrome Viura (see page 298). Sauvignon Blanc, first introduced here in the 1980s for varietal wines, can also be used.

The grape seems to have originated in Rueda, and was widely planted there until phylloxera. Afterwards the growers opted for the higher yields of Viura and it was the Marqués de Riscal and the Marqués de Griñon who in the 1980s rescued it from obscurity. Early examples tended to be clean but too neutral, no doubt in reaction to the grape's tendency to oxidize easily. Yet wine from Verdejo has the structure and balance to age well, and becomes nutty and honeyed with a few years in bottle, though it is usually drunk young for its Sauvignon-like crunchiness of youth. Its high glycerol content quickly rounds the wine out with a little bottle age. Pre-fermentation skin maceration and some barrel fermentation and aging can help the complexity of the wine, though the oak can easily be overdone.

It is also found in the Toro and Cigales DOs, and can be a minority grape in Rioja. Best producers: (Spain) Alvarez y Diez, Belondrade y Lurton, Angel Lorenzo Cachazo, Marqués de Riscal, Vega de la Reina, Angel Rodríguez Vidal, Castilla la Vieja.

VERDELHO

A grape or several grapes found in Portugal. On the island of Madeira, Verdelho is both a grape and a style; since 1993 any Madeira wine calling itself Verdelho must be made with at least 85 per cent of that grape.

As a style of Madeira, Verdelho comes between Sercial and Bual, less dry than the former (though Sercial styles are becoming less dry) but less rich than the latter. It has high, piercing acidity but when young and unaged has more fruit than the other noble Madeira varieties and is the most 'complete'. It is now also being used for table wines in Madeira, sometimes with an admixture of Arnsburger to soften the acidity. These table wines do not at the moment seem to offer great quality potential, partly perhaps because of the difficulty of persuading growers on the island to wait until their grapes are ripe before picking them.

In Portugal, the grape known in Dão as Verdelho is actually Godello.

Verdelho is also proving very successful in Australia, where plantings are relatively small but quality is high. Flavour develops late in the ripening cycle – 'and then one hot night and it shoots through the roof,' says Bruce Tyrrell. The wines have intense flavours of lime cordial and honeysuckle, tending to oiliness at the riper end. It is so susceptible to powdery mildew that one Australian producer suggests that rose growers should plant a Verdelho vine at the end of each row of roses as a warning sign. (Er, that's a joke. Roses are often planted at the ends of vine rows to warn of mildew onset.) The Swan District of Western Australia, Cowra in New South Wales and Langhorne Creek in South Australia all have some. New Zealand's Esk Valley makes a delightful version. Best producers: (Australia) Bleasdale, Chapel Hill, Fox Creek, Lamont, Margan, Moondah Brook, Oakvale, Rothbury, Sandalford, Tyrrells, Upper Reach; (Portugal) Barros e Sousa, Blandy, Cossart Gordon, Esporão, Henriques & Henriques, Leacock.

VERDICCHIO

The best and most characterful white grape of Italy's Marche region, Verdicchio has benefited in recent years from lower yields and better balance and structure. The wine is not aromatic but has a good flavour of nuts and lemons and plenty of weight. It is typically central Italian in that it relies on subtlety and texture rather than flamboyance for its effects – and it goes with food extremely well. The locals like to say that it's a red wine dressed up as a white. Its character tends to come more from winemaking decisions – time of picking (since it is high in phenols, which increase with ripeness), level of ripeness, yields – than from the soil, which in the Marche is relatively homogeneous. Top crus can age extremely well, becoming gloriously honeyed. There is a lot of clonal variation.

There are two Marche DOCs: the larger Castelli dei Jesi, and the small but exciting di Matelica, which is rarely seen but whose wines are richer, rounder, more succulent yet still dry. There is also some *spumante* and some *passito*. You will also find Verdicchio elsewhere in Italy under such names as Trebbiano di Soave and Trebbiano de Lugana. Best producers: (Italy) Belisario, Bisci, Brunori, Bucci, Casal Farneto, Colonnara, Coroncino, Coste del Molino, Fazi Battaglia, Garofoli, Mancinelli, Mecella, La Monacesca, Moncaro, Monte Schiavo, Santa Barbara, Sartarelli, Tavignano, Terre Cortese, Umani Ronchi, Vallerosa-Bonci, Fratelli Zaccagnini.

VERDUZZO

Found mainly in Friuli and Veneto in north-eastern Italy, wine from Verduzzo varies from the pleasantly crisp to the interestingly sweet. Much depends on whether it comes from the Verduzzo Trevigiano grape or from the apparently unrelated Verduzzo Friuliano.

The latter is finer, and has been in Friuli longer, since at least the early 19th century. The former seems to have appeared in the early 20th century and, since it yields both generously and reliably, has inevitably taken a considerable slice of the vineyard for itself. The DOCs in which it features most heavily are Lison Pramaggiore and Piave north of Venice. Other DOCs are Friuli Aquileia, Friuli Grave, Colli Orientali, Friuli Isonzo and Friuli Latisana. Colli Orientali is generally the best: hillside vineyards suit the vine better than ones down on the plains, and besides, the Verduzzo grown here is more likely to be Verduzzo Giallo, which seems to be a superior subvariety of Verduzzo Friuliano. The other subvariety, Verduzzo Verde, is less interesting and more likely to be found in the plains.

Verduzzo Giallo makes good sweet wines, honeyed and quite rich though not with enormous complexity. Even better sweet wines come from Verduzzo Rascie, which is billed as yet another subvariety, this time with looser

Verdicchio growing in Italy's Marche region. This grape is all about structure, weight and texture, and the best can age extremely well, becoming honeyed and nutty while keeping a taut acidity.

clusters. Such details matter: looser clusters are less likely to succumb to rot while the grapes linger on the vine late into the autumn.

Sweet Verduzzo is usually made from late-picked grapes; an alternative is to pick the berries and leave them to dry and shrivel before fermenting them. The wines may then be aged in *barrique*. The DOC of Ramandolo has the reputation of producing the best sweet Verduzzo. Best producers: (Italy) Dario Coos, Dorigo, Giovanni Dri, Filiputti, Lis Neris-Pecorari, Livon, Paolo Rodaro, Ronchi di Manzano, Ronco del Gnemiz, Torre Rosazza.

VERMENTINO

A protean vine found the length of Italy, from Liguria in the northwest to Sardinia. It is first mentioned by name in Piedmont, where it is the same grape as Favorita; it may have been taken there from Liguria, where it is the same grape as Pigato. The best wines come from Tuscany, Sardinia and Liguria, where they have a weight and breadth not found elsewhere. The wines have a typically Italian aroma of lemons, sometimes ripe green apples, and nuts and leaves, combined with racy acidity and robust structure. There is no point in aging them, but when young they have considerable character.

It is also widely and successfully grown in Languedoc-Roussillon, where it generally goes under the name of Rolle (see page 217). It is widely planted on Corsica, where it sometimes takes the name of Malvoisie de Corse. There's a little in Australia, usually softer, riper, but Yalumba's version is pretty racy. The rare Vermentino Nero may be a mutation. Best producers: (Italy) Argiolas, Capichera, Giovanni Cherchi, Gallura co-op, Guado al Tasso, Le Macchiole, Piero Mancini, Pedra Majore, Santadi co-op, Sella & Mosca.

VERNACCIA

Vernaccia wines are found all over Italy, but to try and relate them to each other is often a waste of time. The name has the same root as the word 'vernacular': it simply indicates a local grape. A number of different vines, mostly white, but sometimes red go under this name. The racy Vernaccia grown in San Gimignano for the DOC of the same name is unrelated to the Vernaccia grown on Sardinia for the sherry-like Vernaccia di Oristano, for example; and the name, translated directly into German, becomes Vernatsch, which is the name of a red grape grown in Trentino-Alto Adige. In the Marche region there is sparkling red Vernaccia di Serrapetrona, though this turns out to be Grenache.

Vernaccia di San Gimignano is the best known Vernaccia, not least because it is the local wine of one of Tuscany's most touristy towns. It is usually good and reliable, with crisp acidity and ripe, leafy, citrus fruit. Best producers: (Italy) Le Calcinaie, Casale-Falchini, Vincenzo Cesani, Attilio Contini, La Lastra, Melini, Montenidoli, Giovanni Panizzi, Il Paradiso, Pietrafitta, La Rampa di Fugnano, Guicciardini Strozzi, Teruzzi & Puthod, Casa alle Vacche, Vagnoni.

VERNATSCH

Like Vernaccia (see left), this grape takes its German name from the same root as the word 'vernacular', showing that in Italy's Alto Adige it has long been regarded as home grown. It refers to a group of varieties, whose Italian names are variations on Schiava, meaning either 'Slav', or, more likely, 'slave'. But the name given to it in Germany, Trollinger (see page 284), points to a Tyrolean origin. All this is rather more interesting than the wine itself, which, at best, is a perfectly pleasant light red with fair acidity and, if you're lucky, a flavour of strawberries and smoked ham.

VESPAIOLO

Make Vespaiolo dry and you get attractive, quite characterful white wine, light and startlingly acidic but with no particular aroma. Make it sweet, however, from *passito* or noble-rotted grapes, and you have one of Italy's finest sweet wines. Indeed, the name is said to derive from the wasps, or *vespe*, which are attracted to the grapes as they hang on the vines getting sweeter and sweeter. It is found in the Veneto in the Vicenza region. Here it is sometimes blended with Garganega, but it is in the Breganze DOC that it really shines. As a dry white, it may be blended with Friulano, but it's all by itself in the sweet Passito wines of Torcolato. The grapes are left to dry until January and the wine may be aged in new *barriques*. The acidity is still there in the wine, balancing the intense sweetness, but the flavour is of dried grapes and apricots, honeysuckle and spice. Alternative names for the grape are Vespaiola or Vesparolo. Best producer: (Italy) Maculan.

VESPOLINA

Vespolina is a red grape grown in Piedmont and Lombardy. It is blended with Nebbiolo and sometimes with Bonarda in wines from Gattinara and Ghemme. In Lombardy it is known as Ughetta. Best producer: (Italy) Antichi Vigneti di Cantalupo.

Inniskillin

Vidal doesn't win many plaudits as a dry wine, but every dog has its day, and Canada creates high-quality, quince-flavoured Icewines from Vidal, especially in Ontario.

VIDAL

A French hybrid grown in Canada and New York State, where it is able to withstand the cold winters. Its main fame is as one of the grapes used for Canada's Icewine. Vidal Icewines aren't elegant or long-lived, but they have good concentration, quince jelly richness, intensity and a certain four-square appeal. Best producers: (Canada) Chateau des Charmes, Henry of Pelham, Inniskillin, Konzelmann, Reif Estate; (USA) Mount Hope.

VILANA

The main white grape of Crete. You'll be hard-pressed to find a Cretan white without it. On its own it can be dull if overcropped; it also oxidizes easily. The best wines can be attractively floral, and lemony.

VINHÃO

The main grape grown for red Vinho Verde, where it constitutes some 80 per cent of red plantings. It is currently enjoying a revival as demand for red Vinho Verde increases in the wake of the worldwide fashion for red wine; however, the swing to red is not as great in Vinho Verde as the swing to white was 20 or 30 years ago. There is an obvious reason for this – red Vinho Verde is a shocking, highly acidic, rip-roaring wine not at all in tune with the current New World vogue for soft ripe fruit and oak spice. Plantings of red and white grapes are now divided fairly evenly in Vinho Verde, with perhaps a slight bias towards white. Vinhão is the same as the Sousão of the Douro, where it's used to deepen the colour and sharpen the acidity of some ports. South Africa, California and Australia have a bit. Best producers: (Portugal) Domingos Alves de Sousa, Sogrape.

VIOGNIER

We should be extremely grateful to Viognier for coming along when it did. But then, Viognier should be extremely grateful we came along when *we* did. We as wine drinkers desperately needed a new taste experience that wasn't oaked Chardonnay or green-streaked snappy Sauvignon Blanc. Viognier was in danger of virtual extinction except for a few precious vines at Condrieu in the northern Rhône Valley.

Luckily for Viognier, the vast hordes of wine lovers who had lapped up so much soft, creamy, oaky Chardonnay were beginning to tire of it. They wanted a new white wine experience not dependent on the creaminess of oak, yet many of them weren't at all fond of the high-acid style of Sauvignon or Riesling. They wanted the weight and softness and suppleness of an oaky Chardonnay – but without the oak.

Other grapes that might give weight were too dependent on oak – Chenin Blanc and Sémillon, for instance, demanded oak to give a round, toasty softness to what were usually rather raw, dull wines. On the other hand, wines like Gewürztraminer or Muscat were too overpoweringly fruity and aromatic: people often thought they were sweet, even when they weren't. Wasn't there a middle way?

Well, there was. But it was a difficult route involving a grape that was notoriously difficult to grow and certainly difficult to persuade to give a regular crop – Viognier. But, wow! If you wanted serious, swooning wine, with texture as soft and thick as apricot juice, perfume as optimistic and uplifting as mayblossom, and a savoury sour creamy richness like a dollop of crème fraîche straight from the ladle of a smiling farmer's wife – in other words, a wine which just oozed sex and sensuality – Condrieu, from the Viognier grape, was it.

But Condrieu is a tiny place. The fame of the wine spread and prices soared as new-found enthusiasts fought for the precious bottles. Suddenly the original swooner was being properly valued, and like any good sex goddess she was being far from liberal with her favours and charging mightily for the swish of a sequinned hem, the fleeting caress of her glistening crimson lips. Viognier was in danger of becoming the silver screen sex symbol that many wanted, but few could have.

If a wine can be described as pretty without insulting it, then Viognier is pretty. It tastes deliciously of the apricots that grow along with Viognier on the banks of the Rhône at Condrieu. There is a theory that the vine arrived by boat, along with Syrah, during the Roman occupation, found a landing place at Condrieu and then remained there as a tiny island of distinctive taste until its recent explosion around the world. Château-Grillet, a minute estate within Condrieu, can be glimpsed in the picture behind the apricot fruit and blossom.

But haven't you ever heard of lookalikes, the ones that make you double-take and say – did I see what I think I saw? Well, you didn't. But did it give you a thrill? Did your heart leap for just a second? Sure it did. And what did it cost you? A diamond ring? A champagne dinner for two at the Ritz? Nope. Maybe just what it costs to get your glasses repaired after you walked into a lamp-post. Maybe a new round of drinks for someone as your swivelling elbow knocked over the waiter's drinks tray.

After a small band of well-heeled wine fanatics discovered Viognier, the Condrieu lookalike business was born. It started in California and Australia – two countries that had already proved they could make look-alike Bordeaux and Burgundy reds and whites, and it continued with some succulent beauties from Virginia, of all places. They each came up with good Viognier and a few producers managed to capture that astonishing, elusive mix of sultry sensuality and bright springtime scent that marks out great Condrieu. Virtually all the other New Wave countries followed – South Africa, Argentina, Chile, as well as Spain, Italy, even Greece, and some did make a Condrieu lookalike that could fool all of us devotees until she was so close we could smell the cologne on her neck. But the country that has flooded the wine world with lookalikes – some very good, some mere pale photocopies of a publicity postcard, but almost all affordable – is France itself.

After all, where do you find the most movie-star lookalikes? Hollywood and LA, that's for sure.

Viognier: from Grape to Glass
Geography and History page 290; Viticulture and Vinification page 292; Viognier around the World page 294; Enjoying Viognier page 296

GEOGRAPHY AND HISTORY

We're lucky to have any Viognier left in the world at all. A mere generation ago few wine enthusiasts had ever heard of the grape and even fewer had tasted its wine. Indeed, back in 1965, Condrieu – the tiny Rhône appellation which virtually single-handedly kept Viognier alive – was down to a mere 8ha (20 acres) of measly, impoverished Viognier vines. One vintage produced precisely 1900 litres (417 gallons) of wine. 1900 litres of Viognier to satisfy the world.

Of course, then, the world didn't care. What has become one of the world's most fashionable white grapes was completely unknown and unlamented. But a couple of people were taking note. Josh Jensen of Calera in California decided Viognier was a great grape and he planted some in his mountain vineyard. And Georges Duboeuf, the king of Beaujolais, was desperate to expand out of the Beaujolais Nouveau straitjacket. He planted a great swathe of Viognier in the Ardèche in central southern

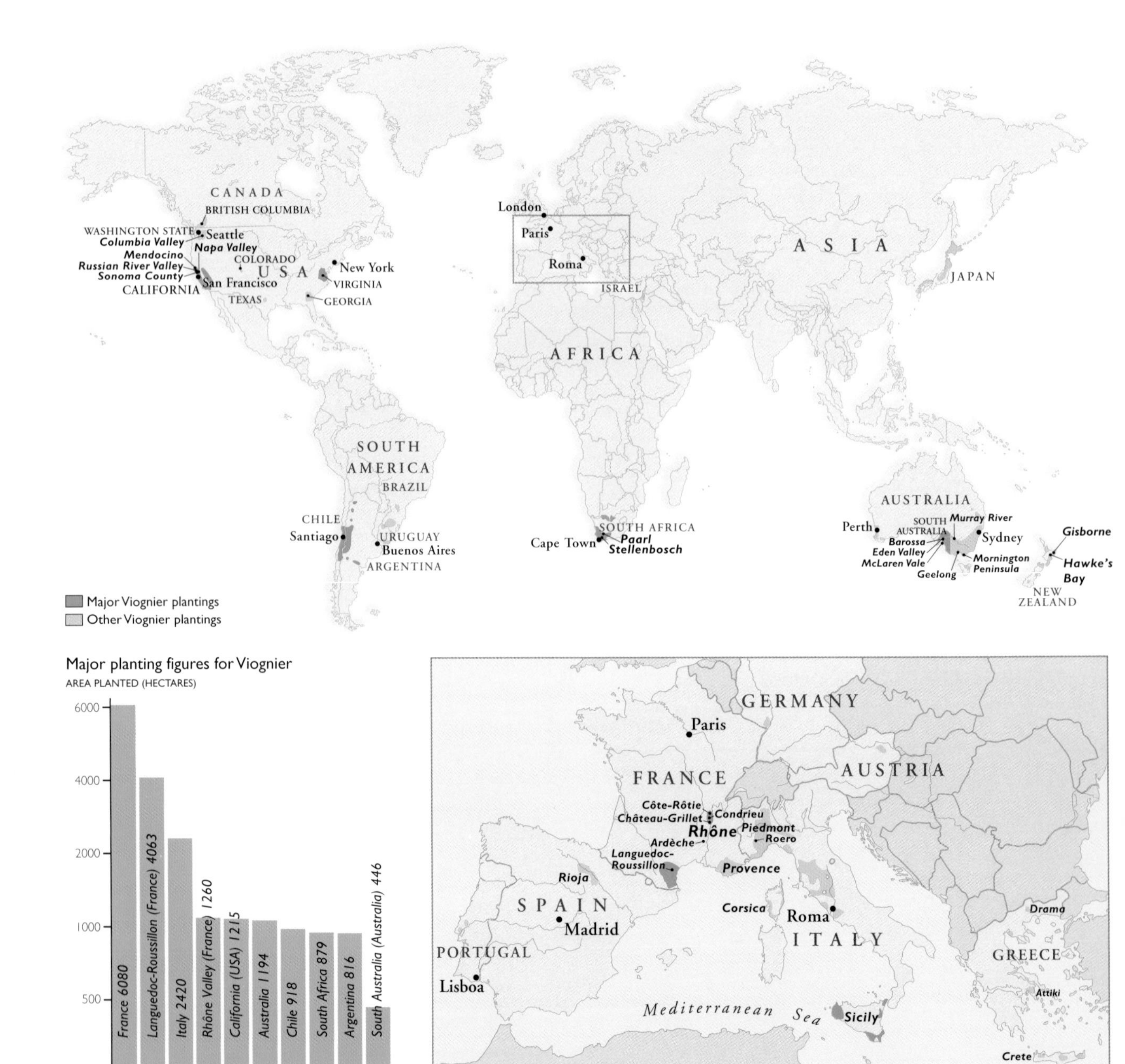

France. By the mid-1990s, not only was Viognier being planted with great enthusiasm in all the New World countries, but, after Georges Duboeuf's lead, Viognier was appearing as a relatively low-priced vin de pays all over the south of France. Just a generation before, the entire world production of one vintage had been precisely 1900 litres (417 gallons).

It hadn't always been this bad. In 1928 the Condrieu vineyards covered 146ha (361 acres). But the slopes high above the Rhône are incredibly steep and difficult to work. A market that was indifferent, and the lure of easier, better-paying jobs in Lyon and Vienne, meant that all but the hardiest growers abandoned their terraces. The fightback from 1965 was painfully slow, but by 1990 plantings had risen to 40ha 100 acres) and by 2013 the figure was 168ha (415 acres). The total plantable area of the appellation is 250ha, so more than half is now planted, including all the best bits.

HISTORICAL BACKGROUND

The origins of Viognier are obscure: the myths that have Syrah travelling up the Rhône in the baggage of Roman legionaries often include Viognier cuttings in the same suitcase. Dalmatia is the region most named as the source of the grape, and it is said to be the Emperor Probus who brought the vine to the Rhône in AD 281. A galley of cargo for the Beaujolais is said to have been captured in the Rhône, conveniently close to Condrieu, and its contents looted by local bandits called *culs de piaux*, after the leather patches on their breeches. It's a suitably inventive story, but alas, invention is what it is. Viognier is related to Syrah, either as a half-sibling or as a grandparent. It's also closely related to Mondeuse Blanche and Freisa. The legionaries had nothing to do with it. Pity.

In more recent times, a new mutant strain has appeared, according to Remington Norman in *Rhône Renaissance*, though this may simply be a more healthy, higher-yielding modern clone from Montpellier. Supposedly this strain looks and tastes quite different and, most worryingly, doesn't seem to possess the fabulous scent of the original. It is found mostly, though not entirely, in the southern Côtes du Rhône. It may have found its way, as a Samsonite clone (see page 24), to other countries, and that could explain why, despite Herculean efforts by really serious New World producers, an awful lot of the resulting wines lack the magic and mystery of true Viognier.

Chester Osborne of D'Arenberg notices a great difference between low-yielding clones originally from Condrieu, and imported into Australia not entirely legally in the 1970s, and the newer, high-yielding Montpellier clone that has made the vine popular in Australia. The latter has an apricot and nectarine character; the former has looser clusters, better acidity and more orange characters. It is now usually known in Australia as the Victorian clone.

The best Condrieu vineyards are perched on terraces above the Rhône, where some of the slopes are as steep as 60%. The ideal soils are mica-based, though soils in the south of Condrieu are more granitic. Shelter from northerly winds at flowering time is crucial. This vineyard is La Maladière, just south of Condrieu itself.

Viognier vines trained one vine to a stake on the steep slopes of Château-Grillet. This is the traditional method of training Viognier in the Rhône.

The La Turque vineyard, almost entirely planted to Syrah, is Guigal's jewel on the Côte Brune in Côte Rôtie. These few boxes represent the 5% which is Viognier.

VITICULTURE AND VINIFICATION

If you're a vine grower and you want an easy life, the last grape you should choose is Viognier. It gives yields that are both low and unpredictable, and in order to get the seductive, exotic flavours of apricots and candied peel, spice and flowers that characterize the wine at its best, you have to pick at precisely the right moment. Pick too early, and you will get sugar ripeness but not flavour development; pick too late and all the beauty and perfume is lost as the wine descends into heavy, oily mediocrity. Acidity is generally low, but picking early to retain acidity doesn't work either. Viognier develops its perfume and richness very late. There is a window of pear blossom aroma at lower ripeness, but most underripe Viognier is bland and flat. And while some oak can be welcome, too many producers are loading on the new oak – which is pointless. Since most such wines end up tasting more of toast and toffee than fruit. Viognier is brimming full of flavour and perfume. Why swamp it?

Climate

The revival of Viognier in the northern Rhône, and its worldwide spread, are both recent, so it is tempting still to look upon many vineyards as experimental. Experience so far suggests that Viognier must be allowed to get very ripe in order to allow its flavour precursors to develop, and that these develop late, after sugar ripeness. It needs a good 13 per cent alcohol to achieve that fine balance between weight and aroma, and though not particularly late-ripening it is generally one of the last varieties to be picked in any region. So it needs warmth: this will hardly come as a surprise to anyone who has visited the northern Rhône in summer.

Condrieu's vineyards tend to be angled towards the south and southeast, where they get plenty of sun in the morning, but are protected from the full strength of the hot evening sun. These steep terraces get as cold in winter as they get hot in summer; the Mediterranean climate does not reach much further north than Valence. The Mistral, too, adds its influence, racing down from the north at speeds of up to 140km/h (87mph) an hour whenever there is high pressure over northern France and low pressure over the western Mediterranean.

Across the world, however, Viognier is grown in a range of climates from cool to warm. It generally produces some varietal characteristics, providing it is well cultivated, but few examples match the finesse of a good Condrieu. Perhaps it is a question of soil; perhaps it is a question of accumulated experience, and subtlety of touch, and perhaps it is a question of new clones having less lushness and scent.

At Mas de Daumas Gassac in the Languedoc, the Guibert family blend Viognier with Chardonnay, Petit Manseng, Chenin Blanc and smaller quantities of other white varieties into an extraordinarily complex cocktail. Few winemakers would be brave enough to blend so many highly aromatic grapes together.

Soil

Condrieu's steep terraces are cut out of heat-retaining granite, with a deep sandy topsoil known as *arzelle*. Erosion can be a terrible problem, in spite of the retaining walls, and since the *arzelle* contributes much to the perfume of the wine, if it washes down the hill the *vigneron* must go down and fetch it back up.

In the South of France, in Languedoc-Roussillon and the Ardèche, where plantings of Viognier are now common, soils are extremely varied, as they are in the Viognier vineyards of the New World. It seems clear that heavy soils can result in a loss of freshness and zest in the end wine. In Australia's Eden Valley, where Yalumba produces some of its best Viognier, the soil is principally grey loam over clay, tightly structured and with low amounts of readily available water. It's very different from Condrieu, but does seem able to produce fragrant, sensual wine.

Cultivation

The exotic flavours of Viognier can prove elusive, so it is hardly a shock that at high yields they disappear altogether. Maximum yields in Condrieu are set at 37hl/ha, with the usual *plafond limite de classement*, the French legal dodge that raises the maximum yield by up to 10 per cent in prolific years. Condrieu's top yield used to be 30hl/ha: it was put up to 37hl/ha during the 1980s, when the appellation rules were revised to take account of the sudden popularity of the wine. There have been discussions since about raising the maximum to 40hl/ha or more in years when the quality is good enough, provided there is a retrospective reduction for the preceding vintage. But most of the people arguing for an increase are the new johnny-come-latelys up on the wind-swept plateau above Condrieu, where Viognier doesn't ripen brilliantly whatever the yield. Since these fertile orchard soils can't produce exciting Viognier, I hope the plateau producers' self-interested plans for an increase are regularly defeated – but when did questions like quality and reputation ever influence the considerations of local politicians?

At the moment, yields are usually under 30hl/ha, and there's no doubt that over 35–40hl/ha the wine loses the gorgeous, sexy lusciousness that makes Condrieu fans get all hot and bothered. In Australia, Yalumba says that yields vary according to the region, the style of wine sought and the price it sells for. Yields for The Virgilius are less than five tonnes per hectare (around 35hl/ha); for Eden Valley Viognier, less than eight tonnes per hectare; for Oxford Landing 10 to 12 tonnes. Remarkably, the Viognier from Oxford Landing is usually succulent and good. In California, yields for the better wines seem to be between three and five tons per acre (approximately 50–85hl/ha).

Trellising systems vary, as well. Some growers of Viognier favour split canopies like Geneva Double Curtain; in Condrieu a single cane is usual, and this is the most widely adopted system elsewhere. The basal buds on Viognier are not very fruitful, so it must be allowed to set fruit higher up the cane.

Poor fruit set is an endemic problem, even if the flowering goes well. The fruit also tends to develop unevenly, leading to what is known as 'hen and chicken' – normal-sized berries and tiny, undeveloped ones growing on the same bunch.

Leaf roll virus is another headache. The first Viognier to be imported into Australia was chosen by Alan Antcliff of the Commonwealth Scientific and Industrial Research Organisation (CSIRO) from the Institut National de Recherche Agronomique (INRA) collection, and came from Montpellier; it was supposedly free of leaf roll. However, later studies have shown that this clone (Montpellier 1968; also known as 642) has leaf roll type one, and more work needs to be done to clean it up. CSIRO has developed a leaf roll-free clone called HT Koolong for Australia. Yalumba is also doing its own work on producing virus-free clones; it says its aim is to redevelop the genetic pool in Australia, as well as checking out potentially exciting massal selection from elsewhere.

The age of the vines is crucial to quality: they really need to be at least 15 years old, and preferably 20, in order to produce good wine. Given that the explosion of interest in Viognier has happened well within the last 30 years, and many plantings are only a decade old, the quality around the world is likely to improve as the vines reach maturity. At the moment, only in the Rhône are there very old vines of 70 years or more.

The terraces of Château-Grillet in its amphitheatre above the Rhône. Ripeness should not be a problem in a sun trap like this, but the key factor with Viognier is to pick at precisely the right time to get all that fat apricot fruit into the wine, yet not allow it to go oily and coarse and lose its haunting, seductive scent.

At the winery

Viognier is made above all in the vineyard: its winemaking is generally characterized by simplicity. In warm climates it is often picked early in the morning, because cool grapes mean clearer juice; the juice may also be allowed to settle for up to 24 hours to fall clear. A few hours of skin contact are sometimes tried, though many winemakers are nervous of getting too many oily, phenolic characters into the wine. Ripe Viognier has soft skins, and phenols are easily extracted.

If the wine is put through the malolactic fermentation, it is to increase weight rather than decrease perceived acidity. In the New World acidification is a possibility. *Bâtonnage*, or lees stirring, is quite common, with the new wine often being left on the fine lees until bottling, generally in the spring or summer following the vintage.

SWEETNESS – AND OAK

Sweet Condrieu is a traditional style which a few growers continue to make on a small scale. The grapes are usually overripe or *passerillé* – shrivelled by the wind and sun – rather than nobly rotten, though botrytized Viogniers are by no means unknown, either in Condrieu or in other countries. The fermentation may be stopped by the addition of sulphur, or it may be allowed to continue until it has practically finished of its own accord. Chilling and sterile filtering may then be necessary to ensure the wine is stable. One argument is that the latter method – allowing the fermentation to finish – gives better aromas, since these are formed towards the end of fermentation.

The leading late-harvest Condrieu is Yves Cuilleron's Récolte Tardive – he wasn't allowed to call it Vendange Tardive because of possible confusion with, of all things, Alsace. He picks the grapes in late October or early November; he does not, however, pick them *à l'assiette*, which was the traditional method. A plate was held underneath the cluster, and the cluster was shaken to make the overripe grapes drop on to the plate. The quantities, one might suspect, were not very large, but I can just see one of the more switched-on Condrieu producers using this *assiette* method in the future, labelling the wine as such – and then selling the handful of bottles for a fortune. I hope it wasn't me that gave them the idea.

Oaked Viogniers are a more common proposition than sweet ones. Some oak, even some new oak, can add complexity and structure, but the perfume of Viognier is easily overwhelmed by too great a dose of vanilla. Fermentation may be in stainless steel, cement or oak but, as with other grapes, greater integration of oak flavours is achieved if the wine is fermented in oak rather than run into oak after fermentation.

Few producers use more than 10 to 30 per cent new oak, with the balance being of various ages. Old oak is just as desirable, since it gives oxygenation and good fruit expression without obvious oaky flavours. Extreme lightness of touch is required in this, as in all aspects of making Viognier.

The lighter your touch, the more you'll be rewarded by lushness of texture and mouthwatering fruit and heady perfume.

VIOGNIER AROUND THE WORLD

Ah, the scent of mayblossom and hothouse apricots, the texture of crème fraîche festooned with apricot and peach and floral fragrance – these are the glories of Condrieu and these are the sensations that inspire Viognier growers all around the world. But few achieve such heights – around the world, or, for that matter, in Condrieu itself.

Condrieu, the Rhône and the South

Condrieu comprises seven communes strung out along a 22km (14 mile) stretch of the Rhône. They are the three original communes named in 1940 when the appellation was created (Condrieu, Chavanay and St-Michel-sur-Rhône) plus Vérin, St-Pierre-de-Boeuf, Malleval and Limony. The soil is relatively homogeneous, but differences do occur and the wines may have aroma or more power according to the balance of *arzelle* and clay. Land just outside the AC boundaries – above 300m (1000m) altitude, on clay plateaux or on the valley floor – may be used to grow vin de pays with considerable success.

Château-Grillet has its own very small appellation (3.08ha/7.6 acres). Its granite soil is marked by a high proportion of decomposed mica, and the wines have higher acidity and are slower to mature than Condrieu. In quality Château-Grillet currently lags behind the best of Condrieu. The vineyard has been under single ownership since 1840 and since 1965 has expanded from 1.75ha (4.3 acres) to its current size; the oldest vines are about 80 years old. Bordeaux super-star Château Latour has recently assumed ownership. This is sure to change the wine's style and bring improvements, but to date much of the criticism hung on the perceived diluteness of the wine (though yields are only slightly higher than in Condrieu), and relatively early picking, lack of new wood and late bottling. The point is not that the wine was poor, but that it was poor value compared to what was obtainable in Condrieu. Indeed,

The whole point about traditional co-fermenting of Syrah and Viognier is that they should both be picked at the same time from the same vineyard, usually in the Côte Blonde in Côte Rôtie. They don't both ripen at the same time, but the Viognier often boasts 1–2% more alcohol at picking, useful for bolstering yet softening the wine. Too much Viognier – some say any more than 5% – starts to flatten the Syrah's flavour. But co-fermentation's success in fixing the Syrah's colour has made the practice popular worldwide, especially in Australia.

I had a bottle recently which showed much more style and perfume than any I've had for a generation. And at Château-Grillet's prices, it's normally only once a generation that I try it.

In Côte-Rôtie, up to 20 per cent Viognier is allowed in the red wine under the AC rules. This may have evolved as a way of adding softness and perfume to the Syrah, or the presence of Viognier might have been an accident. In any case, many growers add no Viognier at all and few add more than 5 per cent. The Viognier must be added to the vat at fermentation. This is tricky, since it ripens earlier than Syrah, and is overripe by the time the Syrah is ripe. So what might start out as 5 per cent in the vineyard is very much less by the time the wine is made.

Nevertheless, its effect is much greater than just an increase in aroma or finesse. Adding it at fermentation is known as co-pigmentation: the different phenolic compounds added by the white grapes help to fix the colour of the red. Most Côte-Rôtie Viognier is on the Côte Blonde, where the limestone-rich soil suits it better than the clay of the Côte Brune.

Viognier is also grown in the southern Rhône, often for blending with Marsanne, Roussanne, Bourboulenc, Clairette and others, and it performs well. But Viognier doesn't like really warm conditions, which is why it was such a success in the warmish Ardèche west of the Rhône, and all its biggest successes in the Languedoc and Provence are in cooler spots.

Australia

Yalumba's first commercial Viognier was planted in 1979, in the Eden Valley, and by the following year the company had propagated enough to plant 1.2ha (2.9 acres) in its Vaughan vineyard at Angaston. This vineyard is now 8ha (19.7 acres) in size; these are the oldest Viognier vines in Australia. Plantings have risen steadily and are scattered between

Delas

Clos Boucher is a tiny vineyard on steep slopes just north of Château-Grillet. Delas are increasingly known for high quality, single vineyard bottlings.

Guigal

Guigal ferments one-third of his Condrieu in new oak to give it the structure he feels the grape lacks. He also has some Viognier in his Côte-Rôtie vineyards.

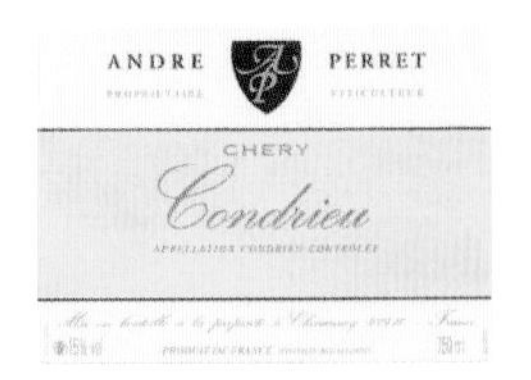

André Perret

André Perret makes very fragrant, lush but properly dry wines from Chéry, a small southern-facing slope of mica and schist near the village of Condrieu.

Murray River, McLaren Vale, Geelong, Central Victoria, Mornington Peninsula, Barossa, Eden Valley, Adelaide Hills, Heathcote and Canberra. Quality varies hugely: the best wines are glorious, the worst dull and lacking in aroma. Says Louisa Rose, senior winemaker at Yalumba, 'The most frustrating thing about Viognier during the ripening season is wondering when the hell it is going to get some flavour. The sugar in the berries may be high, but this is no indication that the flavours are ripe. You have to wait and wait – and then all of a sudden it changes from being dull and boring, and there is a massive explosion of musky apricot character.' Getting the fruit flavours ripe determines the moment of picking; the alcohol level normally turns out between 13 and 14 per cent, but it can go higher and retain its perfume. Tastes in Viognier are becoming more sophisticated, however, and moving away from blowsy fruit to spicy, exotic characters and with the urgent desire to produce lower alcohol wines, Viognier's lower alcohol windows of scent and fruit are being explored. But even people who don't make varietal Viognier want the grape in order to blend it with Shiraz: there's hardly a more fashionable blend. In warm climates like the Barossa it adds complexity and softens the intensity of the Shiraz; and in cool climates it fills out the palate and lifts the perfume of Shiraz.

Virginia has rapidly made a reputation in the wine world by producing some of the most scented, orchard-fresh Viognier grown anywhere on the planet. Given Virginia's warm, humid climate and its susceptibility to the violent weather patterns of America's East Coast, this may seem surprising, but remember, Viognier has a thick skin and resists rot well – and then you experience a view of the Blue Ridge Mountains at harvest time, like this one from above Veritas winery, and you're just glad they've worked out how to do it.

California and Virginia

With nearly 1229ha (3039 acres) in 2013, and a small but steady annual increase, California is cautiously in love with Viognier, largely because people are searching for a white alternative to Chardonnay that can be made as cult wine and sold for a high price. Viognier hasn't quite achieved these early ambitions, and it's no longer necessarily a top-dollar ticket. It was first planted in the early 1980s by Josh Jensen at Calera; he still makes a fantastic example. San Joaquin and Madera in the northern Central Valley, and Sonoma and Santa Barbara counties have the most. The same issues apply in California as everywhere: unsuitable sites, yields, ripeness, skin contact and oak, but the best wines are superb. Yet even with the best treatment in the world, Viognier is not a versatile variety and some of the most dedicated Rhône Rangers are beginning to wonder if there might not be more future in Marsanne and Roussanne.

But one state is making world class Viognier – Virginia on the East Coast, with wines as near to Condrieu as any New World Region gets. It just goes to show you can't predict where Viognier will work because the warm, humid vineyards at the foot of the Blue Ridge Mountains don't resemble Condrieu in the northern Rhône Valley in any way. But Viognier's thick skin and resistance to rot allows it to hang long enough to develop fabulous fragrance.

Rest of the world

Central Italy and Piedmont are having some success with it. In Spain it is grown experimentally in Rioja, and there are also plantings in Greece and the wines, while lacking a little perfume, are fat and full and sensuous. It's showing considerable promise in Bulgaria. In the New World, there is some in Brazil, Argentina, Uruguay and in Chile. South Africa and New Zealand have significant plantings partly for co-fermenting with Syrah. Japan has a bit.

Ascheri
Ascheri are innovative Piedmont producers and planted Viognier as long ago as 1993 on Montalupa Hill just north of Barolo near the Tanaro River.

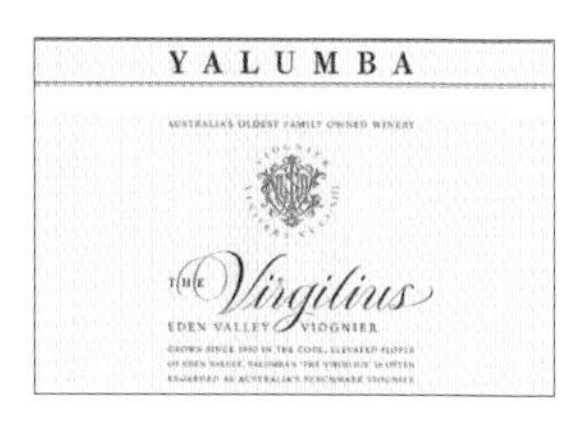

Yalumba
Yalumba was the first winery to plant Viognier in Australia, in 1980. The original plantings in the Eden Valley form the core of this rich, heady wine.

Burgozone
Bulgaria's leading producer of Viognier planted on north-facing vineyards sloping down to the Danube that were once a Roman vineyard area.

ENJOYING VIOGNIER

There are exceptions to every rule, but the general rule for Viognier is that it does not improve with age. The point of the wine is its head-spinning, intoxicating perfume, and, frankly, this is often at its most sexy and seductive at not much more than a year old. Well, I'd say even less. I'd say, if you wheedle your way into a Condrieu cellar, you need to pour yourself a glass straight from the barrel. Viognier is never better than this. If you must age it – well, it can age. I've had 10-year-old examples – deep, brooding, rich and rather exciting, but with none of the eye-spinning thrill of the young model. Four or five years old and a good Viognier will be at the halfway point between self-indulgent youth and corruptly delicious middle age. This youth fetish makes Condrieu an anomaly. High prices usually go hand in hand with ability to age. But the money here flows after youth, making Condrieu the most expensive early-drinking wine in the world. It also seems to change rapidly in bottle according to unknown rules of its own. Certainly, if you age it, each bottle can turn out wildly different from all the others.

The wines that do age are as likely to come from Australia and California as from Condrieu. But beware. Do you really want to give up the exuberant flirtatious perfumes of youth in the half-hope the wine will gain a beeswax rich majesty with age? Are you sure you want to risk it?

The taste of Viognier

Think of every aromatic flower and fruit you can, and throw them together into a glass. Viognier can remind you of honeysuckle, jasmine or primroses, apricots and peaches, candied peel, musk and spice. But that in itself is not sufficient: all these aromas could become top-heavy and wearying without finesse and subtlety on the one hand, and a remarkable texture halfway to crème fraîche on the other.

Acidity is usually low, and if growers pick early for acidity they generally miss out on perfume. It should still be a weighty wine, of at least 13 per cent alcohol, and sometimes 15 per cent. The danger then is of the flavours turning oily and phenolic and lumpen.

Simpler versions – not necessarily inexpensive, because the enormous demand for Viognier means that it is hardly ever very inexpensive – may lack complexity and have a perfectly attractive, though hardly thrilling, aroma of apricots and pears. In that case one might wonder what all the fuss is about. Well, a cheap vin de pays Chardonnay is hardly going to resemble a Corton-Charlemagne, so why should cheap Viognier resemble Condrieu? As it happens, quite a few do have at least a passing similarity – at a fraction of the price. It is a difficult grape to grow and vinify, requiring the same balance as Chardonnay but not possessing any of Chardonnay's obliging nature, and achieving that subtlety of flavour is extremely difficult. It also requires mature vines, which is another reason why only a minority of Viogniers throughout the world are as good as they could be. Or will be: what Viognier mostly needs now, worldwide, is time and experience. As each year passes, it gets a bit more of both.

Virginia has established a remarkable reputation for being able to ripen grapes that other regions find too difficult – Nebbiolo, Petit Verdot and Petit Manseng are examples, but the flagship variety is definitely Viognier. Virginia Viognier manages to blend apricot and Williams Pear ripeness with lemon peel acidity and jasmine scent in a memorable way. Barboursville is one of the best exponents. From the northern Rhône, Yves Cuilleron's Condrieu les Chaillets Vieilles Vignes is one of the most complex, concentrated yet scented wines in the appellation.

MATCHING VIOGNIER AND FOOD

Fresh, young Viognier is at its best drunk as an apéritif. Its subtle flavours have an affinity with such aromatic herbs and spices as rosemary and saffron, and it is also a good partner for mildly spiced Indian dishes, such as chicken korma. In fact, it is not a bad match with any chicken in a creamy sauce, or with rich-tasting seafood such as crab, lobster and scallops. The apricot aroma that typifies even inexpensive Viognier suggests another good pairing: pork or chicken with an apricot stuffing.

CONSUMER INFORMATION

Synonyms & local names

There are none to worry about.

Best producers

FRANCE/Rhône Valley Gilles Barge, P Benetière, P & C Bonnefond, Burgaud, Chapoutier, Château-Grillet, Chêne, Louis Chèze, Clusel-Roch, Colombo, Yves Cuilleron, Delas, Pierre Dumazet, C Facchin, Phillipe Faury, Gilles Flacher, Font de Michelle, Pierre Gaillard, Y Gangloff, Jean-Michel Gérin, les Goubert, Guigal, Jaboulet, Jamet, Jasmin, F Merlin, Monteillet, Mouton, Rémi Niero, Robert Niero, Niero-Pinchon, Alain Paret, André Perret, Christophe Pichon, Philippe Pichon, J Pilon, René Rostaing, Ste-Anne, St-Cosme, Ets L de Vallouit, Georges Vernay, Verrière. Vidal-Fleury, Gérard Villano, François Villard; **Languedoc** l'Aiguelière, l'Arjolle, Aspes, Berlou co-op, Ch. Cazal-Viel, Clovallon, Florensac co-op, Gourgazaud, Charles Guitard, Mas de Daumas Gassac, Pech-Céleyran, Peirière, Pezilla, Terres Falmet, Trois Terroirs; **Provence** Routas; **South-West** du Cèdre.
SWITZERLAND Grillette.
ITALY/Piedmont Giacomo Ascheri; **Tuscany** D'Alessandro/Manzano.
GREECE Gerovassiliou.
USA/California Alban Vineyards, Araujo, Arrowood, Au Bon Climat, Calera, Caymus, Cayuse, Cline Cellars, Fetzer, Jade Mountain, Kunde, La Jota, Landmark, McDowell Valley, Miner Family, Andrew Murray, Phelps, Qupé, Tablas Creek, Zaca Mesa; **Oregon** Pebblestone; **Washington State** Abeja, Cayuse, Cougar Crest, Columbia, McCrea Cellars, Powers, Charles Smith K Vintners; **Virginia** Barboursville, Horton, Keswick, King Family, Valhalla, Veritas, Williamsburg.
AUSTRALIA Belgravia, by Farr, Clonakilla, Domain Day, Dromana Estate, Elgee Park, Haselgrove, Millbrook, Mitchelton, Mr Riggs, Petaluma, Shelmerdine, Stella Bella, Stonehaven, Tahbilk, Trentham Estate, Yalumba (Yalumba, Heggies and Y labels).
NEW ZEALAND Bilancia, Millton. Te Mata, Trinity Hill.
SOUTH AFRICA Backsberg, Bellingham, Buitenverwachtung, Fairview, Rustenberg.

RECOMMENDED WINES TO TRY

Fifteen classic Rhône wines

Guigal Château d'Ampuis Condrieu la Doriane
Gilles Barge Condrieu
Louis Chèze Condrieu Coteau de Brèze
Cuilleron Condrieu les Chaillets Vieilles Vignes
Delas Frères Condrieu Clos Bondes
Pierre Dumazet Condrieu
Pierre Gaillard Condrieu
Jean-Michel Gerin Condrieu Coteau de la Loye
Guigal Condrieu
Domaine du Monteillet Condrieu
Rémi Niero Condrieu
André Perret Condrieu Coteau de Chéry
René Rostaing Condrieu la Bonette
Vernay Condrieu Coteau de Vernon
François Villard Condrieu les Terrasses du Palat

Five other French Viognier-dominated white blends

Domaine de l'Arjolle Viognier-Sauvignon Blanc
Dom. de la Verrière Viognier
Font de Michelle Côtes du Rhône Cépage Viognier
Domaine les Goubert Côtes du Rhône Viognier
Ch. Pech-Céleyran Vin de Pays d'Oc Viognier
Domaine Ste-Anne Côtes du Rhône Viognier

Sixteen New World Viognier wines

Alban Vineyards Edna Valley Viognier (California)
Barboursville Viognier (Virginia)
Clonakilla Viognier (Australia)
Calera Mt Harlan Viognier (California)
Haselgrove McLaren Vale 'H' Viognier (Australia)
Horton Vineyards Viognier (Virginia)
Jade Mountain Mount Veeder Viognier (California)
King Family Viognier (Virginia)
Landmark Viognier (California)
Millton Riverpoint Viognier (New Zealand)
McCrea Cellars Columbia Valley Viognier (Washington State)
Pebblestone Viognier (Oregon)
Te Mata Zara Viognier (New Zealand)
Trinity Hill Gimblett Gravels Viognier (New Zealand)
Veritas Viognier (Virginia)
Yalumba The Virgilius (Australia)

Lovers of Condrieu shouldn't despair at the small quantities available: there is still plenty more unplanted land in the appellation. However, the best sites are already planted and the vines do need to be properly mature to give expressive flavours.

Maturity charts

Occasional Viogniers will age in bottle, even for 10 years or more. The vast majority, however, should be drunk within a few years.

2014 Condrieu

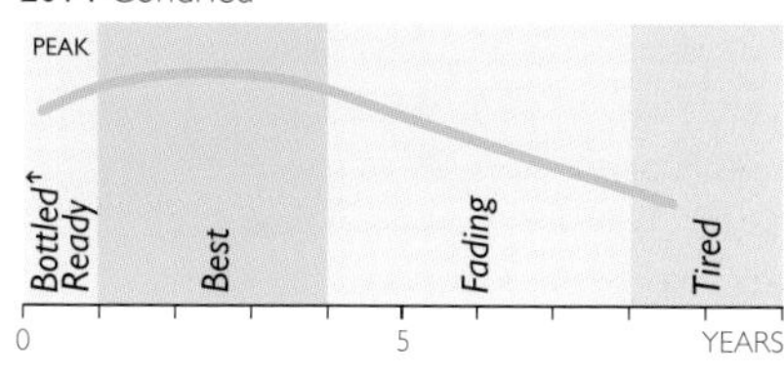

Condrieu is one of the few high-priced whites which really does not need bottle age. Some people even maintain it is at its best straight from the barrel. Mmm.

2014 Languedoc IGP

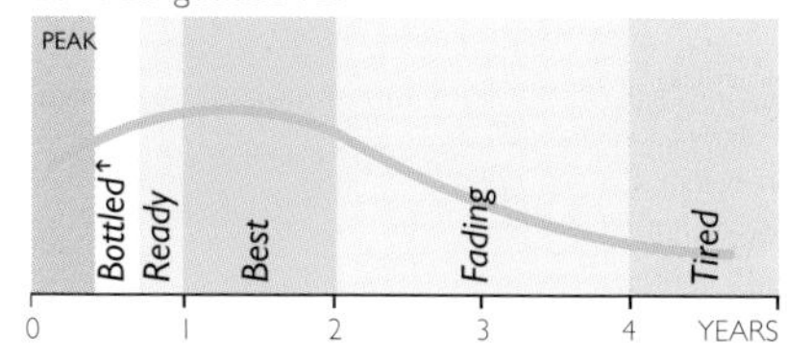

Viognier from Languedoc-Roussillon should be drunk within a couple of years of the harvest. It will not improve in bottle.

2014 Top California

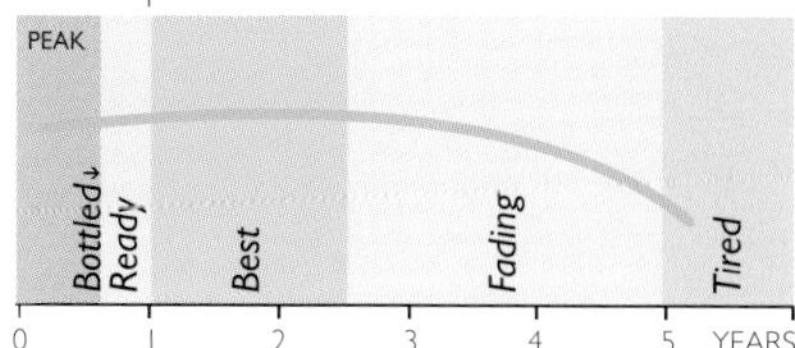

California Viognier is no exception to the 'drink young' rule. Left in bottle too long, it loses its perfume and gains nothing in return.

VIOSINHO

A northern Portuguese grape grown in the Douro and Trás-os-Montes regions – and, as is the way these days, anywhere else an enterprising winemaker decides to try it out. It gives structure and flavour to a blend, and these are valuable qualities in Portugal, which has a wealth of characterful red grapes, but is slightly less well off for good white ones. Best producers: (Portugal) Domingos Alves de Sousa, Sogrape.

VITAL

A usually rather dull low-acid grape grown in central Portugal and the Douro. In the latter it may be called Malvasia Fina do Douro or Malvasia Corada, but it has nothing to do with true Malvasia Fina. Casal Figuera, north of Lisbon, takes it seriously and it repays their efforts with mineral, lemon-fruited wine of some elegance.

VIURA

The name by which Macabeo (see page 129) is known in Rioja in northern Spain, where it probably arrived from Catalunya around the turn of the 20th century, and in Rueda. It used to have a role as a softening grape in red Rioja: many regions employed such strategems in the past, if tannins and acidity were likely to be too high for fun. Nowadays higher yields from the vines have softened up the reds, though diluted their character too, and better viticulture and winemaking are also producing softer results. This has left some of the world's more neutral white grapes in search of a job. (Trebbiano in Chianti is in the same boat – see page 284.)

In Rioja, Viura was often thought to be inferior to Malvasia, but the latter's soft, quick-to-age personality needed the lean, acidic Viura. With better winemaking, there's a lot of good sharp young Viura Rioja, and some of the best long-barrel-aged Riojas are based on Viura, where it develops a remarkable flavour of custard richness and orange blossom scent. It had its day in the white wine boom of the 1980s, when bodegas promoted young, aggressive, grapefruit-scented wines. Viura's neutrality is the reason that much white Rioja tasted of oak, youthful fruit and little else, but it does take obligingly to *barrique* fermentation and lees aging, and if picked early has good acidity. However, early picking means that it tends to lose flavour, so you pays your money and you takes your choice. Nowadays everybody is calling for red, and the last thing that current fashion demands is lighter reds diluted with white, so growers in Rioja are tending to replace Viura with Tempranillo.

At the moment Rioja plantings of Viura (about 7500ha/18,530 acres) constitute some 15 per cent of its plantings in Spain, but new plantings of Viura are being passed over in favour of red varieties, especially Tempranillo. One small bonus, therefore, is that most of the vines are quite old. In Navarra, too, where it forms some 8 per cent of the vineyard and is the principal white grape, red wines are in the ascendant at the expense of white. Best producers: (Spain) Artadi, Bodegas Bretón, Martínez Bujanda, CVNE, Enomar, López de Heredia, Marqués de Caceres, Marqués de Murrieta, Montecillo, La Rioja Alta.

VRANAC

An old variety which hails from Croatia, Serbia and Montenegro and is important in Macedonia. It is related to Zinfandel, alias Tribidrag. The wines are not dissimilar: lots of colour, lots of tannin, lots of alcohol, lots of black fruit flavours, not much acidity.

WEISSBURGUNDER

Sometimes written as Weisser Burgunder, this is a common German and Austrian synonym for Pinot Blanc (see pages 184–185). Best producers: (Germany) Bercher, Bergdolt, Schlossgut Diel, Fürst, Dr Heger, Karl-Heinz Johner, Franz Keller, U Lützkendorf, Müller-Catoir, Rebholz; (Austria) Feiler-Artinger, Walter Glatzer, Hiedler, Lackner-Tinnacher, Hans Pittnauer, Fritz Salomon.

WEISSER RIESLING

Anything called Weisser Riesling or White Riesling is proper Riesling (see pages 204–215), as opposed to the lesser Welschriesling (see below).

WELSCHRIESLING

A grape that is not entitled to its suffix of Riesling, since it is unrelated (and should be called by its Croatian name Grasevina) to the genuine Riesling of Germany. However, one can perhaps see how the name arose. 'Welsch' means 'foreign' in Germanic tongues, and since it is known by the name of Welschriesling in Austria, where it is widely grown, one can surmise that it did not originate there. (It is not grown in Germany.) But the name might possibly derive from Wallachia in Romania: the Slav name for Wallachia is Vlaska, and an alternative name for the grape, used in Slovenia and Vojvodina, is Laski Rizling. Other names include Laski Rizling in Slovenia and Vojvodina, Olasz Rizling in Hungary, Rizling Vlassky in the Czech Republic and Slovakia and Riesling Italico in Italy. Riesling Renano is their name for true, or Rhine Riesling.

In northern Europe it has long had a reputation for extremely poor quality, but this reputation is based on the large quantities of Yugoslav Laski Rizling exported until the end of communism (and indeed until the break-up of Yugoslavia). The low quality of this wine was the result of poor winemaking, poor storage and, perhaps, the determination of the shippers to meet certain price points; certainly the same state-owned winery that produced the worst wine also produced some very enjoyable wine from the same grape.

Good Welschriesling, as made in Austria, the Czech Republic, Croatia and Slovenia, is non-aromatic, nutty and quite weighty, with good roundness and relatively low acidity. It makes excellent sweet wines in Burgenland. However, the fashion for reds means that it is being uprooted to make way for Merlot, Pinot Noir and Austrian varieties like Blaufränkisch. North Italian versions are light and crisp.

It yields generously, and yields must be controlled if the wine is to have the concentration it needs. In the future it may regain its own reputation as a good, perhaps very good grape, but it would help if all countries adopted the Croatian name of Grasevina and gave up trying to pretend a relationship with Riesling. But for now it is still blighted by its notoriety as the grape that has contributed so cynically to the entirely blameless real German Riesling's loss of popularity and respect in much of the wine world. Best producers: (Austria) Feiler-Artinger, Alois Kracher, Velich.

WHITE RIESLING

Another name for the proper Riesling of Germany (see pages 204–215).

WILDBACHER

See Blauer Wildbacher page 48.

XAREL-LO

A high-yielding Spanish grape found especially in Cataluña, where it is one of the mainstays of the Cava blend. Parellada (see page 179) and Macabeo (see page 129) are its traditional blending partners. Chardonnay is also popular, not least because it is richer in the proteins needed by the yeast for the second fermentation. But Xarel-lo is highly regarded by the region's best winemakers, and with good reason: it gives wines of structure, balance and

Naoussa in northern Greece is not an easy place to ripen grapes, which makes their use of Xynomavro – the 'acid black' grape – all the more challenging. Kir-Yianni's vineyards are at 280–330m (900–1 100ft), the highest point in the appellation and the wine greatly benefits from aging.

character – sometimes quite a firm, phenolic character – and it can live for years. Indeed, it only starts to be really expressive after two or three years. It contributes backbone to a blend, whereas Macabeo gives flesh and Parellada gives prettiness. Xarel-lo also expresses its *terroir* rather well, when it's allowed to: the idea of single-vineyard wines in Cava is new, but interesting.

In Alella, north of Barcelona, it takes the name of Pansa Blanca (see page 179) and gains an attractive lime cordial aroma and flavour. Best producers: (Spain) Albet i Noya, Castellblanch, Codorníu, Gramona, Recaredo, Freixenet, Augustí Torelló.

XYNISTERI

If one is to judge the quality of this white Cypriot by that of most of the table wines emerging from Cyprus, one would not give it the time of day, but there are signs that better winemakers are taking an interest in what Cyprus has to offer. Xynisteri is showing itself to be capable of making rather good, zesty table wines, especially if carefully blended with international varieties like Chardonnay or Sémillon and vinified with New World care. It is the best grape (better and more subtle than the black Mavro, that is) for making Commandaria, the dark, sweet, often fortified wine that is Cyprus's claim to vinous fame. Best producer: (Cyprus) Kyperoundas.

XYNOMAVRO

A highly acidic, dark-coloured Greek grape, the name of which means, oddly enough, 'acidic and black'. All that tannin and acidity does mean that the wines age well – between 5 and 12 years. It is the backbone of rich, full-bodied spicy reds from Naoussa and Goumenissa in the heart of the Macedonian region, and is grown widely all over northern Greece. It ripens late, and may be planted at high altitudes for sparkling wine. Best producers: (Greece) Ktima Alpha.

ZALEMA

A dull grape used mostly in Condado de Huelva and Montilla-Moriles in southern Spain. There's a tendency to replace it with Palomino Fino.

ZAMETOVKA

This old Slovenian variety is used mostly for Cvicek, an acidic, light red local speciality. Its other claim to fame is that the oldest vine in the world, over 400 years old and found in the town of Maribor, is Zametovka.

ZETA

The new name for the Hungarian grape Oremus, a Furmint x Bouvier early ripener which likes heavier, colder, clay soils. It gets botrytis easily, produces creamy, aromatic, fat wines with good acidity, and blends happily with Furmint. It's been tried for dry wines, but lacks the verve of Furmint.

ZIBIBBO

The Sicilian name for the less aromatic and less seductive variety of Muscat, Muscat of Alexandria (see pages 156–165), especially on the island of Pantelleria. Best producers: (Italy) Colosi, D'Ancona, De Bartoli, Donnafugata, Pellegrino, Salvatore Murana; (Australia) Brown Bros.

ZIERFANDLER

Austrian grape found in the Thermenregion area south of Vienna, where it is the traditional partner for Rotgipfler (see page 217) in the white wine called Gumpoldskirchner, though sometimes it is also made as a varietal. It makes big, spicy, honeyed, long-lived wines of substance, especially when not fully dry. When dry the high alcohol level can seem too much of a good thing. It makes excellent sweet wines. Best producers: (Austria) Karl Alphart, Gottfried Schellmann, Stadlmann, Richard Thiel.

ZINFANDEL

See pages 300–309.

ZWEIGELT

Austria's most planted red grape, Zweigelt is high-yielding and easy to grow. It buds early and ripens early, so avoiding bad weather at the end of the growing season and, perhaps as a result of its high yields, is often treated as a workhorse grape. It can, however, produce quite rich, cherry-fruited wines with a very attractive pepper tingle which can even take some oak aging if they are sufficiently concentrated. The best can even age for a few years in bottle. My favourite, however, are purple-pink and bristling with cherry fruit and peppery fire.

The vine was produced in 1922 by a Dr Zweigelt, who crossed Sankt Laurent (see page 221) with Blaufränkisch (see page 48); alternative names are Blauer Zweigelt or Zweigeltrebe. It has potential for other cool climates, notably England, Canada, Czech Republic, Slovakia, northern Japan and parts of Germany. Best producers: (Austria) Feiler-Artinger, Gernot Heinrich, Hans Pitnauer, Umathum; (Hungary) Akos Kamocsay.

And that – unless anyone can find us any grapes starting Zx, Zy or Zz – is *IT*.

EL DORADO
OF THE
STATES OF AMERICA
E·R

ZINFANDEL

The way some Californian wine marketeers go on, you might think that Cabernet Sauvignon, Merlot and Chardonnay were Californian grapes. But they're not. No way. They're French classics that the Californians have adopted with a revivalist fervour that threatens to eclipse their origins in Bordeaux and Burgundy.

But the Californians *do* have a grape that they just might claim as their own – indeed a lot of them do just that. And that's the Zinfandel. As Californian as the Hollywood sign, Big Sur and the Golden Gate Bridge. Except that now some smart alec techno-wizards have proved through DNA fingerprinting that Zin ain't Zin. Zinfandel is the same as Primitivo from the south of Italy. And if that idea isn't hard enough for Zin fanatics to swallow, it is also the same as Crljenak Kastelanski from the Croatian coast – which I can't even pronounce. Hey, that's nothing. The latest name the white-coated wizards have come up with is Tribidrag. Funny, the more they get to learn about this grape, the less attractive its names become. A glass of Tribidrag, anyone? I don't think so.

Zinfandel, now known to be the same as southern Italy's Primitivo grape, may have come to the eastern seaboard of the USA with Italian immigrants. From there it made the journey west to California with the Gold Rush – hence the gold pan, the poster advertising the Gold Fields – the new El Dorado – and the nugget of gold. The popularity of the grape has proved to be a more lasting national treasure than the seams of gold which fast ran out.

Now, let's get one thing straight. Zin was nothin', but *nothin'* before the Californians came along. So, even if it is the same as Primitivo, did you ever read a single word (until recently, that is) saying how great Primitivo from the heel of Italy was? How many syllables can you remember being expended on the merits of Croatian Crljenak Kastelanski in the world's library of wine wisdom? And as for Tribidrag, well, that was a new one, even for me. There were precisely nine vines of it in the garden of a very ancient Croatian lady. If she'd popped off, and they'd sold her garden for holiday villas, that's the last we'd have heard of Tribidrag. So no matter what the evidence, I'm brazenly going to wave the Californian flag for the wine that above all others symbolizes the spirit of the Golden West – red, pink or white, heavy or light, sweet or dry – California Zin.

Yes, Italy's Primitivo obviously did stray across the Adriatic sea from Croatia; it's not far as the crow flies. But vines need prophets and producers of genius if we are ever to see how good they are – and California had two. Both very different, with very different objectives, but both of vital importance.

First there was Bob Trinchero of Sutter Home Winery. Now, if you don't live in the USA, the 'white' Zinfandel craze may have passed you by, but during the 1980s and early 1990s this pale pink, sweetish winey glugger was one of the most popular wine styles in America. Bob Trinchero invented it. He says he saved Zinfandel from extinction as California, dollar signs swivelling in its eyes, went crazy for Cabernet and Chardonnay, uprooting all its old vines – many of them Zin – with mindless determination. Perhaps he did. In which case – here's to you, Bob.

But the true prophet, the bearded, bespectacled genius of Zinfandel is Paul Draper of Ridge Vineyards, high in the earthquake-rattled hills above San José in the Santa Cruz Mountains. With some grapes you can point towards a bunch of growers and say – these producers saved the grape, or showed us how great its wine could be. But with Zinfandel, if you're looking for the one person who has raised its wine to an art form, a heady, swirling, perfumed riot of richness and ripeness that never somehow topples over into the banality of stewed fruit and jam, it is the scholarly, amiable Paul Draper. For me, he's the true king of Zin.

And from round the world I hear faint but lusty cries of support – from Cape Mentelle in Western Australia, from Fairview Estate in South Africa, from Mexico and Chile, from Puglia in Italy and even a hoarse but hearty shout from the Croatian coast itself.

Zinfandel: from Grape to Glass

Geography and History page 302; Viticulture and Vinification page 304; Zinfandel around the World page 306; Enjoying Zinfandel page 308

GEOGRAPHY AND HISTORY

To listen to some of the Californian Zin producers talk, you would think that Zinfandel only grows in California, only ever could grow in California, only ever did grow in California and that anyone who dares protest that they too can grow Zinfandel in any less-favoured area than America's Golden West is cruisin' for a bruisin' (or mushin' for a crushin' depending on your preferred winemaking style). Well, folks, it ain't so. Let's start with the USA: 14 States of the Union grow Zin, and some of it's pretty good. And there's more. Zin isn't Californian; it isn't even American. First they thought it was Italian, called Primitivo. Now they reckon it's Croatian and called Crljenak Kastelanski, or Tribidrag, or goodness knows what. Definitely not Californian.

Now, I can feel the hurt, because there's a powerful movement called ZAP (Zinfandel Advocates and Producers) which declares

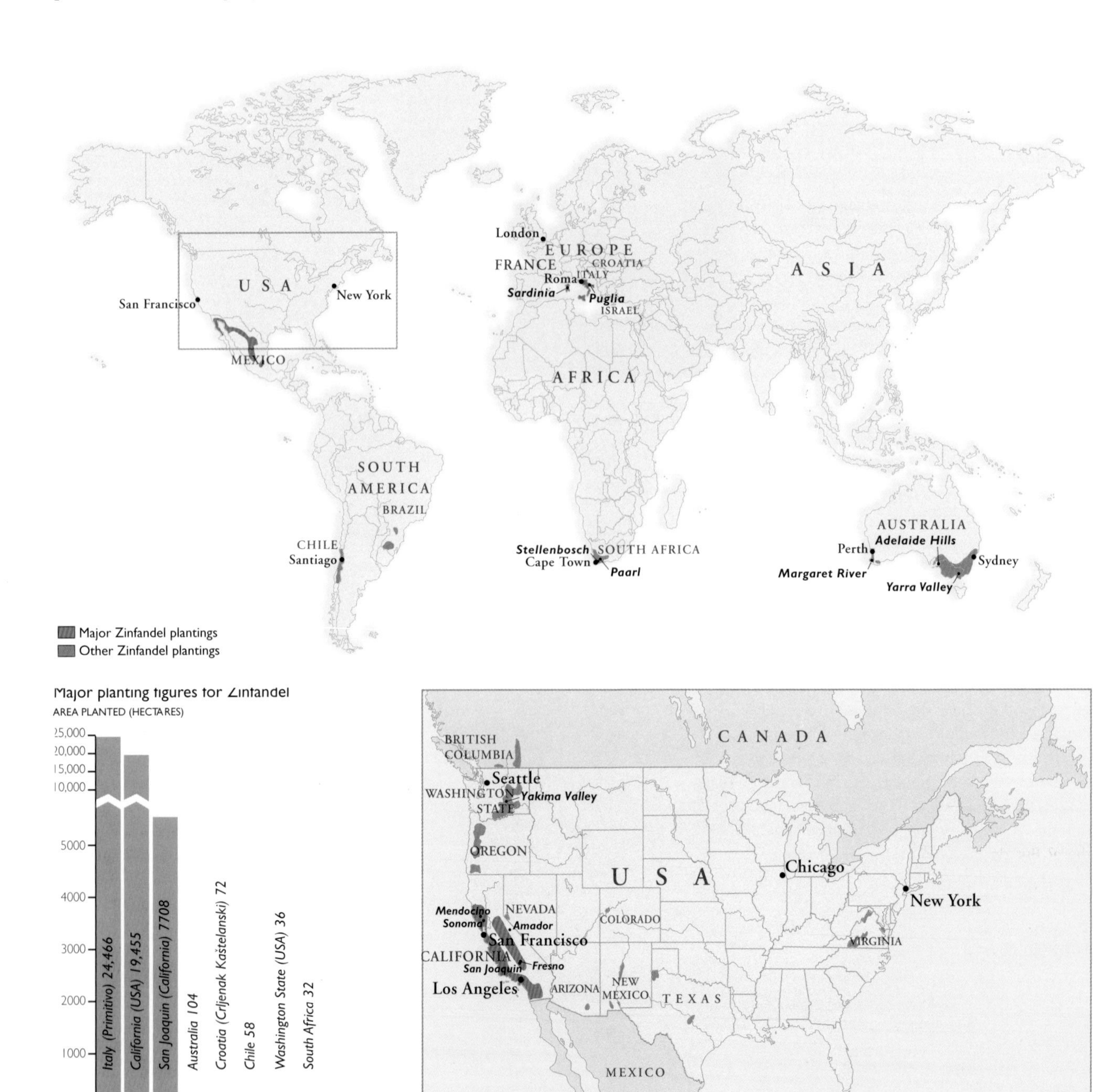

Zinfandel as California's indigenous grape. But California didn't *have* any indigenous *Vitis vinifera*: – at least it was California's own University of California at Davis that proved the grape was the same as southern Italy's Primitivo and Croatia's whatever. Zin flourished in California in the gold rush, providing the forty-niners with cheap hooch, and home winemakers turned to it with enthusiasm during Prohibition. California's first wine boom, in the 1880s, was based on Zinfandel. Then the wine was red. A century later, another Californian wine boom galloped along on the broad shoulders of Zin – but this time it was pink. It can be pink or deepest black, sweet or dry; a properly white mutation even exists. In its protean nature it is the epitome of the American dream, constantly able to reinvent itself to suit the Zeitgeist. Modern America is based on waves of immigrants adapting deftly to the unique conditions of the continent. Zinfandel, the Croatian/Italian immigrant, has achieved that par excellence.

HISTORICAL BACKGROUND

We now know not only that Zinfandel and Italy's Primitivo are the same vine, but that they are identical to Tribidrag and Crljenak Kastelanski – you pronounce the first word 'tsurl-yen-nak'. Yes, really you do. It's a vine found in Croatia, and, more specifically, found there by Dr Carole Meredith, then of the University of California at Davis (UCD). Her researches, using DNA fingerprinting, first of all narrowed the search to islands off the coast of Croatia, where she found some close matches. She then found the identical vine. And judging by the number of its close relatives there, it has been in Croatia for a long time.

The earliest mention of Zinfandel in the USA is in the 1820s, when George Gibbs, a Long Island nurseryman, brought cuttings from the Imperial botanical collection in Vienna. By 1832 'Zinfendal' vines were being sold by a Boston nursery; in subsequent years it was cultivated as a table grape in the northeastern states. From there it is believed to have been taken to California in the 1840s by another nurseryman, Frederick Macondary.

Its name was probably acquired because of confusion with an Austrian vine called Zierfandler; and it probably arrived in Vienna in the first place from Croatia. It has also been suggested that Zinfandel was taken to Italy from the USA by returning emigrants: some 20 per cent of Italian emigrants returned to their homeland. In which case, of course, there should be no problem with Italian Primitivos labelling themselves as Zinfandels, should there? But do you really think that returning emigrés would bring vines back to somewhere like Italy? And anyway, there are records in Puglia of Primitivo arriving there from Dalmatia (now Croatia) in the 18th century.

So, Zin moved from Croatia to California – but it hasn't stopped moving yet. Central and South America have some, as do Australia and South Africa. There's even a Zin vineyard on the hallowed Syrah domain of the hill of Hermitage (for amusement only, I must add).

Early morning mist in the Lytton Springs Vineyard on the eastern side of Dry Creek Valley, Sonoma, California. Zinfandel was first brought to the valley by Italian immigrants in the 1870s. Part of the secret of this region's great Zinfandel is unique, old gravelly soil, known as Dry Creek Conglomerate.

One hundred-year-old vines in Lytton Springs, Dry Creek Valley. Over half the Zinfandel in California is reckoned to be more than 50 years old.

Ridge makes superb wines from the Dusi Ranch in Paso Robles. It's been family-owned for almost a century, and the young 'uns get involved early.

VITICULTURE AND VINIFICATION

Folks do get stuck on Zin. Pinot Noir devotees work themselves into paroxysms of distress over their beloved grape's many shortcomings. But Zin worshippers are a more laid-back crew. Zin does have shortcomings, but these guys seem to welcome them.

It will produce obligingly large crops on poor soil in dry conditions, but it ripens so unevenly that it is not uncommon to find green, ripe and raisined berries all on the same cluster. This, in any other variety, would be regarded with horror for the unbalanced flavours it produces; Zinfandel aficionados seem to treat it with the indulgence of a parent for a spoilt child. Green, ripe and raisined? Why, that sounds pretty balanced to me. And so it often is. If Zin rarely achieves the heights of Shiraz or Cabernet Sauvignon, producers who understand it make some magnificent, mouth-warming, heart-warming stuff.

Climate

What Zinfandel likes is warmth, but a long growing season as well. To get the full flavours of the grape, you seem to need 14 per cent alcohol: Paul Draper of Ridge Vineyards says, 'You must go to 101 per cent ripeness to get the true fruit. Whereas an overripe Cabernet does not work, an overripe Zinfandel does not exist'. Ripeness carried to the ultimate can mean 17 per cent alcohol, and such wines are not everyone's idea of a balanced, appealing drink (and Draper's wines do not go to these levels). Such monster Zins had their greatest vogue during the 1970s, but they are still around – and a few are magnificent.

The vine performs best in a Mediterranean climate with plenty of sun; but let it get too hot and the wine will taste jammy and baked. Dry Creek Valley, California's prime Zinfandel region, points to its hot days (35–38°C, 95–100°F) and its relatively cool nights (7–10°C, 45–50°F) as factors that produce good ripeness and preserve acidity at the same time. The Central Valley, where a lot is planted for sweetish 'blush' (in reality pale pink) Zin, is even hotter, but without the redeeming cool nights.

In the south of Italy, where the quality of Primitivo has yet to rival that of Californian Zinfandel, much of the wine hits 16 per cent alcohol. A few quality-oriented producers are looking at ways of slowing the maturation of this grape – not easy given that it's a variety that used to be famous for ripening before any other grape in Europe. It got its name Primitivo from the Latin for 'early ripener'.

When the grapes are picked is key to the wine style. Earlier picking gives flavours in the strawberry, then cranberry/raspberry range; later picking gives black cherry, blackberry and plum flavours, right through to prune, dates and raisin. Crucially, underripe Zin does not taste vegetal, unlike, say, Cabernet; and many growers like to pick when each cluster contains a mixture of unripe, ripe and overripe berries.

Paul Draper, 'King of Zin', taking a well-earned rest. He arrived at Ridge Vineyards as winemaker in 1969; Ridge Zinfandels, made with grapes from various sources, are famed for their great intensity, concentration and long life.

Up to 23 Brix (the standard US measurement of fermentable sugars in the grape juice strawberry flavours dominate; cherry comes into the picture between 23 and 24, and at 25 the flavours turn blackberryish. Raisined grapes will be even higher in sugar, about 27 Brix; most growers will accept ten per cent or so raisined grapes in the vat. Much blush Zin is picked at 20 Brix: varietal character is not at a premium for such wines, but, interestingly, pink (or so-called white) Zin does have a very recognizable apple flesh and tobacco flavour, dappled with raisin.

Soil

Italian settlers, planting Zinfandel in the 19th century, usually chose hillsides and river benchlands, and soils that were too poor for other crops. Such practices reflected their habits back home; and it turned out that they suited Zinfandel very well. Poor, well-drained but mineral-rich soils produce good results, but good wines also come from the valley floors. As is usual in California, choice of soil is secondary to choice of climate. And in any case, even the now-revered Dry Creek Valley was originally planted for bulk wine. That it turned out to produce excellent quality was as happy a chance.

Cultivation

California abounds in Zinfandel vines of 50 years old or more; which is one of the reasons why the USA regards the variety as a national treasure. (Being generally planted on St George rootstock, rather than the susceptible AXR1, means that they escaped phylloxera in the early 1990s.) Such vines are bush-trained and head-pruned, just as they are in the south of Italy, but bush-trained vines are not necessarily naturally low-yielding: 3 to 5 tons per acre is about the maximum for top quality; 6 tons is all right, but even bush-trained vines can give up to 8 tons per acre – at a low density of perhaps 512 vines per acre – unless they are carefully managed. (The oldest vines may not manage more than 2 tons per acre.)

Clusters need to be thinned, and the second-crop clusters, which Zinfandel is adept at producing because of its uneven flowering, need to be removed. More recent plantings, including those of the higher-yielding, virus-free clones produced by the University of California at Davis (UCD) in the 1970s, are trained on wires, and mechanically cultivated and picked. But some producers try to achieve more consistent ripeness by hand-picking in successive tries.

UCD has been busy preserving the diversity of Zinfandel in California. In its Heritage Vineyard at Oakville in the Napa Valley, it gathered together 90 massal selections from vineyards planted before 1930 from all over California: from Sonoma, Mendocino, Napa, Contra Costa, Sierra Foothills, San Luis Obispo, San Joaquin, Lake, Amador, El Dorado, Calaveras, Alameda,

Autumnal vineyards of Chateau Potelle on the slopes of Mount Veeder AVA in Napa, California. Much of the Zinfandel planted in the 1990s is trained on wires such as these ones. There is also a debate over whether hillside vineyards like this one, or vineyards in the valleys, produce the longest-lived wines. Mount Veeder reds, from whatever variety, are some of the deepest and longest lived in California.

Santa Cruz, San Bernardino and Riverside Counties. These it has narrowed down to 20 selections all grafted on to the same St George rootstock, head trained and spur pruned, and planted in 2.7 x 2.4m (9ft by 8ft) spacing. Such variables as cluster weight, berry size and cluster tightness will be studied, to see which ones are genetic and which are derived from site and circumstances. Some of these vines may be cloned and made commercially available to growers in the future: it's a long-term project.

At the winery

Zinfandel can be treated in almost any way the producer likes, once it gets to the winery. For a start, you can make it like a white wine: run the juice off quickly, ferment it nice and cool. Leave a fair amount of sugar in it and – hey presto – you've got a white Zin, or, more accurately, a Blush Zin since there'll be a shy pink bloom to the wine.

If you're making reds, whole cluster fermentation will give fruitier, strawberryish flavours; a long maceration with the skins will give more colour and tannin and that'll set the wine up for some oak aging. Traditionally, old oak barrels or even just great big redwood vats were used for softening the wine. In the right hands, this produced some blockbuster stuff. But, nowadays, the risk of oxidation and infection isn't seen as acceptable, so small, frequently new barrels are used.

I still prefer it when the oak is kept in check, but the rich vanilla and spice flavours of American oak do seem to marry pretty well with the deep date and raisin richness of Zin. Some French oak is now being used, though I can't see the point for such an over-the-top grape as Zin. Anyway, a good winemaker using well-coopered barrels will make a wine where you don't notice the oak at all.

Late-harvest Zins, made from completely raisined grapes, may be fermented out to dryness, which gives somewhat intimidating alcohol levels of over 17 per cent; Paul Draper at Ridge has also made some sweet Zinfandel Essence from botrytized grapes. Remarkably, the wine kept its dark colour: most botrytized reds end up as a nondescript pink, since one of the effects of *Botrytis cinerea* is to destroy colour.

BLENDING ZINFANDEL

If you go back far enough, traditional Zinfandel came from vineyards in which the vine was interplanted with other varieties such as Petite Sirah, Carignan, Grenache, Mourvèdre, Mission and even Muscat: a hotchpotch of warm-climate grapes which, blended together, gave extra complexity to the exotic flavours of Zinfandel. Carignan would give tartness to such a blend; Petite Sirah colour and tannin; Zinfandel would have provided the richness. It's what's known as a field blend, or mista nera (dark blend), and although such plantings are rare today, they are reputed to survive in some of California's oldest vineyards.

If a single variety is to be blended with Zinfandel, it is usually Petite Sirah. Paul Draper of Ridge Vineyards is particularly keen on this mix, and all his Zins have 10 to 15 per cent of Petite Sirah included in the blend. J Lohr adds a touch of Riesling to its white Zinfandel, to help the aroma – rather mundane employment for such a fine variety.

ZINFANDEL AROUND THE WORLD

We're living in an age when great big gobfuls of super-ripe juicy red fruit win a lot of the plaudits in the wine world. Well, step forward Zin. That's what Zin does best, and that is why growers around the world are having a go, as well as explaining a great increase in respect and enthusiasm for Zinfandel as Primitivo in its homeland of southern Italy.

USA: California

Zinfandel can be found all over California, from the Sierra Foothills to the coastal hills of Paso Robles. And plantings are still rising: almost doubling between 1986 and 2013 to 48,638 acres (19, 683ha). It is now California's second most widely planted red grape, after Cabernet Sauvignon. Zinfandel Advocates and Producers (ZAP), a group formed in 1991 to promote the vine and its wines, now has over 300 members. And that the vine has been proved to be identical to Primitivo has not altered the opinion of the Bureau of Alcohol, Tobacco, Firearms and Explosives (BATF) that Zinfandel is unique to the USA, and that the two wines must be labelled differently when sold in the USA.

Dry Creek Valley in Sonoma County is the region associated above all with the grape. Here it is warm, but not excessively so: it is cooler than Alexander Valley, where the vine also flourishes, though warmer than Russian River Valley, where Zinfandel really needs warm years to excel. Typical Zin flavours from Dry Creek are juicy blackberry and pepper; bright flavours with good acidity. Russian River examples also show acidity – too much so in cool vintages when the grapes don't ripen fully. Sonoma generally gives this character, with relatively low tannin. Santa Cruz wines have complexity and depth – but is that Santa Cruz character, or the skill of Paul Draper at Ridge Vineyards? Its advocates say that Zinfandel reflects its site more than, say, Cabernet, but it reflects its producer as well, and the diversity of styles make it difficult definitively to assign particular characters to particular regions – even if the regions were homogeneous, which they are not. But single-vineyard Zins, often made in tiny quantities, are enjoying a vogue.

Paso Robles Zinfandels often have soft, round flavours without the acidity of Sonoma, and in Napa, where the vine has generally given way to Cabernet Sauvignon, the flavours can be plummy, blackberry-scented and intense. Contra Costa wines can be dusty and dense; Lodi, self-proclaimed Zinfandel Capital of World, is probably the most quality-conscious volume producer in California with fleshy, approachable reds, and a great deal of blush.

Lodi is at the northern end of the vast San Joaquin Valley but its vines, cooled by breezes from San Francisco Bay, are far superior. In particular, Lodi has some of the biggest plantings of mature Zinfandel, allowing rich-flavoured, charismatic Zins to be quite widely available at a fair price.

There is a distinct difference in flavour when the vines are old – and with California so rich in vines of half a century and more, there is not the same devaluation of the term 'old vines' that one sees in most countries. 'Old Vines' on the label still means something here. Old vine Zinfandels are denser, more compact and have a much more thrilling, brooding intensity. Their balance is partly because of their lower yields and partly because they were planted on poorer soils and on hillside sites where maturation is slower and flavours have longer to develop.

Blush, or white, Zinfandel was the great commercial success of the 1980s, although the grape had been used for flavoursome rosé in the 19th century. Bob Trinchero of Sutter Home is the man credited with commercializing the first modern blush Zin: the red wine one year was pallid, so he took the skins from one vat and added them to another, leaving him with some properly deep coloured wine, and some pale pink wine. Well, he wasn't going to throw it away, was he?

In the late 1990s this bland, sweetish style of wine was still outselling red Zinfandel. It is often said that it was blush Zin that saved the vine in California: sales had been falling until then and vineyards were being uprooted. But the vines that produce blush wine are not the same as those that produce top-quality red: blush is nearly always a commodity wine, produced from high-yielding, wire-trained, mechanically harvested vines. Old bush-trained vines cannot be mechanically harvested.

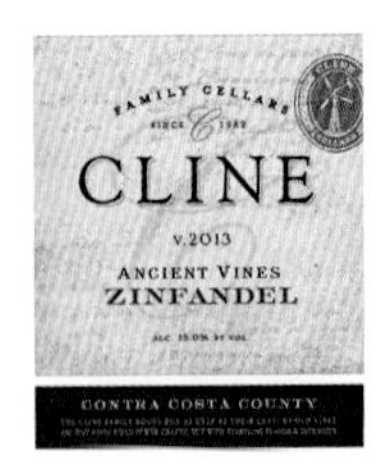

Cline
Spectacular 100-year-old vines, somehow preserved between a railway line and a chemical factory in Oakley, Contra Costa County, make stunning Zin.

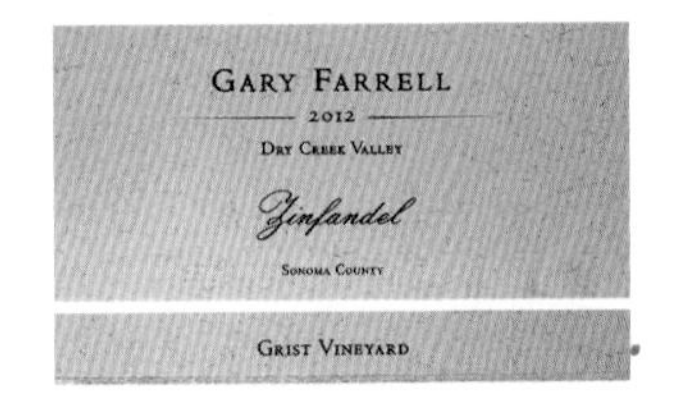

Gary Farrell
Farrell makes good use of old Zinfandel vines, from original plantings in 1900, planted high up on the northwest side of Dry Creek Valley.

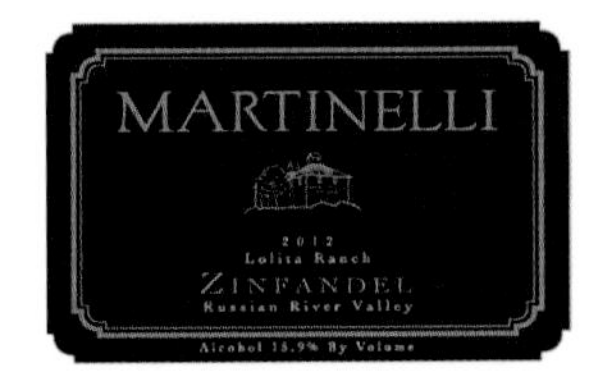

Martinelli
As it enters the redwood forests on its way to the Pacific, the Russian River Valley is relatively cool, but Lolita's Ranch can still ripen its Zin to 15.9%.

Unirrigated Zinfandel vines dating from the 19th century basking in a suntrap in California's Arroyo Grande AVA south of San Luis Obispo. At the coast only 16km (10 miles) away it's so chilly you can barely ripen Chardonnay and Pinot Noir. That's the glory of the Coastal Ranges in California. Pacific maritime influence is everything – southern Rhône to Chablis, as it were, in just 16km (10 miles).

Nevertheless, blush Zin did revive interest in the variety. If nothing else, it paved the way for Zinfandel's current cult status.

Rest of the USA

Other states with Zinfandel include Arizona, Colorado, Illinois, Indiana, Iowa, Massachusetts, Nevada, New Mexico, North Carolina, Ohio, Oregon, Tennessee, Texas and Washington. They rarely achieve the ripeness of California, but I recently tasted a couple of New Mexico and Arizona examples that were unmistakably Zinfandel and deliciously different.

Italy

Primitivo was a widely planted but deeply obscure grape from the south of Italy until its resemblance to Zinfandel was noted in the 1970s. I don't mean to imply that Primitivo wasn't popular – it was. It was tremendously popular in the North, with the red wine producers of Tuscany and Piedmont. Whoops. Wash my mouth out with soap and water for even entertaining such an idea. But, as a very early ripener, with loads of flavour and alcohol, it could be fermented and finished down in the South before the North had even begun picking. And in poor years, a good slug of Primitivo will have turned many an insipid thin northern red into something rather nice. So Primitivo was popular, but, understandably, it was obscure. You don't go round praising a grape's name when it's been blended with something entirely different hundreds of miles away.

Anyway, as soon as the similarity between Zinfandel and Primitivo was noted, obscurity was out. The University of California at Davis proved in 1994 that the two are genetically identical, and now not only are bottlings of Primitivo on the increase (the bulk still goes to the north of the country for blending), but some is being renamed Zinfandel – though not if it is to be sold in the USA.

Varietal DOCs are Primitivo di Manduria and Primitivo di Gioia, both of which are in Puglia. Most wines are somewhat rustic and solid, and the Manduria DOC allows for sweet and fortified versions as well; alcohol in even the unfortified versions commonly tops 16 per cent. However, young and ambitious producers are busy turning out imitations of the California style, with 14 per cent alcohol and oak (often American oak) aging.

Rest of the world

South Africa is beginning to look at the possibilities of Zinfandel, though plantings haven't really taken off yet. Producers generally have to blend it with something like Cinsaut or Carignan to bring the brawny alcohol level down below 16 per cent.

In Chile, too, quantities are still small, but rising and the flavour is absolutely spot on. The producer Cono Sur suggests that the colour of Chilean Zin is lighter than that of California – but then produces a sample that is black as night. Mexico and Brazil also have some, and I'd expect Argentina to give it a try.

In Australia, its most successful plantings so far are in Western Australia's Margaret River region where the flavours are tremendous, even if the alcohol heads unerringly towards 16 per cent most years. South Australia, is also successful, including, surprisingly, cool Adelaide Hills; Victoria's even cooler Yarra Valley less so. As for New Zealand – well, they've got just 4ha (10 acres). In a super-hot vintage the wine could be interesting, but I doubt if it'll do much in a normal year. Even so, the New Zealanders always get excited by super-hot vintages and some guys simply can't resist planting a raft of warm climate varieties. The 2013 vintage should have provided a proper test for them.

Luna Rossa
New Mexico claims it is the USA's original wine state. High altitude and relentless sunshine allow great Zin to be grown in Mimbres Valley west of the Rio Grande.

McHenry Hohnen
Australia's Margaret River is a cool climate region, but David Hohnen's Californian background meant he had to give Zin a try – and successfully, too.

Blaauwklippen
Zinfandel isn't much grown in South Africa, and often ripens to port-like levels when it appears, but this is a positively restrained version at a mere 14.5%.

ENJOYING ZINFANDEL

I've not had a great deal of luck in aging Zin, and it's a lot to do with the nature of the grape. Ripe Zinfandel, even when young, seems to have the seeds of its own maturity lurking in its flavours. Flavours like prune, dates and raisins do friendly battle with the pepper and blackberry of youth, and within a year or two they've usually gained the upper hand as the blackberry browns and the pepper is soothed away.

So why age them? Why indeed. Many Zins are made to be drunk young anyway – and the grape is ideally suited for it. Especially when grown in high-yielding vineyards in hot areas like California's Central Valley, the low tannin and soft raspberry and blackberry fruit is perfect for a young glugger. And even from old vines in a top area like Dry Creek – well, the low tannin and burly brawny fruit makes them so easy to drink young, and few undergo any great sea-change in flavour, just a gradual loss of sweetness and fruit, eventually becoming almost tarry with time. If you like that style, well, fine. Many Zinophiles (as American lovers of the grape call themselves) do like it. Others feel that the point of Zinfandel is the young, spicy, forward fruit, the exotic perfume, the exuberance of its style, and when that fades the best of the wine is gone. Many Zinfandels don't have much tannic structure or complexity, and neither attribute can be magically produced by bottle age if it's not there in the first place.

Paul Draper of Ridge Vineyards maintains that it goes through a dumb phase between losing its primary fruit after eight or 10 years, and becoming more interesting after 12 or 15 years. But remember, he is a genius winemaker – and he always includes a proportion of grapes like Petite Sirah to add backbone and muscle.

The taste of Zinfandel

Zinophiles probably already know about the Zinfandel Aroma Wheel produced by ZAP; they may also know about the website (www.zinfandel.org) from which the Aroma Wheel can be downloaded, and on which they can add to the fund of knowledge via the 'submit-a-smell' option. I love it!

It already includes pretty well all the aromas of which red wine is capable, including faults like TCA (cork taint) and horsiness. Spicy flavours include black pepper, clove, cinnamon and oregano; floral flavours are violets and roses; fruits range from cranberries, strawberries and raspberries (these are found on lighter wines) through to the blackcurrant, black cherry, plums, prunes and raisins of increasing ripeness.

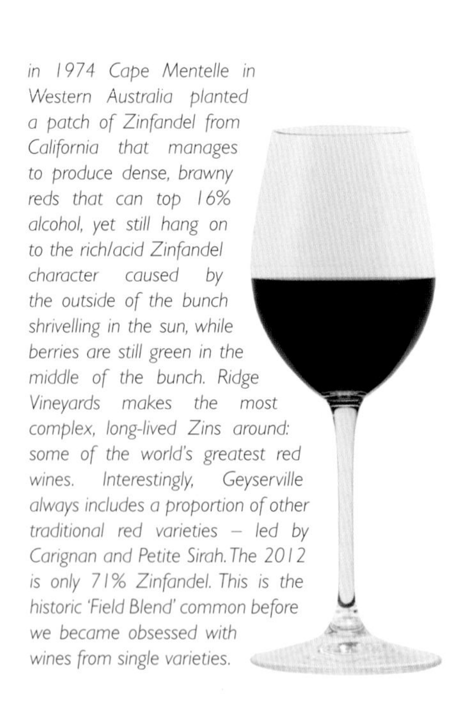

In 1974 Cape Mentelle in Western Australia planted a patch of Zinfandel from California that manages to produce dense, brawny reds that can top 16% alcohol, yet still hang on to the rich/acid Zinfandel character caused by the outside of the bunch shrivelling in the sun, while berries are still green in the middle of the bunch. Ridge Vineyards makes the most complex, long-lived Zins around: some of the world's greatest red wines. Interestingly, Geyserville always includes a proportion of other traditional red varieties – led by Carignan and Petite Sirah. The 2012 is only 71% Zinfandel. This is the historic 'Field Blend' common before we became obsessed with wines from single varieties.

Unripe wines taste of green beans, artichokes, green bell peppers, mint and eucalyptus.

Zinfandel can taste nutty or chocolatey, cedary or of green oak – the latter, one assumes, because of winemaker error. The malolactic fermentation gives creamy tastes; oak gives toast, spice and vanilla flavours.

So that's that, then. Zinfandel is the protean grape, capable of just about any flavour in the world. Well, no. It's just that these Zin freaks get a bit evangelical if you let them. In fact, you're only likely to find such wealth of flavours in wines from great winemakers like Paul Draper of Ridge and from superb, low-yielding, old vines. Otherwise, Zinfandel is blackberryish, raspberryish, peppery when young, with the flavours of dates, raisins, prunes and herbs never far from the surface. And that's not a bad bunch of flavours.

MATCHING ZINFANDEL AND FOOD

California's most versatile grape is used for a bewildering variety of wine styles from bland, slightly sweet pinks to rich, elegant fruity reds. And the good red Zinfandels themselves vary greatly in style. If they aren't too oaky, they are good with barbecued meats, venison and roast chicken. The hefty, old-style wines are a great match with the spicy, mouthfilling San Francisco cuisine. The pale blush style of Zin goes well with tomato-based dishes, such as pizza and pasta, as well as with hamburgers.

CONSUMER INFORMATION

Synonyms & local names

It has now been established by DNA fingerprinting that Zinfandel, Primitivo, Crljenak Kastelanski and Tribidrag are one and the same variety.

Best producers

USA/California Bianchi, Boeger, Brown, Chateau Montelena, Cline Cellars, Dashe, De Loach, Dry Creek, Dutton Goldfield, Gary Farrell, Hartford Family, Kunde, Mariah, Martinelli, Michael-David (Earthquake), Oakville, Peachy Canyon, Preston, Quivira, Rafanelli, Ravenswood, Ridge, Rosenblum, Saddleback Cellars, St Francis, Sausal, Scherrer, Seghesio, Trinitas, Turley, Villa Mt Eden, Wellington, Williams Selyem; **Washington State** Kiona, Sineann; **Arizona** Arizona Stronghold, Callaghan, Pillsbury; **New Mexico** Luna Rossa. **MEXICO** Bibayoff, L A Cetto. **ITALY/Primitivo di Manduria** Felline, Morella, Pervini, Giovanni Soloperto. **CROATIA** Zlatan Plenkovic. **FRANCE/Languedoc** Dom. de l'Arjolle. **AUSTRALIA** Cape Mentelle, Kangarilla Road, McHenry Hohnen, Nepenthe, Smidge, Tscharke. **SOUTH AFRICA** Blaauwklippen, Glen Carlou, Hartenberg. **CHILE** MontGras.

RECOMMENDED WINES TO TRY

Twenty top Californian Zinfandel wines

Amador Foothill Winery Shenandoah Valley Ferrero Vineyard Zinfandel
Boeger El Dorado Zinfandel
Cline Ancient Vines Zinfandel
Dashe Lily Hill Vineyard, Dry Creek Valley
De Loach OFS Russian River Valley Zinfandel
Gary Farrell Dry Creek Valley Zinfandel
Hartford Family Russian River Valley, Hire Wire Vineyard Zinfandel
Kunde Sonoma Valley Century Vines Zinfandel
Martinelli Russian River Valley Jackass Vineyard Zinfandel
Peachy Canyon Paso Robles Dusi Ranch Zinfandel
A Rafanelli Dry Creek Valley Zinfandel
Ravenswood Sonoma Valley Monte Rosso Vineyard Zinfandel
Renwood Shenandoah Valley Grandpère Vineyard Zinfandel
Ridge Lytton Springs Dry Creek Valley Zinfandel
Rosenblum Napa Valley George Hendry Vineyard Zinfandel
Saddleback Cellars Napa Valley Old Vines Zinfandel
Sausal Alexander Valley Zinfandel
Scherrer Alexander Valley Old & Mature Vines Zinfandel
Turley Cellars Paso Robles Dusi Vineyard Zinfandel
Villa Mt Eden Cellar Select Zinfandel

Ten southern Italian Primitivo wines

Leone De Castris Primitivo di Manduria Santera
Felline Primitivo di Manduria
A Mano Primitivo di Puglia
Masseria Pepe Primitivo di Manduria Dunico
Pervini Primitivo di Manduria Archidamo and Primitivo del Tarantino Monili
Pichierri/Vinicola Savese Primitivo di Manduria Mamma Teresa
Sava co-op Primitivo di Manduria Terra di Miele
Sinfarosa Primitivo di Manduria Zinfandel
Torrevento Primitivo del Tarantino I Pastini

Eleven other Zinfandel wines

L A Cetto Zinfandel (Mexico)
Callaghan Vineyards Z13 (Arizona)
Cape Mentelle Zinfandel (Australia)
Hartenberg Zinfandel (South Africa)
Kangarilla Road Zinfandel (Australia)
Luna Rossa Zinfandel (New Mexico)
McHenry Hohnen Zinfandel (Australia)
Mont Gras Single Vineyard Zinfandel (Chile)
Nepenthe Zinfandel (Australia)
Sineann Columbia Valley Old Vine Zinfandel (Washington)
Zlatan Plenkovic Zinfandel (Croatia)

When Zinfandel grapes are really ripe, they look like this: not just black but starting to shrivel. Growers of many other red varieties would take fright at this point, and think their grapes overripe. Zinfandel growers just love them like this. These grapes come from very old vines at Summit Lake Vineyard in the Howell Mountain AVA in Napa.

Maturity charts

Whether you think that old vine Zin actually improves in bottle or merely changes depends on your taste.

2014 California Zinfandel (full-bodied, old vine)

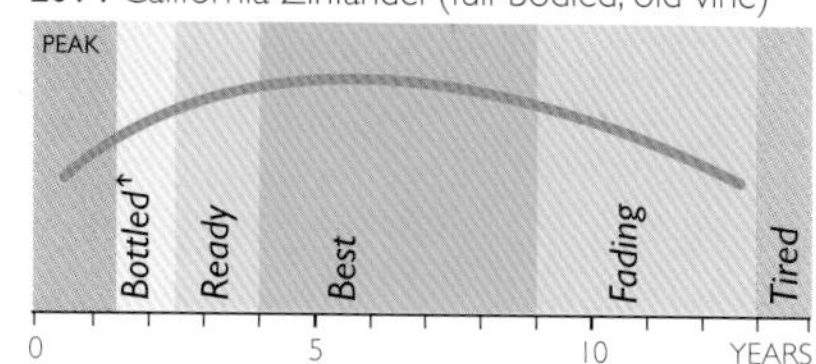

From a Zinfandel specialist such as Ridge, these wines can be beautifully balanced without the common Zinfandel problems of excessive alcohol or oak.

2014 California Zinfandel (fruity red)

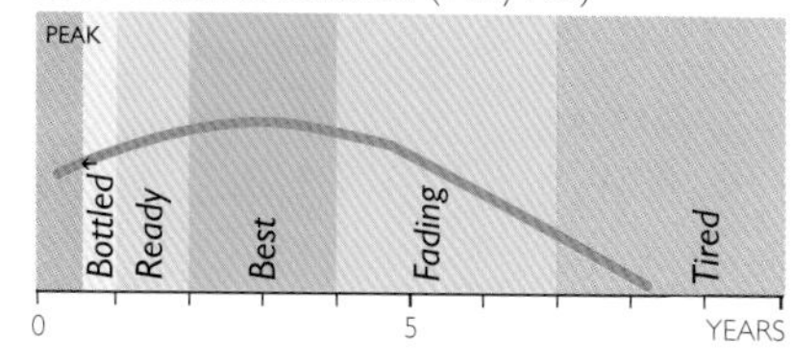

Simple Zinfandel is best drunk young, while it has all its delicious blackberry flavours. Aging this style in bottle does it no favours.

2014 Primitivo di Manduria

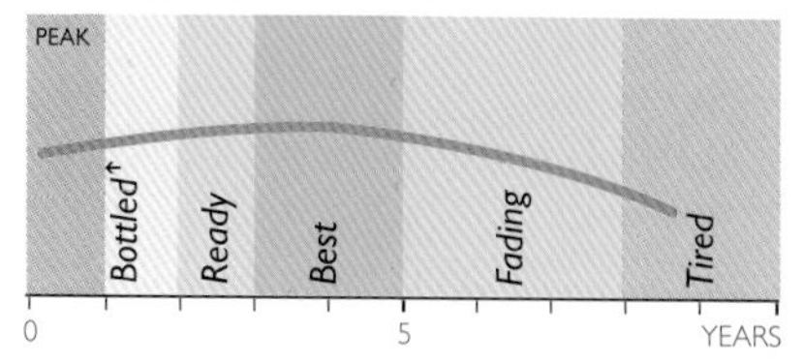

This DOC from southern Italy makes full-bodied wines with ripe, spicy black fruit and typical southern warmth and smoothness.

KEY TO THE CLASSIC GRAPE PAINTINGS

In each painting the artist Lizzie Riches has combined an accurate rendition of the grape in question with images associated with myths and legends surrounding the grape and its most famous wine styles. In so doing she has taken considerable licence with reality – for example, in the painting of Sauvignon Blanc the gooseberry and the grape should not fruit in the same season, nor do the hills of the Marlborough region in New Zealand's South Island lie directly behind the château of Villandry in the Loire Valley. The exact origin of these classic grape varieties is often conjectural, but the myths that remain behind are more evocative than prosaic fact. Here follows a brief explanation of the images chosen to convey the nature and quality of 17 of the world's classic grape varieties and the wonderful wines they can produce.

LIZZIE RICHES

Lizzie Riches, who lives at Norwich in the UK, has been a mainstay of London's Portal Gallery for many years. Largely self-taught, for figurative art was not much in vogue at art school in the 1960s, she has always been drawn to rich detail, exotic subjects and obscure references. Her taste in art is broad, but she has an especial fondness for the glittering surfaces of Veronese and wishes she could paint a feather as well as Bogdani. She was one of the artists selected for the 'Art in the Underground' series and has recently completed a cycle of 16 paintings for P&O. She has had exhibitions in London, Paris and Chicago, in Holland and Germany and is represented in collections worldwide.

Cabernet Sauvignon (Page 52)
Aristocratic and magnificent, Cabernet Sauvignon is represented here by the sunburst, the emblem of France's king Louis XIV, also known as Le Roi Soleil or the Sun King. His brilliant court at the Palace of Versailles was filled with images of Louis' glory. The painting captures Cabernet Sauvignon's self-importance and regal position in the world of wine.

Chardonnay (Page 68)
There seem to be more flavours associated with Chardonnay than with any other grape and it also has a wonderful affinity with new oak barrels. So here carved in the fresh new oak are many of these flavours, including cloves, hazelnuts, warm brioche and a host of different fruits.

Chenin Blanc (Page 82)
Floating on air and water, the château of Chenonceau stretches across the river Cher in Touraine, with the river Loire beyond. Chenin Blanc was first planted here in the heart of the Loire Valley in the 15th century and Anjou and Touraine are still where it produces its most exciting wines, whether sweet or dry, sparkling or still.

Garnacha Tinta/Grenache Noir (Page 102)
High on the skyline loom the gaunt ruins of the castle built by the popes during their sojourn at Avignon in the southern Rhône, instead of Rome. This was their new castle, their château neuf – and the vineyards spread around the castle walls are those of Châteauneuf-du-Pape. Gnarled old Grenache vines grow in a soil covered in large white pebbles or galets roulés that retain the heat of the southern sun long into the night. The pebbles make good homes for lizards, too.

Gewürztraminer (Page 112)
The Gothic-style, carved wooden spice cabinet, stork, and buttery, knot-shaped sweet pretzels suggest Alsace where the grape achieves its highest fame. The pomander, cloves, nutmeg and spice mortar are references to the word 'Gewürz' which means spice in German but Gewürztraminer is a complex story. The wines are far more than just being fat and spicy, famous only for their heady perfume.

Merlot (Page 138)
Merlot's name is said to be derived from merle, French for blackbird, which apparently loves its sweet, early-ripening fruit. Being planted like fury around the world, Merlot has done particularly well in the Napa Valley where the California poppy, as orange as a blackbird's bill, grows wild.

Muscat (Page 156)
The wine-dark sea around the island of Samos, a home of the Muscat grape, lies in the distance. Was this the wine drunk by Dionysus, the Greek god of wine? The ancient Greeks clearly loved wine and all its paraphernalia. Here is a mixing jar, a wine jug and a beautiful kylix, or two-handled wine cup. And goodness me, some hooligan has been scrawling his name on the side of the cup. Disgraceful.

Nebbiolo (Page 166)
Dusty purple grapes and bright red leaves seen against a background of snow and fog – this is the Piedmontese town of Alba, famous for red wine and white truffles, in autumn. Against a backdrop of snow-capped Alps, the town's medieval towers emerge from the late autumn fogs or nebbie that have given the Nebbiolo grape its name. In late autumn truffle hunters and their specially trained dogs set off into the oak forests bordering the vineyards in search of the revered delicacy.

Pinot Noir (Page 188)
Seen through an arched Gothic window from the Hospices de Beaune in Burgundy, Pinot Noir's homeland, is the château at Gevrey-Chambertin, one of the Côte de Nuits' best known wine villages. The moonlit scene is a tribute to the word 'Nuits'. On the windowsill are various references to Pinot Noir – a silver Burgundian tastevin *or tasting cup, a Champagne cork and wire cage, an Oregon pine cone and a Knave of Spades to illustrate Pinot Noir's capriciousness.*

Riesling (Page 204)
On the banks of the river Mosel stands the town of Bernkastel directly behind which climb steep Riesling vineyards, just like the ridged green stem of the traditional Mosel wine glass. The intricate ironwork shop signs typical of the Mosel region were surely originally inspired by coiling vine tendrils.

Sangiovese (Page 222)
Sangiovese means literally the 'Blood of Jove'. Nice to think that this Roman god left off seducing mortals and dropping thunderbolts to give his name to this vine. Almost certainly of Tuscan origin, Sangiovese is still central Italy's most important grape. Florence, the heart of Tuscany, lies beyond and Keats's nightingale has strayed from Provence as this grape seems to give more truly 'a beaker full of the warm south'.

Sauvignon Blanc (Page 232)
A piece of the trellis from the potager garden at the château of Villandry illustrates the Loire Valley's long association with Sauvignon Blanc. The flavours of Sauvignon Blanc are linked to myriad fruit and vegetables but none more so than gooseberries. Behind the trellis stretch the vineyards and hills of the Marlborough region in New Zealand's South Island, the world's new classic area for Sauvignon Blanc.

Sémillon (Page 244)
Seen here in the honeyed autumnal light evocative of its precious golden wine, Château d'Yquem is the supreme example of the majestic sweet wines of Sauternes. Pickers will go through the vineyards four times or more during the harvest, picking fully botrytized grapes or even green ones, according to the constantly changing instructions from the winery.

Syrah/Shiraz (Page 258)
Whether it originated in the northern Rhône, in Syracuse in Sicily or was brought back from Shiraz in Persia, there is something exotic about this grape. It's a nice idea that Syrah reached Europe at roughly the same time as the damask rose, and along the same route. In the distance is the chapel on the Hermitage hill above the Rhône and in the foreground some of the wild violets that seem to perfume the wine. The parrot is an Adelaide Rosella from the Barossa Valley where Shiraz excels.

Tempranillo (Page 270)
The history and grandeur of Tempranillo is conveyed by references to two Spanish artists of the 17th century: first, Diego Velázquez, the great Spanish court painter whose paintings of the royal family, including the Infanta Maria Teresa depicted here, were much admired at the time, and second, Juan Sánchez Cotán who is best known for his still lifes with their detailed realism.

Viognier (Page 288)
If a wine can be described as pretty without insulting it, then Viognier is pretty. It tastes deliciously of the apricots that grow along with Viognier on the banks of the Rhône at Condrieu. There is a theory that the vine arrived by boat, along with Syrah, during the Roman occupation, found a landing place at Condrieu and then remained there as a tiny island of distinctive taste until its recent global explosion. Château-Grillet, a tiny estate in Condrieu, can be glimpsed behind the apricot fruit and blossom.

Zinfandel (Page 300)
Zinfandel, now known to be the same as southern Italy's Primitivo grape, may have come to the eastern seabord of the USA with Italian immigrants. From there it made the journey west to California with the Gold Rush – hence the gold pan, the poster advertising the Gold Fields – the new El Dorado – and the nugget of gold. The popularity of the grape has proved to be a more lasting national treasure than the seams of gold which fast ran out.

WHICH GRAPES MAKE WHICH WINES

Most European wines are named according to where they come from – the appellation – rather than according to the grape variety or varieties in them. This list of the main European appellations and their authorized grape varieties (with local names in brackets) will help you find the grape you want in the A–Z section (pages 40–309). Grape names listed in *italic* are approved minor varieties that are either seldom used nowadays or allowed only in small percentages, sometimes as experimental. Virtually all the grapes listed here have an entry in the A–Z. A few, however, although they are still listed in the regulations do not in practice appear in the vineyards any more and have, therefore, been omitted from the A–Z for reasons of space.

The list does not include varietal wines which state their main grape on the label and also appellations named after their sole or principal grape, for example Muscat de Beaumes-de-Venise and Vernaccia di San Gimignano. If you cannot find the name you are looking for try the Index of Grape Names and their Synonyms on page 322.

Grapes allowed for the following wines:
Red wine; Sparkling white wine
White wine; Sparkling rosé wine
Rosé wine; Sparkling red wine

AUSTRIA
Gumpoldskirchner Rotgipfler, Zierfandler
Mittelburgenland Blaufränkisch
Steiermark Schilcher Blauer Wildbacher
Weinviertal Grüner Veltliner

CYPRUS
Commandaria Mavro, Xynisteri

FRANCE
Ajaccio Sciacarello, Barbarossa, Nielluccio, Vermentino, *Carignan, Cinsaut, Grenache Noir*; Vermentino, Ugni Blanc
Aloxe Corton *Pinot Noir, Pinot Gris (Pinot Beurot), Pinot Liébault, Pinot Blanc, Chardonnay*; Chardonnay
Anjou, Anjou Gamay, Anjou Pétillant Cabernet Franc, Cabernet Sauvignon, Pineau d'Aunis; Cabernet Franc, Cabernet Sauvignon, Pineau d'Aunis, Gamay, Cot, Groslot; Chenin Blanc (Pineau de la Loire), *Chardonnay, Sauvignon Blanc*; **Anjou Coteaux de la Loire** Chenin Blanc (Pineau de la Loire); **Anjou Mousseux** Chenin Blanc (Pineau de la Loire), *Cabernet Sauvignon, Cabernet Franc, Cot, Gamay, Groslot, Pineau d'Aunis*; **Anjou Mousseux** Cabernet Sauvignon, Cabernet Franc, Cot, Gamay, Groslot, Pineau d'Aunis; **Anjou-Villages** Cabernet Franc, Cabernet Sauvignon
Arbois Poulsard Noir (Ploussard), Trousseau, Pinot Noir; Savagnin Blanc (Naturé), Chardonnay (Melon d'Arbois, Gamay Blanc), Pinot Blanc; **Arbois Vin de Paille** Poulsard Noir (Ploussard), Trousseau, Chardonnay (Melon d'Arbois, Gamay Blanc), Savagnin Blanc (Naturé)
Auxey-Duresses Pinot Noir, *Pinot Gris (Pinot Beurot), Pinot Liébault, Pinot Blanc, Chardonnay*; Chardonnay, *Pinot Blanc*
Bandol Mourvèdre, Grenache Noir, Cinsaut, *Syrah, Carignan*; Mourvèdre, Grenache Noir, Cinsaut, *Syrah, Carignan plus any of the following white varieties*; Bourboulenc, Clairette, Ugni Blanc, *Sauvignon Blanc*
Banyuls, Banyuls Grand Cru VDN Grenache Noir, Grenache Gris, Grenache Blanc, Maccabéo, Tourbat (Malvoisie du Roussillon), Muscat Blanc à Petits Grains, Muscat of Alexandria (Muscat Romain), *Carignan Noir, Cinsaut, Syrah*
Barsac Sémillon, Sauvignon Blanc, Muscadelle
Les Baux-de-Provence Grenache Noir, Syrah, Mourvèdre, *Cinsaut, Counoise, Carignan, Cabernet Sauvignon*; Grenache Noir, Syrah, Cinsaut, *Mourvèdre, Counoise, Carignan, Cabernet Sauvignon*
Béarn Tannat, Cabernet Franc (Bouchy), Cabernet Sauvignon, Fer (Pinenc), Manseng Noir, Courbu Noir; Petit Manseng, Gros Manseng, Courbu, Lauzet, Camaralet, Raffiat, Sauvignon Blanc
Beaujolais, Beaujolais Supérieur, Beaujolais-Villages Gamay, *Pinot Noir, Pinot Gris, Chardonnay, Aligoté, Melon de Bourgogne*; Chardonnay, *Aligoté*
Beaune Pinot Noir, *Pinot Gris (Pinot Beurot), Pinot Liébault, Pinot Blanc, Chardonnay*; Chardonnay, *Pinot Blanc*
Bellet Braquet, Folle Noir (Fuella), Cinsaut, *Grenache Noir plus any of the following white varieties*; Braquet, Folle Noir (Fuella), Cinsaut, *Grenache Noir plus any of the following white varieties except Chardonnay and Muscat Blanc à Petits Grains*; Rolle, Roussanne, Spagnol (Mayorquin), *Clairette, Bourboulenc, Chardonnay, Pignerol, Muscat Blanc à Petits Grains*
Bergerac, Bergerac Sec Sémillon, Sauvignon Blanc, Muscadelle, Ondenc, Chenin Blanc, *Ugni Blanc*; **Bergerac, Côtes de Bergerac** Cabernet Sauvignon, Cabernet Franc, Merlot, Malbec (Cot), Fer Servadou, Merille (Périgord)
Blagny Pinot Noir, *Pinot Gris (Pinot Beurot), Pinot Liébault, Pinot Blanc, Chardonnay*; Chardonnay, *Pinot Blanc*
Blanquette de Limoux Mauzac, Chardonnay, Chenin Blanc; **Blanquette Méthode Ancestrale Mousseux** Mauzac
Blaye Cabernet Sauvignon, Cabernet Franc, Merlot, Malbec, *Prolongeau (Bouchalès), Cahors, Béquignol, Petit Verdot*; Ugni Blanc, *Colombard, Sémillon, Sauvignon Blanc, Muscadelle, Chenin Blanc (Pinot de la Loire)*
Bonnes-Mares Pinot Noir, *Pinot Liébault, Pinot Blanc, Pinot Gris, Chardonnay*
Bonnezeaux Chenin Blanc (Pineau de la Loire)
Bordeaux, Bordeaux Clairet, Bordeaux Rosé, Bordeaux Supérieur Cabernet Sauvignon, Cabernet Franc, Merlot, *Carmenère, Malbec, Petit Verdot*;
Bordeaux, Bordeaux Blanc Sec Sémillon, Sauvignon Blanc, Muscadelle, *Merlot Blanc, Colombard, Mauzac, Ondenc, Ugni Blanc*;
Bordeaux-Côtes de Francs Cabernet Franc (Bouchet), Cabernet Sauvignon, Merlot, Malbec (Pressac); Sauvignon Blanc, Sémillon, Muscadelle; **Bordeaux Haut-Benauge** Sémillon, Sauvignon Blanc, Muscadelle;
Bordeaux Mousseux Sémillon, Sauvignon Blanc, Muscadelle, Cabernet Sauvignon, Cabernet Franc, Carmenère, Merlot, Malbec, Petit Verdot, *Ugni Blanc, Merlot Blanc, Colombard, Mauzac, Ondenc*; Cabernet Sauvignon, Cabernet Franc, *Carmenère, Merlot, Malbec, Petit Verdot*
Bourg, Côtes de Bourg, Bourgeais Cabernet Sauvignon, Cabernet Franc, Merlot, Malbec; Sauvignon Blanc, Sémillon, Muscadelle, Merlot Blanc, Colombard
Bourgogne, Bourgogne-Hautes Côtes de Beaune, Bourgogne-Hautes Côtes de Nuits, Bourgogne-Côte Chalonnaise Pinot Noir, *Pinot Gris (Pinot Beurot), Pinot Liébault, Pinot Blanc, Chardonnay*, plus César and Tressot in the Yonne, plus Gamay in Beaujolais; Chardonnay (Beaunois, Aubaine), Pinot Blanc;
Bourgogne Passetoutgrains Gamay, Pinot Noir, *Pinot Blanc, Pinot Gris, Chardonnay*
Bourgueil Cabernet Franc (Breton), *Cabernet Sauvignon*
Bouzeron Aligoté
Brouilly Gamay, *Chardonnay, Aligoté, Melon de Bourgogne*
Buzet Merlot, Cabernet Sauvignon, Cabernet Franc, Malbec (Cot); Sémillon, Sauvignon Blanc, Muscadelle
Cabardès Grenache Noir, Syrah, Cabernet Sauvignon, Merlot, Cabernet Franc, Malbec (Cot), Fer, *Cinsaut*
Cabernet d'Anjou, Cabernet de Saumur Cabernet Sauvignon, Cabernet Franc
Cadillac Sémillon, Sauvignon Blanc, Muscadelle
Cahors Malbec (Cot, Auxerrois), Merlot, Tannat
Canon, Canon-Fronsac Merlot, Cabernet Franc (Bouchet), Cabernet Sauvignon, Malbec (Pressac)
Cassis Grenache Noir, Mourvèdre, Carignan, Cinsaut, Barbaroux, *Terret Noir*; Clairette, Marsanne, *Doucillon, Pascal Blanc, Sauvignon Blanc, Terret, Ugni Blanc*
Cérons Sémillon, Sauvignon Blanc, Muscadelle
Chablis, Chablis Premier Cru, Chablis Grand Cru, Petit Chablis Chardonnay (Beaunois)
Chambertin, Chambertin Clos-de-Bèze, Chapelle-Chambertin, Charmes-Chambertin, Griotte-Chambertin,

Latricières-Chambertin, Mazis-Chambertin, Mazoyères-Chambertin and Ruchottes-Chambertin R Pinot Noir, *Pinot Gris (Pinot Beurot), Pinot Liébault, Pinot Blanc, Chardonnay*

Chambolle-Musigny Pinot Noir, *Pinot Liébault, Pinot Gris (Pinot Beurot), Pinot Blanc, Chardonnay*

Champagne Pinot Noir, Pinot Meunier, Chardonnay

Charlemagne Chardonnay, *Aligoté*

Chassagne-Montrachet Pinot Noir, *Pinot Gris (Pinot Beurot), Pinot Liébault, Pinot Blanc, Chardonnay*; Chardonnay, *Pinot Blanc*

Château-Chalon Savagnin Blanc

Château-Grillet Viognier

Châteaumeillant Gamay, Pinot Gris, Pinot Noir

Châteauneuf-du-Pape Grenache Noir, Syrah, Mourvèdre, Picpoul, Terret Noir, Counoise, Muscardin, Vaccarèse, Picardan, Cinsaut, Clairette, Roussanne, Bourboulenc

Châtillon-en-Diois Gamay, *Syrah, Pinot Noir*; Aligoté, Chardonnay

Chénas Gamay, *Chardonnay, Aligoté, Melon de Bourgogne*

Cheverny Gamay, Pinot Noir, *Cabernet Franc, Cot*; Gamay, Pinot Noir, *Cabernet Franc, Cot, Pineau d'Aunis*; Sauvignon Blanc, *Chardonnay, Arbois (Menu Pineau), Chenin Blanc*

Chinon Cabernet Franc (Breton), *Cabernet Sauvignon*; Chenin Blanc (Pineau de la Loire)

Chiroubles Gamay, *Chardonnay, Aligoté, Melon de Bourgogne*

Chorey-lès-Beaune Pinot Noir, *Pinot Gris (Pinot Beurot), Pinot Liébault, Pinot Blanc, Chardonnay*; Chardonnay, **Pinot Blanc**

Clairette de Die Muscat Blanc à Petits Grains, Clairette Blanche

Clos des Lambrays, Clos de la Roche, Clos St-Denis, Clos de Tart, Clos de Vougeot Pinot Noir, *Pinot Gris (Pinot Beurot), Pinot Liébault, Pinot Blanc, Chardonnay*

Collioure Grenache Noir, Mourvèdre, Syrah, *Carignan, Cinsaut*; Grenache Noir, Mourvèdre, Syrah, *Carignan, Cinsaut, Grenache Gris*; Grenache Blanc, Grenache Gris, *Malvoisie, Macabeu, Marsanne, Roussanne, Vermentino*

Condrieu Viognier

Corbières Carignan, Grenache Noir, Lladoner Pelut, Mourvèdre, Piquepoul Noir, Terret Noir, Syrah, *Cinsaut, Maccabeu, Bourboulenc*; Bourboulenc (Malvoisie), Maccabeu, Grenache Blanc, *Clairette Blanche, Muscat Blanc à Petits Grains, Piquepoul Blanc, Terret Blanc, Marsanne, Roussanne, Vermentino Blanc*

Cornas Syrah

Corton Pinot Noir, *Pinot Gris (Pinot Beurot), Pinot Liébault, Pinot Blanc, Chardonnay*; Chardonnay

Corton-Charlemagne Chardonnay

Costières de Nîmes Carignan, Grenache Noir, Mourvèdre, Syrah, Cinsaut; Carignan, Grenache Noir, Mourvèdre, Syrah, Cinsaut *plus any of the following white varieties*; Clairette Blanche, Grenache Blanc, Bourboulenc Blanc, Ugni Blanc, Roussanne, Rolle, Maccabéo, Marsanne

Côte de Beaune, Côte de Beaune-Villages Pinot Noir, *Pinot Gris (Pinot Beurot), Pinot Liébault, Pinot Blanc, Chardonnay*; Chardonnay, *Pinot Blanc*

Côte de Brouilly Gamay, *Chardonnay, Aligoté, Melon de Bourgogne, Pinot Noir, Pinot Gris*

Côte de Nuits-Villages Pinot Noir, *Pinot Liébault, Chardonnay, Pinot Blanc, Pinot Gris*; Chardonnay, *Pinot Blanc*

Côte Roannaise Gamay

Côte-Rôtie Syrah, *Viognier*

Coteaux d'Aix-en-Provence Cinsaut, Counoise, Grenache Noir, Mourvèdre, Syrah, *Cabernet Sauvignon, Carignan, plus any of the following white varieties*; Bourboulenc, Clairette, Grenache Blanc, Vermentino Blanc, *Ugni Blanc, Sauvignon Blanc, Sémillon*

Coteaux de l'Aubance Chenin Blanc (Pineau de la Loire)

Coteaux Champenois Pinot Noir, Pinot Meunier, Chardonnay

Coteaux de Die Clairette Blanche

Coteaux du Giennois Gamay, Pinot Noir; Sauvignon Blanc

Coteaux du Languedoc Syrah, Mourvèdre, Grenache Noir, Cinsault, Carignan; Grenache Blanc, Clairette Blanche, Bourboulenc, Piquepoul, Marsanne, Roussanne, Rolle, *Maccabéo, Terret Blanc, Carignan Blanc, Ugni Blanc*

Coteaux du Layon Chenin Blanc (Pineau de la Loire)

Coteaux du Loir Chenin Blanc (Pineau de la Loire); Pineau d'Aunis, Cabernet Franc, Cabernet Sauvignon, Gamay, Cot; Pineau d'Aunis, Cabernet Franc, Cabernet Sauvignon, Gamay, Cot, *Groslot*

Coteaux du Lyonnais Gamay; Chardonnay, Aligoté

Coteaux de Pierrevert Grenache Noir, Syrah, *Carignan, Cinsaut, Mourvèdre*; Grenache Noir, Syrah, Carignan, Cinsaut *plus any of the white varieties*; Grenache Blanc, Vermentino, Ugni Blanc, Clairette, Roussanne, Marsanne, Piquepoul

Coteaux de Saumur Chenin Blanc (Pineau de la Loire)

Coteaux du Tricastin Grenache Noir, Cinsaut, Mourvèdre, Syrah, Picpoul Noir, *Carignan plus any of the following white varieties*; Grenache Blanc, Clairette Blanche, Picpoul Blanc, Bourboulenc, *Ugni Blanc, Marsanne, Roussanne, Viognier*

Coteaux Varois Grenache Noir, Syrah, Mourvèdre, *Carignan, Cinsaut, Cabernet Sauvignon*; Grenache Noir, Cinsaut, *Syrah, Mourvèdre, Carignan, Tibouren plus any of the following white varieties*; Clairette, Grenache Blanc, Rolle, *Sémillon, Ugni Blanc*

Coteaux du Vendomois Chenin Blanc, Chardonnay; Pineau d'Aunis, Gamay, Pinot Noir, Cabernet Franc, Cabernet Sauvignon; Pineau d'Aunis, Gamay

Côtes d'Auvergne Gamay, Pinot Noir; Chardonnay

Côtes de Blaye Colombard, *Sémillon, Sauvignon Blanc, Muscadelle, Merlot Blanc, Folle Blanche, Chenin Blanc (Pinot de la Loire)*

Côtes de Bordeaux-St-Macaire Sémillon, Sauvignon Blanc, Muscadelle

Côtes de Bourg, Bourg, Bourgeais Cabernet Franc (Bouchet), Cabernet Sauvignon, Merlot, Malbec (Pressac); Sauvignon Blanc, Sémillon, Muscadelle, *Merlot Blanc, Colombard*

Côtes de Castillon Cabernet Franc (Bouchet), Cabernet Sauvignon, Merlot, Malbec (Pressac)

Côtes de Duras Sauvignon Blanc, Sémillon, Muscadelle, Mauzac, Chenin Blanc, Ondenc, *Ugni Blanc*; Cabernet Sauvignon, Cabernet Franc, Merlot, Malbec (Cot)

Côtes du Forez Gamay

Côtes du Frontonnais Négrette, *Malbec (Cot), Mérille, Fer, Syrah, Cabernet Franc, Cabernet Sauvignon, Gamay, Cinsaut, Mauzac*

Côtes de Gien, Coteaux du Giennois Gamay, Pinot Noir; Chardonnay

Côtes du Jura Poulsard Noir (Ploussard), Trousseau, Pinot Noir (Gros Noirien, Pinot Gris); Savagnin Blanc (Naturé), Chardonnay (Melon d'Arbois, Gamay Blanc); **Mousseux** Poulsard Noir (Ploussard), Trousseau, Chardonnay (Melon d'Arbois, Gamay Blanc), Savagnin Blanc (Naturé)

Côtes du Jura Vin Jaune Savagnin Blanc

Côtes du Lubéron Grenache Noir, Syrah, Mourvèdre, *Carignan, Cinsaut, Counoise Noir, Picpoul Noir, Gamay, Pinot Noir*; Grenache Noir, Syrah, Mourvèdre, *Carignan, Cinsaut, Counoise Noir, Picpoul Noir, Gamay, Pinot Noir plus any of the following white varieties*; Grenache Blanc, Clairette Blanche, Bourboulenc, Ugni Blanc, Vermentino (Rolle), *Roussanne, Marsanne*

Côtes du Marmandais Cabernet Franc, Cabernet Sauvignon, Merlot, *Abouriou, Malbec (Cot), Fer, Gamay, Syrah*; Sauvignon Blanc, *Muscadelle, Ugni Blanc, Sémillon*

Côtes de Montravel Sémillon, Sauvignon Blanc, Muscadelle

Côtes de Nuits-Villages Pinot Noir, *Pinot Gris (Pinot Beurot), Pinot Liébault, Chardonnay, Pinot Blanc*; Chardonnay, Pinot Blanc

Côtes de Provence Cinsaut, Grenache Noir, Mourvèdre, Syrah, Tibouren, *Cabernet Sauvignon, Carignan plus any of the following white varieties*; Clairette, Sémillon, Ugni Blanc, Vermentino Blanc (Rolle)

Côtes du Rhône Grenache Noir, Syrah, Mourvèdre, Terret Noir, *Carignan, Cinsaut, Counoise, Muscardin, Camarèse, Vaccarèse, Picpoul Noir, Terret Noir, Grenache Gris, Clairette Rose plus any of the following white varieties*; Grenache Blanc, Clairette Blanche, Marsanne, Roussanne, Bourboulenc, Viognier, *Ugni Blanc, Picpoul Blanc*

Côtes du Rhône-Villages Grenache Noir, Syrah, Mourvèdre, Cinsaut *plus any of the other red varieties for Côtes du Rhône*; Grenache Noir, Camarèse, Cinsaut, Carignan *plus any of the other red varieties for Côtes du Rhône*; Clairette, Roussanne, Bourboulenc, *Grenache Blanc plus any of the other white varieties for Côtes du Rhône*

Côtes du Roussillon Carignan, Cinsaut, Grenache Noir, Lladoner Pelut Noir, Syrah, Mourvèdre, Maccabeu Blanc; Grenache Blanc, *Maccabeu Blanc*, Tourbat Blanc (Malvoisie du Roussillon), Marsanne, Roussanne, Vermentino

Côtes du Roussillon-Villages Carignan, Grenache Noir, Lladoner Pelut Noir, Syrah, Mourvèdre, *Maccabéo*

Côtes du Ventoux Grenache Noir, Syrah, Cinsaut, Mourvèdre, Carignan, *Picpoul Noir, Counoise, Clairette, Bourboulenc, Grenache Blanc, Roussanne*; Clairette, Bourboulenc, Grenache Blanc, *Roussanne*
Côtes du Vivarais Grenache Noir, Syrah, Cinsaut, Carignan; Grenache Noir, Cinsaut, Syrah Clairette Blanche, Grenache Blanc, Marsanne
Cour-Cheverny Romorantin Blanc
Crémant d'Alsace Pinot Blanc, Pinot Noir, Pinot Gris, Riesling, Muscat, Sylvaner, Chasselas; Pinot Noir
Crémant de Bordeaux Sémillon, Sauvignon Blanc, Muscadelle, Cabernet Sauvignon, Cabernet Franc, Merlot, Carmenère, Malbec, Petit Verdot, *Colombard, Ugni Blanc*; Cabernet Sauvignon, Cabernet Franc, Merlot, Carmenère, Malbec, Petit Verdot
Crémant de Bourgogne Pinot Noir, Pinot Gris, Pinot Blanc, Chardonnay, *Gamay, Aligoté, Melon, Sacy*
Crémant de Die Clairette Blanche
Crémant du Jura Poulsard (Ploussard), Pinot Noir (Gros Noirien), Pinot Gris, Trousseau, Savagnin (Naturé), Chardonnay (Melon d'Arbois, Gamay Blanc)
Crémant de Limoux Mauzac, Chardonnay, Chenin Blanc
Crémant de Loire Chenin Blanc, Cabernet Franc, Cabernet Sauvignon, Pineau d'Aunis, Pinot Noir, Chardonnay, Menu Pineau, *Grolleau Noir, Grolleau Gris*
Crépy Chasselas Rouge, Chasselas Vert
Crozes-Hermitage Syrah, *Marsanne, Roussanne*; Marsanne, Roussanne
Echézeaux Pinot Noir, Pinot Gris *(Pinot Beurot), Pinot Liébault, Pinot Blanc, Chardonnay*
Entre-Deux-Mers, Entre-Deux-Mers Haut-Benauge Sémillon, Sauvignon Blanc, Muscadelle, *Merlot Blanc, Colombard, Mauzac, Ugni Blanc*
L'Étoile Chardonnay, Poulsard (Ploussard), Savagnin; **Vin Jaune** Savagnin
Faugères Carignan, Cinsaut, Grenache Noir, Lladoner Pelut Noir, *Mourvèdre, Syrah*
Fitou Carignan, Grenache Noir, Lladoner Pelut Noir, Mourvèdre, Syrah, *Cinsaut, Maccabéo Blanc, Terret Noir*
Fixin Pinot Noir, Pinot Gris *(Pinot Beurot), Pinot Liébault, Pinot Blanc, Chardonnay*; Chardonnay (Beaunois, Aubaine), Pinot Blanc
Fleurie Gamay, *Chardonnay, Aligoté, Melon de Bourgogne*
Fronsac Merlot, Cabernet Franc (Bouchet), Cabernet Sauvignon, Malbec (Pressac)
Gaillac Duras, Fer Servadou, Syrah, *Cabernet Sauvignon, Cabernet Franc, Merlot, Gamay*; **Gaillac, Gaillac Premières Côtes** Len de l'El, Mauzac, Mauzac Rosé, Muscadelle, Ondenc, Sauvignon Blanc, Sémillon
Gevrey-Chambertin Pinot Noir, *Pinot Gris (Pinot Beurot), Pinot Liébault, Pinot Blanc, Chardonnay*
Gigondas Grenache Noir, *Syrah, Mourvèdre, plus any of the varieties for Côtes du Rhône, except Carignan*; Grenache Noir *plus any of the varieties for Côtes du Rhône, except Carignan*
Givry Pinot Noir, *Pinot Gris (Pinot Beurot), Pinot Liébault, Chardonnay*; Chardonnay, *Pinot Blanc*
La Grande Rue Pinot Noir, *Pinot Blanc, Pinot Gris, Chardonnay*
Grands Echézeaux Pinot Noir, *Pinot Gris (Pinot Beurot), Pinot Liébault, Pinot Blanc, Chardonnay*
Graves Cabernet Sauvignon, Cabernet Franc, Merlot, *Malbec, Petit Verdot*; **Graves, Graves Supérieures** Sauvignon Blanc, Sémillon, Muscadelle; **Graves de Vayres** Cabernet Sauvignon, Cabernet Franc, Merlot, *Carmenère, Malbec, Petit Verdot*; Sémillon, Sauvignon Blanc, Muscadelle, *Merlot Blanc*
Haut-Médoc Cabernet Sauvignon, Cabernet Franc, Merlot, *Carmenère, Malbec, Petit Verdot*
Haut-Montravel Sémillon, Sauvignon Blanc, Muscadelle
Hermitage Syrah, *Marsanne, Roussanne*; Marsanne, Roussanne
Irancy Pinot Noir, César
Irouléguy Cabernet Sauvignon, Cabernet Franc, Tannat; Courbu, Manseng
Jasnières Chenin Blanc (Pineau de la Loire)
Juliénas Gamay, *Chardonnay, Aligoté, Melon de Bourgogne*
Jurançon, Jurançon Sec Petit Manseng, Gros Manseng, Courbu, *Camaralet, Lauzet*
Ladoix Pinot Noir, *Pinot Gris (Pinot Beurot), Pinot Liébault, Pinot Blanc, Chardonnay*; Chardonnay, *Pinot Blanc*
Lalande-de-Pomerol Merlot, Cabernet Franc (Bouchet), Cabernet Sauvignon, *Malbec (Pressac)*
Limoux Mauzac, Chardonnay, Chenin Blanc
Lirac Grenache Noir, Syrah, Mourvèdre, Syrah, Cinsaut, *Carignan*; Grenache Noir, Syrah, Mourvèdre, Syrah, Cinsaut, *Carignan plus any of the following white varieties*; Clairette Blanche, Grenache Blanc, Bourboulenc, *Ugni Blanc, Picpoul, Marsanne, Roussanne, Viognier*
Listrac-Médoc Cabernet Sauvignon, Cabernet Franc, Merlot, *Carmenère, Malbec, Petit Verdot*
Loupiac Sémillon, Sauvignon Blanc, Muscadelle
Lussac-St-Émilion Cabernet Franc (Bouchet), Cabernet Sauvignon, Merlot, *Malbec (Pressac)*
Mâcon, Mâcon Supérieur, Mâcon-Villages or Mâcon plus commune name Chardonnay, Pinot Blanc; Gamay, Pinot Noir, Pinot Gris, *Chardonnay, Aligoté, Gamay Blanc (Melon)*
Madiran Tannat, Cabernet Sauvignon, Cabernet Franc (Bouchy), Fer (Pinenc)
Maranges Pinot Noir, Pinot Gris *(Pinot Beurot), Pinot Liébault, Pinot Blanc, Chardonnay*; Chardonnay
Marcillac Fer Servadou, *Cabernet Sauvignon, Cabernet Franc, Merlot*
Margaux Cabernet Sauvignon, Cabernet Franc, Merlot, *Carmenère, Malbec, Petit Verdot*
Marsannay Pinot Noir, *Pinot Gris (Beurot), Chardonnay, Pinot Blanc*; Chardonnay, *Pinot Blanc*
Maury VDN Grenache Noir, *Grenache Gris, Grenache Blanc, Maccabeu, Carignan Noir, Syrah*; Grenache Blanc, Grenache Gris, Maccabeu, Tourbat (Malvoisie du Roussillon), *Muscat of Alexandria, Muscat Blanc à Petits Grains*
Médoc Cabernet Sauvignon, Cabernet Franc, Merlot, *Carmenère, Petit Verdot, Malbec*
Menetou Salon Sauvignon Blanc; Pinot Noir
Mercurey Pinot Noir, *Pinot Gris (Pinot Beurot), Pinot Liébault, Chardonnay*; Chardonnay
Meursault Pinot Noir, *Pinot Gris (Pinot Beurot), Pinot Liébault, Pinot Blanc, Chardonnay*; Chardonnay, Pinot Blanc
Minervois Grenache Noir, Syrah, Mourvèdre, Lladoner Pelut Noir, Carignan, *Cinsaut, Picpoul Noir, Terret Noir, Aspiran Noir*; Grenache Noir, Syrah, Mourvèdre, Lladoner Pelut Noir, Carignan, *Cinsaut, Picpoul Noir, Terret Noir, Aspiran Noir plus any of the following white varieties*; Grenache Blanc, Bourboulenc (Malvoisie), Maccabeu Blanc, Marsanne, Roussanne, Vermentino (Rolle), *Picpoul Blanc, Clairette Blanche, Terret Blanc, Muscat Blancs à Petits Grains*
Monbazillac Sémillon, Sauvignon Blanc, Muscadelle
Montagne-St-Émilion Cabernet Franc (Bouchet), Cabernet Sauvignon, Merlot, *Malbec (Pressac)*
Montagny Chardonnay (Beaunois, Aubaine)
Monthélie Pinot Noir, *Pinot Gris (Pinot Beurot), Pinot Liébault, Pinot Blanc, Chardonnay*; Chardonnay, *Pinot Blanc*
Montlouis Chenin Blanc (Pineau de la Loire)
Montrachet, Bâtard-Montrachet, Bienvenues-Bâtard-Montrachet, Chevalier-Montrachet, Criots-Bâtard-Montrachet Chardonnay (Beaunois, Aubaine)
Montravel Sémillon, Sauvignon Blanc, Muscadelle, Ondenc, Chenin Blanc, *Ugni Blanc*
Morey St-Denis Pinot Noir, *Pinot Gris (Pinot Beurot), Pinot Liébault, Pinot Blanc, Chardonnay*; Chardonnay (Beaunois, Aubaine), *Pinot Blanc*
Morgon Gamay, *Chardonnay, Aligoté, Melon de Bourgogne*
Moulin-à-Vent Gamay, *Chardonnay, Aligoté, Melon de Bourgogne*
Moulis, Moulis-en-Médoc RCabernet Sauvignon, Cabernet Franc, Merlot, *Carmenère, Malbec, Petit Verdot*
Muscadet, Muscadet-Côtes de Grandlieu, Muscadet-Coteaux de la Loire, Muscadet-Sèvre et Maine Melon de Bourgogne
Musigny Pinot Noir, *Pinot Gris (Pinot Beurot), Pinot Liébault, Pinot Blanc, Chardonnay*; Chardonnay
Nuits-St-Georges Pinot Noir, *Pinot Gris (Pinot Beurot), Pinot Liébault, Pinot Blanc, Chardonnay*; Chardonnay, Pinot Blanc
Pacherenc du Vic Bilh Gros Manseng, Arrufiac, Courbu, Petit Manseng, *Sauvignon Blanc, Sémillon*
Palette Mourvèdre, Grenache Noir, Cinsaut (Plant d'Arles), *Manosquin (Téoulier), Durif, Muscat Noir, Carignan, Syrah, Castets, Brun-Fourcat, Terret Gris, Petit-Brun, Tibourenc, Cabernet Sauvignon plus any of the following white varieties*; Clairette (various forms), Ugni Blanc,

Ugni Rosé, Grenache Blanc, Muscat, Pascal, Terret-Bourret, Piquepoul, Aragnan, Colombard, Tokay
Patrimonio Nielluccio, *Grenache Noir, Sciacarello, Vermentino*; Vermentino, *Ugni Blanc*
Pauillac Cabernet Sauvignon, Cabernet Franc, Merlot, *Carmenère, Malbec, Petit Verdot*
Pécharmant Cabernet Sauvignon, Cabernet Franc, Merlot, Malbec (Cot)
Pernand-Vergelesses Pinot Noir, *Pinot Gris (Pinot Beurot), Pinot Liébault, Pinot Blanc, Chardonnay*; Chardonnay, *Pinot Blanc*
Pessac-Léognan Cabernet Sauvignon, Cabernet Franc, Merlot, *Carmenère, Malbec (Cot), Petit Verdot*; Sauvignon Blanc, Sémillon, Muscadelle
Petit Chablis Chardonnay (Beaunois)
Pomerol Merlot, Cabernet Franc (Bouchet), Cabernet Sauvignon, *Malbec (Pressac)*
Pommard Pinot Noir, Pinot Gris *(Pinot Beurot) Pinot Liébault, Chardonnay, Pinot Blanc*
Pouilly-Fuissé, Pouilly-Vinzelles, Pouilly-Loché Chardonnay (Beaunois)
Pouilly-Fumé, Blanc Fumé de Pouilly Sauvignon Blanc (Blanc Fumé)
Pouilly-sur-Loire Chasselas, Sauvignon Blanc (Blanc Fumé)
Premières Côtes de Blaye Cabernet Sauvignon, Cabernet Franc, Merlot, *Malbec (Cot)*; Sémillon, Sauvignon Blanc, Muscadelle, *Merlot Blanc, Colombard, Ugni Blanc*
Premières Côtes de Bordeaux Cabernet Sauvignon, Cabernet Franc, Merlot, *Carmenère, Malbec (Cot), Petit Verdot*; Sauvignon Blanc, Sémillon, Muscadelle
Puisseguin-St-Émilion Cabernet Franc (Bouchet), Cabernet Sauvignon, Merlot, *Malbec (Pressac)*
Puligny-Montrachet Pinot Noir, *Pinot Gris (Pinot Beurot), Pinot Liébault, Pinot Blanc, Chardonnay*; Chardonnay, *Pinot Blanc*
Quarts de Chaume Chenin Blanc (Pineau de la Loire)
Quincy Sauvignon Blanc
Rasteau VDN Grenache Noir, Grenache Gris, Grenache Blanc
Régnié Gamay
Reuilly Sauvignon Blanc; Pinot Noir, Pinot Gris
Richebourg Pinot Noir, *Pinot Gris (Pinot Beurot), Pinot Liébault, Pinot Blanc, Chardonnay*
Rivesaltes VDN Grenache Noir, Grenache Gris, Grenache Blanc, Maccabeo, Tourbat (Malvoisie du Roussillon), Carignan, Cinsaut, Syrah, Listan; Grenache Gris, Grenache Blanc, Maccabéo, Tourbat (Malvoisie du Roussillon), *Muscat Blanc à Petits Grains, Muscat of Alexandria (Muscat Romain)*
LaRomanée, Romanée-Conti, Romanée-St-Vivant Pinot Noir, *Pinot Gris (Pinot Beurot) Pinot Liébault, Chardonnay, Pinot Blanc*
Rosé d'Anjou Cabernet Franc, Cabernet Sauvignon, Pineau d'Aunis, Gamay, Cot, Groslot
Rosé de Loire Cabernet Franc, Cabernet Sauvignon, Pineau d'Aunis, Pinot Noir, Gamay, Grolleau
Rosé des Riceys Pinot Noir
Rosette Sémillon, Sauvignon Blanc, Muscadelle
Roussette de Savoie Altesse
Rully Pinot Noir, *Pinot Gris (Pinot Beurot), Pinot Liébault, Chardonnay*; Chardonnay
St-Amour Gamay, *Chardonnay, Aligoté, Melon de Bourgogne*
St-Aubin Pinot Noir, *Pinot Gris (Pinot Beurot), Pinot Liébault, Pinot Blanc*, Chardonnay; *Chardonnay, Pinot Blanc*
St-Chinian Grenache Noir, Lladoner Pelut Noir, Mourvèdre, Syrah, Carignan, Cinsaut
St-Émilion, St-Émilion Grand Cru Merlot, Cabernet Franc (Bouchet), Cabernet Sauvignon, *Carmenère, Malbec (Pressac, Cot)*
St-Estèphe Cabernet Sauvignon, Cabernet Franc, Merlot, *Carmenère, Malbec, Petit Verdot*
St-Georges-St-Émilion Cabernet Franc (Bouchet), Cabernet Sauvignon, Merlot, *Malbec (Pressac)*
St-Joseph Syrah, *Marsanne, Roussanne*; Marsanne, Roussanne
St-Julien Cabernet Sauvignon, Cabernet Franc, Merlot, *Carmenère, Malbec, Petit Verdot*
St Nicolas-de-Bourgueil Cabernet Franc (Breton), Cabernet Sauvignon
St-Péray Roussanne (Roussette), Marsanne
St-Pourçain Gamay, Pinot Noir, Gamay Teinturiers; Tressallier, St-Pierre-Doré, Aligoté, Chardonnay, Sauvignon Blanc
St-Romain Pinot Noir, Pinot Gris *(Pinot Beurot), Pinot Liébault, Pinot Blanc, Chardonnay*; Chardonnay, *Pinot Blanc*
St-Véran Chardonnay
Ste-Croix-du-Mont Sémillon, Sauvignon Blanc, Muscadelle
Ste-Foy-Bordeaux Cabernet Sauvignon, Cabernet Franc, Merlot, *Malbec, Petit Verdot*; Sémillon, Sauvignon Blanc, Muscadelle, *Merlot Blanc, Colombard, Mauzac, Ugni Blanc*
Sancerre Sauvignon Blanc; Pinot Noir
Santenay Pinot Noir, *Pinot Gris (Pinot Beurot), Pinot Liébault, Pinot Blanc, Chardonnay*; Chardonnay, *Pinot Blanc*
Saumur Chenin Blanc, Chardonnay, Sauvignon Blanc; Cabernet Franc, Cabernet Sauvignon, Pineau d'Aunis;
Saumur-Champigny Cabernet Franc, Cabernet Sauvignon, Pineau d'Aunis; Chenin Blanc, Cabernet Franc, Cabernet Sauvignon, Cot, Gamay, Grolleau, Pineau d'Aunis, Pinot Noir, *Chardonnay, Sauvignon Blanc*; Cabernet Franc, Cabernet Sauvignon, Cot, Gamay, Grolleau, Pineau d'Aunis, Pinot Noir;
Saussignac Sémillon, Sauvignon Blanc, Muscadelle, Chenin Blanc
Sauternes Sémillon, Sauvignon Blanc, Muscadelle
Savennières Chenin Blanc (Pineau de la Loire)
Savigny-lès-Beaune Pinot Noir, *Pinot Gris (Pinot Beurot), Pinot Liébault, Pinot Blanc, Chardonnay*; Chardonnay, *Pinot Blanc*
Seyssel Roussette; Molette, Chasselas (Bon Blanc, Fendant), Roussette
La Tâche Pinot Noir, *Pinot Gris (Pinot Beurot), Pinot Liébault, Pinot Blanc, Chardonnay*
Tavel Grenache Noir, Cinsaut, Clairette Blanche, Clairette Rose, Picpoul, Calitor, Bourboulenc, Mourvèdre, Syrah, *Carignan*
Touraine Chenin Blanc (Pineau de la Loire), Arbois (Menu Pineau), Sauvignon Blanc, *Chardonnay*; Cabernet Franc (Breton), Cabernet Sauvignon, Cot, Pinot Noir, Pinot Meunier, Pinot Gris, Gamay, Pineau d'Aunis; Cabernet Franc (Breton), Cabernet Sauvignon, Cot, Pinot Noir, Pinot Meunier, Pinot Gris, Gamay, Pineau d'Aunis, Grolleau; Chenin Blanc (Pineau de la Loire), Arbois (Menu Pineau), *Chardonnay, Cabernet Franc (Breton), Cabernet Sauvignon, Pinot Noir, Pinot Gris, Pinot Meunier, Pineau d'Aunis, Cot, Grolleau*; Cabernet Franc (Breton), Cot, Noble, Gamay, Grolleau; Cabernet Franc (Breton)
Vacqueyras Grenache Noir, Syrah, Mourvèdre *plus any of the varieties for Côtes du Rhône, except Carignan*; Grenache Noir, Mourvèdre, Cinsaut *plus any of the varieties for Côtes du Rhône, except Carignan*; Grenache Blanc, Clairette Blanche, Bourboulenc, *Marsanne, Roussanne, Viognier*
Vin de Corse Nielluccio, Sciacarello, Grenache Noir, *Cinsaut, Mourvèdre, Barbarossa, Syrah, Carignan, Vermentino (Malvoisie de Corse)*; Vermentino (Malvoisie de Corse), *Ugni Blanc (Rossola)*
Vin de Savoie Gamay, Mondeuse, Pinot Noir, *Persan, Cabernet Sauvignon, Cabernet Franc, Étraire de la Dui, Servanin, Joubertin*; Aligoté, Altesse, Jacquère, Chardonnay, Velteliner Rouge Précoce, Mondeuse Blanche, Gringet, *Roussette d'Ayze, Chasselas, Marsanne, Verdesse*; the above white grapes plus Gamay, Pinot Noir, Mondeuse, Molette
Viré-Clessé Chardonnay
Volnay, Volnay-Santenots Pinot Noir, *Pinot Gris (Pinot Beurot), Pinot Liébault, Pinot Blanc, Chardonnay*
Vosne-Romanée Pinot Noir, *Pinot Gris (Pinot Beurot), Pinot Liébault, Pinot Blanc, Chardonnay*
Vougeot Pinot Noir, *Pinot Gris (Pinot Beurot), Pinot Liébault, Pinot Blanc, Chardonnay*; Chardonnay (Beaunois. Aubaine), *Pinot Blanc*
Vouvray Chenin Blanc (Pineau de la Loire, Gros Pinot), Petit Pinot (Menu Pinot)

GERMANY

Liebfraumilch Müller-Thurgau, Silvaner, Kerner, and/or most unusually Riesling must constitute 51 per cent.

GREECE

Amindeo Xynomavro
Anhialos Roditis, *Savatiano*
Archanes Kotsifali, Mandelaria
Côtes de Meliton Limnio, Cabernet Sauvignon, Cabernet Franc; Athiri, Roditis, Assyrtico, *Sauvignon Blanc, Ugni Blanc*
Dafnes Liatiko
Goumenissa Xynomavro, *Negoska*
Limnos Muscat of Alexandria
Mantinia Moscophilero, Asproudos
Mavrodaphne of Cephalonia Mavrodaphne
Mavrodaphne of Patras Mavrodaphne
Messenikola Messenikola, Carignan, Syrah
Muscat of Cephalonia Muscat Blanc à Petits Grains
Muscat of Limnos Muscat of Alexandria
Muscat of Patros Muscat Blanc à Petits Grains

Muscat of Rhodes Muscat Blanc à Petits Grains, Muscat Trani
Muscat of Rio of Patras Muscat Blanc à Petits Grains
Naoussa Xynomavro
Nemea Aghiorghitiko
Paros Monemvassia (Malvaria), *Mandelaria*
Patras Roditis
Peza Kotsifali, Mandelaria; Vilana
Rapsani Xynomavro, Krassato, Stavroto
Rhodes Mandelaria; Athiri
Samos Muscat Blanc à Petits Grains
Santorini Assyrtiko, *Athiri, Aidini*
Sitia Liatiko
Zitsa Debina

HUNGARY

Bulls Blood of Eger, Egri Bikaver Kékfrankos, Kadarka, Blauburger, Portugieser, Zweigelt, Kékmedoc, Cabernet Sauvignon, Merlot, Cabernet Franc, Pinot Noir
Tokaji Furmint, Hárslevelü, Muscat Blanc à Petits Grains, Zéta, Kovérszolo

ITALY

Alcamo Catarratto, Ansonica, Inzolia, Grillo, Grecanino, Chardonnay, Müller-Thurgau, Sauvignon Blanc *and others*;
Alcamo Classico Catarratto *and others*; Nerello Mascalese, Calabrese, Nero d'Avola, Sangiovese, Frappato, Perricone, Cabernet Sauvignon, Merlot, Syrah; Calabrese, Nero d'Avola, Frappato, Sangiovese, Perricone, Cabernet Sauvignon, Merlot, Syrah *and others*
Aleatico di Puglia Aleatico, *Negroamaro, Malvasia Nera, Primitivo*
Alto Adige Pinot Bianco, Pinot Nero, Chardonnay
Asti Moscato Bianco
Barbaresco Nebbiolo
Barbera d'Alba Barbera; **Barbera d'Asti, Barbera del Monferrato** R Barbera, *Freisa, Grignolino, Dolcetto*
Bardolino Corvina, Rondinella, Molinara, Negrara, *Rossignola, Barbera, Sangiovese, Garganega*
Barolo Nebbiolo
Bianco di Custoza Trebbiano Toscano (Castelli Romani), Garganega, Tocai Friulano, Cortese (Bianca Fernanda), Malvasia Toscana, Riesling Italico, Pinot Bianco, Chardonnay
Bolgheri Trebbiano Toscano, Vermentino, Sauvignon Blanc *and others*; Cabernet Sauvignon, Merlot, Sangiovese and others;
Bolgheri Sassicaia Cabernet Sauvignon, *Cabernet Franc and others*
Breganze Tocai Friulano, *Pinot Bianco, Pinot Grigio, Riesling Italico, Sauvignon Blanc, Vespaiolo*; Merlot, *Groppello Gentile, Cabernet Franc, Cabernet Sauvignon, Pinot Nero, Freisa*
Brindisi Negroamaro, *Susumaniello, Malvasia Nera, Sangiovese, Montepulciano*
Brunello di Montalcino Sangiovese
Caldaro (Kalterer)/Lago di Caldaro (Kalterersee) Schiava, *Pinot Nero, Lagrein*
Cannonau di Sardegna Garnacha Tinta (Cannonau)
Carema Nebbiolo *and others*
Carmignano, Barco Reale di Carmignano Sangiovese, Canaiolo Nero, Cabernet Franc, Cabernet Sauvignon, *Trebbiano Toscano, Canaiolo Bianco, Malvasia del Chianti and others*
Carso Terrano, *Piccolo Nero, Pinot Noir*
Castel del Monte Pampanuto, Chardonnay, Bombino Bianco *and others*; Uva di Troia, Aglianico, Montepulciano *and others*; Bombino Nero, Aglianico, Uva di Troia *and others*
Chianti, Chianti Classico Sangiovese, *Canaiolo Nero, Trebbiano Toscano, Malvasia del Chianti, Cabernet Sauvignon, Merlot and others*
Cirò Gaglioppo, *Trebbiano Toscano, Greco Bianco*; Greco Bianco, *Trebbiano Toscano*
Colli Berici mostly varietal wines; Pinot Bianco, Pinot Grigio, Chardonnay, Sauvignon Blanc
Colli Bolognesi mostly varietal wines;
Colli Bolognesi Bianco Albana, Trebbiano Romagnolo
Cinqueterre Bosco, Albarola, Vermentino
Colli di Luna Sangiovese, Canaiolo, Pollera Nera, Ciliegiolo Nero; Vermentino, Trebbiano Toscano
Colli Euganei Garganega, Prosecco, Tocai Friulano, Sauvignon Blanc, *Pinella, Pinot Bianco, Riesling Italico, Chardonnay*; Merlot, Cabernet Franc, Cabernet Sauvignon, Barbera, Raboso Veronese
Colli Piacentini Gutturnio Barbera, Croatina (Bonarda)
Donnici Gaglioppo, *Greco Nero, Malvasia Bianco, Pecorello, Greco Bianco*; Montonico Bianco, Greco Bianco, Malvasia Blanca, Pecorello
Est! Est!! Est!!! di Montefiascone Trebbiano Toscano (Procanico), Malvasia Bianca Toscana, *Rossetto (Trebbiano Giallo)*
Falerno del Massico Falanghina; Aglianico, Piedirosso, *Primitivo, Barbera*
Fiano di Avellino Fiano, *Greco, Coda di Volpe Bianca, Trebbiano Toscano*
Franciacorta Pinot Bianco, Chardonnay, Pinot Nero
Frascati Malvasia Bianca di Candia, Trebbiano Toscano, Greco, Malvasia di Lazio *and others*
Garda Riesling Italico, Riesling Renano; Groppello, Sangiovese, Marzemino, Barbera
Gattinara Nebbiolo (Spanna), *Vespolina, Bonarda di Novarese (Uva Rara)*
Gavi, Cortese di Gavi Cortese
Ghemme Nebbiolo (Spanna), Vespolino, Bonarda Novarese (Uva Rara)
Greco di Tufo Greco, *Coda di Volpe Bianca*
Ischia Forastera Bianco, Biancolella; Guarnaccia, Piedirosso
Lambrusco di Sorbara Lambrusco di Sorbara, Lambrusco Salamino
Locorontondo Verdeca, Bianco d'Alessano, **Fiano, Bombino, Malvasia Toscano**
Lugana Trebbiano di Soave (Trebbiano di Lugana) *and others*
Malvasia delle Lipari Malvasia di Lipari, Corinto Nero
Marino Malvasia Bianca di Candia (Malvasia Rossa), Trebbiano Toscano, Trebbiano Romagnolo, Trebbiano Giallo, Trebbiano di Soave, Malvasia del Lazio (Malvasia Puntinata) *and others*
Marsala Grillo, Catarratto, Pignatello, Calabrese, Nerello Mascalese, Inzolia, Nero d'Avola, Damaschino
Montecarlo Trebbiano Toscano, Sémillon, Pinot Grigio, Pinot Bianco, Vermentino, Sauvignon Blanc, Roussanne; Sangiovese, Canaiolo Nero, Ciliegiolo, Colorino, Malvasia Nera, Syrah, Cabernet Franc, Cabernet Sauvignon, Merlot *and others*
Montefalco Sangiovese, Sagrantino *and others*; Grechetto, Trebbiano Toscano *and others*
Montepulciano d'Abruzzo Montepulciano *and others*
Montescudaio Trebbiano Toscano, Malvasia del Chianti, Vermentino *and others*; Sangiovese, Trebbiano Toscano, Malvasia del Chianti *and others*
Morellino di Scansano Sangiovese, *Canaiolo, Malvasia Nera*
Moscato d'Asti Moscato Bianco
Moscato di Pantelleria Muscat of Alexandria (Zibibbo)
Oltrepò Pavese Barbera, Croatina, Uva Rara, Ughetta (Vespolina), Pinot Nero; Pinot Nero,Chardonnay, Pinot Grigio, Pinot Bianco
Orvieto Trebbiano Toscano (Procanico), Verdello, Grechetto, Canaiolo Bianco (Drupeggio), Malvasia Toscana
Piemonte varietal wines; Chardonnay, Pinot Bianco, Pinot Grigio, Pinot Nero
Pomino Sangiovese, Canaiolo, Cabernet Sauvignon, Cabernet Franc, Merlot *and others*; Pinot Bianco, Chardonnay, Trebbiano Toscano *and others*
Prosecco di Conegliano-Valdobbiadene/ Prosecco di Conegliano/Prosecco de Valdobbiadene Prosecco, *Verdiso*
Ramandolo Verduzzo Friulano
Riviera del Garda Gropello, Sangiovese, Barbera, Berzamino (Marzemino)
Riviera del Garda Bresciano Groppello, Sangiovese, Barbera, Marzemino (Berzemino), Barbera *and others*; Riesling Italico, Riesling Renano and others
Riviera Ligure di Ponente Ormeasco Dolcetto *and others*
Roero Nebbiolo, *Arneis and others*; Arneis
Rosso Conero Montepulciano, Sangiovese
Rosso di Montalcino Sangiovese
Rosso di Montepulciano R Sangiovese (Prugnolo Gentile), *Canaiolo Nero, Mammolo, Cabernet Sauvignon, Merlot, Syrah*
Rosso Piceno Montepulciano, Sangiovese *and others*
Salice Salentino Negroamaro, *Malvasia Nera*; *Chardonnay*
San Gimignano Sangiovese *and others*; Sangiovese, Canaiolo Nero, Trebbiano Toscano, Malvasia del Chianti, Vernaccia di San Gimignano and others
Sangiovese di Romagna Sangiovese
Savuto Gaglioppo, Greco *Nero, Nerello Cappuccio, Magliocco Canino, Sangiovese, Malvasia Bianca, Pecorino*
Scavigna Trebbiano Toscano, Chardonnay, Greco Bianco, *Malvasia* Gaglioppo,Nerello Cappuccio, Aglianico
Soave Garganega, Pinot Bianco, Chardonnay, Trebbiano di Soave (Nostrano) *and other Trebbiano varieties*
Taurasi Aglianico *and others*

Teroldego Rotaliano Teroldego
Terre di Franciacorta Chardonnay, Pinot Bianco, Pinot Nero; Cabernet Sauvignon, Cabernet Franc, *Barbera, Nebbiolo, Merlot and others*
Torcolato Vespaiolo
Torgiano Trebbiano Toscano, Grechetto *and others*; Sangiovese, Canaiolo, Trebbiano Toscano *and others*; Chardonnay, Pinot Nero; **Torgiano Rosso Riserva** Sangiovese, Canaiolo, *Trebbiano Toscano, Ciliegiolo, Montepulciano*
Trebbiano d'Abruzzo Trebbiano d'Abruzzo (Bombino Bianco), Trebbiano Toscano *and others*
Trebbiano di Romagna Trebbiano Romagnolo *and others*
Trentino mostly varietal wines; Chardonnay, Pinot Bianco, Sauvignon Blanc, Müller-Thurgau, Incrocio Manzoni; Cabernet Franc, Cabernet Sauvignon, Merlot; Enantio, Schiava, Teroldego, Lagrein
Trento Chardonnay, Pinot Bianco, Pinot Nero, Pinot Meunier
Val di Cornia Trebbiano Toscano, Vermentino *and others*; Sangiovese, Cabernet Sauvignon, Merlot *and others*
Valdadige Pinot Bianco, Pinot Grigio, Riesling Italico, Müller-Thurgau, Chardonnay, *Bianchetta Trevigiana, Trebbiano Toscano, Nosiola, Vernaccia, Garganega*; Schiava, Lambrusco, *Merlot, Pinot Nero, Lagrein, Teroldego, Negrara*
Valle d'Aosta Chambave Petit Rouge, Dolcetto, Gamay, Pinot Nero *and others*; **Valle d'Aosta Donnas** Nebbiolo (Picoutener), *Freisa, Neyret*
Valpolicella, Amarone, Recioto Corvina Veronese, Corvinone, Rondinella, Molinara, *Croatina, Dindarella, Oseletta, Rossignola, Negrara Trentina, Terodola, Cabernet Sauvignon, Cabernet Franc, Merlot*
Valtellina Nebbiolo (Chiavennasca), *Pinot Nero, Merlot, Rossola, Pignola Valtellinese*
Vesuvio Coda di Volpe, Verdeca, Falanghina, Greco; Piedirosso, Sciascinoso, *Aglianico*
Vino Nobile di Montepulciano Sangiovese (Prugnolo Gentile), *Canaiolo Nero, Mammolo, Cabernet Sauvignon, Merlot, Syrah*
Vin Santo Occhio di Pernice Sangiovese

PORTUGAL

Alenquer Camarate, Trincadeira, Periquita, Preto Martinho, Tinta Miúda; Vital, Jampal, Arinto, Fernão Pires
Bairrada Baga, Camarate, Castelão Nacional, Bastardo, Jaén, Alfrocheiro Preto, Touriga Nacional, Trincadeira Preta, Rufete; Arinto, Bical, Cercial Branco, Rabo de Ovelha, Fernão Pires
Bucelas Arinto, Esgana Cão, Rabo de Ovelha
Colares Ramisco; Malvasia
Dão Alfrocheiro Preto, Bastardo, Jaén, Tinta Amarela, Tinta Roriz, Touriga Nacional, Cabernet Sauvignon; Encruzado, Assario Branco, Borrado das Moscas, Cerceal, Verdelho, Arinto, Rabo de Ovelha, Tamarez, Malvasia Fina, Rabo de Ovelha, Terrantez, Uva Cão
Douro Touriga Nacional, Touriga Franca, Tinta Roriz, Tinto Cão, Tinta Barroca, *Tinta Amarela, Mourisco Tinto, Bastardo, Periquita, Rufete, Tinta da Barca, Tinta Francisca*; Couveio, Viosinho, Rabigato, Malvasia Fina, Donzelinho, Esgaña Cão, Folgazao
Madeira varietals from Sercial, Verdelho, Boal or Malvasia; *minor grapes include Tinta Negra Mole, Bastardo, Malvasia Roxa, Verdelho Tinto, Terrantez*
Port Touriga Franca, Touriga Nacional, Tinta Roriz, Tinto Cão, Tinta Barroca, *Bastardo, Mourisco Tinto, Tinta Amarela, Cornifesto, Donzelinho, Malvasia, Periquita, Rufete, Tinta de Barca*; Gouveio Verdelho, Malvasia Fina, Rabigato Rabo di Ovelha, Viosinho, Donzelinho, Códega, Arinto, Boal, Cerceal, Esgaña Cão, Folgasão, Moscatel de Bago Miúdo, Samarrinho, Vital
Setúbal Moscatel de Setúbal (Muscat of Alexandria); Moscatel Roxo
Vinho Verde Vinhão Sousão, Espadeiro, Padeiro Basto, Rabo de Ovelha, Borraçal, Brancelho, Pedral; Loureiro, Trjadura, Padernã, Azal, Batoca, Alvarinho

ROMANIA

Cotnari Grasa, Tamîioasa, Francusa, Feteasca˘ Alba

SPAIN

Alella Garnacha Tinta, Tempranillo (Ull de Llebre), *Merlot, Cabernet Sauvignon, Pansá Rosada, Garnacha Peluda, Garnacha Negra, Pinot Noir*; Xarel-lo (Pansá Blanca), Garnacha Blanca, *Chardonnay, Macabeo, Parellada, Picapoll, Chenin Blanc, Sauvignon Blanc*
Bierzo Mencía, *Garnacha Tintorera*; Godello, Doña Blanca, *Palomino (Jerez), Malvasía*
Calatayud Garnacha Tinta, Tempranillo, Carineña (Mazuelo), *Monastrell, Cabernet Sauvignon, Merlot, Syrah*; Macabeo (Viura), Malvasía, *Garnacha Blanca, Muscat of Alexandria (Moscatel Blanco/Moscatel Romano), Chardonnay*
Campo de Borja Garnacha Tinta, Tempranillo, *Macabeo, Carineña (Mazuelo), Cabernet Sauvignon, Merlot, Syrah*; Macabeo, *Muscat of Alexandria (Moscatel Romano)*
Carineña Garnacha Tinta, Tempranillo, Carineña (Mazuela), *Cabernet Sauvignon, Monastrell, Merlot, Moristel (Juan Ibáñez)*; Macabeo (Viura), *Garnacha Blanca, Muscat of Alexandria (Moscatel Romano), Parellada*
Cava Macabeo (Viura), Parellada, Xarel-lo, *Chardonnay, Subirat Parent (Malvasía Riojana), Pinot Noir*; Macabeo (Viura), Parellada, Xarel-lo, *Chardonnay, Garnacha Tinta, Monastrell, Pinot Noir*
Conca de Barberá Macabeo, Parellada, *Chardonnay, Viognier*; Trepat, Garnacha, Tempranillo (Ull de Llebre), *Cabernet Sauvignon, Merlot*
Costers del Segre Macabeo, Parellada, Chardonnay, Xarel-lo, Garnacha Blanca; Garnacha Tinta, Cabernet Sauvignon, Tempranillo (Ull de Llebre/Gotim Bru), Pinot Noir, Trepat, Monastrell, Merlot, Cariñena
Jerez y Manzanilla Palomino Fino (Listán), Pedro Ximénez, Muscat of Alexandria (Moscatel Romano)
Málaga Pedro Ximénez, Muscat of Alexandria (Moscatel Romano)
La Mancha Tempranillo (Cencibel), Garnacha Tinta, Cabernet Sauvignon, Moravia, Merlot; Airén, Macabeo, Pardilla, *Chardonnay*
Montilla-Moriles Pedro Ximénez, Moscatel Romano, Airén (Lairén), Baladí, Torrontés
Navarra Garnacha Tinta, Tempranillo, Cabernet Sauvignon, Mazuelo, Graciano, *Merlot and other experimental varieties*; Macabeo (Viura), Muscat Blanc à Petits Grains (Moscatel de Frontignan, Moscatel de Grano Menudo), *Chardonnay, Malvasía, Garnacha Blanca and other experimental varieties*
Penedès Tempranillo (Ull de Llebre), Garnacha Tinta, Cabernet Sauvignon, Cariñena, Monastrell, Samsó, *Cabernet Franc, Pinot Noir, Merlot*; Xarel-lo (Pansá Blanca), Macabeo, Parellada, Subirat Parent (Malvasía Riojana), Sauvignon Blanc, Riesling, Chardonnay, Muscat d'Alsace, Gewürztraminer, Chenin Blanc
Priorat Garnacha Tinta, *Cariñena, Garnacha Peluda, Cabernet Sauvignon, Syrah, Merlot*; Garnacha Blanca, Macabeo, Pedro Ximénez, *Parellada, Chenin Blanc*
Rías Baixas Albariño, Loureiro Blanco (Marqués), Treixadura, Caiño Blanco, Torrontés; Caiño Tinto, Brancellao, Espadeiro, Mencía, Loureira Tinta, Sousón
Ribeiro Treixadura, Godello, Lado, Loureiro, Albariño, *Palomino (Jerez), Torrontés, Macabeo, Albillo*; Caiño, Ferrón, Sousón, Brancellao, *Garnacha Tintorera (Alicante), Tempranillo, Mencía*
Ribera del Duero Tempranillo (Tinto Fino, Tinto del País), *Garnacha Tinta (Tinto Aragonés), Cabernet Sauvignon, Merlot, Malbec, Albillo*
Rioja Tempranillo, Garnacha Tinta (Garnacha), Cariñena (Mazuelo), Graciano, *Cabernet Sauvignon, Merlot*; Macabeo (Viura), Malvasía Riojana, Garnacha Blanca, *Chardonnay, Sauvignon Blanc*
Rueda Verdejo, Sauvignon Blanc, Macabeo (Viura), Palomino Fino, *Chardonnay*
Somontano Moristel, Tempranillo, Cabernet Sauvignon, Garnacha Tinta, Parreleta, *Merlot, Pinot Noir and others*; Macabeo, Garnacha Blanca, Chardonnay, Alcañón, Gewürztraminer and others
Tarragona Garnacha Tinta, Cariñena (Mazuela), Tempranillo (Ull de Llebre), *Cabernet Sauvignon, Merlot, Syrah*; Macabeo, Parellada, Xarel-lo, Garnacha Blanca, *Chardonnay, Muscat of Alexandria (Moscatel Romano)*
Toro Tempranillo (Tinto de Toro), *Garnacha Tinta, Cabernet Sauvignon*; Garnacha Tinta; Malvasía, Verdejo Blanco
Valdeorras Mencía, *Garnacha Tintorera (Alicante), Gran Negro, Merenzao (Bastardo), Cabernet Sauvignon, Tempranillo, Merlot*; Godello, *Doña Blanca (Valenciana), Palomino (Jerez)*
Valdepeñas Tempranillo (Cencibel), *Garnacha Tinta, Cabernet Sauvignon, Merlot, Pinot Noir*; Airén, *Macabeo, Chardonnay*

SWITZERLAND

Dôle Pinot Noir, Gamay
L'Oeil-de-Perdrix de Neuchâtel Pinot Noir

GLOSSARY OF TECHNICAL TERMS

Words in SMALL CAPITALS have their own entries elsewhere in the glossary.

Acidity Naturally present in grapes; gives red wine an appetizing 'grip' and whites a refreshing tang. Too much can make a wine seem sharp, but too little and it will be flabby. See also MALIC ACID, TARTARIC ACID.
Aging Essential for many fine wines and for some everyday ones. May take place in vat, barrel or bottle, and may last for months or years. It has a mellowing effect on a wine, but if the wine has too long in storage it may lose its fruit.
Alcoholic content Alcoholic strength, sometimes expressed in degrees, equivalent to the percentage of alcohol in the total volume.
Alcoholic fermentation Biochemical process whereby yeasts convert the grape sugars into alcohol and carbon dioxide, transforming grape juice into wine. It normally stops when all the sugar has been converted or when the alcohol level reaches about 15 per cent.
American Viticultural Area (AVA) American appellation system introduced in the 1980s. AVA status requires that 85 per cent of grapes in a wine come from a specified region. It does not guarantee any standard of quality.
Ampelography The study of grape varieties.
Appellation d'Origine Contrôlée (AC or AOC) Official designation in France guaranteeing a wine by geographical origin, grape variety and production method.
Assemblage French term for blending of wines.
Auslese German and Austrian category for wines made from selected bunches of grapes. The wines will be generally sweet and sometimes touched by noble rot or BOTRYTIS. Sometimes they are fermented dry, making rich, powerful wines.
Barrel aging Time spent maturing in wood, normally oak, during which the wines may take on flavours from the wood if new barrels are used. The gentle oxygenation caused by gaseous penetration through the pores of the wood is important to the aging process.
Barrel fermentation Oak barrels may be used for FERMENTATION instead of stainless steel. If new barrels are used for this, the integration of oak flavour is better than if the wine is merely put into barrels after fermentation.
Barrique The *barrique bordelaise* is the traditional Bordeaux oak barrel of 225 litres (50 gallons) capacity, used for aging and sometimes for fermenting wine.
Bâtonnage A traditional Burgundian practice of stirring the LEES of fine white wines, now increasingly taken up by producers around the world. It is occasionally used for red wines.
Baumé Hydrometric must weight scale which determines the sugar content of grape juice by measuring its density. This indicates the potential ALCOHOLIC CONTENT. Commonly used in France and Australia. See table page 307.
Beerenauslese German and Austrian QmP category for wines made from individually selected berries. Almost always affected by noble rot or botrytis fungus. The wines are sweet to very sweet. Beerenauslese from new, non-Riesling grapes can be dull but Riesling and many a Scheurebe and occasionally a Silvaner will be astonishing.
Bereich German wine regions, of which there are 13, are subdivided into Bereiche.
Bin number Australian system used by wine companies to identify batches of wine. Bin numbers are often used as brand names.
Biodynamic viticulture This approach works with the movement of the planets and cosmic forces to achieve health and balance in the soil and in the vine. Vines are treated with infusions of mineral, animal and plant materials, applied in homeopathic quantities according to the position of the planets.
Blanc de blancs White wine, especially Champagne, made only from white grapes. Blanc de Noirs is white wine from black grapes.
Blending The art of mixing together wines of different origin, styles or age, often to balance out ACIDITY, weight etc.
Bodega Spanish winery or wine firm.
Bordeaux mixture Copper sulphate, slaked lime and water, sprayed on to vines throughout the growing season to prevent downy mildew. It is one of the few chemical treatments permissable in organic viticulture.
Botrytis Botrytis is rot – a fungus which attacks grapes. Often rot is bad news; but under certain circumstances (see page 27) the fungus concerned, *Botrytis cinerea*, has beneficial effects. It is often known as noble rot and is responsible for many of the world's great sweet wines.
Brix A scale used in the USA and New Zealand to measure sugar levels in grape juice. See table page 307.
Brut Term for 'dry', usually seen on Champagne labels and sparkling wines in the New World. In Champagne the term 'Extra Dry' is, in fact, slightly sweeter.
Cal-Ital A group of Californian wineries that promotes Italian grape varieties grown in California. Also refers to the wines.
Canopy The above-ground part of the vine, including stem, leaves and fruit.
Canopy management Term that includes pruning, training and everything that is done to control or alter the shape of the vine's vegetation. It is aimed at regulating the position of the fruit within the canopy, and at controlling the amount of sunlight and shade on the fruit, to produce a crop of optimum size and ripeness.
Carbonic maceration Winemaking method traditional to Beaujolais and now often used elsewhere. Bunches of uncrushed grapes are fermented whole in closed containers to give well-coloured, fruity wine for early drinking.
Cava Spanish fizz made by the traditional method, as used in Champagne.
Cask Wooden (usually oak) barrel used for aging and storing wine. Known in France as *foudres* and Italy as *botti*.
Cépage French for 'grape variety'. Often followed on labels by the name of a single variety, such as Merlot. In the southern Rhône and the Midi the expression *cépage améliorateur* or 'improving variety' refers to the better quality local grapes: Syrah, Grenache or Mourvèdre.
Chai French term for the building in which wine is stored.
Champagne method Traditional way of making sparkling wine by inducing a second FERMENTATION in the bottle in which the wine will be sold. Now known as the traditional method. The term Champagne method is, ludicrously, no longer legal currency.
Chaptalization Addition of sugar during FERMENTATION to raise a wine's alcoholic strength. More necessary in cool climates where lack of sun may mean insufficient natural sugar in the grapes.
Château A wine-producing estate, especially in Bordeaux. Applied to all sizes of property.
Claret English term for red Bordeaux wine.
Clarification Term covering any winemaking process (such as FILTERING or FINING) that involves the removal of solid matter either from the must or the wine.
Classico Italian term for the heartland of a wine zone where its best wines are produced.
Climat French term for a specifically defined area of vineyard, often very small.
Clone Propagating vines by taking cuttings produces clones of the original plant. However, the term is more usually taken to mean laboratory-produced, virus-free clones, selected to produce higher or lower quantity or quality, or selected for resistance to frost or disease.
Clos Term for a vineyard that is (or was) wall-enclosed; traditional to Burgundy.
Cold fermentation Long, slow FERMENTATION at low temperature to produce wines of maximum freshness. Crucial for whites in hot climates.
Consorzio Italian for consortium or association, especially of wine producers. Each DOC has a Consorzio which lays down the rules for its region. The French equivalent is Comité Interprofessionel and in Spain Consejo Regulador.
Corked/corky Wine fault derived from a cork that has become contaminated, usually with Trichloranisole or TCA. Nothing to do with pieces of cork in the wine. The mouldy, stale smell is unmistakable.
Commune A French village and its surrounding area or parish.
Cosecha Spanish for 'vintage'.
Côtes/Coteaux French for 'slopes'. Hillside vineyards often produce better wine than low-lying ones.

Coulure Failure of the fruit to set after flowering, often accentuated by cold, wet or windy weather.
Crémant Traditional-method sparkling wine from French regions other than Champagne, e.g. Crémant de Bourgogne.
Crianza Spanish term used to describe both the process of AGING a wine and the youngest official category of matured wine. A Crianza wine is aged in barrel, tank and/or bottle for at least two years.
Cross, Crossing Grape bred from two *Vitis vinifera* varieties.
Cru French for 'growth'. Used to describe a wine from a single vineyard.
Cru Bourgeois In Bordeaux, a quality ranking immediately below CRU CLASSÉ.
Cru Classé Literally 'Classed Growth', indicating that a vineyard is included in the official ranking system of its region.
Cryoextraction Technique of freezing grapes to remove excess water and increase concentration, adopted in Sauternes in the 1980s to improve lesser vintages.
Cultivar Term mainly used in South Africa for a single grape variety.
Cuve close or Charmat method A bulk process used to make sparkling wines. The second fermentation, which produces the bubbles, takes place in tank rather than in the bottle (as in the superior but more costly TRADITIONAL METHOD).
Cuvée The term usually indicates a blend, which may mean different grape varieties or simply putting together the best barrels of wine.
Degree The alcoholic strength of wine, usually expressed in degrees equivalent to the percentage of alcohol in the total volume. It is used in a broad-brush way to classify regions by the warmth of their climate and determine which vines might succeed.
Degree days A system devised to measure the growth potential of vines in a specific area in terms of the climate.
Demi-sec Confusingly, it means medium tending to sweet, rather than medium-dry.
Denominación de Origen (DO) The main quality classification for Spanish wine. Rules specify each region's boundaries, grape varieties, vine-growing and winemaking methods.
Denominação de Origem Controlada (DOC) Portugal's top quality classification. Rules specify each region's boundaries, grapes, vine-growing and winemaking methods.
Denominacíon de Origen Calificada (DOC) New Spanish quality wine category, one step up from DO. So far only the Rioja DO has been promoted to DOC.
Denominazione di Origine Controllata (DOC) Italian quality wine classification for wines of controlled origin, grape varieties and style.
Denominazione di Origine Controllata e Garantita (DOCG) Top Italian quality wine classification meant to be one notch above DOC, with tighter restrictions on grape varieties, yields and a tasting panel. It is supposed to give recognition to particularly good vineyard sites.
Density The number of vines planted per hectare or acre.
DNA DNA fingerprinting techniques are being used to identify vine varieties. They can also identify the ancestry of vine varieties. Cabernet Sauvignon, for example, is now known to be the offspring of Cabernet Franc and Sauvignon Blanc.
Domaine Estate, especially in Burgundy.
Downy mildew Common vine fungus, also called peronospera, which destroys leaves and shrivels fruit. It can reach epidemic proportions in heavy rain.
Einzellage German for an individual vineyard. The name of the vineyard is generally preceded by that of the village, e.g. Wehlener Sonnenuhr is the Sonnenuhr vineyard in the village of Wehlen.
Eiswein Rare German and Austrian wine made from grapes harvested and pressed while still frozen to remove the slimy, sweet concentrate. The water, in its icy state, stays separate. Known as icewine in Canada.
élevage French term covering all wine-making stages between FERMENTATION and bottling.
Embotellado de/en Origen Spanish term for ESTATE-BOTTLED.
Engarrafado na Origem Portuguese term for ESTATE-BOTTLED.
Enologist/Oenologist Winemaker. The role has become increasingly high profile in recent years.
Espumoso Spanish for 'sparkling'.
Estate-bottled Wine made from grapes grown on the estate's vineyards and then bottled where it has been made. In France, this is indicated on the label as *mis en bouteilles* followed by *au domaine*, *au château*.
Fermentation See ALCOHOLIC FERMENTATION, MALOLACTIC FERMENTATION.
Filtering Removal of YEASTS, solids and any impurities from a wine before bottling.
Fining Method of clarifying wine by adding coagulants, traditionally egg whites, to the surface. As these fall through the wine they collect solids. They are removed and leave nothing of themselves behind.
Flavour or aroma compounds Substances in wine that can be smelled or tasted.
Flor Film of YEAST that grows on the surface of certain wines when in barrel, especially sherry. Protects the wine from oxidation, and imparts a unique taste.
Flying winemaker Term coined in the late 1980s to describe ENOLOGISTS, many of them Australian-trained, brought in to improve quality in many of the world's under-performing wine regions.
Fortified wine Wine which has high-alcohol grape spirit added, either before or after the ALCOHOLIC FERMENTATION is completed.
Foxy Used to describe the very distinctive, perfumed character of many American native and hybrid grapes, especially red ones. Usually a pejorative term.
Frizzante Italian term for lightly sparkling wine.
Garrafeira Portuguese term for high-quality wine with at least half a per cent of alcohol higher than the required minimum, that has had at least three years' aging for reds, and at least one year for whites.
Geographical Indication (GI) Australian term to indicate the origin of a wine.
Grafting Since phylloxera the only sure method of growing grape vines. It involves grafting a cutting of *Vitis vinifera* on to a phylloxera-resistant American rootstock.
Gran Reserva Top-quality, mature Spanish wine from an especially good vintage, with at least five years' aging (cask and bottle) for reds, and four for whites.
Grand Cru 'Great growth'; the top quality classification in Burgundy, Alsace and Champagne. The Grands Crus of Bordeaux may be subdivided into different ranks, according to the region.
Grand vin Term used in Bordeaux to indicate a producer's top wine. Usually bears a CHÂTEAU name.
Hectolitres 100 litres; 22 imperial gallons or 133 standard 75-cl bottles. See Measurements page 305.
Hybrid Grape bred from an interspecific crossing of an American vine species and European *Vitis vinifera*.
Indicação de Proveniência Regulamentada (IPR) Official Portuguese category for wine regions aspiring to DOC status.
Indicazione Geografica Tipica (IGT) A quality level for Italian wines (roughly equivalent to French vin de pays) in between VINO DA TAVOLA and DOC.
Institut National des Appellations d'Origine des Vins et des Eaux-de-Vie (INAO) The organization in charge of administering the French appellation contrôlée system.
Kabinett Lowest level of German QmP wines. Made from ripe grapes, usually lighter in alcohol than ordinary QbA and often delicious. In Austria Kabinett is a sub-division of Qualitätswein.
KMW The Austrians use KMW or Klosterneuburger Mostwaage as a scale to determine the must weight or original sugar in freshly picked grapes. Like OECHSLE degrees in Germany, each quality category of wine sets a minimum number of KMW. See table page 307.
Late harvest Late-harvested grapes contain more sugar and concentrated flavours; the term is often used for sweetish New World wines. Vendange Tardive is the French term.
Lees Sediment – dead YEASTS etc – thrown by wine in a CASK and left behind after RACKING. Some wines stay on the fine lees for as long as possible to take on extra flavour. See BÂTONNAGE.
Lieu-dit Burgundian term for a single vineyard below the rank of PREMIER CRU that may nevertheless be named on the label.
Liquoroso Italian term for wines high in alcohol, often – but not always – fortified.
Maceration An important winemaking process whereby colour, flavour and/or TANNIN are extracted from grape skins before, during or after FERMENTATION.
Macroclimate Refers to the climate of a region. See also MESOCLIMATE, MICROCLIMATE.
Maderization A form of OXIDATION in white wines caused by heating, usually over a period of time. It takes its name from Madeira, which is the epitome of the style. Unintentional maderization, for example in a light white wine, is a fault.
Malic acid One of the two principal acids found in grapes (the other is TARTARIC). Levels of malic acid are significantly higher in cooler climates.
Malolactic fermentation Secondary fermentation whereby sharp, appley-tasting MALIC acid is converted into riper-tasting lactic acid and

carbon dioxide; usually occurs after ALCOHOLIC FERMENTATION. It is encouraged in red wines, [illegible] prevented in whites to preserve a fresh taste, especially in wines made in warm regions, where natural acidity will be lower.

Maturation The beneficial AGING of wine.

Meritage American, primarily Californian, term for red or white wines made from Bordeaux grape varieties.

Mesoclimate Describes the climate of a specific geographical area, be it one vineyard or simply a hillside or valley. See also MACROCLIMATE, MICROCLIMATE.

Microclimate Describes the immediate physical environment of a vine. Often confused with MESOCLIMATE. See also MACROCLIMATE.

Mildew See DOWNY MILDEW and OIDIUM.

Millerandage The failure of some young grapes to develop normally on an otherwise normal bunch of grapes.

Mousseux French term for sparkling wine not made by the TRADITIONAL METHOD.

Musqué A French term meaning both musky and Muscat-like. Some grape varieties, for example Chardonnay, have a Musqué mutation which is particularly aromatic.

Must The mixture of grape juice, skins, pips and pulp produced after crushing (but prior to completion of FERMENTATION), which will eventually become wine.

Must weight An indicator of the sugar content of juice – and therefore the ripeness of grapes. See BAUMÉ, BRIX, KMW, OECHSLE and table page 307.

Négoçiant French term for merchant or shipper who buys in wine from growers, then matures, maybe blends and bottles it for sale.

Noble rot See BOTRYTIS.

Nouveau, novello French and Italian terms for new wine. Wine for drinking very young, from November in year of vintage.

Oak The most common wood for wine CASKS. During aging or fermenting it gives flavours, such as vanilla and TANNIN, to the wines. The newer the wood, the greater its impact. French oak is subtler in flavour than American.

Oechsle In Germany, MUST WEIGHT is measured in degrees Oechsle; in effect, it indicates the level of sweetness in the juice. Each quality category has a minimum required Oechsle degree. See table page 307.

Oidium Common fungal disease, also called powdery mildew, attacking vine leaves, tendrils and shoots.

Organic There is no such thing, properly speaking, as organic wine, only organic viticulture. Term applied to an increasing number of wines which have been subjected to restrictive viticulture and winemaking practices. There are as yet no universally agreed regulations. Unlike BIODYNAMISM, the concept is simply to eliminate the use of chemical fertilizers and pesticides.

Oxidation Over-exposure of wine to air, causing loss of fruit and flavour.

Passito Italian term for strong, sweet wine made from dried or semi-dried grapes.

Pétillant French for semi-sparkling.

Phenolics Chemical compounds found in the pips, stalks, skins, juice and pulp of grapes, especially in [illegible] tannins, colour-giving anthocyanins and flavour compounds.

Phylloxera Vine aphid (*Phylloxera vastatrix*) which devastated viticulture worldwide in the late 19th century onwards. Since then, the vulnerable European Vitis vinifera has been grafted on to phylloxera-resistant American rootstocks. Phylloxera has never reached Chile and parts of Australia, so vines there are ungrafted and can live up to twice as long – if allowed to.

Plafond limité de classement (PLC) A French system whereby the maximum YIELD permitted within an APPELLATION CONTRÔLÉE is increased in abundant years.

Prädikat One of the six German QUALITÄTSWEIN MIT PRÄDIKAT or QmP categories of wine.

Premier Cru 'First growth'; the top quality classification in parts of Bordeaux, but second to Grand Cru in Burgundy. Used in Champagne to designate vineyards just below Grand Cru.

Prohibition 18th Amendment to the US Constitution, passed in 1920, banning the manufacture, sale and transportation of alcoholic beverages; the measure ruined many wineries, but some survived making grape juice, communion and medicinal wines. Repealed in 1933.

Pruning Method of trimming the vine which takes place mainly in the dormant winter months. Also the primary means of controlling the YIELD.

Pulp The flesh of the grape.

Pumping over The process, called *remontage* in French, whereby the fermenting must is drawn over the cap of skins in the vat. Essential for red wines. It helps to extract colour and tannin.

Qualitätswein bestimmter Anbaugebiete (QbA) German wine classification for 'quality wine from designated regions' – the German equivalent, in EU terms, of French AOC and Italian DOC. Most QbA wines are distinctly ordinary in quality; only buy QbA wines from good growers.

Qualitätswein mit Prädikat (QmP) German wine classification for 'quality wine with distinction'. There are six categories, in order of increasing ripeness of the grapes: KABINETT, SPÄTLESE, AUSLESE, BEERENAUSLESE, TROCKENBEERENAUSLESE and EISWEIN. Some wines (usually Kabinett or Spätlese) may be Trocken (dry) or Halbtrocken (half-dry). Austria has a similar system but regards Kabinett as a subdivision of simple Qualitätswein. Other Austrian Prädikats are Ausbruch (between Beerenauslese and Trockenbeerenauslese) and Strohwein.

Quinta Portuguese farm or wine estate.

Racking The transferring of wine from one barrel to another, leaving the LEES or sediment behind. Racking also produces aeration necessary for the aging process and softens TANNINS.

Rancio Style of wine that is deliberately oxidized; either naturally strong or fortified, it is aged in the sun in glass bottles, earthenware jars or wooden barrels.

Récoltant French for 'grower'. They may make their own wine or sell the grapes to a merchant or NÉGOCIANT.

Reserva In Spain, quality wine from a good vintage with at least three years' AGING (cask and bottle) for reds, and [illegible] designation [illegible] higher than the minimum for the region.

Reserve Many New World producers use this term, or similar ones such as Private Reserve and Special Selection, freely on their wine labels to indicate different wine styles or a special selection rather than a better wine. It has no legal meaning.

Reverse osmosis Method of MUST concentration whereby the wine or juice to be concentrated passes through a filter, leaving the water behind. Other methods of must concentration include evaporation in a vacuum and CRYOEXTRACTION.

Rhône Ranger A phrase coined in the 1980s in California to describe local winemakers fascinated with traditional Rhône grape varieties, especially Syrah and Viognier.

Ripasso Valpolicella wine refermented on the LEES of Amarone della Valpolicella to give extra richness.

Riserva Italian term for wines aged for a specific number of years according to DOC(G) laws.

Rootstock The root of the vine on to which the fruiting branches are grafted. Most rootstocks are from PHYLLOXERA-resistant American vines.

Rosado, rosato Spanish, Portuguese and Italian for pink wine or rosé.

Sec French for 'dry'. When applied to Champagne, it actually means medium-dry.

Second wine A CUVÉE put together from wines selected out of a producer's main wine. It may come from young vines or from less favoured parts of the vineyard. Usually lighter and quicker-maturing than the main wine.

Sekt German term for sparkling wine.

Solera Blending system used for sherry and some other FORTIFIED wines. When mature wine is run off a cask for bottling, only a quarter or so of the volume is taken, and the space is filled with similar but younger wine from another cask, which in turn is topped up from an even younger cask, and so on.

Spätlese German QmP category for wines made from 'late-picked' (therefore riper) grapes. Often moderately sweet, though there are now dry versions.

Spumante Italian for sparkling.

Sugar Naturally present in grapes. Transformed during FERMENTATION into alcohol and carbon dioxide.

Sulphur Commonly used during vinification as a disinfectant for equipment; with fresh grapes and wine as an anti-oxidant; and added as sulphur dioxide to the MUST to arrest or delay FERMENTATION.

Supérieur French term for wines with a higher alcohol content and made according to slightly stricter rules than the basic AC.

Superiore Italian term for wines with higher alcohol, maybe more AGING too.

Super-Tuscan English term for high-quality, non-DOC Tuscan wine.

Sur lie French for 'on the LEES', meaning wine bottled direct from the FERMENTATION vat or cask to gain extra flavour from the lees. Muscadet is the most famous example.

Tafelwein German for 'table wine', the most basic quality designation.

Tannin Harsh, bitter element in red wine, derived from grape skins, pips, stems and from aging in oak barrels; softens with time and is essential for a wine's long-term aging. Producing wines with ripe TANNINS that may be drunk earlier is a priority of red winemaking these days.

Tartaric acid One of the two most important acids naturally present in grapes, the other being MALIC. It tends to be the predominant acid in warm areas. It may be added during FERMENTATION to correct LOW ACIDITY.

Teinturier Black *vinifera* grapes with red pulp. All other *vinifera* grapes have colourless pulp. Teinturier grapes are usually inferior in quality.

Terroir A French term used to denote the combination of soil, climate and exposure to the sun that makes each vineyard and region unique. It is the basis of the French APPELLATION D'ORIGINE CONTRÔLÉE or AC/AOC system.

Traditional method The accepted term for what used to be called the CHAMPAGNE METHOD. The Champenois have succeeded in preventing other regions from using the term.

Training Method of vine management using a permanent vine structure, either free-standing, up stakes, along wires or onto a trellis or training system, which will determine the type of PRUNING.

Trocken German for 'dry'.

Trockenbeerenauslese (TBA) German quality wine category for wines made from individually picked single grapes, shrivelled by noble rot – often the highest level of sweetness.

University of California at Davis (UCD) The leading US viticultural research institute and college for aspiring wine-growers and winemakers.

Varietal The character of wine derived from the grape; also wine made from, and named after, a single or dominant grape variety and usually containing at least 75 per cent of that variety. The minimum percentage varies slightly between countries and, in the USA, between states.

Vendange tardive See LATE HARVEST.

Vieilles vignes Wine from old vines. The term has no legal weight, and 'old' may mean 25 or 100 years.

Vigneron French for 'wine-grower'.

Vigour The growth rate of the vine. Vigorous vineyards are often, but not always, associated with high yields. Soils, too, may be regarded as having low or high potential vigour according to the growth of vines planted on them.

Vin Délimité de Qualité Supérieure (VDQS) Second category of French wines, below AC, abbreviated to VDQS.

Vin de garage Wines made on so small a scale they could be made in one's garage. Particularly applies to some wines from Bordeaux's St-Émilion area. Such wines may be made from vineyards of a couple of hectares or less, and are often of extreme concentration. *Vins de garage* are highly fashionable and sell for high prices.

Vin de paille Wine made by drying the grapes on straw (*paille*) before FERMENTATION. This concentrates the sugar in the grapes: the resulting wines are sweet. Mostly from the Jura region of France. Similar wines may be made in other countries, for example Italy and Austria.

Vin de pays French for 'country wine'. Although it is the third category in the official classification of French wines, it includes some first-class wines which don't follow local AC rules.

Vin doux naturel (VDN) French sweet wine fortified with grape spirit. Mostly from Languedoc-Roussillon.

Vin Santo Historic but extremely variable sweet white wine from Tuscany and Italy. Called Vino Santo in Trentino.

Viña Spanish for 'vineyard'.

Vinification The process of turning grapes into wine.

Vino da tavola Italian for 'table wine'. Quality may be basic or exceptional, although many of these latter wines are now being reclassified as INDICAZIONE GEOGRAFICA TIPICA or IGT.

Vino tipico New Italian category for VINO DA TAVOLA with some regional characteristics.

Vintage The year's grape harvest, also used to describe the wine of a single year.

Viticulture Vine-growing and vineyard management.

Vitis vinifera The species of vine, native to Europe and Central Asia, responsible for all the world's fine wine, as opposed to other species such as the native American *Vitis labrusca*, which is still used in the eastern USA to make grape juice and sweetish wines but which is more suited to juice and jelly manufacture.

Wine of Origin (WO) South African system of controlled appellations which certifies the wine's area of origin, grape variety/varieties and vintage.

Yeast Organism which, in the wine process, causes grape juice to ferment. In the New World it is common to start FERMENTATION with cultured yeasts, rather than rely on the natural yeasts, known as ambient yeasts, present in the winery.

Yield The amount of fruit, and ultimately wine, produced from a vineyard. Measured in hectolitres per hectare (hl/ha) in most of Europe and in the New World as tons/acre or tonnes/hectare. Such figures on their own are, however, meaningless unless taken in conjunction with the density of planting. It is the yield per vine that is important, and a yield of 50hl/ha will obviously give very different wine at 2000 vines per hectare than at 10,000 vines per hectare. High yields are traditionally associated with lesser quality, because the larger the quantity of grapes per vine the less chance the vine has of ripening them successfully. However, reducing the yield per vine below a certain point will not improve quality. Modern viticultural techniques such as the production of virus-free clones tend to increase yields, and modern methods of canopy management may enable larger crops to be ripened successfully. Ascertaining the optimum crop for a vineyard so that the ideal size of crop may be brought to ideal ripeness is the aim of many viticulturalists. This is rather overturning the traditional idea that small yields are invariably better than high.

Measurements and conversions

Mass

1 metric tonne = 0.9842 imperial ton

1 imperial ton = 1.016 metric tonne

Surface area

1 hectare (ha) = 10,000 square metres = 2.471 acres

Temperature

To convert Celsius into Fahrenheit, multiply by 1.8 and add 32. To convert Fahrenheit into Celsius, subtract 32 and multiply by 0.5555.

Volume/capacity

1 hectolitre (hl) = 100 litres = 22 gallons (British) or 26 gallons (USA)

1 USA gallon = 3.78 litres

Yields

In Europe these are measured in hl/ha. To convert to tons/acre (the system used in the New World) use the figure of 18hl/ha = 1 ton/acre. This is necessarily an approximate conversion.

EQUIVALENT MUST WEIGHTS

These are the systems used in different parts of the world to measure the ripeness of grapes or their must weight. Equivalents are necessarily approximate.

Baumé (degrees)	Brix (degrees)	Oeschle (degrees)	KMW (degrees)	Potential alcohol (%)
10	18	74	14.8	10
10.55	19	78	15.6	10.55
11	19.8	83	16.6	11
11.55	20.8	88	17.6	11.55
12	21.7	92	18.4	12
12.55	22.6	96	19.2	12.55
13	23.4	100	20	13
13.55	24.4	107	21.4	13.55
14	25.2	108	21.6	14
14.55	26.2	112	22.4	14.55
15	27.1	117	23.4	15

INDEX OF GRAPE NAMES AND THEIR SYNONYMS

Page numbers in **bold** refer to main references in the A-Z.

GENERAL INDEX

Page numbers in *italic* refer to illustration captions.

C

ACKNOWLEDGMENTS

BIBLIOGRAPHY

Nicolas Belfrage *Barolo to Valpolicella* (Faber & Faber, London, 1999)

Stephen Brook *The Wines of California* (Faber & Faber, London, 1999)

Bruce Cass (Ed.) *The Oxford Companion to the Wines of North America* (Oxford University Press, Oxford, 2000)

Pierre Galet *Dictionnaire Encyclopédique des Cépages* (Hachette, Paris, 2000) and *Précis d'Ampélographique Pratique* (J F Impression, St-Jean-de-Védas, Montpellier, France, 1998)

Anthony Hanson MW *Burgundy* (Faber & Faber, London, 1995)

Ian Hutton *The Zinfandel Trail* (IGH Publications, Esher, Surrey, England, 1998)

James Halliday *Wine Atlas of Australia and New Zealand* (HarperCollinsPublishers, Sydney, 1998)

Ron S. Jackson *Wine Science, Principles and Applications* (Academic Press, San Diego, California, 1994)

Hugh Johnson *The Story of Wine* (Mitchell Beazley, London, 1989)

Nicolas Joly *Le Vin du ciel à la terre* (Sang de la Terre, Paris, 1997)

Giles MacDonogh Austria: *New Wines from the Old World* (Österreichischer Agrarverlag, Klosterneuburg, Austria, 1997)

Richard Mayson Portugal's *Wines and Wine Makers, 2nd edition* (The Wine Appreciation Guild, San Francisco, 1998)

Alex McKay, Garry Crittenden Peter Dry, Jim Hardie *Italian Winegrape Varieties in Australia* (Winetitles, Adelaide, 1999)

Remington Norman MW *Rhône Renaissance* (Mitchell Beazley, London, 1995)

John Radford *The New Spain* (Mitchell Beazley, London, 1998)

Jancis Robinson (Ed.) *The Oxford Companion to Wine, 2nd edition* (Oxford University Press, Oxford, 1999)

ACKNOWLEDGMENTS

The authors and publishers would like to thank the countless wineries and individuals all over the world who have given invaluable help and advice with this book, in particular the following:

ARGENTINA Susanna Balbo, Vintage S.A.; Pedro Marchevsky, Nicolas Catena
AUSTRALIA Cape Mentelle; Louise Helmsley-Smith, D'Arenberg; Robert O'Callaghan, Rockford Wines; Louisa Rose, Yalumba; Adam Wynn, Mountadam
AUSTRIA Austrian Wine Marketing Board
BULGARIA Domaine Boyar UK
CANADA British Columbia Wine Institute; Vintners Quality Alliance, Ontario; Wines of Canada
CHILE Wines of Chile
CZECH REPUBLIC CMVVU
CYPRUS Ministry of Commerce, Industry & Tourism – Cyprus
FRANCE Bureau Interprofessionnel des Vins de Bourgogne (BIVB); Conseil Interprofessionnel des Vins d'Alsace (CIVA); Comité Interprofessionnel du Vin de Champagne (CIVC), Comité Interprofessionnel des Vins du Jura; Comité Interprofessionnel des Vins Doux Naturels à Appellations Contrôlées; Comité Intersyndical des Vins de Corse; Conseil Interprofessionnel du Vin de Bordeaux (CIVB); Conseil Interprofessionel des Vins du Roussillon à Appellation d'Origine Contrôlée; Olivier Humbrecht MW; Interprofession des Vins du Val de Loire; Inter Rhône; Maison de la Vigne et du Vin – Gaillac; ONIVINS; Christian Seely, AXA Millésimes; Syndicat des Côtes de Provence; Syndicat Régional des Vins de Savoie; Union Interprofessionnelle des Vins de Beaujolais (UIVB)
GERMANY Dr Bernhard Abend; Geheimer Rat Dr von Basserman-Jordan; Jochen Becker-Köhn, Weingut Robert Weil; Weingut Josef Biffar; Reichsrat von Buhl; Ferdinand Erbgraf zu Castell-Castell, Fürstlich Castell'sches Domänenamt; Forstmeister Geltz-Zilliken; Horst Kolesch, Juliusspital; Franz Künstler; Ernst Loosen, Weingut Dr Loosen; Bernd Philippi, Koehler-Ruprecht; Annegret Reh-Gartner, Reichsgraf von Kesselstatt; Stefan Ress, Balthasar Ress; Dr Joachim Schmid, Geisenheim Viticultural Research Institute; Johannes Selbach, Selbach-Oster; Dr Heinrich Wirsching, Hans Wirsching
GREECE Boutari
HUNGARY Dominique Arangoits, then of Disznókö; Crown Estates (Tokaj Trading House); The Hungarian Food & Wine Bureau; Royal Tokaji Wine Company
ISRAEL Golan Heights Winery
ITALY Dr Alberto Antononi; ARSSA (Abruzzo); Mario Consorte, Sella & Mosca; Consorzio di Tutela Vini DOC Valtellina; Enoteca Italiana, Siena; Enoteca Regionale Emilia-Romagna; Dott. Giancarlo Montaldo; Regione Autonoma Valle d'Aosta; Regione Campania; Regione Marche; Regione Siciliana (Instituto Regionale della Vite e del Vino); Regione Toscana; Trentino Vini
japan Miyoko Stevenson
LEBANON Chateau Musar
MEXICO Mexican Association of Grape Growers
MOLDOVA Premium Brand Corporation
NEW ZEALAND Cloudy Bay; Alan Limmer, Stonecroft; Philip Manson, Winegrowers of New Zealand; Dr Neil McCallum, Dry River; The Wine Institute of New Zealand
PORTUGAL Jacques A Faro da Silva, Madeira Wine Company; Francisco Manuel Machado Albuquerque, Madeira Wine Company; ICEP – Portuguese Trade & Tourism Office – Wines of Portugal; Vasco Magalhaes, Sogrape
SLOVAKIA Ministry of Agriculture of the Slovak Republic
SLOVENIA Ministry of Agriculture, Forestry & Food – Slovenia
SOUTH AFRICA South African Wine Industry Information & Systems; Gyles Webb, Thelema
SPAIN Ministerio de Agricultura, Pesca y Alimentacion; Carlos Read, Moreno Wines; Rioja Wine Group; Bodegas Riojanas
SWITZERLAND Yvon Roduit, Caves Imesch
UNITED KINGDOM Argentine Trade Department, Wine Department; Australian Wine Bureau; Bordeaux Wine Information Service; Ben Campbell-Johnson, J E Fells; Tony Potter, Denbies Wine Estate; English Wine Producers; German Wine Information Service; David Gleave MW; Bill Gunn MW; Margaret Harvey MW; Italian Trade Centre, Wine Department; Geoffrey Kelly; Antony Lacey, Mistral Wines; Paul Lapsley, Southcorp; Della Madison, Madeira Wine Company; Angela Muir MW; Joanna Simon; Westbury Communications; Wines of Chile; Wines from Spain; Wines of South Africa
USA Dr Carole Meredith, University of California at Davis (UCD); New York Wine & Grape Federation; Oregon Wine Advisory Board; Lynn Penner-Ash, Rex Hill; Michael Silacci, Warren and Julia Winiarski, Stag's Leap Wine Cellars; Wine Institute of California; Zinfandel Advocates & Producers (ZAP)
ZIMBABWE Cairns Wineries

PHOTOGRAPHY CREDITS

Bottle photography

The Publishers and Oz Clarke would like to thank the following for their help with sourcing bottles for the photographs: AG Wines, Armit Wines, Berry Bros, Bethel Heights Vineyard, Bibendum Wine Ltd, Cadman Fine Wines Ltd, Concho Y Toro, Cornish Point Wines Ltd, Diamond Creek Vineyards, Ellis of Richmond, Enotria, Handford Wines, Hedonism Wines, John E Fells & Sons Ltd, Justerini & Brooks, Lay & Wheeler Ltd, Lea & Sandeman, Les Caves de Pyrène, Liberty Wines Ltd, Majestic Wines, Moët Hennessy, New Horizon Wines, Treasury Wine Estates, Tyrrell's Wines, Raisin Social Ltd, Waddesdon Wine Ltd, Yapp Brothers. Photography by Martin Norris.

Other photography

Andrew Will Winery 143; Barboursville/Luca Paschina 172; Bien Nacido 173; 'Vintagers and rope makers at work, copy of a wall painting from the tomb of Kha'emwese at Thebes, c. 1450 by Mrs Nina Garis Davies (1881-1965)', from British Museum, London, UK/Bridgeman Art Library; J Dusi Winery 303; California Wine Institute 306; Feiler-Artinger 177; Angelo Gaja 170; Ania Grom-Yoncali 336 (below); Jeff Grosset 212; Kevin Judd 236; Langmeil Winery 109; Ted Lemon/Littorai Wines 196; Dr Loosen 210; Masi 92; Pavilion/Nigel James 336 (centre); le Pin 142; Pizzini Wines 229; Ridge Vineyards 304; Symington Family Estates 277; Tablas Creek Vineyards 219; Thick as Thieves 336 (above); Torres 105 (below right), 275; Josef Umathum 221; Warren Winiarski 10 (left); Doug Wregg 11 (right), 64; Jon Wyand 71 (below left), 72, 144, 191 (below left), 207 (below left), 227, 274; Yalumba Family Wines13.

All other photographs supplied by Cephas Picture Library. All photographs by Mick Rock except: Jerry Alexander 29; Kevin Argue 209; Emma Borg 89; Karine Bossavy 249; Fernando Briones 43 (right); Hervé Champillion 163; Andy Christodolo 18, 25 (centre), 28, 30 (below), 56, 57, 80, 93, 131 (left), 153 (left), 160, 207 (below right), 292; David Copeman 133 (left), 254; David Edwardson 97; Bruce Fleming 22 (right), 59; Dario Fusaro 169 (below right); Kevin Judd 19 (top), 24 (above), 25 (left), 36 (right), 76, 115 (above), 116, 131 (right), 147, 185 (right), 187 (left), 201 (right), 215, 219 (right), 253, 297; Herbert Lehmann 67 (right), 99, 225 (below left and right), 299; Diana Mewes 3, 273 (below left), 291 (below left); Janis Miglavs 13 (below); Matt Molchen 169 (left); R & K Muschenetz 11 (right), 257 (left), 307; Jean-Bernard Nadeau 32, 141 (above), 250; 141 (below left); Alain Proust 12 (all), 13 (top), 87, 201 (left); Ted Stefanski 24 (below), 25 (right), 60, 305; Matt Wilson 20, 67 (left), 77, 146; Stephen Wolfenden 36 (left).

Oz Clarke is one of the world's leading wine experts and is known for his phenomenal palate, irreverent style, accurate predictions and enthusiasm for life in general and wine in particular. His passion for the subject dates from his student days at Oxford University, when he won tasting competitions at a precociously early age. Since then his tasting skills have earned him an international reputation and he is acknowledged as having one of the finest palates of anyone writing about wine today. He has won all the major wine-writing awards both in the UK and the USA, including the Glenfiddich (three times), André Simon, the Wine Guild (three times), the James Beard, the Julia Child, World Food Media, the Prix Lanson (five times) and most recently the Louis Roederer award. In 2015 he was awarded a Lifetime Achievement Award by *Off Licence News* in recognition of his diverse career as TV presenter, columnist and author. Clarke's frequent BBC TV and radio appearances are broadcast around the world.

Margaret Rand is an award-winning wine writer and writes and tastes for *Decanter*, *World of Fine Wine*, *Harpers* and other leading wine and food journals, as well as judging at international wine shows.

Above: It all starts here. Fragrant Pinot Noir from Thick as Thieves in the Yarra Valley begins its long journey to the bottle as a crimson mush.